21st Century Homestead: Non-Timber Forest Products

Contents

Chapter 1

Non-timber forest product

Mahua seeds

Non-timber forest products (**NTFPs**), also **special, non-wood, minor, alternative** and **secondary forest products,** are useful substances, materials and/or commodities obtained from forests which do not require harvesting (logging) trees. They include game animals, fur-bearers, nuts, seeds, berries, mushrooms, oils, foliage, medicinal plants, peat, fuelwood, and forage.*[1]

Research on NTFPs have focused on their commodifiability for rural incomes and markets, as an expression of traditional knowledge or as a livelihood option for rural household needs, and, as a key component of sustainable forest management and conservation strategies. All research promote forest products as valuable commodities and tools that can promote the conservation of forests.

1.1 Definitions

There is a wide variety of NTFPs, including mushrooms, huckleberries, ferns, transplants, seed cones, piñon seeds, tree nuts, moss, maple syrup, cork, cinnamon, rubber, tree oils and resins, and ginseng. The United Kingdom's Forestry Commission defines NTFPs as "any biological re-

Tendu patta (leaf) collection

sources found in woodlands except timber," *[2] and Forest Harvest, part of the Reforesting Scotland project, defines them as "materials supplied by woodlands - except the conventional harvest of timber." *[3] These definitions include wild and managed game, fish and insects.*[4] NTFPs are commonly grouped into categories such as floral greens, decoratives, medicinal plants, foods, flavors and fragrances, fibers, and saps and resins.

Other terms similar to NTFPs include *special, non-wood, minor, alternative* and *secondary forest products*. NTFPs in particular highlight forest products which are of value to local people and communities but have been overlooked in the wake of forest management priorities (for example, timber production and animal forage). In recent decades, interest has grown in using NTFPs as alternatives or supplements to forest management practices. In some forest types, under the right political and social conditions, forests can be managed to increase NTFP diversity and, consequently, to increase biodiversity and potentially economic diversity.

1.2 Uses

The harvest of NTFPs remains widespread throughout the world. People from a wide range of socio-economic, geographical and cultural contexts harvest NTFPs for a number of purposes, including but not limited to: household subsistence, maintenance of cultural and familial traditions, spiritual fulfillment as well as physical and emotional well-being, scientific learning and income.*[4]*[5] Other terms synonymous with harvesting include *wild-crafting, gathering, collecting* and *foraging*. NTFPs also serve as raw materials for industries ranging from large-scale floral greens suppliers and pharmaceutical companies to micro-enterprises centred upon a wide variety of activities (such as basketmaking, woodcarving and the harvest and processing of various medicinal plants).

1.3 Economic importance

It is difficult to estimate the contribution of NTFPs to national or regional economies as there is a lack of broad-based systems for tracking the combined value of the hundreds of products that make up various NTFP industries. One exception to this is the **maple syrup industry**, which in 2002 in the US alone yielded 1.4 million US gallons (5,300 m^3) worth USD$38.3 million.*[6] In temperate forests such as in the US, wild edible mushrooms such as matsutake, medicinal plants such as ginseng, and floral greens such as salal and *sword fern* are multimillion dollar industries. While these high-value species may attract the most attention, a diversity of NTFPs can be found in most forests of the world.

In tropical forests, for example, NTFPs can be an important source of income that can supplement farming and/or other activities. A value-analysis of the Amazon rainforest in Peru found that exploitation of NTFPs could yield higher net revenue per hectare than would timber harvest of the same area, while still conserving vital ecological services.*[7] Their economic, cultural and ecological value, when considered in aggregate, makes managing NTFPs an important component of sustainable forest management and the conservation of biological and cultural diversity.

1.4 Research

Research on NTFPs have focused on three perspectives: NTFPs as a commodity with a focus on rural incomes and markets, as an expression of traditional knowledge or as a livelihood option for rural household needs, and, finally, as a key component of sustainable forest management and conservation strategies. These perspectives promote forest products as valuable commodities and important tools that can promote the conservation of forests. In some contexts, the gathering and use of NTFPs can be a mechanism for poverty alleviation and local development.*[8]*[9]

1.5 See also

- Arid Forest Research Institute
- Biomass
- Biomass (ecology)
- Bioproducts
- Ethnobotany
- Indigenous (ecology)
- Wildness
- Wilderness
- Wildlife

1.6 References

1.6.1 Notes

[1] "Glossary of Forestry Terms in British Columbia" (PDF). Ministry of Forests and Range (Canada). March 2008. Retrieved 2009-04-06.

[2] "Forest Research - Social, cultural and economic values of contemporary non-timber forest products: Wild Harvests". Forestry.gov.uk. Retrieved 2013-11-21.

[3] "non timber forest products in Scotland". ForestHarvest. Retrieved 2013-11-21.

[4] "Forests and non-timber forest products". Cifor.org. Retrieved 2013-11-21.

[5] Kala, CP. (2013). "Harvesting and Supply Chain Analysis of Ethnobotanical Species in the Pachmarhi Biosphere Reserve of India," *American Journal of Environmental Protection* 1(2), pp.20-27.

[6] "SIC 0831 Forest Nurseries and Gathering of Forest Products". *Encyclopedia of American Industries, 5th ed.* Gale. 2008.

[7] Peters, Charles M.; Alwyn H. Gentry; Robert O. Mendelsohn (29 June 1989). "Valuation of an Amazonian rainforest". *Nature* **339**. doi:10.1038/339655a0.

[8] Belcher, B.M. (2003). "What isn't an NTFP?". *International Forestry Review* **5** (2): 161–168. doi:10.1505/IFOR.5.2.161.17408.

[9] Kala, CP 2003. Medicinal Plants of Indian Trans-Himalaya. http://www.cabdirect.org/abstracts/20066710101.html; jsessionid=453D5B8DEE32770EACE0E447E613263E

1.6.2 Bibliography

- Delang, Claudio O. 2006. *The Role of Wild Food Plants in Poverty Alleviation and Biodiversity Conservation in Tropical Countries.* Progress in Development Studies 6(4): 275-286

- Emery, Marla and Rebecca J. McLain; (editors). 2001. *Non-Timber Forest Products: Medicinal Herbs, Fungi, Edible Fruits and Nuts, and Other Natural Products from the Forest.* Food Products Press: Binghamton, New York.

- Guillen, Abraham; Laird, Sarah A.; Shanley, Patricia; Pierce, Alan R. (editors). 2002. *Tapping the Green Market: Certification and Management of Non-Timber Forest Products.* Earthscan

- Jones, Eric T. Rebecca J. McLain, and James Weigand. eds. 2002. *Non Timber Forest Products in the United States.* Lawrence: University Press of Kansas.

- Mohammed, Gina H. 2011. *The Canadian NTFP Business Companion: Ideas, Techniques and Resources for Small Businesses in Non-Timber Forest Products & Services.* Candlenut Books: Sault Ste Marie, Ontario

1.7 External links

- Center of Minor Forest Products
- Center for Livelihoods and Ecology
- FAO Nonwood Forest Products Website
- Non Timber Forest Product United States
- Midwest Special Forest Products
- Nontimber Forest Products in Alaska
- Northern Forest Diversification Centre
- NTFP.org
- Virginia Tech Non-Timber Forest Products
- Alaska mushroom guide : for harvesting morels / compiled and written by Jay Moore. Hosted by the Alaska State Publications Program.
- Nepal NTFP Network (NNN)
- Lao NTFP Wiki
- P&M Technologies - NTFP Page

Chapter 2

Akpeteshie

Akpeteshie is a homebrewed alcoholic spirit produced in Ghana and other West African nations by distilling palm wine or sugar cane juice. Other names for this drink include apio, ogogoro (in Nigeria), sodabi, keley, "hot" or "hot drink" and "kutukù" (in Nzema) or VC10. Use of this high-proof spirit is increasing in West Africa, as is the concern over the social and public health problem increased use might entail.

2.1 History and origins

Before the advent of European colonization of what is today Ghana, the Anlo brewed a local gin also known as "kpótomenui," meaning "something hidden in a coconut mat fence." *[1]

With British colonization of what became known as the Gold Coast, such local brewing was outlawed in the early 1930s. According to a 1996 interview with S.S. Dotse about his life under British colonial rule: "Our contention was that the drink the white man brought is the same as ours. The white men's contention was that ours was too strong...Before the white men came we were using akpeteshie. But when they came they banned it, probably because they wanted to make sales on their own liquor. And so we were calling it kpótomenui. When you had a visitor whom you knew very well, then you ordered that kpótomenui be brought. This is akpeteshie, but it was never referred to by name." *[1]

The name "akpeteshie" was given to the drink with its prohibition: the word comes from the Ga language - Ape te shie - the act of hiding spoken in greater Accra and means "they are hiding," referring to the secretive way in which non-European inhabitants were forced to consume the beverage.*[2] Despite being outlawed, Illicit spirits remained commonplace, with reports that even schoolboys were able to easily obtain akpeteshei through the 1930s. Demand for akpeteshie and the profits to be made from its sale was enough to encourage the spread of sugar cane cultivation in the Anlo region of Ghana.*[3]

Distillation was legalized with decolonization and Ghanaian independence. The first factory was established in the Volta Region, taking advantage of the area's supply of sugar cane plantations.*[3]

2.2 Brewing

local distillation process of Akpeteshie

Akpeteshie is generally distilled from palm wine, Raffia palm wine, or sugarcane.*[4] This sweetened liquid or wine is first fermented in a large barrels, sometimes with the help of yeast.*[5] After this first stage of fermentation, fires are built under the barrels in order to bring the liquid to a boil and pass the resulting vapor through a copper pipe within cooling barrels, where it condenses and drips into sieved jars. The boiled juice then undergoes a second stage of fermentation.*[6] The resulting spirit is between 40 and 50% alcohol by volume.*[7]

4

2.3 Packaging and consumption

Akpeteshie is not professionally bottled or sealed, but instead poured into unlabeled used bottles. The spirit can be bought wholesale from a brewer or by the glass at boutiques and bars. Although not professionally advertised, the drink is very popular. This is partially due to its price, which is lower than that of other professionally bottled or imported drinks. It's relative inexpensive makes it a drink associated more with the poor, but even those who can afford better quality are said to consume the spirit in secret.*[6]

The potency of the liquor heavily affects the bodily senses, providing a feeling likened to that of a knockout punch. Practiced drinkers can be seen acknowledging receipt by blowing out air or pounding their chest.*[6]

2.4 Health

Medical practitioners have been critical of the drink's high concentration of alcohol, particularly the damage it can cause the liver and the risk of alcoholism.*[7]

2.5 References

[1] Akyeampong, Emmanuel Kwaku (2001). *Between the Sea & the Lagoon: An Eco-social History of the Anlo of South-eastern Ghana : C. 1850 to Recent Times.* James Curry Publishers. p. 154.

[2] Peele, Stanton (1999). *Alcohol and Pleasure: A Health Perspective.* Psychology Press. p. 123.

[3] Akyeampong, Emmanuel Kwaku (2001). *Between the Sea & the Lagoon: An Eco-social History of the Anlo of South-eastern Ghana : C. 1850 to Recent Times.* James Curry Publishers. p. 155.

[4] Peele, Stanton (1999). *Alcohol and Pleasure: A Health Perspective.* Psychology Press. p. 123.

[5] Chernoff, John M. (2005). *Exchange is Not Robbery: More Stories of an African Bar Girl.* Chicago: University of Chicago Press. p. 194.

[6] GBC News. "Expert Warns Against "Akpeteshie' Consumption" . Retrieved 8 February 2013.

[7] Luginaaha, Isaac; Crescentia Dakubob (2003). "Consumption and impacts of local brewed alcohol (akpeteshie) in the Upper West Region of Ghana: a public health tragedy" . *Social Science & Medicine* **57** (9). doi:10.1016/s0277-9536(03)00014-5.

Chapter 3

Allspice

Allspice, also called **Jamaica pepper**, **pepper**, **myrtle pepper**, **pimenta**,*[2] **turkish Yenibahar**, **English pepper***[3] or **newspice**, is the dried unripe fruit (berries, used as a spice) of *Pimenta dioica*, a midcanopy tree native to the Greater Antilles, southern Mexico, and Central America, now cultivated in many warm parts of the world.*[4] The name 'allspice' was coined as early as 1621 by the English, who thought it combined the flavour of cinnamon, nutmeg, and cloves.*[5]

Several unrelated fragrant shrubs are called "Carolina allspice" (*Calycanthus floridus*), "Japanese allspice" (*Chimonanthus praecox*), or "wild allspice" (*Lindera benzoin*). Allspice is also sometimes used to refer to the herb costmary (*Tanacetum balsamita*).

Allspice is the dried fruit of the *P. dioica* plant. The fruits are picked when green and unripe and are traditionally dried in the sun. When dry, they are brown and resemble large brown smooth peppercorns. The whole fruits have a longer shelf life than the powdered product and produce a more aromatic product when freshly ground before use.

Fresh leaves are used where available. They are similar in texture to bay leaves and are thus infused during cooking and then removed before serving. Unlike bay leaves, they lose much flavor when dried and stored, so do not figure in commerce. The leaves and wood are often used for smoking meats where allspice is a local crop. Allspice can also be found in essential oil form.

3.1 Preparation/form

Whole allspice berries

3.2 Uses

Allspice is one of the most important ingredients of Caribbean cuisine. It is used in Caribbean jerk seasoning (the wood is used to smoke jerk in Jamaica, although the spice is a good substitute), in *moles*, and in pickling; it is also an ingredient in commercial sausage preparations and curry powders. Allspice is also indispensable in Middle Eastern cuisine, particularly in the Levant, where it is used to flavour a variety of stews and meat dishes. In Palestinian cuisine, for example, many main dishes call for allspice as the sole spice added for flavouring. In the U.S., it is used mostly in desserts, but it is also responsible for giving Cincinnati-style chili its distinctive aroma and flavour. Allspice is commonly used in Great Britain, and appears in many dishes, including cakes. Even in many countries where allspice is not very popular in the household, as in Germany, it is used in large amounts by commercial sausage makers. It is a main flavour used in barbecue sauces. In the West Indies, an allspice liqueur called "pimento*[2] dram" is produced.

Allspice has also been used as a deodorant. Volatile oils found in the plant contain eugenol, a weak antimicrobial agent.*[6]

3.3 Cultivation

Pimenta dioica *leaves in Goa, India*

The allspice tree, classified as an evergreen shrub, can reach 10–18 m (33–59 ft) in height. Allspice can be a small, scrubby tree, quite similar to the bay laurel in size and form. It can also be a tall, canopy tree, sometimes grown to provide shade for coffee trees planted underneath it. It can be grown outdoors in the tropics and subtropics with normal garden soil and watering. Smaller plants can be killed by frost, although larger plants are more tolerant. It adapts well to container culture and can be kept as a houseplant or in a greenhouse.

To protect the pimenta trade, the plant was guarded against export from Jamaica. Many attempts at growing the pimenta from seeds were reported, but all failed. At one time, the plant was thought to grow nowhere except in Jamaica, where the plant was readily spread by birds. Experiments were then performed using the constituents of bird droppings; however, these were also totally unsuccessful. Eventually, passage through the avian gut, whether due to the acidity or the elevated temperature, was found to be essential for germinating the seeds. Today, pimenta is spread by birds in Tonga and Hawaii, where it has become naturalized on Kaua'i and Maui.[7]

3.4 Western history

Allspice (*P. dioica*) was encountered by Christopher Columbus on the island of Jamaica during his second voyage to the New World, and named by Dr. Diego Álvarez Chanca. It was introduced into European and Mediterranean cuisines in the 16th century. It continued to be grown primarily in Jamaica, though a few other Central American countries produced allspice in comparatively small quantities.[8]

3.5 References

[1] "The Plant List: A Working List of All Plant Species". Retrieved 19 August 2015.

[2] The name *pimento*, often substituted when *pimenta* is intended, is also used for a certain kind of large, red, heart-shaped sweet pepper.

[3] In Hebrew, the spice is called פלפל אנגלי, literally: English pepper.

[4] Riffle, Robert L. (1 August 1998). *The Tropical Look: An Encyclopedia of Dramatic Landscape Plants.* Timber Press. ISBN 0-88192-422-9.

[5] *Oxford English Dictionary* (2 ed.). Oxford, UK: Clarendon Press. 1 March 1989. ISBN 0-19-861186-2. Retrieved 12 December 2009.

[6] Yaniv, Zohara; Bacharach, Uriel, eds. (1 April 2005). *Handbook of Medicinal Plants.* Brighamton, New York: Food Products Press and Haworth Medical Press. p. 336. ISBN 1-56022-994-2.

[7] Lorence, David H.; Flynn, Timothy W.; Wagner, Warren L. (1 March 1995). "Contributions to the Flora of Hawai'i III" (PDF). *Bishop Museum Occasional Papers* (Honolulu, Hawaii: Bishop Museum Press) **41**: 19–58. ISSN 0893-1348. Retrieved 12 December 2009.

[8] Nancy Gaifyllia. "About.com Greek Food – Allspice". Archived from the original on 7 July 2011. Retrieved 26 June 2011.

3.6 External links

- Media related to Pimenta dioica at Wikimedia Commons

- Media related to Allspice at Wikimedia Commons

- Data related to Pimenta dioica at Wikispecies

- "*Pimenta dioica*". *Floridata Plant Encyclopedia.*

- "*Pimenta dioica*". *Plants of Hawaii.* Hawaiian Ecosystems at Risk project (HEAR).

- "Allspice". *Gernot Katzer's Spice Pages.*

- "Allspice". Trade Winds Fruit.

- "Allspice". *The Encyclopedia of Spices.* Epicentre.com.

Chapter 4

Bay leaf

Indian bay leaf Cinnamomum tamala

Indonesian bay leaf Syzygium polyanthum

Bay leaf (plural **bay leaves**) refers to the aromatic leaves of several plants used in cooking. These include:

- Bay laurel (*Laurus nobilis*, Lauraceae). Fresh or dried bay leaves are used in cooking for their distinctive flavor and fragrance. The leaves are not meant to be eaten, although it is safe to do so. The leaves are often used to flavor soups, stews, braises and *pâtés* in Mediterranean cuisine. The fresh leaves are very mild and do not develop their full flavor until several weeks after picking and drying.[*][1]

- California bay leaf – the leaf of the California bay tree (*Umbellularia californica*, Lauraceae), also known as California laurel, Oregon myrtle, and pepperwood, is similar to the Mediterranean bay laurel, but has a stronger flavor.

- Indian bay leaf or malabathrum (*Cinnamomum tamala*, Lauraceae) is somewhat similar in appearance to the leaves of bay laurel, but is culinarily quite different, having a fragrance and taste similar to cinnamon (cassia) bark, but milder.

- Indonesian bay leaf or Indonesian laurel (*salam* leaf, *Syzygium polyanthum*, Myrtaceae) is not commonly found outside of Indonesia; this herb is applied to meat and, less often vegetables.[*][2]

- West Indian bay leaf, the leaf of the West Indian bay tree (*Pimenta racemosa*, Myrtaceae), used culinarily and to produce the cologne called bay rum.

- Mexican bay leaf (*Litsea glaucescens*, Lauraceae).

4.1   Taste and aroma

If eaten whole, bay leaves are pungent and have a sharp, bitter taste. As with many spices and flavorings, the fragrance of the bay leaf is more noticeable than its taste. When dried, the fragrance is herbal, slightly floral, and somewhat similar to oregano and thyme. Myrcene, which is a component of many essential oils used in perfumery, can be extracted from the bay leaf. They also contain the essential oil eugenol.[*][3]

4.2 Uses

Bay leaves were used for flavoring by the ancient Greeks.[4] They are a fixture in the cooking of many European cuisines (particularly those of the Mediterranean), as well as in the Americas. They are used in soups, stews, meat, seafood, vegetable dishes, and sauces. The leaves also flavor many classic French dishes. The leaves are most often used whole (sometimes in a *bouquet garni*) and removed before serving (they can be abrasive in the digestive tract). Thai cuisine employs bay leaf (Thai name *bai kra wan*) in a few Arab-influenced dishes, notably massaman curry.[5]

In Indian and Pakistani cuisine, bay laurel leaves are sometimes used in place of Indian bay leaf, although they have a different flavor. They are most often used in rice dishes like *biryani* and as an ingredient in *garam masala*. Bay (laurel) leaves are frequently packaged as *tejpatta* (the Hindi term for Indian bay leaf), creating confusion between the two herbs.

In the Philippines, dried bay laurel leaves are added as a spice in the Filipino dish Adobo.

Bay leaves can also be crushed or ground before cooking. Crushed bay leaves impart more of their desired fragrance than whole leaves, but are more difficult to remove, and thus they are often used in a muslin bag or tea infuser. Ground bay laurel may be substituted for whole leaves, and does not need to be removed, but it is much stronger due to the increased surface area and in some dishes the texture may not be desirable.

Bay leaves can also be used scattered in a pantry to repel meal moths,[6] flies, roaches, mice, and silverfish.

Bay leaves have been used in entomology as the active ingredient in killing jars. The crushed, fresh, young leaves are put into the jar under a layer of paper. The vapors they release kill insects slowly but effectively, and keep the specimens relaxed and easy to mount. The leaves discourage the growth of molds. They are not effective for killing large beetles and similar specimens, but insects that have been killed in a cyanide killing jar can be transferred to a laurel jar to await mounting.[7] It is not clear to what extent the effect is due to cyanide released by the crushed leaves, and to what extent other volatile products are responsible.

4.3 Safety

Some members of the laurel family, as well as the unrelated but visually similar mountain laurel and cherry laurel, have leaves that are poisonous to humans and livestock. While these plants are not sold anywhere for culinary use, their visual similarity to bay leaves has led to the oft-repeated belief that bay leaves should be removed from food after cooking because they are poisonous. This is not true - bay leaves may be eaten without toxic effect. However, they remain very stiff even after thorough cooking, and if swallowed whole or in large pieces, they may pose a risk of scratching the digestive tract or even causing choking. There are multiple cases of intestinal perforations caused by swallowing bay leaves and they should not be swallowed or left in the food before serving to prevent the occurrence of a possibly fatal surgical emergency.[8][9] Thus, most recipes that use bay leaves will recommend their removal after the cooking process has finished.[10]

4.4 References

[1] "Spice Trade: Bay Leaf" . Archived from the original on 12 April 2009. Retrieved 2009-04-11.

[2] "Spice Pages: Indonesian Bay-Leaf" . Retrieved 2012-12-01.

[3] "Encyclopedia of Spices: Bay Leaf" . Archived from the original on 16 April 2009. Retrieved 2009-04-11.

[4] "Ancient Egyptian Plants: Trees" *www.reshafim.org.il* Retrieved October 29, 2013

[5] Tan, Hugh T. W. (2005). *Herbs & Spices of Thailand*. Marshall Cavendish. p. 71.

[6] "How to Repel Grain Moths with Bay Leaves" . Retrieved 2009-04-11.

[7] Smart, John (1963). *British Museum (Natural History) Instructions for Collectors NO. 4A. Insects*. London: Trustees of the British Museum.

[8] Lingenfelser, T; Adams, G; Solomons, D; Marks, I. N. (1992). "Bay leaf perforation of the small bowel in a patient with chronic calcific pancreatitis" . *Journal of clinical gastroenterology* **14** (2): 174–6. PMID 1556436.

[9] Bell, C. D.; Mustard, R. A. (1997). "Bay leaf perforation of Meckel's diverticulum" . *Canadian Journal of Surgery* **40** (2): 146–7. PMC 3952980. PMID 9126131.

[10] "Straight Dope: Are Bay Leaves Poisonous?". Retrieved 2009-04-11.

Chapter 5

Benzoin resin

Not to be confused with benzoin.

Benzoin resin is a balsamic resin obtained from the bark

Kemenyan, *benzoin resin as sold in Gombong, Central Java*

of several species of trees in the genus *Styrax*. It is used in perfumes, some kinds of incense, as a flavoring, and medicine (see tincture of benzoin). Commonly called "benzoin", it is called "benzoin resin" here to distinguish it from the chemical compound benzoin. Benzoin resin does *not* contain this crystalline compound.

Benzoin is also called **gum benzoin** or **gum benjamin**, but "gum" is incorrect as benzoin is not a polysaccharide. Its name came via the Italian from the Arabic *lubān jāwī* (لبان جاوي, "frankincense from Java").[1]

Benzoin resin is also called **styrax balsam** or **styrax resin**, but wrongly, since those resins are obtained from a different plant family, Hamamelidaceae.

Benzoin resin is a common ingredient in incense-making and perfumery because of its sweet vanilla-like aroma and fixative properties. Gum benzoin is a major component of the type of church incense used in Russia and some other Orthodox Christian societies, as well as Western Catholic Churches.[2] Most benzoin is used in Arab States of the Persian Gulf and India, where it is burned on charcoal as an incense. It is also used in the production of Bakhoor (Arabic بخور - scented wood chips) as well as various mixed resin incense in the Arab countries and the Horn of Africa. Benzoin resin is also used in blended types of Japanese incense, Indian incense, Chinese incense (known as Anxi xiang; 安息香), and Papier d'Arménie as well as incense sticks.

There are two common kinds of benzoin resin, benzoin Siam and benzoin Sumatra. Benzoin Siam is obtained from *Styrax tonkinensis*, found across Thailand, Laos, Cambodia, and Vietnam. Benzoin Sumatra is obtained from *Styrax benzoin*, which grows predominantly on the island of Sumatra.[3] Unlike Siamese benzoin, Sumatran benzoin contains cinnamic acid in addition to benzoic acid.[4] In the United States, Sumatra benzoin (Styrax benzoin and Styrax paralleoneurus) is more customarily used in pharmaceutical preparations, Siam benzoin (Styrax tonkinensis et al.) in the flavor and fragrance industries.[5]

In perfumery, benzoin is used as a fixative, slowing the dispersion of essential oils and other fragrance materials into the air.[3] Benzoin resin is used in cosmetics, veterinary medicine, and scented candles.[4] It is used as a flavoring in alcoholic and nonalcoholic beverages, baked goods, chewing gum, frozen dairy, gelatins, puddings, and soft candy.[6]

5.1 References

[1] A. Dietrich (1986), "LUBĀN", *The Encyclopaedia of Islam* **5** (2nd ed.), Brill, p. 786a

[2] St. Alban Blend

[3] Karl-Georg Fahlbusch; et al. (2007), "Flavors and Fragrances", *Ullmann's Encyclopedia of Industrial Chemistry* (7th ed.), Wiley, p. 87

[4] Klemens Fielbach; Dieter Grimm (2007), "Resins, Natural", *Ullmann's Encyclopedia of Industrial Chemistry* (7th ed.), Wiley, p. 4

[5] James A. Duke (2008), "Benzoin (Styrax benzoin Dryander.)", *Duke's Handbook of Medicinal Plants of the Bible*, Taylor & Francis, p. 446

[6] George A. Burdock (2010), "Benzoin Resin", *Fenaroli's Handbook of Flavor Ingredients* (6th ed.), Taylor & Francis, pp. 139–140

5.2 External links

Chapter 6

Berry

For botanical usage, see Berry (botany). For other uses, see Berry (disambiguation).

In everyday language, a **berry** is a small, pulpy and often edible fruit. Berries are usually juicy, rounded, brightly colored, sweet or sour, and do not have a stone or pit, although many pips or seeds may be present.[1] Common examples are strawberries, raspberries, blueberries; and red- and blackcurrants.[2] In Britain **soft fruit** is a horticultural term for such fruits.[3][4][5]

In scientific terminology, a botanical berry is a fruit produced from the ovary of a single flower in which the outer layer of the ovary wall develops into an edible fleshy portion (botanically the pericarp). The definition includes many fruits that are not commonly known as berries, such as grapes, tomatoes, cucumbers, eggplants (aubergines) and bananas. Fruits excluded by the botanical definition include strawberries and raspberries. A plant bearing berries is said to be *bacciferous* or *baccate*.

Many berries are edible, but some are poisonous to humans, such as the fruits of the potato, the deadly nightshade and pokeweed, and can cause harm. Others, such as the white, red mulberry, and elderberry are poisonous when unripe, but are edible in their ripe form.[6]

Berries are eaten worldwide and often used in jams, preserves, cakes or pies. Some berries are commercially important. The berry industry varies from country to country as do types of berries cultivated or growing in the wild. Many berries such as raspberries and strawberries have been bred for thousands of years and are distinct from their wild counterparts, while some berries such as lingonberries and cloudberries grow almost exclusively in the wild.

6.1 History

6.1.1 Biological

6.1.2 Human

Berries have been valuable as a food source since before the start of agriculture to humans, and remain among the primary food sources of other primates. They were a seasonal staple for early hunter-gatherers for thousands of years, and wild berry gathering remains a popular activity in Europe and North America today. In time, humans learned to store berries so that they could be used in the winter, and they may be made into fruit preserves, and among Native Americans, mixed with meat and fats as pemmican.[7]

Berries also began to be cultivated in Europe and other countries. Some species of blackberries and raspberries of the genus *Rubus* have been cultivated since the 17th century, while smooth-skinned blueberries and cranberries of the genus *Vaccinium* have been cultivated in the United States for over a century.[7] In Japan, between the 10th and 18th centuries, the term "*ichibigo*" (which later became "*ichigo*") referred to many berry crops. The most widely cultivated berry of modern times, however, is the strawberry which is produced globally at twice the amount of all other berry crops combined.[8] Strawberry was mentioned by ancient Romans who thought it had medicinal properties,[9] but it was then not a staple of agriculture.[10] Wood strawberry began to be grown in garden in France in the 14th century, musky-flavored strawberry (*F. moschata*) in late 16th century in European gardens, and later Virginia strawberry in Europe and United States.[11] The most commonly consumed strawberry, the garden strawberry, is an accidental hybrid of Virginia strawberry and a Chilean variety *Fragaria chiloensis* first noted by a French gardener around the mid 18th century after *F. moschata* and *F. virginiana* were planted in between rows of *F. chiloensis*. Antoine Nicolas Duchesne began to study the breeding of strawberries,[12] and hundreds of cultivars have since been produced through the breeding of strawberries.[10]

6.1.3 Etymology

A form of the word "berry" is found in all the Germanic languages; for example, Old English *berie* compares with Old Saxon and Old High German *beri*, and Old Norse *ber*. These forms point to the Old German **bazjo-m*, which has been traced to the Old German **bazo-z* (which also leads to the English word *bare*, as in "a bare fruit"). In Old English, the word was mainly applied to grapes, but has since grown to its current definition.

6.2 Botanical definition

Main article: Berry (botany)

In botanical terminology, a berry is a simple fruit with seeds and pulp produced from the ovary of a single flower. It is fleshy throughout, except for the seeds. It does not have a special "line of weakness" along which it splits to release the seeds when ripe (i.e. it is indehiscent).[13] A berry may develop from an ovary with one or more carpels (the female reproductive structures of a flower). The seeds are usually embedded in the fleshy interior of the ovary, but there are some non-fleshy examples such as peppers, with air rather than pulp around their seeds. The differences between the everyday and botanical uses of "berry" result in three categories: those fruits which are berries under both definitions; those fruits which are botanical berries but not commonly known as berries; and those parts of plants commonly known as berries which are not botanical berries, and may not even be fruits.

Different kinds of berries

Lingonberries – berries under both definitions

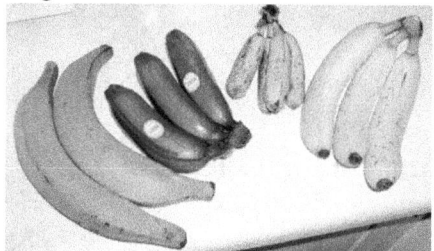

Bananas – botanically berries, but not commonly described as such

Blackberries – botanically aggregate fruits

Sloe berries – botanically stone fruits or drupes

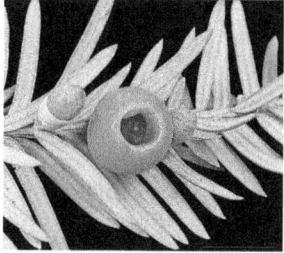

Yew berry – botanically a modified seed-bearing conifer cone

Berries under both definitions include blueberries, cranberries, lingonberries, and the fruits of many other members of the heather family, as well as gooseberries, goji berries and elderberries. The fruits of some "currants" (*Ribes* species), such as blackcurrants, redcurrants and white currants, are botanical berries, and are treated as horticultural berries (or as soft fruit in the UK), even though their most commonly used names do not include the word "berry".

Botanical berries not commonly known as berries include bananas,[14][15] tomatoes,[1] grapes, eggplants or aubergines, persimmons, watermelons and pumpkins.

There are several different kinds of fruits which are commonly called berries but are not botanical berries. Blackberries, raspberries and strawberries are kinds of aggregate fruits;[1] they contain seeds from different ovaries of a single flower. In aggregate fruits like blackberries, the individual "fruitlets" making up the fruit can be clearly seen. The fruits of blackthorn may be called "sloe berries",[16] but botanically are small stone fruits or drupes, like plums or apricots. Junipers and yews are commonly said to have berries, but these differ from botanical berries. They are highly modified seed-bearing cones. In juniper berries, used to flavour gin, the cone scales, which are hard and woody in most conifers, are instead soft and fleshy when ripe. The bright red berries of yews consist of a fleshy outgrowth (aril) almost enclosing the poisonous seed.

6.3 Cultivation

Strawberries have been grown in gardens for a long time in Europe. Blueberries were domesticated starting in 1911 with the first commercial crop in 1916.[*][17] Huckleberries of all varieties are not fully domesticated but domestication was attempted from 1994-2010 for the economically significant western huckleberry.[*][18][*][19] Many other varieties of *Vaccinium* are likewise not domesticated, with some being of commercial importance.

6.3.1 Agricultural methods

Like most other food crops, berries are commercially grown with both conventional pest management and integrated pest management (IPM) practices. Organically certified berries are becoming more widely available.[*][20][*]:5

Many soft fruit berries require a period of temperatures between 0 °C and 10 °C for breaking dormancy, in general: strawberries require 200–300 hours, blueberries 650–850 hours, blackberries 700 hours, raspberries 800–1700 hours, currants and gooseberries 800–1500 hours, and cranberries 2000 hours.[*][21] However too low a temperature will also kill the crops: blueberries do not tolerate temperatures below −29 °C, raspberries, depending on variety, may tolerate as low as −31 °C, and blackberries are injured at less than −20 °C.[*][21] Spring frosts are, however, much more damaging to berry crops than low winter temperatures causing sites with moderate slopes (3-5%) and north or east facing in the northern hemisphere near large bodies of water which regulate spring temperature to be considered ideal in preventing spring frost injury to the new leaves and flowers.[*][21] All berry crops have shallow root systems.[*][21] Many land-grant university extension offices suggest that strawberries should not be planted more than five years on the same site due to the danger of black root rot (though many other illnesses go by the same name), which is controlled in major commercial production by annual methyl bromide fumigation.[*][22][*][23][*][24][*][25][*][26][*][27][*][28][*][29] As well as years in production soil compaction, frequency of fumigation, and usage of herbicides increases the appearance of black root rot in strawberries.[*][29] Raspberries, blackberries, strawberries, and many other berries are susceptible to verticillium wilt. Blueberries and cranberries grow poorly if the clay or silt content of the soil is higher than 20%, while most other berries tolerate a wide range of soil types.[*][21] For most berry crops the ideal soil is well drained sandy loam with a pH of 6.2-6.8 with a moderate to high organic content; however, blueberries have an ideal pH of 4.2-4.8 and can be grown on muck soils and blueberries and cranberries prefer poorer soils with lower cation exchange, lower calcium, and lower levels of phosphorus.[*][21]

Growing most berries organically requires the usage of proper crop rotation, the right mix of cover crops, and the cultivation of the correct beneficial microorganisms in the soil.[*][29] As blueberries and cranberries thrive in soils that are not hospitable to most other plants and conventional fertilizers are toxic to them, the primary concern when growing them organically is bird management.[*][29]

Post-harvest small fruit berries are generally stored at 90-95% relative humidity and 0 °C.[*][30] Cranberries are however frost sensitive and should be stored at 3 C.[*][30] Berries do not respond to ethylene, except blueberries but flavor does not improve after harvest so they require the same treatment as other berries: removal of ethylene may reduce disease and spoilage in all berries.[*][30] Precooling within one to two hours post-harvest to storage temperature, generally 0 °C, via forced air cooling increases the storage life of berries by about a third.[*][30] Under optimum storage conditions raspberries and blackberries last for two to five days, strawberries 7–10 days, blueberries two to four weeks, and cranberries two to four months.[*][30] Berries can be shipped under high carbon dioxide or modified atmosphere of 10-15% carbon dioxide for high carbon dioxide or 15-20% carbon dioxide and 5-10% oxygen for a modified atmosphere container to increase shelf life and prevent grey mold rot.[*][30]

6.3.2 Breeding

6.3.3 Horticultural soft fruit berries

Some fruit not commonly referred to as berries and not always botanically berries are included by land-grant university extension offices in their guides for cultivation of berries, or in guides of identifying local wild edible and non-edible berries. Examples include beach plums,[*][31] American persimmons, pawpaws, Pacific crabapples, and prickly pears.[*][32]

6.4 Commercial production

One source suggests that in the year 2005, there were 1.8 million acres of land worldwide cultivating berries, with 6.3 million tons produced.[*][20][*]:4

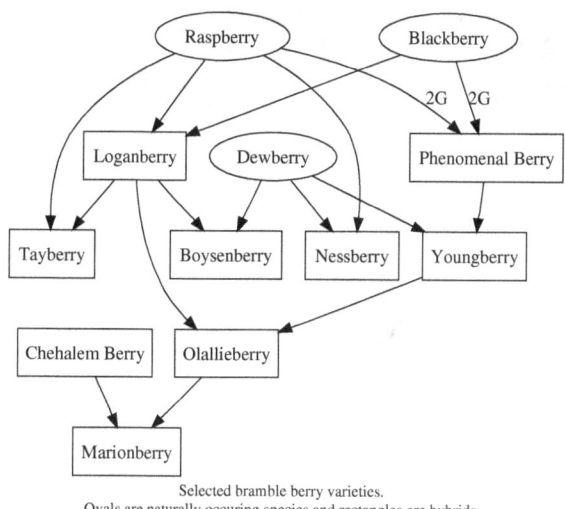

Selected bramble berry varieties.
Ovals are naturally occuring species and rectangles are hybrids.
"2G" denotes a second-generation hybrid.

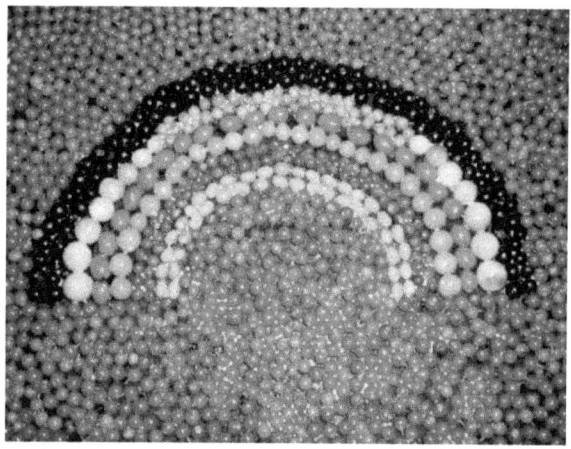

Example of color contrast in (mostly inedible) wild berries

Brambles have been cultivated for thousands of years and been crossed back and forth to create the range of edible Rubus *which we have today.*

6.4.1 Economics

According to figures from Global Berry Congress 2015 in the US over 6 million dollars of soft fruit are sold accounting for 19% of total supermarket revenue, more than bananas (a botanical berry) at 11%, and apples (14%); with continued rapid growth in the market expected.*[33]

In certain regions berrypicking can be a large part of the economy, and it is becoming increasingly common for western European countries such as Sweden and Finland to import cheap labor from Thailand or Bulgaria to do the berry picking.*[34]*[35] This practice has come under scrutiny in the past years because of low wages and living standard for the "berry-pickers" as well as lack of worker safety.*[34]

6.5 Color and potential health benefits

Berries are typically of a contrasting color to their background (often of green leaves), making them visible and attractive to frugivorous animals and birds. This assists the wide dispersal of the plants' seeds.

Berry colors are due to natural plant pigments, such as anthocyanins, together with other flavonoids localized mainly in berry skins, seeds and leaves.*[36]*[37]*[38] Although berry pigments have antioxidant properties *in vitro*,*[39] there is no physiological evidence established to date that berry pigments have actual antioxidant or any other functions within the human body. Consequently, it

is not permitted to claim that foods containing polyphenols have antioxidant health value on product labels in the United States or Europe.*[40]*[41]

6.6 Culinary significance

6.6.1 Use in baked goods

A slice of blueberry pie

Berries are commonly used in pies or tarts, such as grape pie, blueberry pie, blackberry pie, and strawberry pie.

Berries are often used in baking blueberry muffins, blackberry muffins, berry cobblers, berry crisps, berry cakes, berry buckles, berry crumb cakes, berry tea cakes, and berry cookies.*[42] Berries are commonly incorporated whole into the batter for baking and care is often taken so as to not burst the berries; frozen or dried berries may be preferable for some baked berry products.*[43]*[44]*[45] Fresh berries are also often incorporated into baked berry desserts, sometimes with cream, either as a filling to the dessert or as a topping.[*][42]

6.6.2   Beverages

Berries are often added to water and/or juiced as in cranberry juice, which accounts for 95% of cranberry crop usage,[*][46] blueberry juice, raspberry juice, goji berry juice, acai juice, aronia berry juice, and strawberry juice.[*][47][*][48] Wine is the principal fermented beverage made from berries (grapes). Fruit wines are commonly made out of other berries. In most cases sugars must be added to the berry juices in the process of Chaptalization to increase the alcohol content of the wine. Examples of fruit wines made from berries include: elderberry wine, strawberry wine, blueberry wine, blackberry wine, redcurrant wine, huckleberry wine, goji wine and cranberry wine.[*][49][*][50][*][51][*][52]

6.6.3   Dried

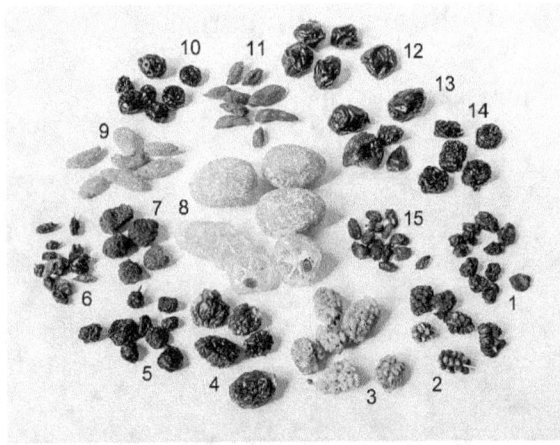

Various dried berries

Currants, raisins and sultanas are examples of dried grape berries, and many other commercially important berries are available in dried form.

6.6.4   Fruit preserves

Main article: Fruit preserves

Berries are perishable fruits with a short shelf life and are often preserved by drying, freezing, pickling or making fruit preserves. Berries such as blackberry, blueberry, boysenberry, lingonberry, loganberry,[*][53] raspberry and strawberry are often used in jams and jellies. In the United States, Native Americans were "the first to make preserves from blueberries."[*][54]

Elderberry jam on bread

6.6.5   Other usages

Chefs have created quick pickled soft fruit such as blackberries,[*][55] strawberries,[*][56] and blueberries.[*][57] Strawberries can be battered and quickly fried in a deep fryer.[*][58][*][59] Sauces made from berries, such as cranberry sauce, can be frozen until hard, battered and deep fried.[*][60] Cranberry sauce is a traditional food item for Thanksgiving, and similar sauces can be made from many other berries such as blueberries, raspberries, blackberries, and huckleberries.[*][61][*][62][*][63][*][64][*][65]

6.7   Cultural significance

6.7.1   Dyeing

Berries have been used in some cultures for dyeing. Many berries contain juices which can easily stain, affording use as a natural dye. For example, blackberries are useful for making dyes, especially when ripe berries can easily release juice to produce a colourfast effect.[*][66][*][67][*][68] *Rubus* berries, such as blackberry, raspberry, black raspberry, dewberry, loganberry and thimbleberry, all produce dye colours once used by Native Americans.[*][68][69] In Hawaii, the native raspberry called 'akala' was used to dye tapa cloth with lavender and pink hues, whereas berries from the dianella lily were used for blue coloration, and berries from the black nightshade were used to produce green coloration.[*][70]

In Swaziland, several berry species are used as a dye.[*][71]

6.8   See also

- List of culinary fruits

- List of inedible fruits

6.9 References

[1] "Berry (Plant reproductive body)". *Encyclopædia Britannica*. Retrieved 16 August 2015.

[2] "Berry". *Merriam-Webster*.

[3] "soft fruit". *Collins English Dictionary – Complete & Unabridged 10th Edition*. HarperCollins. Retrieved 11 August 2015.

[4] "Soft Fruit List: 2014–15". Royal Horticultural Society. Archived from the original on 11 August 2015. Retrieved 11 August 2015.

[5] "Berry". *The Free Dictionary*. Retrieved 10 August 2015.

[6] "ELDERBERRY (SAMBUCUS SPECIES)". *The Poison Plant Patch*. Novia Scotia Museum. Retrieved 13 August 2015.

[7] Kenneth F. Kiple, ed. (2000). *The Cambridge World History of Food, Volume 2*. Cambridge University Press. pp. 1731–1732. ISBN 978-0521402156.

[8] Aaron Liston, Richard Cronn and Tia-Lynn Ashman (2014). "Fragaria: A genus with deep historical roots and ripe for evolutionary and ecological insights". *American Journal of Botany* **101** (10): 1686–99. doi:10.3732/ajb.1400140. PMID 25326614.

[9] Jack Staub (2008). *75 Remarkable Fruits for Your Garden*. Gibbs Smith. p. 213. ASIN B001PGX05K.

[10] Chittaranjan Kole, ed. (2011). *Wild Crop Relatives: Genomic and Breeding Resources: Temperate Fruits*. Springer. pp. 22–23. ASIN B008CN2MQC.

[11] Vern Grubinger. "History of the Strawberry". University of Vermont.

[12] George M. Darrow (1966). *The strawberry; history, breeding, and physiology* (PDF). New York Holt Rinehart and Winston. pp. 38–43.

[13] Kiger, Robert W. & Porter, Duncan M. (2001). "Find term 'berry'". *Categorical Glossary for the Flora of North America Project*. Retrieved 2015-08-14.

[14] "Banana from *Fruits of Warm Climates* by Julia Morton". Purdue University. Archived from the original on 15 April 2009. Retrieved 16 April 2009.

[15] Armstrong, Wayne P. "Identification of Major Fruit Types". Wayne's Word: An On-Line Textbook of Natural History. Retrieved 17 August 2013.

[16] Shilling, Jane (20 August 2014). "Why these bitter berries are summer's sweetest fruit: Mixed bag of weather results in an early burst of the sloe". *MailOnline*. Retrieved 15 August 2015.

[17] "Blueberries – Celebrating 100 Years". *Blueberry Council*. Retrieved 11 August 2015.

[18] Russell, Betsy Z. "Wild huckleberry nearly tamed". *idahoptv*. Retrieved 11 August 2015.

[19] Pittaway, Jenna. "Dr Barney Interview on the Western Huckleberry". *wildhuckleberry*. Retrieved 11 August 2015.

[20] Yanyun Zhao (6 June 2007). *Berry Fruit: Value-Added Products for Health Promotion*. CRC Press. ISBN 978-1-4200-0614-8.

[21] Pritts, Dr. Marvin. "Site and Soil requirements for small fruit crops" (PDF). *Cornell Fruit*. Retrieved 11 August 2015.

[22] Handley, David T. "Growing Strawberries". *University of Maine Extension*. Retrieved 13 August 2015.

[23] "Growing Strawberries". *University of Illinois Extension*. Retrieved 13 August 2015.

[24] Whiting, David. "Growing Strawberries in Colorado Gardens". *Colorodo State University Extension*.

[25] Gao, Gary. "Strawberries are an Excellent Fruit for the Home Garden". *Ohio State University Extension*. Retrieved 13 August 2015.

[26] Kluepfel, Marjan; Polomski, Bob. "Growing Strawberries". *Clemson Cooperative Extension*. Retrieved 13 August 2015.

[27] "Strawberry Production Systems". *Maine Organic Farmers and Gardners Association*. Retrieved 13 August 2015.

[28] Ruttan, Denise. "Plant strawberries and boost your health". *Oregon State University Extension Service*. Retrieved 13 August 2015.

[29] Pritts, Dr. Marvin. "Key Features of Organic Berry Crop Production" (PDF). *Cornell Fruit*. Retrieved 11 August 2015.

[30] DeEll, Dr. Jennifer. "Postharvest Handling and Storage of Berries". *omafra*. Retrieved 12 August 2015.

[31] Whitlow, Dr. Thomas. "Beach Plum". *Cornell*. Retrieved 13 August 2015.

[32] "Edible Berries of the Pacific Northwest". *Northern Bushcraft*. Retrieved 13 August 2015.

[33] "Retail revenue soft fruit in US bigger than bananas or apples". *freshplaza*. Retrieved 11 August 2015.

[34] "Berrypickers, unite!". *The Economist*. ISSN 0013-0613. Retrieved 12 August 2015.

[35] Teivainen, Aleksi. "Record number of Thai berry pickers to arrive in Finland". *helsinkitimes.fi*. Retrieved 12 August 2015.

[36] Wrolstad, Ronald E. (2001). "The Possible Health Benefits of Anthocyanin Pigments and Polyphenolics". Linus Pauling Institute, Oregon State University, Corvallis. Archived from the original on 7 July 2014. Retrieved 7 July 2014.

[37] Mattivi F, Guzzon R, Vrhovsek U, Stefanini M, Velasco R (2006). "Metabolite profiling of grape: Flavonols and anthocyanins". *J Agric Food Chem* **54** (20): 7692–702. PMID 17002441.

[38] González CV, et al. (2015). "Fruit-localized photoreceptors increase phenolic compounds in berry skins of field-grown Vitis vinifera L. cv. Malbec". *Phytochemistry* **110**: 46–57. doi:10.1016/j.phytochem.2014.11.018. PMID 25514818.

[39] Wu X, Beecher GR, Holden JM, Haytowitz DB, Gebhardt SE, Prior RL; Beecher; Holden; Haytowitz; Gebhardt; Prior (June 2004). "Lipophilic and hydrophilic antioxidant capacities of common foods in the United States". *Journal of Agricultural and Food Chemistry* **52** (12): 4026–37. doi:10.1021/jf049696w. PMID 15186133.

[40] Guidance for Industry, Food Labeling; Nutrient Content Claims; Definition for "High Potency" and Definition for "Antioxidant" for Use in Nutrient Content Claims for Dietary Supplements and Conventional Foods U.S. Department of Health and Human Services, Food and Drug Administration, Center for Food Safety and Applied Nutrition, June 2008

[41] EFSA Panel on Dietetic Products, Nutrition and Allergies (NDA)2, 3 (2010). "Scientific Opinion on the substantiation of health claims related to various food(s)/food constituent(s) and protection of cells from premature aging, antioxidant activity, antioxidant content and antioxidant properties, and protection of DNA, proteins and lipids from oxidative damage pursuant to Article 13(1) of Regulation (EC) No 1924/2006" (PDF). *EFSA Journal* (Parma, Italy: European Food Safety Authority) **8** (10): 1752. doi:10.2903/j.efsa.2010.1752.

[42] "60 Berry desserts". *Martha Stewart*. Retrieved 13 August 2015.

[43] "Baking with Blueberries". *U.S Highbush Blueberry Council*. Retrieved 13 August 2015.

[44] Gordon, Megan. "Frozen Berries In Off-Season Baking: Should You Thaw Before Using?". *The Kitchn*. Retrieved 13 August 2015.

[45] "Fresh Fruit vs Frozen Fruit in baking recipes". *Baking Bites*. Retrieved 13 August 2015.

[46] Geisler, Malinda. "Cranberries Profile". *AgMRC*. Retrieved 13 August 2015.

[47] Beck, Margery A. "Aronia berry gaining market foothold in U.S.". *USA Today*. Retrieved 13 August 2015.

[48] "Fruit Juices". *Agriculture and Agri-Food Canada*. Retrieved 13 August 2015.

[49] Wright, John. "How to make Blackberry Wine and Whisky". *The Gaurdian*. Retrieved 13 August 2015.

[50] Kime, Robert. "Strawberry Wine" (PDF). *Berry Resources Cornell*. Retrieved 13 August 2015.

[51] "Bring on the Blueberry Wine". *Wine Mag*. Retrieved 13 August 2015.

[52] Rudebeck, Clare. "A berry nice vintage: It's time to rediscover the ancient art of fermenting fruit wines". *.independent.co.uk*. Retrieved 13 August 2015.

[53] *The Jam Book*. Taylor & Francis. 2014. p. 121. ISBN 978-1-317-84605-5.

[54] Grotto, D. (2007). *101 Foods That Could Save Your Life*. Random House Publishing Group. p. 53. ISBN 978-0-553-90451-2.

[55] Satterfield, Steven. "Spiced and Pickled Blackberries". *Food and Wine*. Retrieved 11 August 2015.

[56] O'Brady, Tara. "Pickled Strawberry Preserves". *david lebovitz*. Retrieved 11 August 2015.

[57] Kord, Tyler. "Pickled Blueberries". *Saveur*. Retrieved 11 August 2015.

[58] "Deep Fried Strawberries". *Driscolls*. Retrieved 31 August 2015.

[59] Fortune, Fia. "Deep-Fried Cheesecake-Stuffed Strawberries". *Forkable*. Retrieved 31 August 2015.

[60] Deen, Paula. "Cranberry Sauce fritters recipe". *Foodnetwork.com*. Retrieved 31 August 2015.

[61] Deen, Paula. "Leopold's Huckleberry Sauce". *Food Network.com*. Retrieved 31 August 2015.

[62] Currah, Allice. "Simple Homemade Blackberry Sauce". *PBS.org*. Retrieved 31 August 2015.

[63] Lagasse, Emeril. "Raspberry Sauce". *Food Network.com*. Retrieved 31 August 2015.

[64] "Perfect Cranberry Sauce". *Food Network.com*. Retrieved 31 August 2015.

[65] Garten, Ina. "Baked Blintzes with Fresh Blueberry Sauce". *Food Network.com*. Retrieved 31 August 2015.

[66] "Dyeing with blackberries". Retrieved 12 August 2015.

[67] "Culturally and Economically Important Nontimber Forest Products of Northern Maine: Blueberry". *US Forest Service*. Retrieved 12 August 2015.

[68] "Native Plant Dyes". *US Forest Service*. Retrieved 12 August 2015.

[69] Mahady, G.B.; Fong, H.H.S.; Farnsworth, N.R. (2001). *Botanical Dietary Supplements:*. Taylor & Francis. p. 47. ISBN 978-90-265-1855-3.

[70] Krohn-Ching, V. (1980). *Hawaii Dye Plants and Dye Recipes*. University Press of Hawaii. p. 13. ISBN 978-0-8248-0698-9.

[71] Shujaa, M.J.; Shujaa, K.J. (2015). *The SAGE Encyclopedia of African Cultural Heritage in North America*. SAGE Publications. p. 382. ISBN 978-1-4833-4638-0.

6.10 Further reading

- Bowling, B.L. (2005). *The Berry Grower's Companion*. Timber Press. ISBN 978-0-88192-726-9

6.11 External links

- United States National Berry Crops Initiative

Chapter 7

Birch bark

A man with a hat made from birch bark in Hankasalmi, Central Finland

Birch bark or **birchbark** is the bark of several Eurasian and North American birch trees of the genus *Betula*.

The strong and water-resistant cardboard-like bark can be easily cut, bent, and sewn, which made it a valuable building, crafting, and writing material, since pre-historic times. Even today birch bark remains a popular type of wood for various handicrafts and arts.

Birch bark also contains substances of medicinal and chemical interest. Some of those products (such as betulin) also

have fungicidal properties that help preserve bark artifacts, as well as food preserved in bark containers.

A Russian birch bark letter (14th century).

7.1 Collection and storage

Birch bark can be removed fairly easily from the trunk or branches, living or recently dead, by cutting a slit lengthwise through the bark and pulling or prying it away from the wood. The best time for collection is spring or early summer, as the bark is of better quality and most easily removed.

Removing the outer (light) layer of bark from the trunk of a living tree may not kill it, but probably weakens it and makes it more prone to infections. Removal of the inner (dark) layer, the phloem, kills the tree by preventing the flow of sap to the roots.

To prevent it from rolling up during storage, the bark should be spread open and kept pressed flat.

7.2 Working

Birch bark can be cut with a sharp knife, and worked like cardboard. For sharp bending, the fold should be scored (scratched) first with a blunt stylus.

Fresh bark can be worked as is; bark that has dried up (be-

Birchbark box with lid and bottom of birch wood

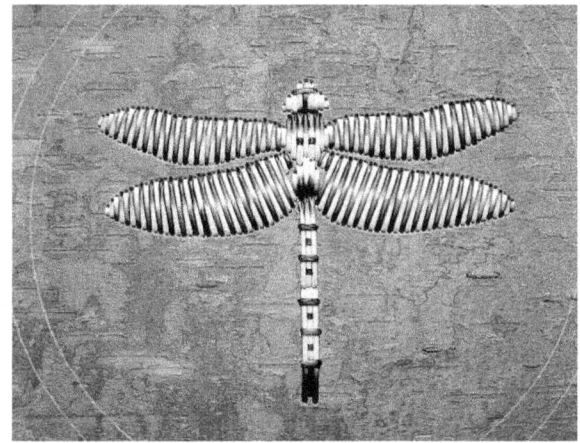

Contemporary quillwork design on birch bark, by Ferdy Goode

Finnish fishing net weights made out of birch bark and stones

North American birchbark canoe

Birchbark knife handle

fore or after collection) should be softened by steaming, by soaking in warm water, or over a fire.

7.3 Uses

Birch bark was a valuable construction material in any part of the world where birch trees were available. Containers like wrappings, bags, baskets, boxes, or quivers were made by most societies well before pottery was invented. Other uses include:

- In various Asian countries (including Siberia) birch bark was used to make storage boxes, paper, tinder, canoes, roof coverings, tents, and waterproof covering for composite bows, such as the Mongol bow, the Chinese bow, Korean bow, Turkish bows, Assyrian bow, the Perso-Parthian bow....etc. It is still being used. More than one variety of birch is used.

- In North America, the native population used birch bark for canoes,[1] wigwams, scrolls, ritual art (birch bark biting), maps (including the oldest maps of North America[2]), torches, fans, musical instruments, clothing, and more.

- In Scandinavia and Finland, it was used as the substratum of sod roofs and birch-bark roofs, for making boxes, casks and buckets, fishing implements, and shoes (as used by the Egtved Girl), etc..

- In Russia, many birch bark manuscripts have survived from the Middle Ages.

- Birch bark knife handles are popular tools to be made

currently.

- In India, birch-bark, along with dried palm leaves, replaced parchment as the primary writing medium. The oldest known Buddhist manuscripts (some of the Gandharan Buddhist Texts), from Afghanistan, were written on birch bark.

Birch bark also makes an outstanding tinder, as the inner layers will stay dry even through heavy rainstorms. To render birch bark useless as tinder, it must be soaked for an extended period of time.

7.4   See also

- Birch bark manuscript
- Mazinibaganjigan (Ojibwa birch bark decorative designs)
- Wiigwaasabak (Ojibwa birch bark scrolls)
- "Wiigwaas" entry in Wiktionary

7.5   References

[1] Tom Vennum, Charles Weber, Earl Nyholm (Director) (1999). *Earl's Canoe: A Traditional Ojibwe Craft.* Smithsonian Center for Folklife Programs and Cultural Studies. Retrieved 2012-12-03.

[2] Hayes, Derek. Historical Atlas of Canada: Canada's History Illustrated with Original Maps. Vancouver: Douglas & McIntyre Ltd, 2002. p. 152.

- *The Algonquin Birchbark Canoe*, by David Gidmark.

7.6   Further reading

- McPhee, John, *The Survival of the Bark Canoe*, Farrar, Straus and Giroux, New York, 1975.

- Adney, Edwin Tappan and Howard Chapelle, *Bark Canoes and Skin Boats of North America*, Skyhorse Publishing, Inc., 2007, 2014.

- Jennings, John, *Bark Canoes: The Art and Obsession of Tappan Adney*, Firefly Books Ltd., 2004.

- Behne, C. Ted, editor, *The Travel Journals of Tappan Adney, 1887-1890*, Estate of Tappan Adney, 2010.

- Goode, F.W., *Ojibwe Birch Bark Canoes: Anishinaabe Wigwassi-Jiimaan*, Beaver Bark Canoes, 2012.

Winter bark etching on canoe

7.7   External links

- The Birch Bark Torch, a *Wilderness Way Magazine's* article by Kevin Finney. Archived April 3, 2007 at the Wayback Machine

- Birchbark articles from the *NativeTech* site.

- Birch and Birch Bark, an article by John Zasada at a University of Minnesota site.

- Birch Bark Canoe Building Courses at the North House Folk School, Minnesota.

- Birch Bark Canoe page on the site of the Algonquins of Pikwàganagàn.

- Watch a documentary on how to build a Birch bark canoe

- Bureau of Catholic Indian Missions Digital Image Collection at Marquette University; keyword: birch bark.

Chapter 8

Birch beer

Working Birch Beer still at the Kutztown Folk Festival. Sign reads "Birch oil is distilled from the sap of the Black Birch tree..."

Birch beer in its most common form is a carbonated soft drink made from herbal extracts, usually from birch bark, although in the colonial era birch beer was made with herbal extracts of oak bark.*[1] It has a taste similar to root beer. There are dozens of brands of birch beer available.*[2]

Various types of birch beer made from birch sap are available as well, distinguished by color. The color depends on the species of birch tree from which the sap is extracted (though enhancements via artificial coloring are common presently). Popular colors include brown, red, blue and clear (often called white birch beer), though others are possible. This drink is most commonly found in the Northeastern United States, and Newfoundland in Canada. After the sap is collected, it is distilled to make birch oil. The oil is added to the carbonated drink to give it the distinctive flavor, reminiscent of teaberry. Black birch is the most common source of extract. In the dairy country of southeastern and central Pennsylvania, an ice cream soda made with vanilla ice cream and birch beer is called a Birch

Beer Float, while chocolate ice cream and birch beer makes a Black Cow.

Alcoholic birch beer, in which the birch sap is fermented rather than reduced to an oil, has been known from at least the seventeenth century. The following recipe is from 1676:

> To every Gallon whereof, add a pound of refined Sugar, and boil it about a quarter or half an hour; then set it to cool, and add a very little Yest to it, and it will ferment, and thereby purge itself from that little dross the Liquor and Sugar can yield: then put it in a Barrel, and add thereto a small proportion of Cinnamon and Mace bruised, about half an ounce of both to ten Gallons; then stop it very close, and about a month after bottle it; and in a few days you will have a most delicate brisk Wine of a flavor like unto Rhenish. Its Spirits are so volatile, that they are apt to break the Bottles, unless placed in a Refrigeratory, and when poured out, it gives a white head in the Glass. This Liquor is not of long duration, unless preserved very cool. Ale brewed of this Juice or Sap, is esteem'd very wholesome. *[3]

8.1 Commercial brands

- A-Treat
- Adirondack
- Boylan Bottling Company
- Crush
- Fanta
- Foxon Park
- Frostop
- Hank's Birch Beer

- Hosmer Mountain Soda

- Izze

- Mercury Brewing Company[*][4]

- Pennsylvania Dutch Birch Beer

- Polar Beverages

- Shurfine (white and dark varieties; store brand)

- Sioux City

- Stewart's Fountain Classics

- White Rock Beverages

8.2   See also

- Birch syrup

- Sarsaparilla (soft drink)

- Sassafras soda

8.3   References

[1]   "Hands on History: Colonial Cooking" .

[2]   Anthony's Root Beer Barrel - Birch Beer reviews

[3]   Vinetum Britannicum, p. 176, London, England 1676.

[4]   http://www.ipswichalebrewery.com/products/soda_pop

Chapter 9

Birch syrup

Several bottles of birch syrup

Birch syrup is a savory mineral tasting syrup made from the sap of birch trees, and produced in much the same way as maple syrup. It is seldom used for pancake or waffle syrup, more often it is used as an ingredient paired with pork or salmon dishes in sauces, glazes, and dressings, and as a flavoring in ice cream, beer, wine, and soft drinks. It is condensed from the sap, which has about 0.5-2% percent sugar content, depending on the species of birch, location, weather, and season. The finished syrup is 66% sugar or more to be classified as a syrup. Birch sap sugar is about 42–54% fructose and 45% glucose, with a small amount of sucrose and trace amounts of galactose. The flavor of birch syrup has a distinctive and mineral-rich caramel-like taste that is not unlike molasses or balsamic condiment or some types of soy, with a hint of spiciness. Different types of birch will produce slightly different flavour profiles; some more copper, others with hints of wildflower honey. Many people remark that while Birch syrup has the same sugar content of maple it is far more savory than sweet.

9.1   Method

Making birch syrup is more difficult than making maple syrup, requiring about 100-150 liters of sap to produce one liter of syrup (more than twice that needed for maple syrup). The tapping window for birch is generally shorter than for maple, primarily because birches live in more northerly climates. It also happens later in the year than maple tapping. The trees are tapped and their sap collected in the spring (generally mid- to late April, about two to three weeks before the leaves appear on the trees). The common belief is that while birches have a lower trunk and root pressure than maples, pipeline or tubing method of sap collection used in large maple sugaring operations is not as useful in birch sap collection. However Rocky Lake Birchworks in The Pas Manitoba is successfully using the tubing method along with a vacuum system for collection of birch sap.

The sap is reduced in the same way as maple sap, using reverse osmosis machines and evaporators in commercial production. While maple sap may be boiled down without the use of reverse osmosis, birch syrup is difficult to produce this way: the sap is more temperature sensitive than is maple sap because fructose burns at a lower temperature than sucrose, the primary sugar in maple sap. This means that boiling birch sap to produce syrup can much more easily result in a scorched taste.

9.2   Production

Most birch syrup is produced in Russia, Alaska and Canada from Paper Birch or Alaska Birch sap (*Betula papyrifera* var. *humilis* and *neoalaskana*). These trees are found primarily in interior and south central Alaska. The Kenai birch (*Betula papyrifera* var. *kenaica*), which is also used, grows most abundantly on the Kenai Peninsula, but is also found in the south central part of the state and hybridizes with *humilis*. The southeast Alaska variety is the Western paper birch, (*Betula papyrifera* var. *commutata*) and has a lower

A birch tree being tapped

sugar content. One litre of syrup from these trees requires evaporation of approximately 130–150 litres of sap.*[1]

Sap dripping from a tapped birch tree

Total production of birch syrup in Alaska is approximately 3,800 liters (1,000 U.S. gallons) per year, with smaller quantities made in other U.S. states and Canada (also from Paper Birch), Russia, Belarus, Ukraine, and Scandinavia (from other species of birch). Because of the higher sap-to-syrup ratio and difficulties in production, birch syrup is more expensive than maple syrup, up to five times the price.

9.3 See also

- Birch beer

- Xylitol, a sugar alcohol extracted from birch

9.4 References

[1] "Haines birch syrups attract gourmet following", Margaret Baumann, Alaska Journal of Commerce, May 29, 2005

9.5 External links

- Kispiox Creations Production and recipes for pure Birch Syrup (no sugar added) made in British Columbia Canada

- Petition to US Food and Drug Administration for establishment of Standard of Identity for birch syrup, including the Alaska Birch Syrupmakers' Association Best Practices. July 18, 2005.

- Birch Boy Gourmet Syrups' educational articles on birch and other syrups

- Crooked Chimney Syrups research page research on sugar content of birch sap

- Birch: white gold in the boreal forest. (pdf download) 2004. Deirdre Helfferich. Agroborealis 35:2, pp. 4-12.

- Forbes Wild Foods - Commercialization of birch syrup and other birch products.

- recipes using birch syrup

Listening

- "Alaska Sap Suckers" (A story from National Public Radio's *All Things Considered* program, May 29, 2001)

- edibletoronto Article on Birch Syrup

-

Chapter 10

Birch tar

Birch bark pitch made in a single pot: The birch bark is heated under airtight conditions, the final product consists of tar and the ashes of the bark.

Birch tar or birch pitch is a substance (liquid when heated) derived from the dry distillation of the bark of the birch tree.

10.1 Compounds

It is compounded of phenols such as guaiacol, cresol, xylenol and creosol.

10.2 Uses

Birch tar was used widely as an adhesive as early as the late Paleolithic or early Mesolithic era. It has also been used as a disinfectant, in leather dressing, and in medicine.

Ends of fletching of arrows were fastened with birch-tar and birch-tar-and-rawhide lashings were used to fix the blade of axes in the Mesolithic period.

Russia leather is a water-resistant leather, oiled with birch oil after tanning. This leather was a major export good from 17th and 18th century Russia, as the availability of birch oil

Modern way of producing birch bark tar in a single pot: The birch bark is heated under airtight conditions, the final product consists of tar and the ashes of the bark.

limited its geographical production.*[1] The oil impregnation also deterred insect attack and gave a distinctive and pleasant aroma that was seen as a mark of quality in leather.

Birch tar oil is an effective repellent of gastropods.*[2] The repellent effect lasts about two weeks.*[2] The repellent effect of birch tar oil mixed with petroleum jelly applied to a fence lasts up to several months.*[2]

Birch tar oil is also used in perfumery as a base note to impart leather, tar, smoky, and wintergreen notes.

10.3 References

[1] "Production of Russia Leather" (PDF). The Honourable Cordwainers' Company. 1807.

[2] Lindqvist I., Lindqvist B., Tiilikkala K., Hagner M., Penttinen O.-P., Pasanen T. & Setälä H. (2010). "Birch tar oil is an effective mollusc repellent: field and laboratory experiments using *Arianta arbustorum* (Gastropoda: Helicidae) and *Arion lusitanicus* (Gastropoda: Arionidae)". *Agricultural and Food Science* **19**(1): 1-12. doi:10.2137/145960610791015050.

10.4 External links

- Non-video demonstration

Chapter 11

Black pepper

"Peppercorn" redirects here. For other uses, see Peppercorn (disambiguation).

Black pepper (*Piper nigrum*) is a flowering vine in the family Piperaceae, cultivated for its fruit, which is usually dried and used as a spice and seasoning. When dried, the fruit is known as a peppercorn. When fresh and fully mature, it is approximately 5 millimetres (0.20 in) in diameter, dark red, and, like all drupes, contains a single seed. Peppercorns, and the ground pepper derived from them, may be described simply as pepper, or more precisely as **black pepper** (cooked and dried unripe fruit), **green pepper** (dried unripe fruit) and **white pepper** (ripe fruit seeds).

Black pepper is native to south India, and is extensively cultivated there and elsewhere in tropical regions. Currently Vietnam is the world's largest producer and exporter of pepper and producing 34% of the world's *Piper nigrum* crop as of 2008.

Dried ground pepper has been used since antiquity for both its flavour and as a traditional medicine. Black pepper is the world's most traded spice. It is one of the most common spices added to European cuisine and its descendants. The spiciness of black pepper is due to the chemical piperine, not to be confused with the capsaicin that gives fleshy peppers theirs. It is ubiquitous in the modern world as a seasoning and is often paired with salt.

11.1 Etymology

The word "pepper" has its roots in the Dravidian word for long pepper, *pippali*.[2][3][4] Ancient Greek and Latin turned *pippali* into the Latin *piper*, which was used by the Romans to refer both to black pepper and long pepper, as the Romans erroneously believed that both of these spices were derived from the same plant.[5] Today's "pepper" derives from the Old English *pipor*. The Latin word is also the source of Romanian *piper*, Italian *pepe*, Dutch *peper*, German *Pfeffer*, French *poivre*, and other similar forms.

In the 16th century, *pepper* started referring to the unrelated New World chili pepper as well. "Pepper" was used in a figurative sense to mean "spirit" or "energy" at least as far back as the 1840s; in the early 20th century, this was shortened to *pep*.[6]

11.2 Varieties

Black and white peppercorns

11.2.1 Black pepper

Black pepper is produced from the still-green, unripe drupes of the pepper plant. The drupes are cooked briefly in hot water, both to clean them and to prepare them for drying. The heat ruptures cell walls in the pepper, speeding the work of browning enzymes during drying. The drupes are dried in the sun or by machine for several days, during which the pepper around the seed shrinks and darkens into a thin, wrinkled black layer. Once dried, the spice is called black peppercorn. On some estates, the berries are separated from the stem by hand and then sun-dried without the boiling process.

Once the peppercorns are dried, pepper spirit and oil can be extracted from the berries by crushing them. Pepper spirit is used in many medicinal and beauty products. Pepper oil is also used as an ayurvedic massage oil and used in certain beauty and herbal treatments.

- Black pepper (*Piper nigrum*) essential oil in a clear glass vial

- Ground black pepper and a plastic pepper shaker

- Roughly cracked black peppercorns, also known as *mignonette* or *poivre mignonette*

The six variants of pepper

11.2.2 White pepper

"White pepper" redirects here. For the Ween album, see White Pepper.

White pepper consists of the seed of the pepper plant

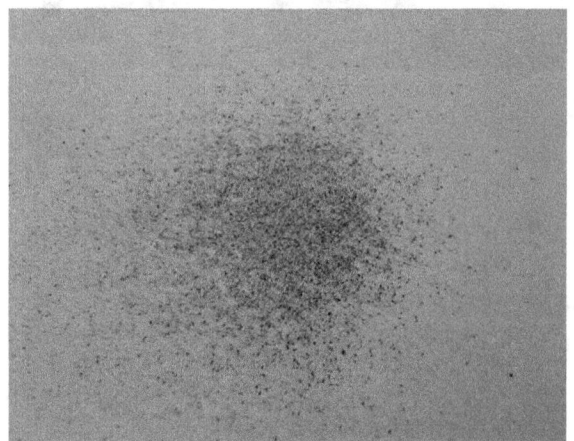

White pepper grains

alone, with the darker-coloured skin of the pepper fruit removed. This is usually accomplished by a process known

as retting, where fully ripe red pepper berries are soaked in water for about a week, during which the flesh of the pepper softens and decomposes. Rubbing then removes what remains of the fruit, and the naked seed is dried. Sometimes alternative processes are used for removing the outer pepper from the seed, including removing the outer layer through mechanical, chemical or biological methods.*[7]

Ground white pepper is often used in cream sauces, Chinese and Thai cuisine, and dishes like salad, light-coloured sauces and mashed potatoes, where black pepper would visibly stand out. White pepper has a slightly different flavour from black pepper, due to the lack of certain compounds present in the outer fruit layer of the drupe, but not found in the seed. A slightly sweet version of white pepper from India is sometimes called *safed golmirch* (Hindi), *shada golmorich* (Bengali), or *safed golmirch* (Punjabi).

*Black, green, pink (*Schinus terebinthifolius*), and white peppercorns*

11.2.3 Green pepper

Green pepper, like black, is made from the unripe drupes. Dried green peppercorns are treated in a way that retains the green colour, such as treatment with sulphur dioxide, canning or freeze-drying. Pickled peppercorns, also green, are unripe drupes preserved in brine or vinegar. Fresh, unpreserved green pepper drupes, largely unknown in the West, are used in some Asian cuisines, particularly Thai cuisine.*[8] Their flavour has been described as spicy and fresh, with a bright aroma.*[9] They decay quickly if not dried or preserved.

11.2.4 Wild pepper

Wild pepper grows in the Western Ghats region of India. Into the 19th century, the forests contained expansive wild pepper vines, as recorded by the Scottish physician Francis Buchanan (also a botanist and geographer) in his book *A journey from Madras through the countries of Mysore, Canara and Malabar* (Volume III).[10] However, deforestation resulted in wild pepper growing in more limited forest patches from Goa to Kerala, with the wild source gradually decreasing as the quality and yield of the cultivated variety improved. No successful grafting of commercial pepper on wild pepper has been achieved to date.[10]

11.2.5 Orange pepper and red pepper

Orange pepper or red pepper usually consists of ripe red pepper drupes preserved in brine and vinegar. Ripe red peppercorns can also be dried using the same colour-preserving techniques used to produce green pepper.[11]

11.2.6 Pink pepper and other plants used as pepper

Pink pepper from *Piper nigrum* is distinct from the more-common dried "pink peppercorns", which are actually the fruits of a plant from a different family, the Peruvian pepper tree, *Schinus molle*, or its relative the Brazilian pepper tree, *Schinus terebinthifolius*. A pink peppercorn (French: baie rose, "pink berry") is a dried berry of the shrub Schinus molle, commonly known as the Peruvian peppertree. As they are members of the cashew family, they may cause allergic reactions including anaphylaxis for persons with a tree nut allergy.

The bark of *Drimys winteri* ("Canelo" or "Winter's Bark") is used as a substitute for pepper in cold and temperate regions of Chile and Argentina where it is easily available.

In New Zealand the seeds of Kawakawa (*Macropiper excelsum*), a relative of black pepper, are sometimes used as pepper and the leaves of *Pseudowintera colorata* (mountain horopito) are another replacement for pepper.

Several plants in the United States are used also as pepper substitutes, such as *Lepidium campestre*, *Lepidium virginicum*, shepherd's purse, horseradish, and field Pennycress.

11.2.7 Region of origin

Peppercorns are often categorized by their place of origin. Two types come from India's Malabar Coast: *Malabar* and *Tellicherry*. Tellicherry comes from grafted Malabar plants grown on Mount Tellicherry.[12]

Sarawak pepper is native to the Malaysian portion of Borneo. White Muntok pepper comes from Indonesia and Lampung hails its island of Sumatra. Vietnam produces

both white and black pepper in the provinces of Bà Rịa–Vũng Tàu, Chu Se District, Bình Phước, and Phú Quốc Island in Kiên Giang Province.[13]

Kampot Pepper is native to Kampot, Cambodia and received Geographical indication (GI) status in 2008. This pepper is grown in a limited geographical region in four varieties: black, green, red, and white.[14]

11.3 Plant

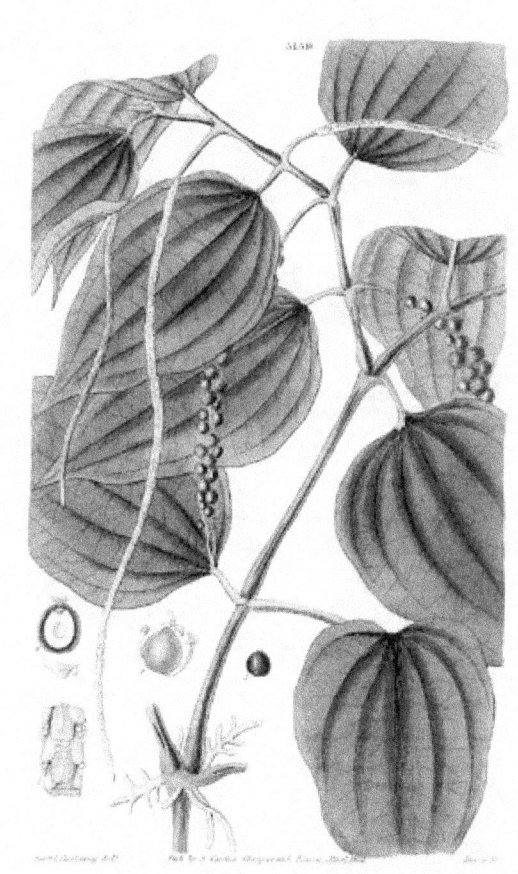

Piper nigrum *from an 1832 print*

The pepper plant is a perennial woody vine growing up to 4 metres (13 ft) in height on supporting trees, poles, or trellises. It is a spreading vine, rooting readily where trailing stems touch the ground. The leaves are alternate, entire, 5 to 10 centimetres (2.0 to 3.9 in) long and 3 to 6 centimetres (1.2 to 2.4 in) across. The flowers are small, produced on pendulous spikes 4 to 8 centimetres (1.6 to 3.1 in) long at the leaf nodes, the spikes lengthening up to 7 to 15 centimetres (2.8 to 5.9 in) as the fruit matures.[15] The fruit of the black pepper is called a drupe and when dried is known as

Unripe drupes of Black Pepper (Piper nigrum*) at Trivandrum, Kerala, India*

a peppercorn.

Pepper can be grown in soil that is neither too dry nor susceptible to flooding, moist, well-drained and rich in organic matter (the vines do not do too well over an altitude of 900 m (3,000 ft) above sea level). The plants are propagated by cuttings about 40 to 50 centimetres (16 to 20 in) long, tied up to neighbouring trees or climbing frames at distances of about 2 metres (6 ft 7 in) apart; trees with rough bark are favoured over those with smooth bark, as the pepper plants climb rough bark more readily. Competing plants are cleared away, leaving only sufficient trees to provide shade and permit free ventilation. The roots are covered in leaf mulch and manure, and the shoots are trimmed twice a year. On dry soils the young plants require watering every other day during the dry season for the first three years. The plants bear fruit from the fourth or fifth year, and typically continue to bear fruit for seven years. The cuttings are usually cultivars, selected both for yield and quality of fruit.

A single stem will bear 20 to 30 fruiting spikes. The harvest begins as soon as one or two fruits at the base of the spikes begin to turn red, and before the fruit is fully mature, and

still hard; if allowed to ripen completely, the fruit lose pungency, and ultimately fall off and are lost. The spikes are collected and spread out to dry in the sun, then the peppercorns are stripped off the spikes.[*][15]

Black pepper is either native to Southeast Asia[*][16] or South Asia.[*][17] Within the genus *Piper*, it is most closely related to other Asian species such as *Piper caninum*.[*][17]

- *Piper nigrum* on tree support in Goa, India

- Pepper vine, Tiruvannamalai, Tamil Nadu, India

11.4 History

Pepper in Kerala, India

Pepper before ripening

Pepper is native to South Asia and Southeast Asia and has been known to Indian cooking since at least 2000 BCE.[*][18] J. Innes Miller notes that while pepper was grown in southern Thailand and in Malaysia, its most important source was India, particularly the Malabar Coast,

Peppercorn close-up

in what is now the state of Kerala*[19] Peppercorns were a much-prized trade good, often referred to as "black gold" and used as a form of commodity money. The legacy of this trade remains in some Western legal systems which recognize the term "peppercorn rent" as a form of a token payment made for something that is in fact being given.

The ancient history of black pepper is often interlinked with (and confused with) that of long pepper, the dried fruit of closely related *Piper longum*. The Romans knew of both and often referred to either as just "piper". In fact, it was not until the discovery of the New World and of chili peppers that the popularity of long pepper entirely declined. Chili peppers, some of which when dried are similar in shape and taste to long pepper, were easier to grow in a variety of locations more convenient to Europe.

Before the 16th century, pepper was being grown in Java, Sunda, Sumatra, Madagascar, Malaysia, and everywhere in Southeast Asia. These areas traded mainly with China, or used the pepper locally.*[20] Ports in the Malabar area also served as a stop-off point for much of the trade in other spices from farther east in the Indian Ocean. Following the British hegemony in India, virtually all of the black pepper found in Europe, the Middle East, and North Africa was traded from Malabar region.

11.4.1 Ancient times

Black peppercorns were found stuffed in the nostrils of Ramesses II, placed there as part of the mummification rituals shortly after his death in 1213 BCE.*[21] Little else is known about the use of pepper in ancient Egypt and how it reached the Nile from South Asia.

Pepper (both long and black) was known in Greece at least as early as the 4th century BCE, though it was probably an uncommon and expensive item that only the very rich could afford. Trade routes of the time were by land, or in ships which hugged the coastlines of the Arabian Sea. Long pepper, growing in the north-western part of India, was more accessible than the black pepper from further south; this trade advantage, plus long pepper's greater spiciness, probably made black pepper less popular at the time.

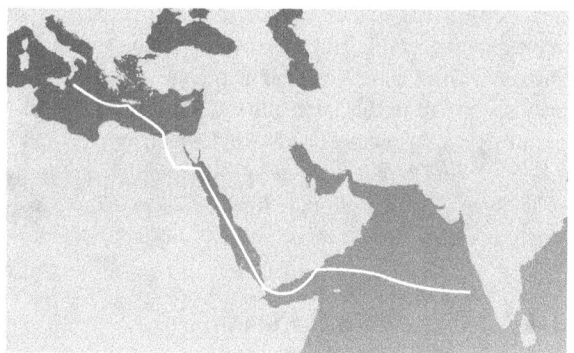

A Roman era trade route from India to Italy

By the time of the early Roman Empire, especially after Rome's conquest of Egypt in 30 BCE, open-ocean crossing of the Arabian Sea direct to southern India's Malabar Coast was near routine. Details of this trading across the Indian Ocean have been passed down in the *Periplus of the Erythraean Sea*. According to the Roman geographer Strabo, the early Empire sent a fleet of around 120 ships on an annual one-year trip to China, Southeast Asia, India and back. The fleet timed its travel across the Arabian Sea to take advantage of the predictable monsoon winds. Returning from India, the ships travelled up the Red Sea, from where the cargo was carried overland or via the Nile-Red Sea canal to the Nile River, barged to Alexandria, and shipped from there to Italy and Rome. The rough geographical outlines of this same trade route would dominate the pepper trade into Europe for a millennium and a half to come.

With ships sailing directly to the Malabar coast, black pepper was now travelling a shorter trade route than long pepper, and the prices reflected it. Pliny the Elder's *Natural History* tells us the prices in Rome around 77 CE: "Long pepper ... is fifteen denarii per pound, while that of white pepper is seven, and of black, four." Pliny also complains "there is no year in which India does not drain the Roman Empire of fifty million sesterces," and further moralizes on pepper:

It is quite surprising that the use of pepper has come so much into fashion, seeing that in other substances which we use, it is sometimes their sweetness, and sometimes their appearance that has attracted our notice; whereas, pepper has nothing in it that can plead as a recommendation to either fruit or berry, its only desirable quality being a certain pungency; and yet it is for this that we import it all the way from India! Who was the first to make trial of it as an article of food? and who, I wonder, was the man that was not content to prepare himself by hunger only for the satisfying of a greedy appetite? (*N.H.* 12.14)*[22]

Black pepper was a well-known and widespread, if expensive, seasoning in the Roman Empire. Apicius' De re coquinaria, a 3rd-century cookbook probably based at least partly on one from the 1st century CE, includes pepper in a majority of its recipes. Edward Gibbon wrote, in *The History of the Decline and Fall of the Roman Empire*, that pepper was "a favorite ingredient of the most expensive Roman cookery".

11.4.2 Postclassical Europe

Pepper was so valuable that it was often used as collateral or even currency. In the Dutch language, "pepper expensive" (*peperduur*) is an expression for something very expensive. The taste for pepper (or the appreciation of its monetary value) was passed on to those who would see Rome fall. Alaric the Visigoth included 3,000 pounds of pepper as part of the ransom he demanded from Rome when he besieged the city in 5th century.*[23] After the fall of Rome, others took over the middle legs of the spice trade, first the Persians and then the Arabs; Innes Miller cites the account of Cosmas Indicopleustes, who travelled east to India, as proof that "pepper was still being exported from India in the sixth century".*[24] By the end of the Early Middle Ages, the central portions of the spice trade were firmly under Islamic control. Once into the Mediterranean, the trade was largely monopolized by Italian powers, especially Venice and Genoa. The rise of these city-states was funded in large part by the spice trade.

A riddle authored by Saint Aldhelm, a 7th-century Bishop of Sherborne, sheds some light on black pepper's role in England at that time:

I am black on the outside, clad in a wrinkled cover,
Yet within I bear a burning marrow.
I season delicacies, the banquets of kings, and the luxuries of the table,
Both the sauces and the tenderized meats of the kitchen.
But you will find in me no quality of any worth,
Unless your bowels have been rattled by my gleaming marrow.*[1]

1. ^ Translation from Turner, p 94. The riddle's answer

is of course *pepper*.

It is commonly believed that during the Middle Ages, pepper was used to conceal the taste of partially rotten meat. There is no evidence to support this claim, and historians view it as highly unlikely: in the Middle Ages, pepper was a luxury item, affordable only to the wealthy, who certainly had unspoiled meat available as well.*[25] In addition, people of the time certainly knew that eating spoiled food would make them sick. Similarly, the belief that pepper was widely used as a preservative is questionable: it is true that piperine, the compound that gives pepper its spiciness, has some antimicrobial properties, but at the concentrations present when pepper is used as a spice, the effect is small.*[26] Salt is a much more effective preservative, and salt-cured meats were common fare, especially in winter. However, pepper and other spices certainly played a role in improving the taste of long-preserved meats.

A depiction of Calicut, India published in 1572 during Portugal's control of the pepper trade

Its exorbitant price during the Middle Ages and the monopoly on the trade held by Italy was one of the inducements which led the Portuguese to seek a sea route to India. In 1498, Vasco da Gama became the first person to reach India by sailing around Africa (see Age of Discovery); asked by Arabs in Calicut (who spoke Spanish and Italian) why they had come, his representative replied, "we seek Christians and spices". Though this first trip to India by way of the southern tip of Africa was only a modest success, the Portuguese quickly returned in greater numbers and eventually gained much greater control of trade on the Arabian sea. It was given additional legitimacy (at least from a European imperialistic perspective) by the 1494 Treaty of Tordesillas, which granted Portugal exclusive rights to the half of the world where black pepper originated.

The Portuguese proved unable to maintain their stranglehold on the spice trade for long. The old Arab and Venetian trade networks successfully 'smuggled' enormous quantities of spices through the patchy Portuguese blockade, and pepper once again flowed through Alexandria and Italy, as well as around Africa. In the 17th century, the Portuguese lost almost all of their valuable Indian Ocean trade to the Dutch and the English who, taking advantage from the Spanish ruling over Portugal (1580–1640), occupied by force almost all Portuguese dominations in the area. The pepper ports of Malabar began to trade increasingly with the Dutch in the period 1661–1663.

Pepper harvested for the European trader, from a manuscript Livre des merveilles de Marco Polo *(The book of the marvels of Marco Polo)*

Pepper mill

As pepper supplies into Europe increased, the price of pepper declined (though the total value of the import trade generally did not). Pepper, which in the early Middle Ages had been an item exclusively for the rich, started to become more of an everyday seasoning among those of more average means. Today, pepper accounts for one-fifth of the world's spice trade.[27]

11.4.3   China

It is possible that black pepper was known in China in the 2nd century BCE, if poetic reports regarding an explorer named Tang Meng (唐蒙) are correct. Sent by Emperor Wu to what is now south-west China, Tang Meng is said to have come across something called *jujiang* or "sauce-betel" . He was told it came from the markets of Shu, an area in what is now the Sichuan province. The traditional view among historians is that "sauce-betel" is a sauce made from betel leaves, but arguments have been made that it actually refers to pepper, either long or black.[28]

In the 3rd century CE, black pepper made its first definite appearance in Chinese texts, as *hujiao* or "foreign pepper" . It does not appear to have been widely known at the time, failing to appear in a 4th-century work describing a wide variety of spices from beyond China's southern border, including long pepper.[29] By the 12th century, however, black pepper had become a popular ingredient in the cuisine of the wealthy and powerful, sometimes taking the place of China's native Sichuan pepper (the tongue-numbing dried fruit of an unrelated plant).

Marco Polo testifies to pepper's popularity in 13th-century China when he relates what he is told of its consumption in the city of Kinsay (Hangzhou): "... Messer Marco heard it stated by one of the Great Kaan's officers of customs that the quantity of pepper introduced daily for consumption into the city of Kinsay amounted to 43 loads, each load being equal to 223 lbs." [30] Marco Polo is not considered a very reliable source regarding China, and this second-hand data may be even more suspect, but if this estimated 10,000 pounds (4,500 kg) a day for one city is anywhere near the truth, China's pepper imports may have dwarfed Europe's.

During the course of the treasure voyages in the early 15th century, Admiral Zheng He and his expeditionary fleets returned with such a large amount of black pepper that the once-costly luxury became a common commodity.[31]

11.5   Phytochemicals, folk medicine and research

Like many eastern spices, pepper was historically both a seasoning and a folk medicine. Long pepper, being stronger, was often the preferred medication, but both were used. Black pepper (or perhaps long pepper) was believed to cure illness such as constipation, diarrhoea, earache, gangrene, heart disease, hernia, hoarseness, indigestion, insect bites, insomnia, joint pain, liver problems, lung disease, oral abscesses, sunburn, tooth decay, and toothaches.[32] Various sources from the 5th century onward also recommend pepper to treat eye problems, often by applying salves or poultices made with pepper directly to the eye. There is no current medical evidence that any of these treatments has any benefit; pepper applied directly to the eye would be quite uncomfortable and possibly damaging.[33] Nevertheless, black pepper, either powdered or its decoction, is widely used in traditional Indian medicine and as a home remedy for relief from sore throat, throat congestion, cough, etc.

'There's certainly too much pepper in that soup!' Alice said to herself, as well as she could for sneezing. Alice in Wonderland (1865). Chapter VI: Pig and Pepper. Note the cook's pepper mill.

Pepper is known to cause sneezing. Some sources say that piperine, a substance present in black pepper, irritates the nostrils, causing the sneezing.[34] Few, if any, controlled studies have been carried out to answer the question.

Piperine is under study for its potential to increase absorption of selenium, vitamin B, beta-carotene and curcumin as well as other nutrients.[35] As a folk medicine, pepper appears in the Buddhist Samaññaphala Sutta, chapter five, as one of the few medicines allowed to be carried by a monk.[36]

Pepper contains phytochemicals,[37] including amides, piperidines, pyrrolidines and trace amounts of safrole which may be carcinogenic in laboratory rodents.[38]

Piperine is under study for a variety of possible physiological effects,[39] although this work is preliminary and mechanisms of activity for piperine in the human body remain unknown.

11.5.1 Nutrition

One tablespoon (6 grams) of ground black pepper contains moderate amounts of vitamin K (13% of the daily value or DV), iron (10% DV) and manganese (18% DV), with trace amounts of other essential nutrients, protein and dietary fibre.[40]

11.6   Flavor

Pepper gets its spicy heat mostly from piperine derived both from the outer fruit and the seed. Black pepper con-

Handheld pepper mills

Black pepper grains

tains between 4.6% and 9.7% piperine by mass, and white pepper slightly more than that.[41] Refined piperine, by weight, is about one percent as hot as the capsaicin found in chili peppers.[42] The outer fruit layer, left on black pepper, also contains important odour-contributing terpenes including pinene, sabinene, limonene, caryophyllene, and linalool, which give citrusy, woody, and floral notes. These scents are mostly missing in white pepper, which is stripped of the fruit layer. White pepper can gain some different odours (including musty notes) from its longer fermentation stage.*[43] The aroma of pepper is attributed to rotundone (3,4,5,6,7,8-Hexahydro-3α,8α-dimethyl-5α-(1-methylethenyl)azulene−1(2H)-one), a sesquiterpene originally discovered in the tubers of cyperus rotundus, which can be detected in concentrations of 0.4 nanograms/L in water and in wine: rotundone is also present in marjoram, oregano, rosemary, basil, thyme, and geranium, as well as in some Shiraz wines.*[44]

Pepper in Kolli Hills in India

Pepper loses flavour and aroma through evaporation, so airtight storage helps preserve its spiciness longer. Pepper can also lose flavour when exposed to light, which can transform piperine into nearly tasteless isochavicine.*[43] Once ground, pepper's aromatics can evaporate quickly; most culinary sources recommend grinding whole peppercorns immediately before use for this reason. Handheld pepper mills or grinders, which mechanically grind or crush whole peppercorns, are used for this, sometimes instead of pepper shakers that dispense pre-ground pepper. Spice mills such as pepper mills were found in European kitchens as early as the 14th century, but the mortar and pestle used earlier for crushing pepper have remained a popular method for centuries as well.*[45]

11.7 World trade

Peppercorns (dried black pepper) are, by monetary value, the most widely traded spice in the world, accounting for 20 percent of all spice imports in 2002. The price of pepper can be volatile, and this figure fluctuates a great deal year to year; for example, pepper made up 39 percent of all spice imports in 1998.*[46] By weight, slightly more chili peppers are traded worldwide than peppercorns.

The International Pepper Exchange is located in Kochi, India. Participation in the IPE however is domestic with regulatory restrictions on international membership on local exchanges; something common to almost all Asian commodity exchanges.

As of 2008, Vietnam is the world's largest producer and exporter of pepper, producing 34% of the world's *Piper nigrum*. Other major producers include India (19%), Brazil (13%), Indonesia (9%), Malaysia (8%), Sri Lanka (6%), China (6%), and Thailand (4%). Global pepper production peaked in 2003 with over 355,000 t (391,000 short tons), but has fallen to just over 271,000 t (299,000 short tons) by 2008 due to a series of issues including poor crop management, disease and weather. Vietnam dominates the export market, using almost none of its production domestically; however its 2007 crop fell by nearly 10% from the previous year to about 90,000 t (99,000 short tons). Similar crop yields occurred in 2007 across the other pepper producing nations as well.*[47] Nowadays, in England, industrial buyers mix Peppers of different origin to maintain a balance between price, taste and other factors. Malabar black peppers are used for weight and taste, Sumatra for colour, and Penang for strength.*[48]

11.8 See also

- Peppercorn sauce

- Salt

11.9 Notes and references

[1] "Piper nigrum information from NPGS/GRIN" . www.ars-grin.gov. Retrieved 2 March 2008.

[2] Dravidian India - T.R. Sesha Iyengar - Google Books. Books.google.com. Retrieved on 31 October 2012.

[3] Intercourse Between India and the Western World - H. G. Rawlinson - Google Books. Books.google.com. Retrieved on 31 October 2012.

[4] Antiquities of India: An Account of the History and Culture of Ancient Hindustan - Lionel D. Barnett - Google Books. Books.google.com. Retrieved on 31 October 2012.

[5] "Pepper" . Tamilnadu.com. 30 October 2012.

[6] Douglas Harper's *Online Etymology Dictionary* entries for *pepper* and *pep*. Retrieved 13 November 2005.

[7] "Cleaner technology for white pepper production" . *The Hindu Business line*. 27 March 2008. Retrieved 29 January 2009.

[8] See Thai Ingredients Glossary. Retrieved 6 November 2005.

[9] Ochef, Using fresh green peppercorns. Retrieved 6 November 2005.

[10] Manjunath Hegde, Bomnalli (19 October 2013). "Meet the pepper queen" (Bangalore). Deccan Herald. Retrieved 22 January 2015.

[11] Katzer, Gernot (2006). Pepper. Gernot Katzer's Spice Pages. Retrieved 2 December 2012.

[12] Peppercorns, from Penzeys Spices. Retrieved 17 October 2006.

[13] Pepper varieties information from A Cook's Wares. Retrieved 6 November 2005.

[14] *Cambodia*. Lonely Planet. 1988. p. 225. GGKEY: ALKFLS6LY8Y.

[15] "Black Pepper Cultivation and Harvest" . Thompson Martinez. Retrieved 14 May 2014.

[16] "Piper nigrum Linnaeus" . *Flora of China*.

[17] Jaramillo, M. Alejandra; Manos (2001). "Phylogeny and Patterns of Floral Diversity in the Genus Piper (Piperaceae)". *American Journal of Botany* **88** (4): 706–16. doi:10.2307/2657072. PMID 11302858.

[18] Davidson & Saberi 178

[19] J. Innes Miller, *The Spice Trade of the Roman Empire* (Oxford: Clarendon Press, 1969), p. 80

[20] Dalby p. 93.

[21] Stephanie Fitzgerald (8 September 2008). *Ramses II, Egyptian Pharaoh, Warrior, and Builder*. Compass Point Books. p. 88. ISBN 0-7565-3836-X. Retrieved 29 January 2008.

[22] From Bostock and Riley's 1855 translation. Text online.

[23] J. Norwich, Byzantium: The Early Centuries, 134

[24] Innes Miller, *The Spice Trade*, p. 83

[25] Dalby p. 156; also Turner pp. 108–109, though Turner does go on to discuss spices (not pepper specifically) being used to disguise the taste of partially spoiled wine or ale.

[26] H. J. D. Dorman and S. G. Deans (2000). "Antimicrobial agents from plants: antibacterial activity of plant volatile oils" . *Journal of Applied Microbiology* **88** (2): 308–16. doi:10.1046/j.1365-2672.2000.00969.x. PMID 10736000.. Full text at Blackwell website; purchase required. "Spices, which are used as integral ingredients in cuisine or added as flavouring agents to foods, are present in insufficient quantities for their antimicrobial properties to be significant."

[27] Jaffee, p. 10.

[28] Dalby pp. 74–75. The argument that *jujiang* was long pepper goes back to the 4th century CE botanical writings of Ji Han; Hui-lin Li's 1979 translation of and commentary on Ji Han's work makes the case that it was *piper nigrum*.

[29] Dalby p. 77.

[30] Yule, Henry; Cordier, Henri, Translation from *The Travels of Marco Polo: The Complete Yule-Cordier Edition*, Vol. 2, *Dover. ISBN 0-486-27587-6. p. 204.*

[31] Finlay, Robert (2008). "The Voyages of Zheng He: Ideology, State Power, and Maritime Trade in Ming China" . *Journal of the Historical Society* **8** (3): 337. doi:10.1111/j.1540-5923.2008.00250.x.

[32] Turner p. 160.

[33] Turner p. 171.

[34] U.S. Library of Congress Science Reference Services "Everyday Mysteries" , Why does pepper make you sneeze?. Retrieved 12 November 2005.

[35] Dudhatra, GB; Mody, SK; Awale, MM; Patel, HB; Modi, CM; Kumar, A; Kamani, DR; Chauhan, BN (2012). "A comprehensive review on pharmacotherapeutics of herbal bioenhancers". *The Scientific World Journal* **2012** (637953): 637953. doi:10.1100/2012/637953. PMC 3458266. PMID 23028251.

[36] Thanissaro Bhikkhu (30 November 1990). *Buddhist Monastic Code II*. Cambridge University Press. ISBN 0-521-36708-5. Retrieved 29 January 2008.

[37] Dawid, Corinna; Henze, Andrea; Frank, Oliver; Glabasnia, Anneke; Rupp, Mathias; Büning, Kirsten; Orlikowski, Diana; Bader, Matthias; Hofmann, Thomas (2012). "Structural and Sensory Characterization of Key Pungent and Tingling Compounds from Black Pepper (*Piper nigrum* L.)". *Journal of Agricultural and Food Chemistry* **60** (11): 2884–2895. doi:10.1021/jf300036a. PMID 22352449.

[38] James A. Duke (16 August 1993). *CRC Handbook of Alternative Cash Crops*. CRC Press. p. 395. ISBN 0-8493-3620-1. Retrieved 29 January 2009.

[39] Srinivasan K (2007). "Black pepper and its pungent principle-piperine: a review of diverse physiological effects" . *Crit Rev Food Sci Nutr* **47** (8): 735–48. doi:10.1080/10408390601062054. PMID 17987447.

[40] "Nutrition facts for black pepper, one tablespoon (6 g); USDA Nutrient Database, version SR-21" . Conde Nast. 2014. Retrieved 25 October 2014.

[41] Pepper. Tis-gdv.de. Retrieved on 31 October 2012.

[42] Lawless, Harry T.; Heymann, Hildegarde (2010). *Sensory Evaluation of Food: Principles and Practices.* Springer. p. 43. ISBN 1441964886.

[43] McGee p. 428.

[44] Siebert, Tracey E.; Wood, Claudia; Elsey, Gordon M.; Alan (2008). "Determination of Rotundone, the Pepper Aroma Impact Compound, in Grapes and Wine" . *J. Agric. Food Chem* **56** (10): 3745–3748. doi:10.1021/jf800184t. PMID 18461962.

[45] Montagne, Prosper (2001). *Larousse Gastronomique.* Hamlyn. p. 726. ISBN 0-600-60235-4. OCLC 47231315 50747863 83960122. "Mill" .

[46] Jaffee p. 12, table 2.

[47] "Karvy's special Reports Seasonal Outlook Report Pepper" (PDF). Karvy Comtrade Limited. 15 May 2008. Retrieved 29 January 2008.

[48] "Black Pepper" . Regency as China Business Limited. 1 January 2014. Retrieved 20 May 2014.

11.10 Bibliography

- Dalby, Andrew (2002). *Dangerous Tastes.* Berkeley: University of California Press. ISBN 0-520-23674-2.

- Davidson, Alan (2002). *Wilder Shores of Gastronomy: Twenty Years of the Best Food Writing from the Journal Petits Propos Culinaires.* Berkeley: Ten Speed Press. ISBN 978-1-58008-417-8.

- Jaffee, Steven (2004). "Delivering and Taking the Heat: Indian Spices and Evolving Process Standards" (PDF). *An Agriculture and Rural Development Discussion Paper* (Washington: World Bank).

- McGee, Harold (2004). "Black Pepper and Relatives" . *On Food and Cooking (Revised Edition).* Scribner. pp. 427–429. ISBN 0-684-80001-2. OCLC 56590708.

- Turner, Jack (2004). *Spice: The History of a Temptation.* London: Vintage Books. ISBN 0-375-70705-0. OCLC 61213802.

11.11 Further reading

- Black Pepper Chemical List (Dr. Duke's Databases)

- "Black Pepper" from Plant Cultures, a collaboration between NYKRIS and Kew Gardens

- Ravindran, P.N. (2000). *Black pepper: piper nigrum.* Amsterdam: Harwood Academic, CRC. ISBN 978-90-5702-453-5

11.12 External links

- Media related to Piper nigrum at Wikimedia Commons

- Data related to Piper nigrum at Wikispecies

- Pepper at Wikibook Cookbooks

Chapter 12

Boscia senegalensis

Boscia senegalensis, or **hanza**, is a member of the family Capparaceae.

The plant originated from West Africa. Still a traditional food plant in Africa, this little-known fruit has potential to improve nutrition, boost food security, foster rural development and support sustainable landcare.*[1]

B. senegalensis is a perennial woody plant species of the *Boscia* genus in the caper (Capparaceae) family.*[2] This plant is classified as a dicot. Native to the Sahel region in Africa, this evergreen shrub can grow anywhere from 2 to 4 m (6 ft 7 in to 13 ft 1 in) in height under favourable conditions. The leaves of the plant are small and leathery, reaching 12 cm × 4 cm (4.7 in × 1.6 in).*[2] *B. senegalensis* produces fruits, clustered in small bunches, in the form of yellow spherical berries, up to 1.5 cm (0.59 in) in diameter. These fruits contain 1–4 seeds, which are a greenish hue when mature.

B. senegalensis is recognized as a potential solution to hunger and a buffer against famine in the Sahel region due to the variety of useful products it yields. It produces products for consumption, household needs, and medicinal and agricultural uses.

Other common names include: *aizen* (Mauritania), *mukheit* (Arabic), *hanza* (Hausa), *bere* (Bambara), *ngigili* (Fulani), and *mandiarha* (Berber). The fruits are also known as *dilo* (Hausa), *bokkhelli* (Arabic), *gigile* (Fulani).*[1]

12.1   History, geography and ethnography

B. senegalensis is a wild species, native to the Sahel region in Africa. It has not yet been domesticated. It currently grows in: Algeria, Benin, Burkina Faso, Cameroon, Central African Republic, Chad, Ghana, Guinea, Kenya, Mali, Mauritania, Niger, Nigeria, Senegal, Somalia, Sudan, and Togo.*[2]

Ethnobotanical indigenous knowledge contributes to the importance of this plant to the Hausa peoples of Niger and Fulani herders in West Africa. During the famine of 1984–1985, it was reported that *B. senegalensis* was the most widely consumed famine food in both Sudan and Darfur, relied on by over 94% of people in northern Darfur.*[1]

12.2   Growing conditions

B. senegalensis grows in altitudes of 60–1,450 m (200–4,760 ft), in temperatures between 22–30 °C (72–86 °F) and with rainfall conditions of 100–500 mm (3.9–19.7 in) annually. It can be found growing in marginal soils: rocky, lateritic, clay stony hills, sand dunes, and sand-clay plains.*[2] These characteristics make it a highly resilient species, able to grow without expensive inputs even in the extremely hot and dry desert region of the Sahel. Herein lies its significance for poor farmers – in times of severe drought and famine, when many other crops have failed, *B. senegalensis* can still survive and provide useful products.

12.3   Other farming issues

Boscia senegalensis - Occurrence in field

B. senegalensis can benefit farmers because it keeps soil from laying bare and thus prevents soil erosion and degradation. It also buffers against wind, stabilizes sand dunes, offers shade to surrounding plants and cycles nutrients.[*][3] In Niger, the trees are often cut or burned down by farmers in the dry season, in order to make space on the field for staple crops as millet or sorghum. However, due to the strong surviving character of the tree, it reappears after the first rains and continues growing as a small bush.

12.4   How consumed and uses

Fruits are ready for human consumption at the beginning of the rainy season, when most crops are just being planted, and there is little other food available. Fruits can be consumed raw and cooked. Raw fruits initially contain a sweet pulp that then dries out to a sugary solid, difficult to separate from seed. Fruits are often cooked prior to consumption. Juice can also be extracted and boiled down into a butter-like consistency that can be mixed with millet and milk to make cakes.[*][1] In Sudan, the fruit is fermented into a beer.[*][2]

The seeds of *B. senegalensis* are also important sources of nutrition, especially during times of famine.[*][4] To gain access to the seeds, fruits are dried in the sun, pounded to remove the outer seed coat and soaked in water for several days, changing the water every day.[*][5] The seed soaking process, also known as *debittering*, is essential to remove bitter and potentially toxic components. Seeds are usually cooked prior to consumption. Cooked seeds are texturally similar to a chickpea and can be used as a cereal substitute in stews, soups and porridges. Additionally, seeds can be re-dried and stored for later use or ground into a flour that can be used to make porridge. Roasted seeds can also serve as a substitute for coffee.[*][1]

Hanza bread, cookies and cooked hanza, Zinder, Republic of Niger

Modern uses of *B. senegalensis* seeds are being developed in Niger Republic. They include cakes, cookies, bread, canned and popped seeds. These products from natural, wild *B. senegalensis* were recognised with the innovation award at an international food fair in Niamey, Niger, 2012.[*][6]

Leaf extracts contain carbohydrate hydrolase enzymes that are useful for the production of cereal-based flour and for reducing the bulk of cereal porridges.[*][7] Due to their proven biocidal activities, leaves are also added to granaries to protect cereals against pathogens. Leaves have many medicinal properties, notably anti-parasitic, fungicidal, anti-inflammatory and wound healing properties.[*][2] Leaves, although not pleasant to taste, can be used as emergency forage for animals.[*][1]

Young roots can be ground and boiled down into a thick, sweet porridge.

Wood can be used for home construction as well as for cooking fuel in times of dire need.[*][1]

B. senegalensis contains natural coagulants that can be used to clarify water sources. Components of the plant (bark, twigs, leaves, fruits) can be added to a bucket of murky water, and the natural coagulants will cause clay and other particulates to compact and sink the bottom, allowing clear water to be obtained from the top.[*][1]

12.5   Nutritional information

Fruits are a significant source of carboydrates, as they contain 66.8% carbohydrates.[*][2]

The seeds are sufficiently nutritious, although they do lack some essential nutrients, notably lysine and threonine. The seeds have significant levels of protein (25% of dry matter) and carbohydrates (60%). In these regards, seeds outperform local staple cereals such as sorghum and millet. Additionally, seeds are rich in zinc, iron, methionine, tryptophan, B-vitamins and linoleic acid (essential fatty acid).[*][5] Seeds contain 3.6 times the World Health Organization (WHO) ideal level of tryptophan.[*][8]

Leaves have high antioxidant capacity (nearly 1.5 times that of spinach) and are high in calcium, potassium, manganese and iron.[*][8] The bioavailability of these compounds, however, is not very well known.[*][9]

12.6   Economics

Leaves, seeds and fruits of *B. senegalensis* are traded in many small markets in the Sahel region.[*][1] Some opportunities to add value are: roasting seeds to be sold as a coffee bean substitute, fermenting fruit into beer, processing fruit and seeds into prepared food, or processing leaves into medicinal applications. It can help raise incomes of the poor by protecting their stored cereals from pests and by substituting for other purchases from the market.

12.7 Gender Issues

Women in rural areas usually have the responsibility of gathering and preparing *B. senegalensis* for consumption. This process can create an extra work burden for women, however, their dominion over this process may result in increased access to this food source and thus contribute to improving their nutritional status.*[10]

12.8 Constraints to wider adoption

A major constraint to the wider adoption of *B. senegalensis* is the recalcitrant nature of its seeds. Seeds of this type are not well suited for ex-situ conservation, as they rapidly lose viability, and embryos are killed when seeds are dried.*[3] This creates a barrier to widespread growth, as it is difficult to propagate large numbers of plants for large-scale genetic selection and breeding. Other drawbacks to consumption include the issue of toxicity and the associated need to use scarce water resources and additional labour to leach out toxins during the debittering process.*[11]

12.9 Practical information

One intervention with the potential to help poor farmers is the creation of cool temperature storage facilities – as *B. senegalensis* seeds can be stored for up to 2 months at 15 °C (59 °F).*[11]

It is imperative to spread knowledge of the wide range of benefits that *B. senegalensis* provides, in order to encourage small farmers to plant it. New plantings would offer increased protection to the soil as well as provide food and other resources in times of famine.

It is recommended that the techniques of grafting and generating hybrids (wide-crosses) with related species be explored, as both techniques have the potential to increase harvests and/or improve the fruits.*[1] Promising preliminary research is being conducted using in vitro tissue culture technologies to propagate *B. senegalensis*.*[3] Additionally, direct seedling trials are recommended and being advanced by the Eden Foundation.*[12]

12.10 References

[1] National Research Council (2008-01-25). "Aizen (Mukheit)". *Lost Crops of Africa: Volume III: Fruits*. Lost Crops of Africa **3**. National Academies Press. ISBN 978-0-309-10596-5. Retrieved 2008-07-25.

[2] Booth, F E M; Wickens, G E (1988). "Boscia Senegalensis" . *Non-timber Uses of Selected Arid Zone Trees and Shrubs in Africa*. Rome: Food and Agricultural Organization. ISBN 9789251027455.

[3] Khalafalla, M M; Daffalla, H M; Abdellatef, E; Agabna, E; El-Shemy, H A (April 2011). "Establishment of an in vitro micropropagation protocol for *Boscia senegalensis* (Pers.) Lam. ex Poir" . *Journal of Zhejiang University Science Biomedicine & Biotechnology* **12** (4): 303–312. doi:10.1631/jzus.B1000205. PMC 3072594. PMID 21462387.

[4] Eden Foundation (2006), *When Endemic Malnutrition is Labeled as Famine*

[5] Salih, O M; Nour, A M; Harper, D B (1991). "Chemical and nutritional composition of two famine food sources used in Sudan, Mukheit (*Boscia senegalensis*) and Maikah (*Dobera roxburghi*)". *Journal of the Science of Food and Agriculture* **57** (3): 367–377. doi:10.1002/jsfa.2740570307. (subscription required)

[6] "Aridité Prospère" . Retrieved 2013-04-04.

[7] Dicko, M H; Leeuwen, M S; Traore, A S; Hilhorst, R; Beldman, G (2001). "Polysaccharide hydrolases from leaves of *Boscia senegalensis*: properties of endo-(1-3)-β-D-glucanase" (PDF). *Applied Biochemistry and Biotechnology* **94** (3): 225–241. PMID 11563825.

[8] Cook, J A; VanderJagt, D J; Dasgupta, A; Mounkaili, G; Glew, R S; Blackwell, W; Glew, R H (1998). "Use of the Trolox assay to estimate the antioxidant content of seventeen edible wild plants of Niger" . *Life Sciences* **63** (2): 105–110. doi:10.1016/S0024-3205(98)00245-8. PMID 9674944. (subscription required)

[9] Cook, J A; VanderJagt, D J; Pastuszym, A; Mounkaila, G; Glew, R S; Millson, M; Glew, R H (2000). "Nutrient and chemical composition of 13 wild plant foods of Niger" . *Journal of Food Composition and Analysis* **13** (1): 83–92. doi:10.1006/jfca.1999.0843. (subscription required)

[10] Becker, B (1983). "The contribution of wild plants to human nutrition in the Ferlo (northern Senegal)" (PDF). *Agroforestry Systems* **1** (3): 257–267. doi:10.1007/BF00130611.

[11] Danthu, P.; Gueye, A; Boye, A; Bauwens, D; Sarr, A (2000). "Seed storage behaviour of four Sahelian and Sudanian tree species". *Seed Science Research* **10** (2): 183–187. (subscription required)

[12] "Eden Foundation" . Retrieved 2012-12-16.

12.11 External links

- Photograph of aizen fruit

- Activities based on *Boscia senegalensis* seeds

- *Boscia senegalensis* in West African plants – A Photo Guide.

- Crop of the Week: Hanza (*Boscia senegalensis*)

- TalTV report on the use of Hanza

Chapter 13

Camphor

For other uses, see Camphor (disambiguation).

Camphor (/ˈkæmfər/) is a waxy, flammable, white or transparent solid with a strong aromatic odor.[*][5] It is a terpenoid with the chemical formula $C_{10}H_{16}O$. It is found in the wood of the **camphor laurel** (*Cinnamomum camphora*), a large evergreen tree found in Asia (particularly in Sumatra, Indonesia and Borneo) and also of the unrelated *kapur tree*, a tall timber tree from the same region. It also occurs in some other related trees in the laurel family, notably *Ocotea usambarensis*. Dried rosemary leaves (*Rosmarinus officinalis*), in the mint family, contain up to 20% camphor. Camphor can also be synthetically produced from oil of turpentine. It is used for its scent, as an ingredient in cooking (mainly in India), as an embalming fluid, for medicinal purposes, and in religious ceremonies. A major source of camphor in Asia is camphor basil (the parent of African blue basil).

Norcamphor is a camphor derivative with the three methyl groups replaced by hydrogen.

13.1 Etymology

The word camphor derives from the French word *camphre*, itself from Medieval Latin *camfora*, from Arabic *kafur*, from Sanskrit, कर्पूरम् / *karpūram*.[*][6] The term ultimately was derived from Old Malay *kapur barus* which means "the chalk of Barus" . Barus was the name of an ancient port located near modern Sibolga city on the western coast of Sumatra island (today North Sumatra Province, Indonesia).[*][7] This port was initially built prior to the Indian–Batak trade in camphor, benzoin and spices. Traders from India, East Asia and the Middle East would use the term *kapur barus* to buy the dried extracted ooze of camphor laurel trees (*Cinnamonum camphora*) from local Batak tribesmen; the camphor tree itself is natively found in that region. In the proto-Malay-Austronesian language, it is also known as *kapur Barus*. Even now, the local tribespeople and In-

donesians in general refer to aromatic naphthalene balls and moth balls as *kapur Barus*.

13.2 Production

A sample of sublimed camphor

In the 19th century, it was known that with nitric acid, camphor could be oxidized into camphoric acid. Haller and Blanc published a semisynthesis of camphor from camphoric acid, which, although demonstrating its structure, would not prove it. The first complete total synthesis for camphoric acid was published by Gustaf Komppa in 1903. Its starting materials were diethyl oxalate and 3,3-dimethylpentanoic acid, which reacted by Claisen condensation to give diketocamphoric acid. Methylation with methyl iodide and a complicated reduction procedure produced camphoric acid. William Perkin published another synthesis a short time later. Previously, some organic compounds (such as urea) had been synthesized in the laboratory as a proof of concept, but camphor was a scarce natural product with a worldwide demand. Komppa realized this and began industrial production of camphor in Tainionkoski, Finland, in 1907.

Camphor can be produced from alpha-pinene, which is abundant in the oils of coniferous trees and can be distilled from turpentine produced as a side product of chemical pulping. With acetic acid as the solvent and with catalysis by a strong acid, alpha-pinene readily rearranges into camphene, which in turn undergoes Wagner-Meerwein rearrangement into the isobornyl cation, which is captured by acetate to give isobornyl acetate. Hydrolysis into isoborneol followed by oxidation gives racemic camphor. By contrast, camphor occurs naturally as D-camphor, the (*R*)-enantiomer.

13.3 Biosynthesis

In biosynthesis, camphor is produced from geranyl pyrophosphate, via cyclisation of linaloyl pyrophosphate to bornyl pyrophosphate, followed by hydrolysis to borneol and oxidation to camphor.

Biosynthesis of camphor from geranyl pyrophosphate

13.4 Reactions

Typical camphor reactions are

- bromination,

- oxidation with nitric acid,

- conversion to isonitrosocamphor.

Camphor can also be reduced to isoborneol using sodium borohydride.

In 1998, K. Chakrabarti and coworkers from the Indian Association for the Cultivation of Science, Kolkata, prepared diamond thin film using camphor as the precursor for chemical vapor deposition.[8]

In 2007, carbon nanotubes were successfully synthesized using camphor in chemical vapor deposition process.[9]

13.5 Uses

The sublimating capability of camphor gives it several uses.

13.5.1 Explosives

Camphor is used as a plasticizer for nitrocellulose, an ingredient for fireworks and explosive munitions.

13.5.2 Pest deterrent and preservative

Camphor is believed to be toxic to insects and is thus sometimes used as a repellent.[10] Camphor is used to make mothballs. Camphor crystals are sometimes used to prevent damage to insect collections by other small insects. Some folk remedies state camphor will deter snakes and other reptiles due to its strong odor. It is kept in clothes used on special occasions and festivals, and also in cupboard corners as a cockroach repellent.

Camphor is also used as an antimicrobial substance. In embalming, camphor oil was one of the ingredients used by ancient Egyptians for mummification.[11]

Solid camphor releases fumes that form a rust-preventative coating and is therefore stored in tool chests to protect tools against rust.[12]

13.5.3 Culinary

In ancient and medieval Europe, camphor was used as an ingredient in sweets. It was used in a wide variety of both savory and sweet dishes in medieval Arabic language cookbooks, such as *al-Kitab al-Ṭabikh* compiled by ibn Sayyâr al-Warrâq in the 10th century,[13] and an anonymous Andalusian cookbook of the 13th century.[14] It also appears in sweet and savory dishes in a book written in the late 15th century for the sultans of Mandu, the *Ni'matnama*.[15] An early international trade in it made camphor widely known throughout Arabia in pre-Islamic times, as it is mentioned in the Quran 76:5 as a flavoring for drinks.[16] By the 13th century, it was used in recipes everywhere in the Muslim world, ranging from main dishes such as *tharid* and stew to desserts.[14]

Currently, camphor is used as a flavoring, mostly for sweets, in Asia. It is widely used in cooking, mainly for dessert dishes, in India where it is known as *kachha karpooram* or "pachha karpoora" ("crude/raw camphor"), in (Telugu:పచ్చ కర్పూరం), (Tamil:பச்சைக் கற்பூரம்), (Kannada:ಪಚ್ಚ ಕರ್ಪೂರ), and is available in Indian grocery stores where it is labeled as "edible camphor".

13.5.4 Medicinal

Camphor is readily absorbed through the skin producing either a coolness or warmth sensation,[17][18] and acts as slight local anesthetic and antimicrobial substance. There are anti-itch gels and cooling gels with camphor as the active ingredient. It is used carefully and in low dosage in baby oil for its calming effects.

Camphor is an active ingredient (along with menthol) in vapor-steam products, such as Vicks VapoRub. It is used as a cough suppressant[19] and as a decongestant.[19] It is also used for aromatherapy.

Camphor may also be administered orally in small quantities (50 mg) for minor heart symptoms and fatigue.[20] Through much of the 1900s this was sold under the trade name Musterole; production ceased in the 1990s.

Camphor was used in ancient Sumatra to treat sprains, swellings, and inflammation.[21] Camphor is a component of paregoric, an opium/camphor tincture from the 18th century. Also in the 18th century, camphor was used by Auenbrugger in the treatment of mania.[22] Based on Hahnemann's writings, camphor (dissolved in alcohol) was also successfully used to treat the 1854-1855 cholera epidemics in Naples.[23]

Small dose

Its effects on the body include tachycardia, vasodilation in skin (flushing), slower breathing, reduced appetite, increased secretions and excretions such as perspiration, diuretic. [24]

The sensation of heat or cold that camphor produces is caused by activating the ion channel TRPV3.[18][25]

Large dose toxicity

Camphor is poisonous in large doses. It produces symptoms of irritability, disorientation, lethargy, muscle spasms, vomiting, abdominal cramps, convulsions, and seizures.[26][27][28] Lethal doses in adults are in the range 50–500 mg/kg (orally). Generally, two grams cause serious toxicity and four grams are potentially lethal.[29]

Regulation

In 1980, the US Food and Drug Administration set a limit of 11% allowable camphor in consumer products, and totally banned products labeled as camphorated oil, camphor oil, camphor liniment, and camphorated liniment (except "white camphor essential oil", which contains no significant amount of camphor). Since alternative treatments exist, medicinal use of camphor is discouraged by the FDA, except for skin-related uses, such as medicated powders, which contain only small amounts of camphor.

13.5.5 Hindu religious ceremonies

Camphor is widely used in Hindu religious ceremonies. Hindus worship a holy flame by burning camphor, which forms an important part of many religious ceremonies. Camphor is used in the Mahashivratri celebrations of Shiva, the Hindu god of destruction and (re)creation. As a natural pitch substance, it burns cool without leaving an ash residue, which symbolizes consciousness. Most temples in southern India have stopped lighting paraffin wax camphor in the main *sanctum sanctorum* because of the heavy carbon deposits it produces; however, they still burn it in open areas.

In Gujarat, Tamil Nadu, Andhra Pradesh, Uttar Pradesh, Madhya Pradesh, Maharashtra, Karnataka, Kerala & Andamans, camphor is the primary ingredient in any holy ritual. At the end of a holy ritual, camphor flame (called *aarti*) is burned for the deities.

In some Hindu puja ceremonies, camphor is burned in a ceremonial plate for performing *aarti*. This type of camphor, the processed white crystalline paraffin wax kind, is sold at Indian grocery stores.

13.6 See also

- 1,4-Dichlorobenzene

- Citral

- Eucalyptol

- Lavender

- Vaporizer

13.7   References

[1] *The Merck Index*, 7th edition, Merck & Co., Rahway, New Jersey, USA, 1960

[2] *Handbook of Chemistry and Physics*, CRC Press, Ann Arbor, Michigan, USA

[3] "NIOSH Pocket Guide to Chemical Hazards #0096" . National Institute for Occupational Safety and Health (NIOSH).

[4] "Camphor (synthetic)". National Institute for Occupational Safety and Health (NIOSH). 4 December 2014. Retrieved 19 February 2015.

[5] Mann JC, Hobbs JB, Banthorpe DV, Harborne JB (1994). *Natural products: their chemistry and biological significance*. Harlow, Essex, England: Longman Scientific & Technical. pp. 309–11. ISBN 0-582-06009-5.

[6] Camphor at the Online Etymology Dictionary

[7] Drakard, Jane (1989). "An Indian Ocean Port: Sources for the Earlier History of Barus" . *Archipel* **37**: 53-82. Retrieved 9 October 2015.

[8] Chakrabarti K,Chakrabarti R, Chattopadhyay KK, Chaudhuri S, Pal AK (1998). "Nano-diamond films produced from CVD of camphor" . *Diam Relat Mater* **7** (6): 845–52. Bibcode:1998DRM.....7..845C. doi:10.1016/S0925-9635(97)00312-9.

[9] Kumar M, Ando Y (2007). "Carbon Nanotubes from Camphor: An Environment-Friendly Nanotechnology" . *J Phys Conf Ser.* **61**: 643–6. Bibcode:2007JPhCS..61..643K. doi:10.1088/1742-6596/61/1/129.

[10] The Housekeeper's Almanac, or, the Young Wife's Oracle! for 1840!. No. 134. New-York: Elton, 1840. Print.

[11] http://www.newscientist.com/article/ dn1475-mummymaking-complexity-revealed.html# .VKVwGMksoow

[12] Tips for Cabinet Making Shops

[13] Nasrallah, Nawal (2007). *Annals of the Caliphs' Kitchens: Ibn Sayyâr al-Warrâq's Tenth-century Baghdadi Cookbook*. Islamic History and Civilization, 70. Leiden, The Netherlands: Brill. ISBN 978-0-415-35059-4.

[14] An Anonymous Andalusian cookbook of the 13th century, translated from the original Arabic by Charles Perry

[15] Titley, Norah M. (2004). *The Ni'matnama Manuscript of the Sultans of Mandu: The Sultan's Book of Delights*. Routledge Studies in South Asia. London, UK: Routledge. ISBN 978-0-415-35059-4.

[16] *[Quran 76:5]

[17] Moqrich, A.; Hwang, Sun Wook; Earley, Taryn J.; Petrus, Matt J.; Murray, Amber N.; Spencer, Kathryn S. R.; Andahazy, Mary; Story, Gina M.; Patapoutian, Ardem (2005). "Impaired Thermosensation in Mice Lacking TRPV3, a Heat and Camphor Sensor in the Skin" . *Science* **307** (5714): 1468–72. Bibcode:2005Sci...307.1468M. doi:10.1126/science.1108609. PMID 15746429.

[18] Green, B. G. (1990). "Sensory characteristics of camphor" . *The Journal of investigative dermatology* **94** (5): 662–6. PMID 2324522.

[19] http://www.drugs.com/cdi/camphor-liquid.html

[20] Lääketietokeskus. *Lääkevalmisteet Pharmaca Fennica 1996*, p. 814.

[21] Miller, Charles. *History of Sumatra : An account of Sumatra*. p. 121.

[22] Pearce, J.M.S. (2008). "Leopold Auenbrugger: Camphor-Induced Epilepsy – Remedy for Manic Psychosis" . *European Neurology* **59** (1–2): 105–7. doi:10.1159/000109581. PMID 17934285.

[23] Bayes (1866). "Cholera, as Treated by Dr. Rubini" . *The American Homoeopathic Review* **6** (11–12): 401–3.

[24] Church, John (1797). *An inaugural dissertation on camphor: submitted to the examination of the Rev. John Ewing, S.S.T.P. provost ; the trustees & medical faculty of the University of Pennsylvania, on the 12th of May, 1797 ; for the degree of Doctor of Medicine*. University of Philadelphia: Printed by John Thompson. Retrieved January 18, 2013.

[25] Moqrich, A.; Hwang, Sun Wook; Earley, Taryn J.; Petrus, Matt J.; Murray, Amber N.; Spencer, Kathryn S. R.; Andahazy, Mary; Story, Gina M.; Patapoutian, Ardem (2005). "Impaired Thermosensation in Mice Lacking TRPV3, a Heat and Camphor Sensor in the Skin" . *Science* **307** (5714): 1468–72. Bibcode:2005Sci...307.1468M. doi:10.1126/science.1108609. PMID 15746429.

[26] "Camphor overdose" . *Medline*. NIH. Retrieved January 19, 2012.

[27] Martin D, Valdez J, Boren J, Mayersohn M (Oct 2004). "Dermal absorption of camphor, menthol, and methyl salicylate in humans" . *Journal of Clinical Pharmacology* **44** (10): 1151–7. doi:10.1177/0091270004268409. PMID 15342616.

[28] Uc A, Bishop WP, Sanders KD (Jun 2000). "Camphor hepatotoxicity" . *Southern Medical Journal* **93** (6): 596–8. doi:10.1097/00007611-200006000-00011. PMID 10881777.

[29] "Poisons Information Monograph: Camphor". International Programme on Chemical Safety.

13.8   External links

- INCHEM at IPCS (International Programme on Chemical Safety)

- NIOSH Pocket Guide to Chemical Hazards - Camphor at Centers for Disease Control and Prevention

Chapter 14

Carnauba wax

Carnauba wax

Carnauba palm

Carnauba (/kɑːrˈnɔːbə/ ʊɪ /kɑːrˈnaʊbə/, *carnaúba*, Portuguese pronunciation: [kaʁnɐˈubɐ]), also called **Brazil wax** and **palm wax**, is a wax of the leaves of the palm *Copernicia prunifera* (Synonym: *Copernicia cerifera*), a plant native to and grown only in the northeastern Brazilian states of Piauí, Ceará, and Rio Grande do Norte.[*][1] It is known as "queen of waxes" [*][2] and in its pure state, usually comes in the form of hard yellow-brown flakes. It is obtained from the leaves of the carnauba palm by collecting and drying them, beating them to loosen the wax, then refining and bleaching the wax.

14.1 Composition

Carnauba consists mostly of aliphatic esters (40 wt%), diesters of 4-hydroxycinnamic acid (21.0 wt%), ω-hydroxycarboxylic acids (13.0 wt%), and fatty acid alcohols (12 wt%). The compounds are predominantly derived from acids and alcohols in the C26-C30 range. Distinctive for carnauba wax is the high content of diesters as well as methoxycinnamic acid.[*][3]

Carnauba wax is sold in several grades, labeled T1, T3, and T4, depending on the purity level. Purification is accomplished by filtration, centrifugation, and bleaching.

Candy coated with carnauba wax

14.2 Properties

Carnauba wax can produce a glossy finish and as such is used in automobile waxes, shoe polishes, dental floss, food products such as sweets, instrument polishes, and floor and furniture waxes and polishes, especially when mixed with beeswax and with turpentine. Use for paper coatings is the most common application in the United States. It was commonly used in its purest form as a coating on speedboat hulls in the early 1960s to enhance speed and aid in handling in salt water environments. It is also the main ingredient in surfboard wax, combined with coconut oil.

Because of its hypoallergenic and emollient properties as well as its shine, carnauba wax appears as an ingredient in many cosmetics formulas where it is used to thicken lipstick, eyeliner, mascara, eye shadow, foundation, deodorant, various skin care preparations, sun care preparations, etc. It is also used to make cutler's resin.

It is the finish of choice for most briar tobacco or smoking pipes. It produces a high gloss finish when buffed on to wood. This finish dulls with time rather than flaking off (as is the case with most other finishes used).

Although too brittle to be used by itself, carnauba wax is often combined with other waxes (principally beeswax) to treat and waterproof many leather products where it provides a high-gloss finish and increases leather's hardness and durability.

It is also used in the pharmaceutical industry as a tablet-coating agent. Adding the carnauba wax aids in the swallowing of tablets for patients. A very small amount (less than a hundredth of one percent by weight, i.e., 30 grams for a 300 kg batch) is sprinkled onto a batch of tablets after they have been sprayed and dried. The wax and tablets are then tumbled together for a few minutes before being discharged from the tablet-coating machine.

In 1890, Charles Tainter patented the use of carnauba wax on phonograph cylinders as a replacement for a mixture of paraffin and beeswax.

Carnauba wax may be used as a mold release agent for manufacture of fibre-reinforced plastics. An aerosol mold release agent is formed by dissolving carnauba wax in a solvent. Unlike silicone or PTFE, carnauba is suitable for use with liquid epoxy, epoxy molding compounds (EMC), and some other plastic types and generally enhances their properties . Carnauba wax is not very soluble in chlorinated or aromatic hydrocarbons.*[4] Carnauba is used in melt/castable explosives to produce an insensitive explosive formula such as Composition B, which is a blend of RDX and TNT.

14.3 Production and export

In 2006, Brazil produced 22,409 tons of carnauba wax, of which 14% was solid wax, and 86% was in powder form. *[5] There are 20-25 exporters of carnauba wax in Brazil who buy the carnauba wax from middlemen or directly from farmers. The exporters refine the wax before exporting it to the rest of the world. The four largest exporters of carnauba wax are Pontes, Brasil Ceras, Foncepi, and Carnauba do Brasil, who together account for around €25 million of the export market.

According to the Brazilian Ministry of Development, Industry and Foreign Trade, the major destinations for exported carnauba wax are:

- USA (25%)
- Japan (15-25%)
- Germany (10-15%)
- Netherlands (5%)
- Italy (5%)
- other destinations (18%)

14.4 Technical characteristics

- INCI name is *Copernicia cerifera (carnauba) wax*
- E Number is E903.
- Melting point: 82–86 °C (180–187 °F), among the highest of natural waxes, higher than beeswax, 62–64C.
- Relative density is about 0.97
- It is among the hardest of natural waxes.
- It is practically insoluble in water, soluble on heating in ethyl acetate and in xylene, and practically insoluble in ethyl alcohol.

14.5 References

[1] Steinle, J. Vernon (September 1936). "Carnauba wax: an expedition to its source" . *Industrial & Engineering Chemistry* **28** (9): 1004–1008. doi:10.1021/ie50321a003.

[2] Parish, Edward J.; Terrence L. Boos; Shengrong Li (2002). "The Chemistry of Waxes and Sterols". In Casimir C. Akoh, David B. Min. *Food lipids: chemistry, nutrition, and biochemistry* (2nd ed.). New York: M. Dekker. p. 103. ISBN 0-8247-0749-4.

[3] Uwe Wolfmeier,Hans Schmidt, Franz-Leo Heinrichs, Georg Michalczyk, Wolfgang Payer,Wolfram Dietsche, Klaus Boehlke, Gerd Hohner, Josef Wildgruber "Waxes" in Ullmann's Encyclopedia of Industrial Chemistry, Wiley-VCH, Weinheim, 2002. doi:10.1002/14356007.a28_103.

[4] Apps, E. A. (1958). *Printing Ink Technology*. London: Leonard Hill Books ltd. p. 86.

[5] , Report on the production of carnauba wax.

14.6   External links

- Botanical description - from the Mildred E. Mathias Botanical Garden

- Carnauba wax data sheet - from the UN Food and Agriculture Organization

- Carnauba Wax Background Paper - published report from field work.

Chapter 15

Chicle

For the tree species, see Manilkara chicle.

Chicle (/ˈtʃɪkəl/) is a natural gum traditionally used

A chiclero *bleeding a tree for chicle, Belize 1917*

in making chewing gum and other products. It is collected from several species of Mesoamerican trees in the *Manilkara* genus, including *M. zapota*, *M. chicle*, *M. staminodella*, and *M. bidentata*.[1][2]

The tapping of the gum is similar to the tapping of latex from the rubber tree: zig-zag gashes are made in the tree trunk and the dripping gum is collected in small bags. It is then boiled until it reaches the correct thickness. Locals who collect chicle are called *chicleros*.

15.1 Etymology

The word *chicle* comes from the Nahuatl word for the gum, *tziktli* ([ˈt͡sikt͡ɬi]), which can be translated as "sticky stuff". Alternatively, "chichle" may have come from the Mayan word *tsicte*.[3] Chicle was well known to the Nahuatl-Aztecs and to the Maya, and early European settlers prized it for its subtle flavor and high sugar content. The ancient word is still used in the Americas, *chicle* being a common term for chewing gum in Spanish and *chiclete* being the Portuguese term (both in Brazil and in parts of Portugal). The word has also been exported to other languages such as Greek, which refers to chewing gum as 'tsichla'.

15.2 History

Historically, the Adams Chewing Gum Company was a prominent user of this ingredient in the production of chewing gum.

In response to a land reform law passed in Guatemala in 1952, which ended feudal work relations and expropriated unused lands and sold them to the indigenous and peasants, the Wrigley Gum Company discontinued buying Guatemalan chicle. Since it was the sole buyer of Guatemalan chicle, the government was forced to create a massive aid program for growers.[4]

By the 1960s, most chewing gum companies had switched from using chicle to butadiene-based synthetic rubber which was cheaper to manufacture. The only U.S. gum companies still using chicle are Glee Gum, Simply Gum, and Tree Hugger Gum.[5]

15.3 References

[1] Mathews, Jennifer P. (2009). *Chicle: The Chewing Gum of the Americas, From the Ancient Maya to William Wrigley*. Tucson: University of Arizona Press. pp. 19–21. ISBN 0816528217.

[2] Chicle, Merriam-Webster.com. Retrieved March 17, 2011.

[3] Mexicolore article on chicle

[4] LaFeber, Walter (1993). *Inevitable revolutions: the United States in Central America*. New York: W.W. Norton. p. 119. ISBN 0-393-30964-9.

[5] Burks, Raychelle (6 August 2007). "Chewing Gum: Popular confection began as a not-so-sweet treat from trees". *Chemical and Engineering News* **85** (32): 36.

Chapter 16

Cinnamon

For other uses, see Cinnamon (disambiguation).

Cinnamon (/ˈsɪnəmən/ *SIN-ə-mən*) is a spice obtained

Raw cinnamon

from the inner bark of several trees from the genus *Cinnamomum* that is used in both sweet and savoury foods. While *Cinnamomum verum* is sometimes considered to be **"true cinnamon"**, most cinnamon in international commerce is derived from related species, which are also referred to as **"cassia"** to distinguish them from "true cinnamon" .*[1]*[2]

Cinnamon is the name for perhaps a dozen species of trees and the commercial spice products that some of them produce. All are members of the genus *Cinnamomum* in the family Lauraceae. Only a few of them are grown commercially for spice.

16.1   Name

The English word *cinnamon* and "cassia" , attested in English since the 15th century, derives from the Greek κιννάμωμον *kinnámōmon* (later *kínnamon*), via Latin and medieval French intermediate forms. The Greek in turn was borrowed from a Phoenician word, which would have been akin to the related Hebrew *qinnamon*.*[3]

The name of *cassia*, first recorded in English around 1000 AD, was borrowed via Latin and ultimately derives from Hebrew *q'tsīʿāh*, a form of the verb *qātsaʿ* 'strip off bark'.*[4]

Early Modern English also used the name *canel* or *canella*, akin to the current names of cinnamon in several other European languages, which are derived from the Latin word *cannella*, a diminutive of *canna*, 'tube', from the way it curls up as it dries.*[5]

16.2   History

Cinnamomum verum, from Koehler's Medicinal-Plants *(1887)*

In the classical times, four types of cinnamon were distinguished (and often confused):

- Cassia (Hebrew קציעה *qəṣi`â*), the bark of *Cinnamomum iners* from Arabia and Ethiopia, literally "the peel of the plant" which is scraped off the tree[6]

- True cinnamon (Hebrew קִנָּמוֹן *qinnamon*), the bark of *C. verum* (also called *C. zeylanicum*) from Sri Lanka

- Malabathrum or malobathrum (from Sanskrit तमालपत्त्रम्, *tamālapattram*, literally "dark-tree leaves"), several species including *C. tamala* from the north of India

- Serichatum, *C. cassia* from Seres, that is, China

Cinnamon has been known from remote antiquity. It was imported to Egypt as early as 2000 BC, but those who report it had come from China confuse it with cassia.[7] Cinnamon was so highly prized among ancient nations that it was regarded as a gift fit for monarchs and even for a god: a fine inscription records the gift of cinnamon and cassia to the temple of Apollo at Miletus.[8] Though its source was kept mysterious in the Mediterranean world for centuries by the middlemen who handled the spice trade, to protect their monopoly as suppliers, cinnamon is native to Bangladesh, Sri Lanka, the Malabar Coast of India, and Burma.[9]

The first Greek reference to *kasia* is found in a poem by Sappho in the seventh century BC. According to Herodotus, both cinnamon and cassia grew in Arabia, together with incense, myrrh, and ladanum, and were guarded by winged serpents. The phoenix was reputed to build its nest from cinnamon and cassia. Herodotus mentions other writers who believed the source of cassia was the home of Dionysos, located somewhere east or south of Greece.

The Greeks used *kásia* or *malabathron* to flavour wine, together with absinth wormwood (*Artemisia absinthium*). While Theophrastus gives a good account of the plants, he describes a curious method for harvesting: worms eat away the wood and leave the bark behind.

Egyptian recipes for *kyphi*, an aromatic used for burning, included cinnamon and cassia from Hellenistic times onward. The gifts of Hellenistic rulers to temples sometimes included cassia and cinnamon as well as incense, myrrh, and Indian incense (*kostos*), so one might conclude that the Greeks used it for similar purposes.

The Hebrew Bible makes specific mention of the spice many times: first when Moses is commanded to use both sweet cinnamon (Hebrew: קִנָּמוֹן, *qinnāmôn*) and cassia in the holy anointing oil;[10] in Proverbs where the lover's bed is perfumed with myrrh, aloes, and cinnamon;[11]

and in Song of Solomon, a song describing the beauty of his beloved, cinnamon scents her garments like "the smell of Lebanon." [12] Cassia was also part of the *ketoret*, the consecrated incense described in the Hebrew Bible and Talmud. It was offered on the specialized incense altar in the time when the Tabernacle was located in the First and Second Jerusalem temples. The *ketoret* was an important component of the temple service in Jerusalem. Psalm 45:8 mentions the garments of the king (or of Torah scholars) that smell of myrrh, aloes, and cassia.

Pliny[13] gives an account of the early spice trade across the Red Sea that cost Rome 100 million sesterces each year. Cinnamon was brought around the Arabian peninsula on "rafts without rudders or sails or oars," taking advantage of the winter trade winds.[14] Pliny also mentions cassia as a flavouring agent for wine.[15]

According to Pliny, a Roman pound (327 grams (11.5 oz)) of cassia, cinnamon, or serichatum cost up to 300 *denarii*, the wage of ten months' labour. Diocletian's Edict on Maximum Prices[16] from 301 AD gives a price of 125 *denarii* for a pound of cassia while an agricultural labourer earned 25 *denarii* per day. Cinnamon was too expensive to be commonly used on funeral pyres in Rome, but the Emperor Nero is said to have burned a year's worth of the city's supply at the funeral for his wife Poppaea Sabina in AD 65.[17]

Malabathrum leaves (*folia*) were used in cooking and for distilling an oil used in a caraway sauce for oysters by the Roman gourmet Gaius Gavius Apicius.[18] Malabathrum is among the spices that, according to Apicius, any good kitchen should contain.

The famous Commagenum unguent produced in Commagene, in present-day eastern Turkey, was made from goose fat aromatised with cinnamon oil and spikenard. Malobathrum from Egypt (Dioscorides I, 63) was based on beef fat and contained cinnamon, as well; one pound cost 300 *denari*. The Roman poet Martial (VI, 55) made fun of Romans who drip unguents, smell of cassia and cinnamon taken from a bird's nest, and look down on a man who does not smell at all.

Through the Middle Ages, the source of cinnamon was a mystery to the Western world. From reading Latin writers who quoted Herodotus, Europeans had learned that cinnamon came up the Red Sea to the trading ports of Egypt, but where it came from was less than clear. When the Sieur de Joinville accompanied his king to Egypt on crusade in 1248, he reported – and believed – what he had been told: that cinnamon was fished up in nets at the source of the Nile out at the edge of the world (i.e., Ethiopia). Marco Polo avoided precision on the topic.[19] Herodotus and other authors named Arabia as the source of cinnamon: they recounted that giant cinnamon birds collected the cinnamon sticks from an unknown land where the cinnamon trees grew and used them to construct their nests, and that the Arabs employed a trick to obtain the sticks. Pliny the Elder wrote in the first century that traders had made this up to charge more, but the story remained current in Byzantium as late as 1310.

The first mention that the spice grew in Sri Lanka was in Zakariya al-Qazwini's *Athar al-bilad wa-akhbar al- 'ibad* ("Monument of Places and History of God's Bondsmen") about 1270.[20] This was followed shortly thereafter by John of Montecorvino in a letter of about 1292.[21]

Indonesian rafts transported cinnamon directly from the Moluccas to East Africa (see also Rhapta), where local traders then carried it north[22][23][24] to Alexandria in Egypt. Venetian traders from Italy held a monopoly on the spice trade in Europe, distributing cinnamon from Alexandria. The disruption of this trade by the rise of other Mediterranean powers, such as the Mamluk sultans and the Ottoman Empire, was one of many factors that led Europeans to search more widely for other routes to Asia.

When Portuguese traders landed in Ceylon (Sri Lanka), they restructured the traditional production and management of cinnamon by the Sinhalese. They established a fort on the island in 1518 and protected Ceylon as their cinnamon monopoly for over a hundred years. Later, Sinhalese held the monopoly for cinnamon in Ceylon.

Dutch traders finally dislodged the Portuguese by allying with the inland Kingdom of Kandy. They established a trading post in 1638, took control of the manufactories by 1640, and expelled the remaining Portuguese by 1658. "The shores of the island are full of it," a Dutch captain reported, "and it is the best in all the Orient. When one is downwind of the island, one can still smell cinnamon eight leagues out to sea." [25]:15 The Dutch East India Company continued to overhaul the methods of harvesting in the wild and eventually began to cultivate its own trees.

In 1767, Lord Brown of the British East India Company established Anjarakkandy Cinnamon Estate near Anjarakkandy in Cannanore (now Kannur) district of Kerala, and this estate became Asia's largest cinnamon estate. The British took control of Ceylon from the Dutch in 1796. However, the importance of the monopoly of Ceylon was already declining, as cultivation of the cinnamon tree spread to other areas, the more common cassia bark became more acceptable to consumers, and coffee, tea, sugar, and chocolate began to outstrip the popularity of traditional spices.

Leaves from a wild cinnamon tree

16.3   Cultivation

Global annual production of cinnamon and cassia amounts to 27,500–35,000 tons. *Cinnamomum verum* accounts for 7,500–10,000 tons of production, with the remainder produced by other species.[1] Sri Lanka produces 80–90% of the world's supply of *C. verum*, but that is the only species grown there; *C. verum* is also cultivated on a commercial scale in Seychelles and Madagascar.[1] Global production of the other species averages 20,000–25,000 tons, of which Indonesia produces around two-thirds of the total, with significant production in China. India and Vietnam are also minor producers.[1]

Cinnamon is cultivated by growing the tree for two years, then coppicing it, i.e., cutting the stems at ground level. The following year, about a dozen new shoots will form from the roots, replacing those that were cut. A number of pests such as *Colletotrichum gloeosporioides*, *Diplodia* spp., and *Phytophthora cinnamomi* (stripe canker) can affect that growing plants, sometimes leading to death.[26]

The stems must be processed immediately after harvesting while the inner bark is still wet. The cut stems are processed by scraping off the outer bark, then beating the branch evenly with a hammer to loosen the inner bark, which is then pried off in long rolls. Only 0.5 mm (0.02 in) of the inner bark is used; the outer, woody portion is discarded, leaving metre-long cinnamon strips that curl into rolls ("quills") on drying. The processed bark will dry completely in four to six hours, provided it is in a well-ventilated and relatively warm environment. Once dry, the bark is cut into 5- to 10-cm (2- to 4-in) lengths for sale. A less than ideal drying environment encourages the proliferation of pests in the bark, which may then require treatment by fumigation. Fumigated bark is not considered to be of the same premium quality as untreated bark.

Sri Lanka cinnamon has a very thin, smooth bark with a light-yellowish brown colour and a highly fragrant aroma. In recent years in Sri Lanka, mechanical devices have been developed to ensure premium quality and worker safety and health, following considerable research by the universities in that country, led by the University of Ruhuna.

16.3.1 Grading

See also: Food grading

The Sri Lankan grading system divides the cinnamon quills into four groups:

- Alba, less than 6 mm (0.24 in) in diameter

- Continental, less than 16 mm (0.63 in) in diameter

- Mexican, less than 19 mm (0.75 in) in diameter

- Hamburg, less than 32 mm (1.3 in) in diameter

These groups are further divided into specific grades. For example, Mexican is divided into M00 000 special, M000000, and M0000, depending on quill diameter and number of quills per kilogram.

Any pieces of bark less than 106 mm (4.2 in) long are categorized as quillings. Featherings are the inner bark of twigs and twisted shoots. Chips are trimmings of quills, outer and inner bark that cannot be separated, or the bark of small twigs.

16.4 Species

Cinnamon sticks, powder, and dried flowers

A number of species are often sold as cinnamon:[27]

- *Cinnamomum cassia* (cassia or Chinese cinnamon, the most common type)

*Ceylon cinnamon (*Cinnamomum verum*) on the left, and Indonesian cinnamon (*Cinnamomum burmannii*) quills*

- *C. burmannii* (Korintje, Padang cassia, or Indonesian cinnamon)

- *C. loureiroi* (Saigon cinnamon, Vietnamese cassia, or Vietnamese cinnamon)

- *C. verum* (Sri Lanka cinnamon or Ceylon cinnamon)

Cassia is the strong spicy flavour associated with cinnamon rolls and other such baked goods, as it handles baking conditions well. Chinese cinnamon is generally a medium to light reddish brown, hard and woody in texture, and thicker (2–3 mm (0.079–0.118 in) thick), as all of the layers of bark are used. Ceylon cinnamon, using only the thin inner bark, has a lighter brown colour, a finer, less dense and more crumbly texture, and is considered to be subtler and more aromatic in flavour than cassia, losing much of its flavour during cooking.

Levels of the blood-thinning agent coumarin in Ceylon cinnamon are much lower than those in cassia.[28][29]

The barks, when whole, are easily distinguished, and their microscopic characteristics are also quite distinct. Ceylon cinnamon sticks (quills) have many thin layers and can easily be made into powder using a coffee or spice grinder, whereas cassia sticks are much harder. Indonesian cinnamon is often sold in neat quills made up of one thick layer, capable of damaging a spice or coffee grinder. Saigon cinnamon (*C. loureiroi*) and Chinese cinnamon (*C. cassia*) are always sold as broken pieces of thick bark, as the bark is not supple enough to be rolled into quills. The powdered bark is harder to distinguish, but if it is treated with tincture of iodine (a test for starch[30]), little effect is visible with pure Ceylon cinnamon, but when Chinese cinnamon is present, a deep-blue tint is produced.[31][32]

16.5   Flavour, aroma and taste

Cinnamomum verum *bark essential oil*

Cinnamon bark

The flavour of cinnamon is due to an aromatic essential oil that makes up 0.5 to 1% of its composition. This essential oil is prepared by roughly pounding the bark, macerating it in sea water, and then quickly distilling the whole. It is of a golden-yellow colour, with the characteristic odour of cinnamon and a very hot aromatic taste. The pungent taste and scent come from cinnamic aldehyde or cinnamaldehyde (about 90% of the essential oil from the bark) and, by reaction with oxygen as it ages, it darkens in colour and forms resinous compounds. Other chemical components of the essential oil include ethyl cinnamate, eugenol (found mostly in the leaves), beta-caryophyllene, linalool, and methyl chavicol.

16.6   Uses

Cinnamon bark is used as a spice. It is principally employed in cookery as a condiment and flavouring material. It is used in the preparation of chocolate, especially in Mexico, which is the main importer of cinnamon.*[33] It is also

Besides use as flavourant and spice in foods, cinnamon-flavoured tea, also flavoured with cardamom, is consumed as a hot beverage in Bangladesh, India and Pakistan.

used in many dessert recipes, such as apple pie, doughnuts, and cinnamon buns as well as spicy candies, coffee, tea, hot cocoa, and liqueurs. In the Middle East, cinnamon is often used in savoury dishes of chicken and lamb. In the United States, cinnamon and sugar are often used to flavour cereals, bread-based dishes, such as toast, and fruits, especially apples; a cinnamon-sugar mixture is even sold separately for such purposes. It is also used in Turkish cuisine for both sweet and savoury dishes. Cinnamon can also be used in pickling. Cinnamon powder has long been an important spice in enhancing the flavor of Persian cuisine, used in a variety of thick soups, drinks, and sweets.*[34]*:10–12

Cinnamon

16.6.1   Use as an alcohol flavorant

Cinnamon is a popular flavoring in numerous alcoholic beverages.*[35]

Cinnamon brandy concoctions, called "Cinnamon liqueur" and made with distilled alcohol, are popular in parts of Greece. In Europe, popular examples of such beverages are *Maiwein* (white wine with woodruff) and *Żubrówka* (vodka flavoured with bison grass).

16.7   Traditional medicine

Cinnamon has a long history of use in traditional medicine, but there is no evidence that it is useful to treat any medical condition.*[36]

16.8   Toxicity

The European Food Safety Authority in 2008 considered toxicity of coumarin, known to cause liver and kidney damage in high concentrations and a significant component of cinnamon, and metabolic effect on humans with CYP2A6 polymorphism, and confirmed a maximum recommended Tolerable Daily Intake (TDI) of 0.1 mg of coumarin per kg of body weight.*[37]*[38] The European Union set a guideline for maximum coumarin content in foodstuffs of 50 mg per kg of dough in seasonal foods, and 15 mg per kg in everyday baked foods.*[39]

16.9   Nutritional information

Ten grams (about 2.1 teaspoons) of ground cinnamon contain:*[40]

- Energy: 103.4 kJ (24.7 kcal)
- Fat: 0.12 g
- Carbohydrates: 8.06 g (of which - fibres: 5.31 g, sugars: 0.2 g)
- Protein: 0.4 g

16.10   See also

- *Canella*, a plant known as "wild cinnamon"
- Cinnamomea, a New Latin adjective meaning "cinnamon-coloured"
- Cinnamon challenge

16.11   References

[1] Iqbal, Mohammed (1993). "International trade in non-wood forest products: An overview" . *FO: Misc/93/11 - Working Paper*. Food and Agriculture Organization of the United Nations. Retrieved November 12, 2012.

[2] "Cassia, also known as cinnamon or Chinese cinnamon is a tree that has bark similar to that of cinnamon but with a rather pungent odour," Bell, Maguelonne Toussaint-Samat ; translated by Anthea (2009). *A history of food* (New expanded ed.). Chichester, West Sussex, U.K.: Wiley-Blackwell. ISBN 978-1405181198.

[3] "cinnamon" . *Oxford English Dictionary* (2nd ed.). Oxford University Press. 1989.; also Harper, Douglas. "cinnamon" . *Online Etymology Dictionary*..

[4] "cassia" . *Oxford English Dictionary* (2nd ed.). Oxford University Press. 1989.; also Harper, Douglas. "cassia". *Online Etymology Dictionary*..

[5] "canella; canel" . *Oxford English Dictionary* (2nd ed.). Oxford University Press. 1989..

[6] Klein, Ernest, *A Comprehensive Etymological Dictionary of the Hebrew Language for Readers of English*, University of Haifa, Carta, Jerusalem, p.589

[7] "The Indians obtained cassia from China" (Toussaint-Samat 2009, p. 437).

[8] Toussaint-Samat 2009, p. 437

[9] "Cinnamon" . *Encyclopaedia Britannica*. 2008. ISBN 1-59339-292-3. (species Cinnamomum zeylanicum), bushy evergreen tree of the laurel family (Lauraceae) native to Bangladesh, Sri Lanka (Ceylon), the neighboring Malabar Coast of India, and Myanmar (Burma), and also cultivated in South America and the West Indies for the spice consisting of its dried inner bark.

[10] Exodus 30:22-25

[11] Proverbs 7:17

[12] Song of Solomon 4:11-14

[13] Pliny, (nat. 12, 86-87)

[14] Pliny the Elder; Bostock, J.; Riley, H.T. (1855). "42, Cinnamomum. Xylocinnamum". *Natural History of Pliny, book XII, The Natural History of Trees* **3**. London: Henry G. Bohn. pp. 137–140.

[15] Pliny, nat. 14, 107f.

[16] ER Graser. A text and translation of the Edict of Diocletian, in An Economic Survey of Ancient Rome Volume V: Rome and Italy of the Empire. Johns Hopkins Press 1940 ISBN 978-0374928483

[17] Toussaint-Samat 2009, p. 437f.

[18] *De re coquinaria*, I, 29, 30; IX, 7

[19] Toussaint-Samat 2009, p. 438 discusses cinnamon's hidden origins and Joinville's report.

[20] Tennent, Sir James Emerson. "Account of the Island of Ceylon". Retrieved 8 November 2014.

[21] Yule, Col. Henry. "Cathay and the Way Thither". Retrieved July 15, 2008.

[22] "The life of spice; cloves, nutmeg, pepper, cinnamon | UNESCO Courier | Find Articles at BNET". Findarticles.com. 1984. Retrieved August 18, 2010.

[23] Independent Online. "News - Discovery: Sailing the Cinnamon Route (Page 1 of 2)". Iol.co.za. Retrieved August 18, 2010.

[24] Gray, E. W.; Miller, J. I. (1970). "The Spice Trade of the Roman Empire 29 B.C.-A.D. 641". *The Journal of Roman Studies* **60**: 222–224. doi:10.2307/299440. JSTOR 299440.

[25] Braudel, Fernand (1984). *The Perspective of the World* **3**. University of California Press. p. 699. ISBN 0-520-08116-1.

[26] https://www.plantvillage.com/en/topics/cinnamon/infos/diseases_and_pests_description_uses_propagation

[27] Culinary Herbs and Spices, The Seasoning and Spice Association. Retrieved August 3, 2010.

[28] High daily intakes of cinnamon: Health risk cannot be ruled out. BfR Health Assessment No. 044/2006, 18 August 2006

[29] "Espoo daycare centre bans cinnamon as "moderately toxic to liver"". Retrieved September 5, 2010.

[30] http://www.webexhibits.org/causesofcolor/6AC.html

[31] "Iodine test for cassia".

[32] Pereira, Jonathan (1854). *The Elements of materia medica and therapeutics* **2**. p. 390.

[33] "Trade and Sustainable Forest Management -Impacts and Interactions". Fao.org. September 26, 2003. Retrieved August 18, 2010.

[34] Fred Czarra. Spices: A Global History. Reaktion Books (May 1, 2009) ISBN 978-1861894267

[35] Haley Willard for The Daily Meal. December 16, 2013 11 Cinnamon-Flavored Liquors for the Holidays

[36] National Center for Complementary and Alternative Medicine (NCCIH). Created: October 2011. Updated: April 2012 Herbs at a Glance: Cinnamon NCCIH Publication No.: 463

[37] Harris, Emily. "German Christmas Cookies Pose Health Danger". National Public Radio. Retrieved May 1, 2007.

[38] "Coumarin in flavourings and other food ingredients with flavouring properties - Scientific Opinion of the Panel on Food Additives, Flavourings, Processing Aids and Materials in Contact with Food (AFC)". doi:10.2903/j.efsa.2008.793.

[39] Guardian newspaper:Cinnamon sparks spicy debate between Danish bakers and food authorities, 20 December 2013

[40] "USDA nutritional information for ground cinnamon". United States Department of Agriculture. Retrieved November 8, 2014.

16.12   Further reading

- Charles Corn. The Scents of Eden: A History of the Spice Trade. Kodansha New York. 1999 ISBN 1-56836-249-8

- Wijesekera R O B, Ponnuchamy S, Jayewardene A L, "Cinnamon" (1975) monograph published by CISIR, Colombo, Sri Lanka

16.13   External links

- BBC News - In pictures: Sri Lanka's spice of life

Chapter 17

Clove

This article is about the spice. For other uses, see Clove (disambiguation).

Cloves are the aromatic flower buds of a tree in the family Myrtaceae, *Syzygium aromaticum*. They are native to the Maluku Islands in Indonesia, and are commonly used as a spice. Cloves are commercially harvested primarily in Indonesia, India, Madagascar, Zanzibar, Pakistan, Sri Lanka and Tanzania.

The clove tree is an evergreen tree that grows up to 8–12 m tall, with large leaves and sanguine flowers grouped in terminal clusters. The flower buds initially have a pale hue, gradually turn green, then transition to a bright red when ready for harvest. Cloves are harvested at 1.5–2.0 cm long, and consist of a long calyx that terminates in four spreading sepals, and four unopened petals that form a small central ball.

Clove model of a proa

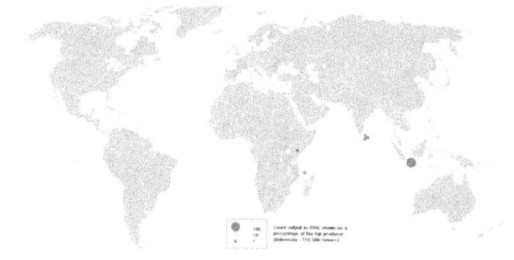

Clove output in 2005

17.1 Uses

Dried cloves

Cloves are used in the cuisine of Asian, African, and the Near and Middle East countries, lending flavor to meats, curries, and marinades, as well as fruit such as apples, pears or rhubarb. Cloves may be used to give aromatic and flavor qualities to hot beverages, often combined with other ingredients such as lemon and sugar. They are a common element in spice blends such as pumpkin pie spice and speculoos spices.

In Mexican cuisine, cloves are best known as *clavos de olor*, and often accompany cumin and cinnamon.[2]

A major component of clove taste is imparted by the chemical eugenol,[3] and the quantity of the spice required is typically small. It pairs well with cinnamon, allspice, vanilla, red wine and basil, as well as onion, citrus peel, star anise, or peppercorns.

17.1.1   Non-culinary uses

The spice is used in a type of cigarette called *kretek* in Indonesia.*[1] They have been smoked throughout Europe, Asia, and the United States. In 2009, clove cigarettes (as well as fruit- and candy-flavored cigarettes) were outlawed in the US. Cigarettes containing clove are now classified as cigars when sold in the US.*[4]

Due to the bioactive chemicals of clove, the spice may be used as an ant repellent.*[5]

They can be used to make a fragrance pomander when combined with an orange. When given as a gift in Victorian England, such a pomander indicated "warmth of feeling."

17.1.2   Traditional medicinal uses

Cloves are used in Indian Ayurvedic medicine, Chinese medicine, and western herbalism and dentistry where the essential oil is used as an anodyne (painkiller) for dental emergencies. Cloves are used as a carminative, to increase hydrochloric acid in the stomach and to improve peristalsis. Cloves are also said to be a natural anthelmintic.*[6] The essential oil is used in aromatherapy when stimulation and warming are needed, especially for digestive problems. Topical application over the stomach or abdomen are said to warm the digestive tract. Applied to a cavity in a decayed tooth, it also relieves toothache.*[7]

In Chinese medicine, cloves or *ding xiang* are considered acrid, warm, and aromatic, entering the kidney, spleen and stomach meridians, and are notable in their ability to warm the middle, direct stomach *qi* downward, to treat hiccough and to fortify the kidney *yang*.*[8] Because the herb is so warming, it is contraindicated in any persons with fire symptoms and according to classical sources should not be used for anything except cold from *yang* deficiency. As such, it is used in formulas for impotence or clear vaginal discharge from *yang* deficiency, for morning sickness together with ginseng and patchouli, or for vomiting and diarrhea due to spleen and stomach coldness.*[8]

Cloves may be used internally as a tea and topically as an oil for hypotonic muscles, including for multiple sclerosis. This is also found in Tibetan medicine.*[9] Some recommend avoiding more than occasional use of cloves internally in the presence of *pitta* inflammation such as is found in acute flares of autoimmune diseases.*[10]

17.1.3   Potential medicinal uses

The U.S. Food and Drug Administration (FDA) has reclassified eugenol (one of the chemicals contained in clove oil), downgrading its effectiveness rating. The FDA now be-

lieves not enough evidence indicates clove oil or eugenol is effective for toothache pain or a variety of other types of pain.*[11]

Studies to determine its effectiveness for fever reduction, as a mosquito repellent, and to prevent premature ejaculation have been inconclusive.*[11] It remains unproven whether clove may reduce blood sugar levels.*[12]

In addition, clove oil is used in preparation of some toothpastes and Clovacaine solution, which is a local anesthetic used in oral ulceration and inflammation. Eugenol (or clove oil generally) is mixed with zinc oxide to form a temporary tooth cavity filling.*[13]

Clove oil can be used to anesthetize fish, and prolonged exposure to higher doses (the recommended dose is 400 mg/l) is considered a humane means of euthanasia.*[14]

17.2   Adulteration

Clove stalks are slender stems of the inflorescence axis that show opposite decussate branching. Externally, they are brownish, rough, and irregularly wrinkled longitudinally with short fracture and dry, woody texture.

Mother cloves (anthophylli) are the ripe fruits of cloves that are ovoid, brown berries, unilocular and one-seeded. This can be detected by the presence of much starch in the seeds.

Brown cloves are expanded flowers from which both corollae and stamens have been detached.

Exhausted cloves have most or all the oil removed by distillation. They yield no oil and are darker in color.*[15]

17.3   History

Archeologists have found cloves in a ceramic vessel in Syria, with evidence that dates the find to within a few years of 1721 BCE.*[16] In the third century BCE, a Chinese leader in the Han Dynasty required those who addressed him to chew cloves to freshen their breath.*[17] Cloves were traded by Muslim sailors and merchants during the Middle Ages in the profitable Indian Ocean trade, the clove trade is also mentioned by Ibn Battuta and even famous *Arabian Nights* characters such as Sinbad the Sailor are known to have bought and sold cloves from India.*[18]

Until modern times, cloves grew only on a few islands in the Maluku Islands (historically called the Spice Islands), including Bacan, Makian, Moti, Ternate, and Tidore.*[16] In fact, the clove tree that experts believe is the oldest in the world, named *Afo*, is on Ternate. The tree is between 350 and 400 years old.*[19] Tourists are told that seedlings from this very tree were stolen by a Frenchman named Poivre in 1770, transferred to France, and then later to Zanzibar, which was once the world's largest producer of cloves.*[19]

Until cloves were grown outside of the Maluku Islands, they were traded like oil, with an enforced limit on exportation.*[19] As the Dutch East India Company consolidated its control of the spice trade in the 17th century, they sought to gain a monopoly in cloves as they had in nutmeg. However, "unlike nutmeg and mace, which were limited to the minute Bandas, clove trees grew all over the Moluccas, and the trade in cloves was way beyond the limited policing powers of the corporation." *[20]

17.4 Chemical compounds

The compound eugenol is responsible for most of the characteristic aroma of cloves.

Eugenol comprises 72-90% of the essential oil extracted from cloves, and is the compound most responsible for clove aroma.*[3] Other important essential oil constituents of clove oil include acetyl eugenol, beta-caryophyllene and vanillin, crategolic acid, tannins such as bicornin,*[3]*[21] gallotannic acid, methyl salicylate (painkiller), the flavonoids eugenin, kaempferol, rhamnetin, and eugenitin, triterpenoids such as oleanolic acid, stigmasterol, and campesterol, and several sesquiterpenes.*[22]

Eugenol is toxic in relatively small quantities; for example, a dose of 5-10 ml has been reported as a near fatal dose for a 2 year old child.*[23]

17.5 See also

- *Cinnamomum cassia*

- Gallic acid

- Insect repellent

17.6 References

[1] "*Syzygium aromaticum* (L.) Merr. & L. M. Perry". *Germplasm Resources Information Network (GRIN) online database*. Retrieved June 9, 2011.

[2] Dorenburg, Andrew and Page, Karen. *The New American Chef: Cooking with the Best Flavors and Techniques from Around the World*, John Wiley and Sons Inc., 2003

[3] Kamatou GP1, Vermaak I, Viljoen AM (2012). "Eugenol--from the remote Maluku Islands to the international market place: a review of a remarkable and versatile molecule". *Molecules* **17** (6): 6953–81. doi:10.3390/molecules17066953.

[4] "Flavored Tobacco". FDA.gov. Retrieved September 7, 2012.

[5] "Get Rid of Ants 24". getridofanst24.

[6] Balch, Phyllis and Balch, James. *Prescription for Nutritional Healing*, 3rd ed., Avery Publishing, 2000, p. 94

[7] Alqareer A, Alyahya A, Andersson L. (May 24, 2012). "The effect of clove and benzocaine versus placebo as topical anesthetics". *Journal of dentistry* **34** (10): 747–50. doi:10.1016/j.jdent.2006.01.009. PMID 16530911.

[8] *Chinese Herbal Medicine: Materia Medica*, Third Edition by Dan Bensky, Steven Clavey, Erich Stoger, and Andrew Gamble 2004

[9] "Question: Multiple Sclerosis". TibetMed. Retrieved September 7, 2012.

[10] Tillotson, Alan (April 3, 2005). "Special Diets for Illness". Oneearthherbs.squarespace.com. Retrieved September 7, 2012.

[11] "Clove". MedlinePlus, U.S. National Library of Medicine and National Institutes of Health. 2014. Retrieved August 18, 2014.

[12] "Clove (Eugenia aromatica) and Clove oil (Eugenol)". *National Institutes of Health, Medicine Plus*. nlm.nih.gov. February 15, 2012. Retrieved September 7, 2012.

[13] Youngken, H.W. (1950). *Text book of pharmacognosy* (6th ed.).

[14] Monks, Neale. "Aquarium Fish Euthanasia: Euthanizing and disposing of aquarium fish.". FishChannel.com. Retrieved August 1, 2011.

[15] Bisset, N.G. (1994). *Herbal drugs and phyotpharmaceuticals, Medpharm*. Stuttgart: Scientific Publishers.

[16] Turner, Jack (2004). *Spice: The History of a Temptation*. Vintage Books. pp. xxvii–xxviii. ISBN 0-375-70705-0.

[17] Andaya, Leonard Y. (1993). "1: Cultural State Formation in Eastern Indonesia". In Reid, Anthony. *Southeast Asia in the early modern era: trade, power, and belief*. Cornell University Press. ISBN 978-0-8014-8093-5.

[18] "The Third Voyage of Sindbad the Seaman - The Arabian Nights - The Thousand and One Nights - Sir Richard Burton translator". Classiclit.about.com. April 10, 2012. Retrieved September 7, 2012.

[19] Worrall, Simon (June 23, 2012). "The world's oldest clove tree". BBC News Magazine. Retrieved June 24, 2012.

[20] Krondl, Michael. *The Taste of Conquest: The Rise and Fall of the Three Great Cities of Spice*. New York: Ballantine Books, 2007.

[21] Li-Ming Bao, Eerdunbayaer, Akiko Nozaki, Eizo Takahashi, Keinosuke Okamoto, Hideyuki Ito and Tsutomu Hatano (2012). "Hydrolysable Tannins Isolated from Syzygium aromaticum: Structure of a New C-Glucosidic Ellagitannin and Spectral Features of Tannins with a Tergalloyl Group.". *Heterocycles* **85** (2): 365–81. doi:10.3987/COM-11-12392.

[22] *Chinese Herbal Medicine: Materia Medica*, Third Edition by Dan Bensky, Steven Clavey, Erich Stoger, and Andrew Gamble. 2004

[23] Hartnoll, G; Moore, D; Douek, D (1993). "Near fatal ingestion of oil of cloves". *Archives of Disease in Childhood* **69** (3): 392–3. doi:10.1136/adc.69.3.392. PMC 1029532. PMID 8215554.

17.7   Further reading

Liu, Bin-Bin; Liu, Luo; Liu, Xiao-Long; Geng, Di; Li, Cheng-Fu; Chen, Shao-Mei; Chen, Xue-Mei; Yi, Li-Tao; Liu, Qing (February 2015). "Essential Oil of Syzygium aromaticum Reverses the Deficits of Stress-Induced Behaviors and Hippocampal p-ERK/p-CREB/Brain-Derived Neurotrophic Factor Expression". *Planta Medica* **81** (3): 185–192. doi:10.1055/s-0034-1396150. Retrieved 27 April 2015.

Chapter 18

Cocoa bean

For other uses of "Cacao", see Cacao (disambiguation). The **cocoa bean**, also **cacao bean**[1] or simply **cocoa**

Cocoa pods in various stages of ripening

(/ˈkoʊ.koʊ/) or **cacao** (/kəˈkaʊ/), is the dried and fully fermented fatty seed of *Theobroma cacao*, from which cocoa solids and cocoa butter are extracted.[2] They are the basis of chocolate, as well as many Mesoamerican foods such as mole sauce and tejate.

18.1 Etymology

Aztec sculpture with cocoa pod

The word *Cocoa* derives [3] from the Spanish word *cacao*, derived from the Nahuatl word *cacahuatl*. [4] The Nahautl word, in turn, ultimately derives from the reconstructed Proto Mije-Sokean word *kakaw~*kakawa. [5]

Cocoa can often also refer to the drink commonly known as hot chocolate; [6] to cocoa powder, the dry powder made by grinding cocoa seeds and removing the cocoa butter from the dark, bitter cocoa solids; or to a mixture of cocoa powder and cocoa butter. [7][8]

18.2   History

The cacao tree is native to the Americas. It may have originated in the foothills of the Andes in the Amazon and Orinoco basins of South America, current day Colombia and Venezuela, where today, examples of wild cacao still can be found. However, it may have had a larger range in the past, evidence for which may be obscured because of its cultivation in these areas long before, as well as after, the Spanish arrived. New chemical analyses of residues extracted from pottery excavated at an archaeological site at Puerto Escondido in Honduras indicate that it was here where cocoa products were first consumed between 1400 and 1500 BC. The new evidence also indicates that, long before the flavor of the cacao seed (or bean) became popular, it was the sweet pulp of the chocolate fruit, used in making a fermented (5% alcohol) beverage, which first drew attention to the plant in the Americas. [9] The cocoa bean was a common currency throughout Mesoamerica before the Spanish conquest. [10]

Cacao trees will grow in a limited geographical zone, of approximately 20 degrees to the north and south of the Equator. Nearly 70% of the world crop today is grown in West Africa. The cacao plant was first given its botanical name by Swedish natural scientist Carl Linnaeus in his original classification of the plant kingdom, who called it *Theobroma* ("food of the gods") *cacao*.

Cocoa was an important commodity in pre-Columbian Mesoamerica. A Spanish soldier who was part of the conquest of Mexico by Hernán Cortés tells that when Moctezuma II, emperor of the Aztecs, dined, he took no other beverage than chocolate, served in a golden goblet. Flavored with vanilla or other spices, his chocolate was whipped into a froth that dissolved in the mouth. It is reported that no fewer than 60 portions each day may have been consumed by Moctezuma II, and 2,000 more by the nobles of his court. [11]

Chocolate was introduced to Europe by the Spaniards, and became a popular beverage by the mid 17th century. [12] They also introduced the cacao tree into the West Indies and the Philippines. It was also introduced into the rest of Asia and into West Africa by Europeans. In the Gold Coast, modern Ghana, cacao was introduced by an African, Tetteh Quarshie.

18.3   Production

Cocoa beans in a freshly cut cocoa pod

18.3.1   Cocoa pod

A cocoa pod (fruit) has a rough and leathery rind about 2 cm (0.79 in) to 3 cm (1.2 in) thick (this varies with the origin and variety of pod). It is filled with sweet, mucilaginous pulp (called 'baba de cacao' in South America) with a lemonade like taste enclosing 30 to 50 large seeds that are fairly soft and a pale lavender to dark brownish purple color. Due to heat buildup in the fermentation process, cacao beans lose most of the purplish hue and become mostly brown in color, with an adhered skin which includes the dried remains of the fruity pulp. This skin is released easily after roasting by winnowing. White seeds are found in some rare varieties, usually mixed with purples, and are considered of higher value. [13][14][15] Historically, white cacao was cultivated by the Rama people of Nicaragua. [16]

18.3.2   Varieties

There are three main varieties of cocoa plant: Forastero, Criollo, and Trinitario. The first is the most widely used, comprising 95% of the world production of cocoa. Cocoa beans of the Criollo variety are rarer and considered a delicacy. [17] Criollo plantations have lower yields than those of Forastero, and also tend to be less resistant to several diseases that attack the cocoa plant, hence very few countries still produce it. One of the largest producers of Criollo beans is Venezuela (Chuao and Porcelana). Trinitario (from

Three main varieties of cocoa: Criollo, Trinitario and Forastero

Trinidad) is a hybrid between Criollo and Forastero varieties. It is considered to be of much higher quality than Forastero, but has higher yields and is more resistant to disease than the former.

18.3.3 Harvesting

Cocoa trees grow in hot, rainy tropical areas within 20° of latitude from the equator.[18] Cocoa harvest is not restricted to one period per year and a harvest typically occurs over several months. In fact, in many countries cocoa can be harvested at any time of the year.[19] Pesticides are often applied to the trees to combat capsid bugs and fungicides to fight black pod disease.[20]

Immature cocoa pods have a variety of colours but most often are green, red, or purple, and as they mature their colour tends towards yellow or orange, particularly in their creases.[19][21] Unlike most fruiting trees, the cacoa pod grows directly from the trunk or large branch of a tree rather than from the end of a branch, similar to jackfruit. This makes harvesting by hand easier as most of the pods will not be up in the higher branches. The pods on a tree do not ripen together; harvesting needs to be done periodically through the year.[19] Harvesting occurs between three and four times weekly during the harvest season.[19] The ripe and near-ripe pods, as judged by their colour, are harvested from the trunk and branches of the cocoa tree with a curved knife on a long pole. Care must be used when cutting the stem of the pod to avoid damaging the junction of the stem with the tree, as this is where future flowers and pods will emerge.[19][22] It is estimated that one person can harvest 650 pods per day.[20][23]

18.3.4 Harvest processing

The harvested pods are opened typically with a machete to expose the beans.[19][20] The pulp and cocoa seeds are removed and the rind is discarded. The pulp and seeds are then piled in heaps, placed in bins, or laid out on grates for several days. During this time, the seeds and pulp undergo "sweating", where the thick pulp liquefies as it ferments. The fermented pulp trickles away, leaving cocoa seeds behind to be collected. Sweating is important[24] for the quality of the beans, which originally have a strong bitter taste. If sweating is interrupted, the resulting cocoa may be ruined; if underdone, the cocoa seed maintains a flavor similar to raw potatoes and becomes susceptible to mildew. Some cocoa producing countries distill alcoholic spirits using the liquefied pulp.[25]

A typical pod contains 20 to 50 beans[18] and about 400 dried beans are required to make one pound - or 880 per kilogram - of chocolate.[18] Cocoa pods weigh an average of 400 grams (0.88 lb) and each one yields 35 to 40 grams (1.2 to 1.4 oz) dried beans (this yield is 40–44% of the total weight in the pod).[20] It is estimated one person can separate the beans from 2000 pods per day.[20][23]

Cocoa beans drying in the sun.

The wet beans are then transported to a facility so they can be fermented and dried.[20] They are fermented for four to seven days and must be mixed every two days.[20][23] They are dried for five to fourteen days, depending on the climate conditions.[20][23] The fermented beans are dried by spreading them out over a large surface and constantly raking them. In large plantations, this is done on huge trays under the sun or by using artificial heat. Small plantations may dry their harvest on little trays or on cowhides. Finally, the beans are trodden and shuffled about (often using bare human feet) and sometimes, during this process, red clay mixed with water is sprinkled over the beans to obtain a finer color, polish, and protection against molds during shipment to factories in the United States, the Netherlands, United Kingdom, and other countries. Drying in the sun is preferable to drying by artificial means, as no extraneous flavors such as smoke or oil are introduced which might otherwise taint the flavor.

The beans should be dry for shipment (usually by sea). Traditionally exported in jute bags, over the last decade, beans are increasingly shipped in 'Mega-Bulk' bulk parcels of several thousand tonnes at a time on ships, or in smaller lots of around 25 tonnes in 20 foot containers. Shipping in bulk significantly reduces handling costs; shipment in bags, however, either in a ship's hold or in containers, is still common.

Throughout Mesoamerica where they are native, cocoa beans are used for a variety of foods. The harvested and fermented beans may be ground to-order at *tiendas de chocolate*, or chocolate mills. At these mills, the cocoa can be mixed with a variety of ingredients such as cinnamon, chili peppers, almonds, vanilla and other spices to create drinking chocolate.[26] The ground cocoa is also an important ingredient in *tejate* and a number of savory foods, such as *mole*.

18.3.5 World production

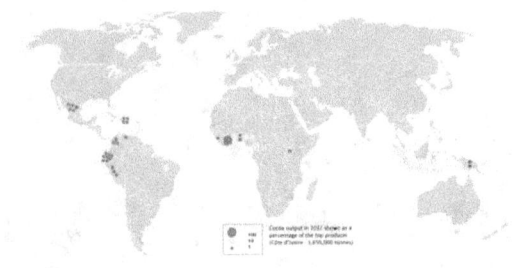

Cocoa bean output in 2012

Nearly 5,000,000 tonnes (4,900,000 long tons; 5,500,000 short tons) of cocoa are produced each year.

The historical global production was

> 1974: 1,556,484 tons,
> 1984: 1,810,611 tons,
> 1994: 2,672,173 tons,
> 2004: 3,607,052 tons.

The production increased by 131.7% in 30 years, representing a compound annual growth rate of 2.9%.

There were 3.54 million tonnes of cocoa beans produced in the 2008–2009 growing year,[18] which runs from October to September.[27] Of this total, African nations produced 2.45 million tonnes (69%), Asia and Oceania produced 0.61 million tonnes (17%) and the Americas produced 0.48 million tonnes (14%).[18] Two African nations, Côte d'Ivoire and Ghana, produce more than half of

the world's cocoa, with 1.23 and 0.73 million tonnes respectively (35% and 21%, respectively).[18] In 2012, the largest cocoa-bean producing countries in the world are as follows.

18.3.6 Consumption

There are different metrics used for chocolate consumption. The Netherlands has the highest monetary amount of cocoa bean imports (US$2.1 billion); it is also one of the main ports into Europe.[18] The United States has highest amount of cocoa powder imports ($220 million); the US has a large amount of cocoa complementary products.[18] The United Kingdom has the highest amount of retail chocolate ($1.3 billion) and is one of the biggest chocolate consumption per capita markets.[18]

Cocoa and its products (including chocolate) are used worldwide. Per capita consumption is poorly understood, with numerous countries claiming the highest: various reports state that Switzerland, Belgium, and the UK have the highest consumption. However, since there is no clear mechanism to determine how much of a country's production is consumed by residents and how much by visitors, any data with respect to consumption remains purely speculative.

18.4 Chocolate production

Main article: Chocolate Production
To make 1 kg (2.2 pounds) of chocolate, about 300 to 600

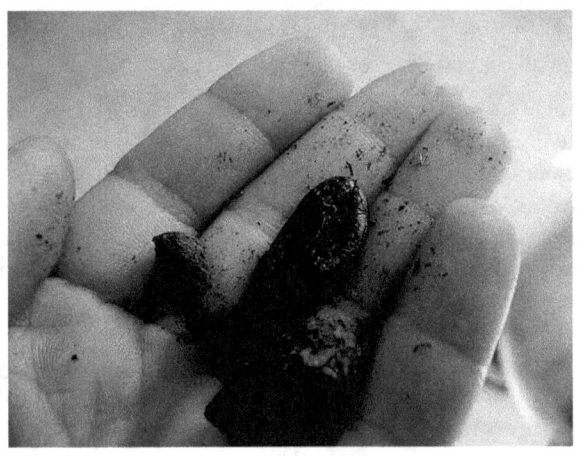

A roasted cocoa bean, the papery skin rubbed loose.

beans are processed, depending on the desired cocoa content. In a factory, the beans are roasted. Next they are cracked and then de-shelled by a "winnower". The resulting pieces of beans are called nibs. They are sometimes sold in small packages at specialty stores and markets to be used in cooking, snacking, and chocolate dishes. Since nibs are directly from the cocoa tree, they contain high amounts of theobromine. Most nibs are ground, using various methods, into a thick creamy paste, known as chocolate liquor or cocoa paste. This "liquor" is then further processed into chocolate by mixing in (more) cocoa butter and sugar (and sometimes vanilla and lecithin as an emulsifier), and then refined, conched and tempered. Alternatively, it can be separated into cocoa powder and cocoa butter using a hydraulic press or the Broma process. This process produces around 50% cocoa butter and 50% cocoa powder. Standard cocoa powder has a fat content of approximately 10–12 percent. Cocoa butter is used in chocolate bar manufacture, other confectionery, soaps, and cosmetics.

Treating with alkali produces Dutch process cocoa powder, which is less acidic, darker and more mellow in flavor than what is generally available in most of the world. Regular (non-alkalized) cocoa is acidic, so when cocoa is treated with an alkaline ingredient, generally potassium carbonate, the pH increases.*[28] This process can be done at various stages during manufacturing, including during nib treatment, liquor treatment or press cake treatment.

Another process that helps develop the flavor is roasting. Roasting can be done on the whole bean before shelling or on the nib after shelling. The time and temperature of the roast affect the result: A "low roast" produces a more acid, aromatic flavor, while a high roast gives a more intense, bitter flavor lacking complex flavor notes.*[29]

18.5 Health benefits of cocoa consumption

In general, cocoa is considered to be a rich source of antioxidants such as procyanidins and flavanoids, which may impart anti-aging properties.*[2]*[30] Cocoa also contain a high level of flavonoids, specifically epicatechin, which may have beneficial cardiovascular effects on health.*[31]*[32]*[33]

The stimulant activity of cocoa comes from the compound theobromine which is less diuretic as compared to theophylline found in tea.*[2] Prolonged intake of flavanol-rich cocoa has been linked to cardiovascular health benefits,*[31]*[32]*[34] though it should be noted that this refers to raw cocoa and to a lesser extent, dark chocolate, since flavonoids degrade during cooking and alkalizing processes.*[35] Studies have found short term benefits in LDL cholesterol levels from dark chocolate consumption.*[36] The addition of whole milk to milk chocolate reduces the overall cocoa content per ounce while increasing saturated fat levels. Although one study*[37] has concluded that milk

Chocolate

impairs the absorption of polyphenolic flavonoids, e.g. epicatechin, a followup*[38] failed to find the effect.

Hollenberg and colleagues of Harvard Medical School studied the effects of cocoa and flavanols on Panama's Kuna people, who are heavy consumers of cocoa. The researchers found that the Kuna People living on the islands had significantly lower rates of heart disease and cancer compared to those on the mainland who do not drink cocoa as on the islands. It is believed that the improved blood flow after consumption of flavanol-rich cocoa may help to achieve health benefits in hearts and other organs. In particular, the benefits may extend to the brain and have important implications for learning and memory.*[39]*[40]*[41]*[42]

Foods rich in cocoa appear to reduce blood pressure but drinking green and black tea may not, according to an analysis of previously published research in the April 9, 2007 issue of Archives of Internal Medicine,*[31] one of the JAMA/Archives journals.*[43]

A 15-year study of elderly men*[44] published in the *Archives of Internal Medicine* in 2006 found a 50 percent reduction in *cardiovascular* mortality and a 47 percent reduction in *all-cause* mortality for the men regularly consuming the most cocoa, compared to those consuming the least cocoa from all sources.

18.6 Child labor

Main articles: Children in cocoa production and Harkin-Engel Protocol

The first allegations that child slavery is used in cocoa production appeared in 1998.*[45] In late 2000 a BBC documentary reported the use of enslaved children in the production of cocoa in West Africa.*[45]*[46]*[47] Other media followed by reporting widespread child slavery and child trafficking in the production of cocoa.*[48]*[49] According to a report by the International Labour Organization (ILO), in 2002, more than 109,000 children were working on cocoa farms in Côte d'Ivoire (Ivory Coast), some of them in "the worst forms of child labour".*[50] The ILO later reported that 200,000 children were working in the cocoa industry in Côte d'Ivoire in 2005.*[51] The 2005 ILO report failed to fully characterize this problem, but estimated that up to 6% of the 200,000 children involved in cocoa production could be victims of human trafficking or slavery.*[51] The cocoa industry was accused of profiting from child slavery and trafficking.*[52] The Harkin-Engel Protocol is an effort to end these practices.*[53] It was signed and witnessed by the heads of eight major chocolate companies, Harkin, Engel, Senator Herb Kohl, the ambassador of the Ivory Coast, the director of the International Programme on the Elimination of Child Labor, and others.*[53] It has, however, been criticized by some groups including the International Labor Rights Forum as an industry initiative which falls short.*[54]*[55]*[56]

18.7   Fairtrade

- There are Fairtrade cocoa producer groups in Belize, Bolivia, Cameroon, The Congo,*[57] Costa Rica, Dominican Republic,*[58] Ecuador, Ghana, Haiti, India, Côte d'Ivoire, Nicaragua, Panama, Peru, Sierra Leone and Sao Tome & Principe.

- As of 2014, less than 1% of the chocolate market was Fair Trade.*[59]

- Cadbury, one of the world's largest chocolate companies, has begun certifying its Dairy Milk bars as Fair Trade; according to Cadbury, in 2010 "around one quarter of ... global sales" of these bars will be Fair Trade.*[60]

18.8   Environmental impact

Main article: Environmental impact of cocoa production

The relative poverty of many cocoa farmers means that environmental consequences such as deforestation are given little significance. For decades, cocoa farmers have encroached on virgin forest, mostly after the felling of trees by logging companies. This trend has decreased as many governments and communities are beginning to protect their remaining forested zones. In general, the use of chemical fertilizers and pesticides by cocoa farmers is limited. When cocoa bean prices are high, farmers may invest in their crops, leading to higher yields which, in turn tends to result in lower market prices and a renewed period of lower investment.

Cocoa production is likely to be affected in various ways by the expected effects of global warming. Specific concerns have been raised concerning its future as a cash crop in West Africa, the current centre of global cocoa production. If temperatures continue to rise, West Africa could simply become unfit to grow the coveted beans.*[61]*[62]

18.8.1   Agroforestry

Cocoa beans may be cultivated under shaded conditions, e.g. agroforestry. Agroforestry can reduce the pressure on existing protected forests for resources, such as firewood, and conserve biodiversity.*[63] Agroforests act as buffers to formally protected forests and biodiversity island refuges in an open, human dominated landscape. Research of their shade-grown coffee counterparts has shown that greater canopy cover in plots is significantly associated to greater mammal species richness and abundance*[64] The amount of diversity in tree species are fairly comparable between shade-grown cocoa plots and primary forests.*[65] Farmers can grow a variety of fruit-bearing shade trees to supplement their income to help cope with the volatile cocoa prices.*[66] Though cocoa has been adapted to grow under a dense rainforest canopy, agroforestry does not significantly further enhance cocoa productivity.*[67]

18.9   Cocoa trading

Cocoa beans, cocoa butter and cocoa powder are traded on two world exchanges: ICE Futures U.S. and NYSE Liffe Futures and Options. The London market is based on West African cocoa and New York on cocoa predominantly from Southeast Asia. Cocoa is the world's smallest soft commodity market.

The future price of cocoa butter and cocoa powder is determined by multiplying the bean price by a ratio. The combined butter and powder ratio has tended to be around 3.5. If the combined ratio falls below 3.2 or so, production ceases to be economically viable and some factories cease extraction of butter and powder and trade exclusively in cocoa liquor.

Cocoa beans can be held in storage for several years in bags or in bulk, during which the ownership can change several times, as the cocoa is traded much the same as metal or other commodities, to gain profit for the owner.

18.10   See also

- Cash crop

- Catechin and epicatechin, flavonoids present in cocoa

- Coenraad Johannes van Houten for Dutch process

- Domingo Ghirardelli for Broma process

- Ghana Cocoa Board

- International CoCoa Farmers Organization

- *Tejate*

- Theobromine, an alkaloid present in cocoa

18.11   References

[1] "Cacao" . Free Dictionary. Retrieved February 17, 2015.

[2] Pharmacognosy and Health Benefits of Cocoa Seeds, Cocoa Powder (Chocolate)

[3] http://www.etymonline.com/index.php?term=cocoa

[4] Ann Bingham; Jeremy Roberts (2010). *South and Meso-American Mythology A to Z*. Infobase Publishing. p. 19. ISBN 978-1-4381-2958-7.

[5] Terrence Kaufman; John Justeson (2006). "History of the Word for 'Cacao' and Related Terms in Ancient Meso-America" . In Cameron L. McNeil. *Chocolate in Mesoamerica: A Cultural History of Cacao*. University Press of Florida. p. 121. ISBN 978-0-8130-3382-2.

[6] "Chocolate Facts" . 2005-06-11. Retrieved 2007-11-12.

[7] Sorting Out Chocolate - Fine Cooking Article

[8] "Cacao Vs. Cocoa: Updating Your Chocolate Vocabulary" . Retrieved 2007-11-12.

[9] http://www.penn.museum/press-releases/739-the-earliest-chocolate-drink-of-the-new-world.html

[10] Wood, G.A.R.; Lass, R.A. (2001). *Cocoa* (4th ed.). Oxford: Blackwell Science. p. 2. ISBN 063206398X.

[11] Díaz del Castillo, Bernal (2005) [1632]. Historia verdadera de la conquista de la Nueva España. Felipe Castro Gutiérrez (Introduction). Mexico: Editores Mexicanos Unidos, S.A.. ISBN 968-15-0863-7. OCLC 34997012

[12] "Chocolate History Time Line" . Retrieved 2007-11-08.

[13] Fabricant, Florence (2011-01-11). "Rare Cacao Beans Discovered in Peru" . *The New York Times*. Retrieved 2014-02-01.

[14] Zipperer, Paul (1902). *The manufacture of chocolate and other cacao preparations* (3 ed.). Berlin: Verlag von M. Krayn. p. 14. white cacao, ... Ecuador ... rare ... In Trinidad also

[15] US patent 5395635, Akira Inoue, Hideo Sasai, Kazuji Yanamoto, "Method of producing white cacao nibs and food using white cacao nibs" , issued 1995-03-07, assigned to Ezaki Glico Kabushiki Kaisha

[16] "Cocoa Beans" . *brainresearchsupplement*. Retrieved 31 August 2015.

[17] http://www.exploratorium.edu/exploring/exploring_chocolate/choc_2.html

[18] "Cocoa Market Update" (PDF). World Cocoa Foundation. May 2010. Retrieved 11 December 2011.

[19] Wood, G. A. R.; Lass, R. A. (2001). *Cocoa*. Tropical agriculture serie (4 ed.). John Wiley and Sons. ISBN 0-632-06398-X.

[20] Olivia Abenyega and James Gockowski (2003). *Labor practices in the cocoa sector of Ghana with a special focus on the role of children*. International Institute of Tropical Agriculture. pp. 10–11. ISBN 978-131-218-1.

[21] Hui, Yiu H. (2006). *Handbook of food science, technology, and engineering* 4. CRC Press. ISBN 0-8493-9849-5.

[22] Dand, Robin (1999). *The international cocoa trade* (2 ed.). Woodhead Publishing. ISBN 1-85573-434-6.

[23] J. Gockowski and S. Oduwole (2003). *Labor practices in the cocoa sector of southwest Nigeria with a focus on the role of children*. International Institute of Tropical Agriculture. pp. 11–15. ISBN 978-131-215-7.

[24] "Yeasts key for cacao bean fermentation and chocolate quality" . Confectionery News. Retrieved 2014-02-02.

[25] "FAQ : Products that can be made from cocoa" . International Cocoa Organization. Retrieved 2014-01-31.

[26] http://food.theatlantic.com/artisans/mexican-chocolate-rustic-strong-better.php

[27] "ICCO Press Releases" . International Cocoa Organization. 30 November 2011. Retrieved 11 December 2011.

[28] Emily Nolan (2002). *Baking For Dummies*. For Dummies. p. 27. ISBN 978-0-7645-5420-9.

[29] "Cocoa: From Bean to Bar," Urbanski, John, Food Product Design, May 2008

[30] Gressner, Olav A (October 2012). "Chocolate Shake and Blueberry Pie...... or why Your Liver Would Love it" . *Journal of Gastroenterology and Hepatology Research* **1** (9): 171–195.

[31] Taubert D, Roesen R, Schömig E (April 2007). "Effect of cocoa and tea intake on blood pressure: a meta-analysis". *Arch. Intern. Med.* **167** (7): 626–34. doi:10.1001/archinte.167.7.626. PMID 17420419.

[32] Schroeter H, Heiss C, Balzer J; et al. (January 2006). "(-)-Epicatechin mediates beneficial effects of flavanol-rich cocoa on vascular function in humans". *Proc. Natl. Acad. Sci. U.S.A.* **103** (4): 1024–9. doi:10.1073/pnas.0510168103. PMC 1327732. PMID 16418281.

[33] "Why Cocoa May Help Heart Health". WebMD. Retrieved February 17, 2015.

[34] 1743-7075-3-2.fm

[35] "Cocoa nutrient for 'lethal ills'". *BBC News.* 2007-03-11. Retrieved 2010-04-30.

[36] http://circ.ahajournals.org/content/119/10/1433.full

[37] Mauro Serafini, Rossana Bugianesi, Giuseppe Maiani, Silvia Valtuena, Somone De Santis, Ala Crozier: "Plasma antioxidants from chocolate", *Nature* **424**(2003)1013. Downloaded from http://eprints.gla.ac.uk/131/01/Crozier, A_2003.pdf

[38] J.B. Keogh, J. McInerney, and P.M. Clifton: "The Effect of Milk Protein on the Bioavailability of Cocoa Polyphenols", *Journal of Food Science* **72**(3)S230-S233, 2007. Downloaded from http://onlinelibrary.wiley.com/doi/10.1111/j.1750-3841.2007.00314.x/abstract

[39] "Flavanols in cocoa may offer benefits to the brain". *International Journal of Medical Sciences.* Nov 2, 2007. Retrieved Nov 1, 2014.

[40] Bayard V, Chamorro F, Motta J, Hollenberg NK (2007). "Does flavanol intake influence mortality from nitric oxide-dependent processes? Ischemic heart disease, stroke, diabetes mellitus, and cancer in Panama". *Int J Med Sci* **4** (1): 53–8. doi:10.7150/ijms.4.53. PMC 1796954. PMID 17299579.

[41] Messerli FH. "Chocolate Consumption, Cognitive Function, and Nobel Laureates". *N Engl J Med* **367**: 1562–1564. doi:10.1056/NEJMon1211064.

[42] Ingham, Richard; Agence France-Presse (5 February 2007). "Cocoa clue to reversing age-related memory loss". *Japan Times.* Retrieved 9 May 2011.

[43] Cocoa, But Not Tea, May Lower Blood Pressure

[44] Buijsse B, Feskens EJ, Kok FJ, Kromhout D (February 2006). "Cocoa intake, blood pressure, and cardiovascular mortality: the Zutphen Elderly Study". *Arch. Intern. Med.* **166** (4): 411–7. doi:10.1001/archinte.166.4.411. PMID 16505260.

[45] Sudarsan Raghavan and Sumana Chatterjee (24 June 2001). "Slaves feed world's taste for chocolate: Captives common in cocoa farms of Africa". Milwaukee Journal Sentinel. Archived from the original on 17 September 2006. Retrieved 25 April 2012.

[46] "Combating Child Labour in Cocoa Growing" (PDF). International Labor Organization. 2005. Retrieved 26 April 2012.

[47] David Wolfe and Shazzie (2005). *Naked Chocolate: The Astonishing Truth about the World's Greatest Food.* North Atlantic Books. p. 98. ISBN 1-55643-731-5. Retrieved 15 December 2011.

[48] Humphrey Hawksley (12 April 2001). "Mali's children in chocolate slavery". BBC News. Retrieved 2 January 2010.

[49] Humphrey Hawksley (4 May 2001). "Ivory Coast accuses chocolate companies". BBC News. Retrieved 4 August 2010.

[50] U.S. Department of State Country Reports on Human Rights Practices, 2005 Human Rights Report on Côte d'Ivoire

[51] ilo.law.cornell.edu

[52] Payson Center for International Development and Technology Transfer (30 September 2010). "Fourth Annual Report: Oversight of Public and Private Initiatives to Eliminate the Worst Forms of Child Labor in the Cocoa Sector of Côte d'Ivoire and Ghana" (PDF). Tulane University. p. 26. Retrieved 23 April 2012.

[53] "Protocol for the growing and processing of cocoa beans and their derivative products in a manner that complies with ILO Convention 182 concerning the prohibition and immediate action for the elimination of the worst forms of child labor" (PDF). International Cocoa Initiative. 2001. Retrieved 25 April 2012.

[54] Tricia Escobedo (19 September 2011). "The Human Cost of Chocolate". CNN. Retrieved 28 April 2012.

[55] Karen Ann Monsy (24 February 2012). "The bitter truth". *Khaleej Times.* Retrieved 28 April 2012.

[56] Payson Center for International Development and Technology Transfer (31 March 2011). "Oversight of Public and Private Initiatives to Eliminate the Worst Forms of Child Labor in the Cocoa Sector of Côte d'Ivoire and Ghana" (PDF). Tulane University. pp. 7–12. Retrieved 26 April 2012.

[57] "GOURMET GARDENS: CONGOLESE FAIR TRADE AND ORGANIC COCOA". befair.be.

[58] "CONACADO: National confederation of cocoa producers".

[59] "The News on Chocolate is Bittersweet: No Progress on Child Labor, but Fair Trade Chocolate is on the Rise". Global Exchange June 2005 (8 pages). retrieved 1 July 2010.

[60] "Fairtrade Cadbury Dairy Milk Goes Global as Canada, Australia, and New Zealand take Fairtrade Further Into Mainstream". Cadbury PLC 2010. Retrieved 1 July 2010.

[61] Climate Change Could Melt Chocolate Production; A new study shows that cocoa will suffer under climate change by Tiffany Stecker and ClimateWire Scientific American October 3, 2011

[62] Climate change: Will chocolate become a costly luxury? If temperatures continue to rise, a new report suggests, West Africa, source of half the world's chocolate, will be unfit to grow the coveted beans posted on The Week September 30, 2011

[63] Bhagwat, Shonil A.; Willis, Katherine J.; Birks, H. John B.; Whittaker, Robert J. "Agroforestry: a refuge for tropical biodiversity?". *Trends in Ecology & Evolution* **23** (5): 261– 267. doi:10.1016/j.tree.2008.01.005.

[64] Caudill, S. Amanda; DeClerck, Fabrice J.A.; Husband, Thomas P. "Connecting sustainable agriculture and wildlife conservation: Does shade coffee provide habitat for mammals?". *Agriculture, Ecosystems & Environment* **199**: 85–93. doi:10.1016/j.agee.2014.08.023.

[65] Vebrova, Hana, Bohdan Lojka, Thomas P. Husband, Maria E.C Zans, Patrick Van Damme, Alexandr Rollo, and Marie Kalousova. "Tree Diversity in Cacao Agroforests in San Alejandro, Peruvian Amazon." Springer Link. Springer Netherlands, 09 Nov. 2013. Web. Feb. 2015.

[66] Oke, D.O.; Odebiyi, K.A. (2007). "Traditional cocoa-based agroforestry and forest species conservation in Ondo State, Nigeria, Agriculture". *Ecosystems & Environment* **122** (3): 305–311. doi:10.1016/j.agee.2007.01.022.

[67] Pedelahore, Philippe. "Farmers Accumulation Strategies and Agroforestry Systems Intensification: The Example of Cocoa in the Central Region of Cameroon over the 1910-2010 Period." Springer Link. Springer Netherlands, 20 Feb. 2014. Web. Retrieved February 2015.

18.12   External links

- International Cocoa Organization (ICCO)

- International CoCoa Farmers Organization (ICCFO)

- Roundtable for a Sustainable Cocoa Economy (RSCE) - working with all stakeholders from the cocoa economy towards a world cocoa economy that is economically viable, ecologically sound and socially acceptable

- Harvard Study on Medical Aspects of Cocoa

- Cocoa Producers' Alliance (COPAL)

- Upcocoa project in Cameroon - A multi-stakeholder initiative on upgrading the capacities of cocoa farmers and their organisations.

- Articles on cocoa trade at the *Agritrade* web site.

Chapter 19

Coconut

For other uses, see Coconut (disambiguation).
"Coconut Tree" redirects here. For the Mohombi song,
see Coconut Tree (song).

The **coconut tree** (*Cocos nucifera*) is a member of the
family Arecaceae (palm family).

It is the only accepted species in the genus *Cocos*.[2] The
term **coconut** can refer to the entire **coconut palm**, the
seed, or the fruit, which, botanically, is a drupe, not a nut.
The spelling **cocoanut** is an archaic form of the word.[3]
The term is derived from the 16th-century Portuguese and
Spanish word *coco* meaning "head" or "skull", from the
three indentations on the coconut shell that resemble facial
features.[4]

The coconut is known for its great versatility as seen in the
many uses of its different parts and found throughout the
tropics and subtropics.[5] Coconuts are part of the daily diets of many people. Coconuts are different from any other
fruits because they contain a large quantity of "water" and
when immature they are known as tender-nuts or jelly-nuts
and may be harvested for drinking. When mature, they still
contain some water and can be used as seednuts or processed to give oil from the kernel, charcoal from the hard
shell and coir from the fibrous husk. The endosperm is
initially in its nuclear phase suspended within the coconut
water. As development continues, cellular layers of endosperm deposit along the walls of the coconut, becoming the edible coconut "flesh".[6] When dried, the coconut flesh is called copra. The oil and milk derived from
it are commonly used in cooking and frying; coconut oil is
also widely used in soaps and cosmetics. The clear liquid
coconut water within is potable. The husks and leaves can
be used as material to make a variety of products for fur-
nishing and decorating. The coconut also has cultural and
religious significance in many societies that use it.

Coconut flowers

19.1 Description

19.1.1 Plant

Cocos nucifera is a large palm, growing up to 30 m (98 ft)
tall, with pinnate leaves 4–6 m (13–20 ft) long, and pinnae
60–90 cm long; old leaves break away cleanly, leaving the
trunk smooth. Coconuts are generally classified into two
general types: tall and dwarf.[7] On very fertile land, a tall
coconut palm tree can yield up to 75 fruits per year, but

more often yields less than 30, mainly due to poor cultural practices.[8][9][10] Given proper care and growing conditions coconut palms produce their first fruit in six to ten years, it takes 15 – 20 years to reach peak production.[11]

19.1.2 Fruit

Botanically, the coconut fruit is a drupe, not a true nut.[12] Like other fruits, it has three layers: the exocarp, mesocarp, and endocarp. The exocarp and mesocarp make up the "husk" of the coconut. Coconuts sold in the shops of non-tropical countries often have had the exocarp (outermost layer) removed. The mesocarp is composed of a fiber, called coir, which has many traditional and commercial uses. The shell has three germination pores (stoma) or "eyes" that are clearly visible on its outside surface once the husk is removed.

A full-sized coconut weighs about 1.44 kg (3.2 lb). It takes around 6,000 full-grown coconuts to produce a tonne of copra.[13]

19.1.3 Roots

Unlike some other plants, the palm tree has neither a tap root nor root hairs, but has a fibrous root system.[14]

The coconut palm root system[15] consists of an abundance of thin roots that grow outward from the plant near the surface. Only a few of the roots penetrate deep into the soil for stability. The type of root system is known as fibrous or adventitious, and is a characteristic of grass species. Other types of large trees produce a single downward-growing tap root with a number of feeder roots growing from it.

Coconut palms continue to produce roots from the base of the stem throughout its life. The number of roots produced depends on the age of the tree and the environment, with more than 3,600 roots possible on a tree that's 60 to 70 years old.

Roots are usually less than about 3 inches in diameter and uniformly thick from the tree trunk to the root tip.

19.1.4 Inflorescence

The palm produces both the female and male flowers on the same inflorescence; thus, the palm is monoecious.[14] Other sources use the term polygamomonoecious.[16] The female flower is much larger than the male flower. Flowering occurs continuously. Coconut palms are believed to be largely cross-pollinated, although some dwarf varieties are self-pollinating.

19.2 Etymology

A dehusked coconut shell from Ivory Coast showing the face-like markings at the base

A cut coconut shell

One of the earliest mentions of the coconut dates back to the One Thousand and One Nights story of Sinbad the Sailor; he is known to have bought and sold coconuts during his fifth voyage.[17] *Tenga*, its Malayalam and Tamil name, was used in the detailed description of coconut found in *Itinerario* by Ludovico di Varthema published in 1510 and also in the later *Hortus Indicus Malabaricus*.[18] Even earlier, it was called *nux indica*, a name used by Marco Polo in 1280 while in Sumatra, taken from the Arabs who called it جوز هندي *jawz hindī*. Both names translate to "Indian nut".[19] In the earliest description of the coconut palm known, given by Cosmos of Alexandria in his *Topographia Christiana* written about 545 AD, there is a reference to the argell tree and its drupe.[18][20]

Historical evidence favors the European origin of the name "coconut", for no name is similar in any of the languages of India, where the Portuguese first found the fruit;

and indeed Barbosa, Barros, and Garcia, in mentioning the Tamil/Malayalam name *tenga*, and Canarese *narle*, expressly say, "we call these fruits *quoquos*", "our people have given it the name of coco", and "that which we call coco, and the Malabars temga".

The OED states: "Portuguese and Spanish authors of the 16th c. agree in identifying the word with Portuguese and Spanish *coco* "grinning face, grin, grimace", also "bugbear, scarecrow", cognate with *cocar* "to grin, make a grimace"; the name being said to refer to the face-like appearance of the base of the shell, with its three holes. According to Losada, the name came from Portuguese explorers, the sailors of Vasco da Gama in India, who first brought them to Europe. The coconut shell reminded them of a ghost or witch in Portuguese folklore called *coco* (also *côca*).[21][22] The first known recorded usage of the term is 1555.[23][24]

The specific name *nucifera* is Latin for "nut-bearing".

19.3   Origin, domestication, and dispersal

19.3.1   Origin

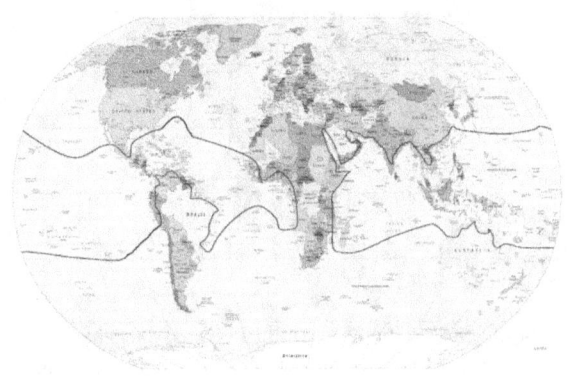

The range of the natural habitat of the coconut palm tree delineated by the red line (based on information in Werth 1933,[25] slightly modified by Niklas Jonsson)

The origin of the plant is the subject of debate.[26][27][28] O.F. Cook was one of the earliest modern researchers to draw conclusions about the location of origin of *Cocos nucifera* based on its current-day worldwide distribution.[29] He hypothesized that the coconut originated in the Americas, based on his belief that American coconut populations predated European contact and because he considered pan-tropical distribution by ocean currents improbable. Thor Heyerdahl later used this hypothesis of the American origin of the coconut to support his theory that the Pacific Islanders originated in

South America.[30] However, more evidence exists for an Indo-Pacific origin either around Melanesia and Malesia or the Indian Ocean.[26][27][28] The oldest fossils known of the modern coconut dating from the Eocene period from around 37 to 55 million years ago were found in Australia and India. However, older palm fossils such as some of nipa fruit have been found in the Americas.[28] Since 1978, the work on tracing the probable origin and dispersal of *Cocos nucifera*[31] has only recently been augmented by a publication on the germination rate of the coconut seednut [32] and another on the importance of the coral atoll ecosystem.[33] Briefly, the coconut originated in the coral atoll ecosystem without human intervention and required a thick husk and slow germination to survive and disperse.

19.3.2   Domestication

Manual harvesting of coconuts

Coconuts could not reach inland locations without human intervention (to carry seednuts, plant seedlings, etc.) and it was early germination on the palm (vivipary) that was important,[34] rather than increasing the number or size of the edible parts of a fruit that was already large enough. Human cultivation of the coconut selected, not for larger size, but for thinner husks and increased volume of endosperm, the solid "meat" or liquid "water" that provides the fruit its food value. Although these modifications for domestication would reduce the fruit's ability to float, this ability would be irrelevant to a cultivated population.

Among modern *C. nucifera*, two major types or variants: a thick-husked, angular fruit and a thin-husked, spherical fruit with a higher proportion of endosperm reflect a trend of cultivation in *C. nucifera*: the first coconuts were of the *niu kafa* type, with thick husks to protect the seed, an angular, highly ridged shape to promote buoyancy during ocean dispersal, and a pointed base that allowed fruits to dig into the sand, preventing them from being washed away during germination on a new island. As early human communities began to harvest coconuts for eating and planting, they (perhaps unintentionally) selected for a larger endosperm to husk ratio and a broader, spherical base, which rendered the fruit useful as a cup or bowl, thus creating the *niu vai* type. The decreased buoyancy and increased fragility of this spherical, thin-husked fruit would not matter for a species that had started to be dispersed by humans and grown in plantations. Harries' adoption of the Polynesian terms *niu kafa* and *niu vai* has now passed into general scientific discourse, and his hypothesis is generally accepted.*[35]*[36]

Variants of *C. nucifera* are also categorized as Tall (var. *typical*) or Dwarf (var. *nana*).*[37] The two groups are genetically distinct, with the Dwarf variety showing a greater degree of artificial selection for ornamental traits and for early germination and fruiting.*[31]*[38] The Tall variety is outcrossing while Dwarf palms are incrossing, which has led to a much greater degree of genetic diversity within the Tall group. It is believed that the Dwarf subgroup mutated from the Tall group under human selection pressure.*[39]

19.3.3 Dispersal

Main article: Genomics of domestication

It is often stated that coconuts can travel 110 days, or 3,000 miles (4,800 km), by sea and still be able to germinate.*[40] This figure has been questioned based on the extremely small sample size that forms the basis of the paper that makes this claim.*[41] Thor Heyerdahl provides an alternative, and much shorter, estimate based on his first-hand experience crossing the Pacific Ocean on the raft Kon-Tiki: "The nuts we had in baskets on deck remained edible and capable of germinating the whole way to Polynesia. But we had laid about half among the special provisions below deck, with the waves washing around them. Every single one of these was ruined by the sea water. And no coconut can float over the sea faster than a balsa raft moves with the wind behind it." *[30] He also notes that several of the nuts began to germinate by the time they had been ten weeks at sea, precluding an unassisted journey of 100 days or more. However, it is more than likely that the coconut variety Heyerdahl chose for his long sea voyage was of the large, fleshy, spherical *niu vai* type, which Harries observed to have a significantly shorter germination type and worse buoyancy than the uncultivated *niu kafa* type.*[31] Therefore, Heyerdahl's observations cannot be considered conclusive when it comes to determining the independent dispersal ability of the uncultivated coconut.

Drift models based on wind and ocean currents have shown that coconuts could not have drifted across the Pacific

unaided.*[41] This provides some circumstantial evidence that Austronesian peoples carried coconuts across the ocean and that they could not have dispersed worldwide without human agency. More recently, genomic analysis of cultivated coconut (*Cocos nucifera L.*) has shed light on the movements of Austronesian peoples. By examining 10 microsatellite loci, researchers found two genetically distinct subpopulations of coconut one originating in the Indian Ocean, the other in the Pacific Ocean. However, admixture, the transfer of genetic material, evidently occurred between the two populations. Given that coconuts are ideally suited for ocean dispersal, individuals from one population possibly could have floated to the other. However, the locations of the admixture events are limited to Madagascar and coastal east Africa, and exclude the Seychelles. This pattern coincides with the known trade routes of Austronesian sailors. Additionally, a genetically distinct subpopulation of coconut on the Pacific coast of Latin America has undergone a genetic bottleneck resulting from a founder effect; however, its ancestral population is the Pacific coconut. This, together with their use of the South American sweet potato, suggests that Austronesian peoples may have sailed as far east as the Americas.*[42]

19.3.4 Distribution

The coconut has spread across much of the tropics, probably aided in many cases by seafaring people. Coconut fruit in the wild are light, buoyant and highly water resistant, and evolved to disperse significant distances via marine currents.*[43] Specimens have been collected from the sea as far north as Norway.*[44] In the Hawaiian Islands, the coconut is regarded as a Polynesian introduction, first brought to the islands by early Polynesian voyagers from their homelands in Oceania.*[19] They have been found in the Caribbean and the Atlantic coasts of Africa and South America for less than 500 years, but evidence of their presence on the Pacific coast of South America predates Christopher Columbus's arrival in the Americas.*[27] They are now almost ubiquitous between 26°N and 26°S except for the interiors of Africa and South America.

19.4 Natural habitat

The coconut palm thrives on sandy soils and is highly tolerant of salinity. It prefers areas with abundant sunlight and regular rainfall (1500 mm to 2500 mm annually), which makes colonizing shorelines of the tropics relatively straightforward.*[45] Coconuts also need high humidity (70–80%+) for optimum growth, which is why they are rarely seen in areas with low humidity, like the southeastern Mediterranean or Andalusia (Spain), even where tem-

Coconut germinating on Black Sand Beach, Island of Hawaii

peratures are high enough (regularly above 24 °C or 75.2 °F). However, they can be found in humid areas with low annual precipitation such as in Karachi, Pakistan, which receives only about 250 mm (9.8 in) of rainfall per year, but is consistently warm and humid.

Coconut palms require warm conditions for successful growth, and are intolerant of cold weather. Some seasonal variation is tolerated, with good growth where mean summer temperatures are between 28 and 37 °C (82 and 99 °F), and survival as long as winter temperatures are above 4–12 °C (39–54 °F); they will survive brief drops to 0 °C (32 °F). Severe frost is usually fatal, although they have been known to recover from temperatures of –4 °C (25 °F).[45] They may grow but not fruit properly in areas with insufficient warmth, such as Bermuda.

The conditions required for coconut trees to grow without any care are:

- Mean daily temperature above 12–13 °C (54–55 °F) every day of the year

- Mean annual rainfall above 1,000 mm (39 in)

- No or very little overhead canopy, since even small trees require direct sun

The main limiting factor for most locations which satisfy the rainfall and temperature requirements is canopy growth, except those locations near coastlines, where the sandy soil and salt spray limit the growth of most other trees.

19.4.1 Diseases

Main article: List of coconut palm diseases

Coconuts are susceptible to the phytoplasma disease lethal yellowing. One recently selected cultivar, the Maypan, has been bred for resistance to this disease.

19.4.2 Pests

The coconut palm is damaged by the larvae of many Lepidoptera (butterfly and moth) species which feed on it, including *Batrachedra* spp.: *B. arenosella*, *B. atriloqua* (feeds exclusively on *C. nucifera*), *B. mathesoni* (feeds exclusively on *C. nucifera*), and *B. nuciferae*.

Brontispa longissima (coconut leaf beetle) feeds on young leaves, and damages both seedlings and mature coconut palms. In 2007, the Philippines imposed a quarantine in Metro Manila and 26 provinces to stop the spread of the pest and protect the $800 million Philippine coconut industry.[46]

The fruit may also be damaged by eriophyid coconut mites (*Eriophyes guerreronis*). This mite infests coconut plantations, and is devastating: it can destroy up to 90% of coconut production. The immature seeds are infested and desapped by larvae staying in the portion covered by the perianth of the immature seed; the seeds then drop off or survive deformed. Spraying with wettable sulfur 0.4% or with neem-based pesticides can give some relief, but is cumbersome and labor-intensive.

In Kerala (India), the main coconut pests are the coconut mite, the rhinoceros beetle, the red palm weevil and the coconut leaf caterpillar. Research into countermeasures to these pests has as of 2009 yielded no results; researchers from the Kerala Agricultural University and the Central Plantation Crop Research Institute, Kasaragode continue to work on countermeasures. The Krishi Vigyan Kendra, Kannur under Kerala Agricultural University has developed an innovative extension approach called the compact area group approach (CAGA) to combat coconut mites.

19.5 Production and cultivation

Coconut palms are grown in more than 90 countries of the world, with a total production of 62 million tonnes per year (table).[*][47] Most of the world production is in tropical Asia, with Indonesia, the Philippines and India accounting collectively for 73% of the world total (table).

19.5.1 Cultivation

Coconut trees are hard to establish in dry climates, and cannot grow there without frequent irrigation; in drought conditions, the new leaves do not open well, and older leaves may become desiccated; fruit also tends to be shed.[*][45]

The extent of cultivation in the tropics is threatening a number of habitats, such as mangroves; an example of such damage to an ecoregion is in the Petenes mangroves of the Yucatán.[*][48]

19.5.2 Harvesting

In some parts of the world (Thailand and Malaysia), trained pig-tailed macaques are used to harvest coconuts. Training schools for pig-tailed macaques still exist both in southern Thailand and in the Malaysian state of Kelantan.[*][49] Competitions are held each year to find the fastest harvester.

19.5.3 India

Coconuts being sold on a street in India

Coconut plucking in Kerala, India

Green coconut fruit strands on the tree are featured on each Maldivian rufiyaa banknote

Traditional areas of coconut cultivation in India are the states of Kerala, Tamil Nadu, Karnataka, Puducherry, Andhra Pradesh, Goa, Maharashtra, Odisha, West Bengal and the islands of Lakshadweep and Andaman and Nicobar. Four southern states combined account for almost 92% of the total production in the country: Kerala (45.22%),

Coconut trees are among the most common sights throughout Kerala

Tamil Nadu (26.56%), Karnataka (10.85%), and Andhra Pradesh (8.93%).*[50] Other states, such as Goa, Maharashtra, Odisha, West Bengal, and those in the northeast (Tripura and Assam) account for the remaining 8.44%. Kerala, which has the largest number of coconut trees, is famous for its coconut-based products coconut water, copra, coconut oil, coconut cake (also called coconut meal, copra cake, or copra meal), coconut toddy, coconut shell-based products, coconut wood-based products, coconut leaves, and coir pith.

Various terms, such as copra and coir, are derived from the native Malayalam language. In Kerala, the coconut tree is called "Thengu" also termed as *kalpa vriksham*, which essentially means all parts of a coconut tree is useful some way or other.

19.5.4 Maldives

The coconut is the national tree of the Maldives and is considered the most important plant in the country. A coconut tree is also included in the country's national emblem or coat of arms. Coconut trees are grown on all the islands. Before modern construction methods were introduced, coconut leaves were used as roofing material for many houses in the islands, while coconut timber was used to build houses and boats.

19.5.5 Middle East

The main coconut-producing area in the Middle East is the Dhofar region of Oman, but they can be grown all along the Persian Gulf, Arabian Sea and Red Sea coasts, because these seas are tropical and provide enough humidity (through seawater evaporation) for coconut trees to grow. The young coconut plants need to be nursed and irrigated with drip pipes until they are old enough (stem bulb de-

velopment) to be irrigated with brackish water or seawater alone, after which they can be replanted on the beaches. In particular, the area around Salalah maintains large coconut plantations similar to those found across the Arabian Sea in Kerala. The reasons why coconut are cultivated only in Yemen's Al Mahrah and Hadramaut governorates and in the Sultanate of Oman, but not in other suitable areas in the Arabian Peninsula, may originate from the fact that Oman and Hadramaut had long dhow trade relations with Burma, Malaysia, Indonesia, East Africa and Zanzibar, as well as southern India and China. Omani people needed the coir rope from the coconut fiber to stitch together their traditional high seas-going dhow vessels in which nails were never used. The 'know how' of coconut cultivation and necessary soil fixation and irrigation may have found its way into Omani, Hadrami and Al-Mahra culture by people who returned from those overseas areas.

Coconut trees line the beaches and corniches of Oman.

The coconut cultivars grown in Oman are generally of the drought-resistant Indian "West Coast tall" (WC Tall) variety. Unlike the UAE, which grows mostly non-native dwarf or hybrid coconut cultivars imported from Florida for ornamental purposes, the slender, tall Omani coconut cultivars are relatively well-adapted to the Middle East's hot dry seasons, but need longer to reach maturity. The Middle East's hot, dry climate favors the development of coconut mites, which cause immature seed dropping and may cause brownish-gray discoloration on the coconut's outer green fiber.

The ancient coconut groves of Dhofar were mentioned by the medieval Moroccan traveller Ibn Battuta in his writings, known as *Al Rihla*.*[51] The annual rainy season known locally as *Khareef* or monsoon makes coconut cultivation easy on the Arabian east coast.

Coconut trees also are increasingly grown for decorative purposes along the coasts of the UAE and Saudi Arabia with the help of irrigation. The UAE has, however, imposed strict laws on mature coconut tree imports from other countries to reduce the spread of pests to other native palm trees, as the mixing of date and coconut trees poses a risk of cross-species palm pests, such as rhinoceros beetles and red palm weevils.*[52] The artificial landscaping adopted in Florida may have been the cause for lethal yellowing, a viral coconut palm disease that leads to the death of the tree. It is spread by host insects, that thrive on heavy turf grasses. Therefore, heavy turf grass environments (beach resorts and golf courses) also pose a major threat to local coconut trees. Traditionally, dessert banana plants and local wild beach flora such as *Scaevola taccada* and *Ipomoea pes-caprae* were used as humidity-supplying green undergrowth for coconut trees, mixed with sea almond and sea hibiscus. Due to growing sedentary life styles and heavy-handed landscaping, there has been a decline in these traditional farming and soil-fixing techniques.

19.5.6 Sri Lanka

An early mention of the planting of coconuts is found in the *Mahavamsa* during the reign of Agrabodhi II around 589 AD.*[18] Coconuts are common in the Sri Lankan diet and the main source of dietary fat.*[53]

19.5.7 United States

The only places in the United States where coconut palms can be grown and reproduced outdoors without irrigation are Hawaii, southern and central Florida,*[54] and the territories of Puerto Rico, Guam, American Samoa, the U.S. Virgin Islands, and the Commonwealth of the Northern Mariana Islands.

Coconut palms will grow from coastal Pinellas County and St. Petersburg southwards on Florida's west coast, and Melbourne southwards on Florida's east coast. The occasional coconut palm is seen north of these areas in favoured microclimates in Tampa and Clearwater, as well as around Cape Canaveral and Daytona Beach on the east coast. They reach fruiting maturity, but can be damaged or killed by the occasional winter freezes in these areas. In South Texas they may also be grown in favoured microclimates around the Rio Grande Valley near Brownsville, and as far north as Corpus Christi , however more severe cold snaps keep them from producing viable fruit. While coconut palms flourish in southern Florida, rare cold snaps can injure coconut palms there, as well. Only the Florida Keys and the distant southern Atlantic coastlines near Miami provide safe havens from the cold for growing coconut palms on the mainland.

19.5.8 Australia

Coconuts are commonly grown around the northern coast of Australia, and in some warmer parts of New South Wales.

19.5.9 Bermuda

Most of the tall mature coconut trees found in Bermuda were shipped to the island as seedlings on the decks of ships. In more recent years, the importation of coconuts was prohibited, therefore, a large proportion of the younger trees have been propagated from locally grown coconuts.

In the winter months, the growth rate of coconut trees declines due to cooler temperatures and people have commonly attributed this to the reduced yield of coconuts in comparison to tropical regions. However, whilst cooler winter temperatures may be a factor in reducing fruit production, the primary reason for the reduced yield is a lack of water. Bermuda's soil is generally very shallow (1.5 to 3 feet) and much of a coconut tree's root mass is found in the porous limestone underneath the soil. Due to the porosity of the limestone, Bermuda's coconut trees do not generally have a sufficient supply of water with which they are able to support a large number of fruit as rain water quickly drains down through the limestone layer to the water table which is far too deep for a coconut's roots to reach. This typically leads to a reduction in fruit yield (sometimes as little as one or two mature fruits) as well as a reduced milk content inside the coconut that often causes the fruit to be infertile.

Conversely, trees growing in close proximity to the sea almost universally yield a much greater volume of fruit as they are able to tap directly into the sea water which permeates the limestone in such areas. Not only do these trees produce a significantly higher yield, but also the fruit itself tends to be far more fertile due to the higher milk content. Trees found growing in Bermuda's marshy inland areas enjoy a similar degree of success as they are also able to tap directly into a constant supply of water.

19.5.10 Europe

The southern Mediterranean islands of Lampedusa and Linosa are the only locations in Europe having a climate favorable for coconuts to grow. Coconuts also grow in the Spanish territory of the Canary Islands and in the Portuguese territory of Madeira, both of which belong geographically to the African continent.

19.5.11 Cooler climates

In cooler climates (but not less than USDA Zone 9), a similar palm, the queen palm (*Syagrus romanzoffiana*), is used in landscaping. Its fruits are very similar to the coconut, but much smaller. The queen palm was originally classified in the genus *Cocos* along with the coconut, but was later reclassified in *Syagrus*. A recently discovered palm, *Beccariophoenix alfredii* from Madagascar, is nearly identical to the coconut, more so than the queen palm and can also be grown in slightly cooler climates than the coconut palm. Coconuts can only be grown in temperatures above 18 °C (64 °F) and need a daily temperature above 22 °C (72 °F) to produce fruit.

Green coconuts

19.6 Overview of uses

Coconut trees used for landscaping along a coastal road in Kota Kinabalu, Sabah, Malaysia.

The coconut palm is grown throughout the tropics for decoration, as well as for its many culinary and nonculinary uses; virtually every part of the coconut palm can be used by humans in some manner and has significant economic value. Coconuts' versatility is sometimes noted in its naming. In Sanskrit, it is *kalpa vriksha* ("the tree which provides all the necessities of life"). In the Malay language, it is *pokok seribu guna* ("the tree of a thousand uses"). In the Philippines, the coconut is commonly called the "tree of life".[*][55]

19.7 Culinary use

The various parts of the coconut have a number of culinary uses. The seed provides oil for frying, cooking, and making margarine. The white, fleshy part of the seed, the coconut meat, is used fresh or dried in cooking, especially in confections and desserts such as macaroons. Desiccated coconut

Coconut water drink

or coconut milk made from it is frequently added to curries and other savory dishes. Coconut flour has also been developed for use in baking, to combat malnutrition.[*][56] Coconut chips have been sold in the tourist regions of Hawaii and the Caribbean. Coconut butter is often used to describe solidified coconut oil, but has also been adopted as a name by certain specialty products made of coconut milk solids or puréed coconut meat and oil. Dried coconut is also used as the filling for many chocolate bars. Some dried coconut is purely coconut but others are manufactured with other in-

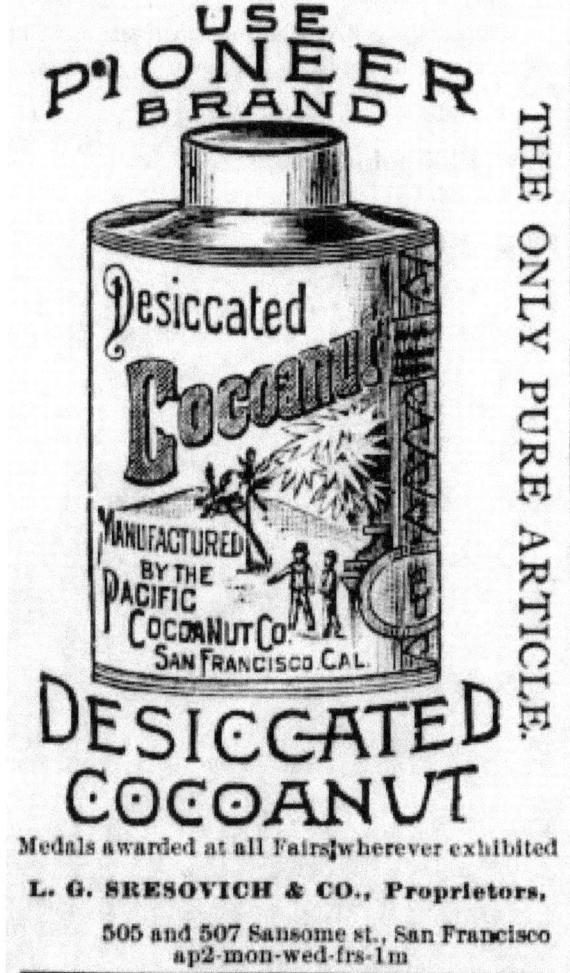

1890 newspaper advertisement showing tin of dried coconut

Coconut water serves as a suspension for the endosperm of the coconut during its nuclear phase of development. Later, the endosperm matures and deposits onto the coconut rind during the cellular phase.[6] It is consumed throughout the humid tropics, and has been introduced into the retail market as a processed sports drink. Mature fruits have significantly less liquid than young, immature coconuts, barring spoilage. Coconut water can be fermented to produce coconut vinegar.

Per 100 gram (100 ml) serving, coconut water contains 19 calories and no significant content of essential nutrients.

Coconut milk

Main article: Coconut milk
Coconut milk, not to be confused with coconut water,

Coconut milk (kakang gata) *from 15 coconuts (Philippines)*

gredients, such as sugar, propylene glycol, salt, and sodium metabisulfite. Some countries in South East Asia use special coconut mutant called Kopyor (in Indonesian) or macapuno (in Philippines) as a dessert drinks.

19.7.1 Nutrition

Per 100 gram serving with 354 calories, raw coconut meat supplies a high amount of total fat (33 grams), especially saturated fat (89% of total fat) and carbohydrates (24 grams) (table). Micronutrients in significant content include the dietary minerals, manganese, iron, phosphorus and zinc (table).

Coconut water

Main article: Coconut water

is obtained primarily by extracting juice by pressing the grated coconut white kernel or by passing hot water or milk through grated coconut, which extracts the oil and aromatic compounds. It has a total fat content of 24%, most of which (89%) is saturated fat, with lauric acid as a major fatty acid.[57] When refrigerated and left to set, coconut cream will rise to the top and separate from the milk. The milk can be used to produce virgin coconut oil by controlled heating and removal of the oil fraction.

A protein-rich powder can be processed from coconut milk following centrifugation, separation and spray drying.[58]

Coconut oil

Main article: Coconut oil

Another byproduct of the coconut is coconut oil. It is commonly used in cooking, especially for frying. It can be used in liquid form as would other vegetable oils, or in solid form as would butter or lard.

Toddy and nectar

The sap derived from incising the flower clusters of the coconut is drunk as *neera*, also known as toddy or *tuba* (Philippines), *tuak* (Indonesia and Malaysia) or *karewe* (fresh and not fermented, collected twice a day, for breakfast and dinner) in Kiribati. When left to ferment on its own, it becomes palm wine. Palm wine is distilled to produce *arrack*. In the Philippines, this alcoholic drink is called *lambanog* or "coconut vodka".[59]

The sap can be reduced by boiling to create a sweet syrup or candy such as *te kamamai* in Kiribati or *dhiyaa hakuru* and *addu bondi* in the Maldives. It can be reduced further to yield coconut sugar also referred to as palm sugar or jaggery. A young, well-maintained tree can produce around 300 liters (66 imp gal; 79 U.S. gal) of toddy per year, while a 40-year-old tree may yield around 400 liters (88 imp gal; 110 U.S. gal).[60]

Heart of palm and coconut sprout

Apical buds of adult plants are edible, and are known as "palm cabbage" or heart of palm. They are considered a rare delicacy, as harvesting the buds kills the palms. Hearts of palm are eaten in salads, sometimes called "millionaire's salad". Newly germinated coconuts contain an edible fluff of marshmallow-like consistency called coconut sprout, produced as the endosperm nourishes the developing embryo.

19.7.2 Indonesia

Coconut is an indispensable ingredient in Indonesian cooking. Coconut meat, coconut milk and coconut water are often used in main courses, desserts and soups throughout the archipelago. In the island of Sumatra, the famous Rendang, the traditional beef stew from West Sumatra, chunks of beef are cooked in coconut milk along with other spices for hours until thickened. In Jakarta, "Soto Babat" or beef tripe soup also uses coconut milk. In the island of Java, the sweet and savoury "Tempe Bacem" is made by cooking tempeh with coconut water, coconut sugar and other spices until thickened. "Klapertart" is the famous Dutch-influenced dessert from Manado, North Celebes, that uses young coconut meat and coconut milk. In 2010, Indonesia increased its coconut production. It is now the world's second largest producer of coconuts. The gross production was 15 million tonnes.[61] A sprouting coconut seed is the logo for Gerakan Pramuka Indonesia, the Indonesian Scouting organization. It can be seen on all the scouting paraphernalia that elementary (SMA) school children wear as well as on the scouting pins and flags.

19.7.3 Philippines

Harvesting coconuts in the Philippines is done by workers who climb the trees using notches cut into the trunk.

From left to right: grated, fresh, mature coconut meat; seed interior; oil, rare two-eyed coconut shell; and more grated meat (Philippines)

The Philippines is one of the world's largest producer of coconuts; the production of coconuts plays an important role in the economy. Coconuts in the Philippines are usually used in making main dishes, refreshments and desserts. Coconut juice is also a popular drink in the country. In the Philippines, particularly Cebu, rice is wrapped in coconut leaves for cooking and subsequent storage; these packets are called *puso*. Coconut milk, known as *gata*, and grated coconut flakes are used in the preparation of dishes such as *laing, ginataan, bibingka, ube halaya, pitsi-pitsi, palitaw, buko* and coconut pie. Coconut jam is made by mixing muscovado sugar with coconut milk. Coconut sport fruits are also harvested. One such variety of coconut is known as *macapuno*. Its meat is sweetened, cut into strands and sold in glass jars as coconut strings, sometimes labeled as "gelatinous mutant coconut". Coconut water can be fermented to make a different product *nata de coco* (coconut gel).

19.7.4 Vietnam

In Vietnam, coconut is grown abundantly across Central and Southern Vietnam, and especially in Bến Tre Province, often called the "land of the coconut". It is used to make coconut candy, caramel, and jelly. Coconut juice and coconut milk are used, especially in Vietnam's southern style of cooking, including *kho, chè* and curry (*cà ri*).

19.7.5 India

In southern India, most common way of cooking vegetables is to add grated coconut and then steam them with spices fried in oil. People from southern India also make chutney, which involves grinding the coconut with salt, chillies, and whole spices. *Uruttu chammanthi* (granulated chutney) is eaten with rice or *kanji* (rice gruel). It is also invariably the main side dish served with *idli, vadai*, and *dosai*. Coconut ground with spices is also mixed in *sambar* and other various lunch dishes for extra taste. Dishes garnished with grated coconut are generally referred to as *poduthol* in North Malabar and *thoran* in rest of Kerala. *Puttu* is a culinary delicacy of Kerala and Tamil Nadu, in which layers of coconut alternate with layers of powdered rice, all of which fit into a bamboo stalk. Recently, this has been replaced with a steel or aluminium tube, which is then steamed over a pot. Coconut (Tamil: தேங்காய்) is regularly broken in the middle-class families in Tamil Nadu for food. Coconut meat can be eaten as a snack sweetened with jaggery or molasses. In Karnataka sweets are prepared using coconut and dry coconut "copra"., Like Kaie Obattu, Kobri mitai etc.

19.8 Commercial, industrial, and household use

Coconuts drying before being processed into copra in the Solomon Islands.

19.8.1 Cultivars

Coconut has a number of commercial and traditional cultivars. They can be sorted mainly into tall cultivars, dwarf cultivars and hybrid cultivars (hybrids between talls and dwarfs). Some of the dwarf cultivars such as *Malayan dwarf* has shown some promising resistance to lethal yellowing while other cultivars such as *Jamaican tall* is highly affected by the same plant disease. Some cultivars are more drought resistant such as *West coast tall* (India) while others such as *Hainan Tall* (China) are more cold tolerant. Other aspects such as seed size, shape and weight and copra thickness are also important factors in the selection of new cultivars. Some cultivars such as *Fiji dwarf* form a large bulb at the lower stem and others are cultivated to produce very sweet coconut water with orange coloured husks (king coconut) used entirely in fruit stalls for drinking (Sri Lanka, India).

19.8.2 Coir

Main article: coir

Coir (the fiber from the husk of the coconut) is used in ropes, mats, door mats, brushes, sacks, caulking for boats, and as stuffing fiber for mattresses.[62] It is used in horticulture in potting compost, especially in orchid mix.

19.8.3 Coconut fronds

The stiff mid-ribs of coconut leaves are used for making brooms in India, Indonesia (*sapu lidi*), Malaysia, the Mal-

A wall made from coconut husks

Toys from coconut leaves

Extracting the fiber from the husk (Sri Lanka)

leaves can be burned to ash, which can be harvested for lime. In India, the woven coconut leaves are used as *pandals* (temporary sheds) for marriage functions especially in the states of Kerala, Karnataka, and Tamil Nadu.

19.8.4   Copra

Main articles: Copra and Coconut oil

Copra is the dried meat of the seed and after processing produces coconut oil and coconut meal. Coconut oil, aside from being used in cooking as an ingredient and for frying, is used in soaps, cosmetics, hair-oil, and massage oil. Coconut oil is also a main ingredient in Ayurvedic oils. In Vanuatu coconut palms for copra production are generally spaced 9 meters apart, allowing a tree density of 100–160 trees per hectare.

19.8.5   Husks and shells

The husk and shells can be used for fuel and are a source of charcoal.[64] Activated carbon manufactured from coconut shell is considered extremely effective for the removal of impurities. The coconut's obscure origin in foreign lands led to the notion of using cups made from the shell to neutralise poisoned drinks. The cups were frequently engraved and decorated with precious metals.[65]

A dried half coconut shell with husk can be used to buff floors. It is known as a *bunot* in the Philippines and simply a "coconut brush" in Jamaica. The fresh husk of a brown coconut may serve as a dish sponge or body sponge.

In Asia, coconut shells are also used as bowls and in the manufacture of various handicrafts, including buttons carved from dried shell. Coconut buttons are often used for

dives and the Philippines (*walis tingting*). The green of the leaves (lamina) are stripped away, leaving the veins (wood-like, thin, long strips) which are tied together to form a broom or brush. A long handle made from some other wood may be inserted into the base of the bundle and used as a two-handed broom. The leaves also provide material for baskets that can draw well water and for roofing thatch; they can be woven into mats, cooking skewers, and kindling arrows, as well. Two leaves (especially the younger, yellowish shoots) woven into a tight shell the size of the palm are filled with rice and cooked to make *ketupat.*[63] Dried coconut

Coconut buttons in Dongjiao Town, Hainan, China

The base of an old coconut palm

Coconut Palace, Manila, Philippines, built entirely out of coconut and local materials

Hawaiian aloha shirts. *Tempurung* as the shell is called in the Malay language can be used as a soup bowl and if fixed with a handle a ladle. In Thailand, the coconut husk is used as a potting medium to produce healthy forest tree saplings. The process of husk extraction from the coir bypasses the retting process, using a custom-built coconut husk extractor designed by ASEAN–Canada Forest Tree Seed Centre (ACFTSC) in 1986. Fresh husks contains more tannin than old husks. Tannin produces negative effects on sapling growth.*[66] In parts of South India, the shell and husk are burned for smoke to repel mosquitoes.

Half coconut shells are used in theatre Foley sound effects work, banged together to create the sound effect of a horse's hoofbeats. Dried half shells are used as the bodies of musical instruments, including the Chinese *yehu* and *banhu*, along with the Vietnamese *đàn gáo* and Arabo-Turkic *rebab*. In the Philippines, dried half shells are also used as a music instrument in a folk dance called *maglalatik*.

In World War II, coastwatcher scout Biuki Gasa was the first of two from the Solomon Islands to reach the shipwrecked and wounded crew of Motor Torpedo Boat PT-109 commanded by future U.S. president John F. Kennedy. Gasa suggested, for lack of paper, delivering by dugout canoe a message inscribed on a husked coconut shell. This coconut was later kept on the president's desk, and is now in the John F. Kennedy Library.

19.8.6 Coconut trunk

Coconut trunks are used for building small bridges and huts; they are preferred for their straightness, strength, and salt resistance. In Kerala, coconut trunks are used for house construction. Coconut timber comes from the trunk, and is increasingly being used as an ecologically sound substitute for endangered hardwoods. It has applications in furniture and specialized construction, as notably demonstrated in Manila's Coconut Palace.

Hawaiians hollowed the trunk to form drums, containers, or small canoes. The "branches" (leaf petioles) are strong and flexible enough to make a switch. The use of coconut branches in corporal punishment was revived in the Gilbertese community on Choiseul in the Solomon Islands in 2005.*[67]

19.8.7   Coconut roots

The roots are used as a dye, a mouthwash, and a medicine for diarrhea and dysentery.[8] A frayed piece of root can also be used as a toothbrush.

19.8.8   Use in beauty products

Coconuts are used in the beauty industry in moisturisers and body butters because coconut oil, due to its chemical structure, is readily absorbed by the skin. The coconut shell may also be ground down and added to products for exfoliation of dead skin. Coconut is also a source of lauric acid, which can be processed in a particular way to produce sodium lauryl sulfate, a detergent used in shower gels and shampoos.[68] The nature of lauric acid as a fatty acid makes it particularly effective for creating detergents and surfactants.

19.9   Role in culture and religion

See also: Coconut Religion

In the Ilocos region of northern Philippines, the Ilocano people fill two halved coconut shells with *diket* (cooked sweet rice), and place *liningta nga itlog* (halved boiled egg) on top of it. This ritual, known as *niniyogan*, is an offering made to the deceased and one's ancestors. This accompanies the *palagip* (prayer to the dead).

A coconut (Sanskrit: nalikera) is an essential element of rituals in Hindu tradition. Often it is decorated with bright metal foils and other symbols of auspiciousness. It is offered during worship to a Hindu god or goddess. Irrespective of their religious affiliations, fishermen of India often offer it to the rivers and seas in the hopes of having bountiful catches. Hindus often initiate the beginning of any new activity by breaking a coconut to ensure the blessings of the gods and successful completion of the activity. The Hindu goddess of well-being and wealth, Lakshmi, is often shown holding a coconut.[69] In the foothills of the temple town of Palani, before going to worship Murugan for the Ganesha, coconuts are broken at a place marked for the purpose. Every day, thousands of coconuts are broken, and some devotees break as many as 108 coconuts at a time as per the prayer. In tantric practices, coconuts are sometimes used as substitutes for human skulls.

In Hindu wedding ceremonies, a coconut is placed over the opening of a pot, representing a womb. Coconut flowers are auspicious symbols and are fixtures at Hindu and Buddhist weddings and other important occasions. In Ker- ala, coconut flowers must be present during a marriage ceremony. The flowers are inserted into a barrel of unhusked rice (paddy) and placed within sight of the wedding ceremony. Similarly in Sri Lanka, coconut flowers, standing in brass urns, are placed in prominent positions.

The Zulu Social Aid and Pleasure Club of New Orleans traditionally throws hand-decorated coconuts, the most valuable of Mardi Gras souvenirs, to parade revelers. The "Tramps" began the tradition *circa* 1901. In 1987, a "coconut law" was signed by Gov. Edwards exempting from insurance liability any decorated coconut "handed" from a Zulu float.

The coconut is also used as a target and prize in the traditional British fairground game "coconut shy". The player buys some small balls which he throws as hard as he can at coconuts balanced on sticks. The aim is to knock a coconut off the stand and win it.

It was the main food of adherents of the now discontinued Vietnamese religion Đạo Dừa in Bến Tre.

19.9.1   Myths and legends

Some South Asian, Southeast Asian and Pacific Ocean cultures have origin myths in which the coconut plays the main role. In the Hainuwele myth from Maluku, a girl emerges from the blossom of a coconut tree.[70] In Maldivian folklore one of the main myths of origin reflects the dependence of the Maldivians on the coconut tree.[71]

According to an urban legend, there are more deaths caused by falling coconuts than by sharks annually.

19.10   Other uses

Making a rug from coconut fiber

The leftover fiber from coconut oil and coconut milk production, coconut meal, is used as livestock feed. The dried calyx is used as fuel in wood-fired stoves. Coconut water is traditionally used as a growth supplement in plant tissue culture/micropropagation.[72] The smell of coconuts comes from the 6-pentyloxan-2-one molecule, known as delta-decalactone in the food and fragrance industries.[73]

19.10.1   Tool and shelter for animals

Researchers from the Melbourne Museum in Australia observed the octopus species *Amphioctopus marginatus* use of tools, specifically coconut shells, for defense and shelter. The discovery of this behavior was observed in Bali and North Sulawesi in Indonesia between 1998 and 2008.[74][75][76] *Amphioctopus marginatus* is the first invertebrate known to be able to use tools.[75][77]

A coconut can be hollowed out and used as a home for a rodent or small birds. Halved, drained coconuts can also be hung up as bird feeders, and after the flesh has gone, can be filled with fat in winter to attract tits.

19.11   Allergies

19.11.1   Food allergies

Coconut can be a food allergen although its prevalence varies from country to country. While coconut is one of the top-five food allergies in India where it is a common food source,[78] such allergies to coconut are considered rare in Australia, the UK, and the United States.[79] As a result, commercial extracts of coconut are not currently available for skin prick testing in Australia or New Zealand.[80]

Despite a low prevalence of allergies to coconut in the United States, the U.S. Food and Drug Administration (FDA) began identifying coconuts in October 2006.[79] Based on FDA guidance and federal U.S. law, coconut must be disclosed as an ingredient.[81]

19.11.2   Topical allergies

Coconut-derived products can cause contact dermatitis. They can be present in cosmetics, including some shampoos, moisturizers, soaps, cleansers and hand washing liquids. Those known to cause contact dermatitis include: coconut diethanolamide, cocamide sulphate, cocamide DEA, CDEA, sodium laureth sulfate, sodium lauroyl sulfate, ammonium laureth sulfate, ammonium lauryl sulfate, sodium lauroyl sarcosinate, sodium cocoyl sarcosinate, potassium coco hydrolysed collagen, triethanolamine laureth sulfate, caprylic/capric triglycerides, triethanolamine lauryl or cocoyl sarcosime, disodium oleamide sulfocuccinate, laureth sulfasuccinate, and disodium dioctyl sulfosuccinate.[80]

19.12   See also

- Coir Board of India

- Death by coconut

- Ravanahatha

- *Voanioala gerardii* – forest coconut, the closest relative of the modern coconut

- Coconut production in Kerala

19.13   References

[1]   Hahn, William J. (1997). Arecanae: The palms. Retrieved April 4, 2011 from the Tree of Life Web Project website.

[2]   Royal Botanic Gardens, Kew. *Cocos*. World Checklist of Selected Plant Families.

[3]   J. Pearsall, ed. (1999). "Coconut" . *Concise Oxford Dictionary* (10th ed.). Oxford: Clarendon Press. ISBN 0-19-860287-1.

[4]   Dalgado, Sebastião. "Glossário luso-asiático" . *google.com* 1. p. 291.

[5]   Source: James A. Duke, Handbook of Energy Crops, unpublished (1983). "*Cocos nucifera* L." . Purdue University, NewCROP, the New Crop Resource Online Program. Retrieved 4 June 2015.

[6]   Paniappan S (December 12, 2002). "The Mystery Behind Coconut Water" . *The Hindu*. Retrieved January 16, 2012.

[7]   T. Pradeepkumar, B. Sumajyothibhaskar, and K.N. Satheesan. (2008). *Management of Horticultural Crops* (Horticulture Science Series Vol.11, 2nd of 2 Parts). New India Publishing. pp. 539–587. ISBN 978-81-89422-49-3.

[8]   Grimwood 1975, p. 18

[9]   Sarian, Zac B. (August 18, 2010). New coconut yields high. *The Manila Bulletin*. Retrieved April 21, 2011.

[10] Ravi, Rajesh. (March 16, 2009). Rise in coconut yield, farming area put India on top. *The Financial Express*. Retrieved April 21, 2011.

[11] "How Long Does It Take for a Coconut Tree to Get Coconuts?". *Home Guides - SF Gate*.

[12] Coconut, Plant of Many Uses. From UCLA course on Economic Botany.

[13] Bourke, R. Michael and Tracy Harwood (Eds.). (2009). *Food and Agriculture in Papua New Guinea*. Australian National University. p. 327. ISBN 978-1-921536-60-1.

[14] Thampan, P.K. (1981). *Handbook on Coconut Palm*. Oxford & IBH Publishing Co.

[15] http://www.agroforestry.net/tti/Cocos-coconut.pdf

[16] Willmer, Pat. (2011). *Pollination and Floral Ecology*. Princeton University Press. p. 57. ISBN 978-0-691-12861-0.

[17] "The Fifth Voyage of Sindbad the Seaman – The Arabian Nights – The Thousand and One Nights – Sir Richard Burton translator". Classiclit.about.com. November 2, 2009. Retrieved February 14, 2012.

[18] Grimwood 1975, p. 1.

[19] Elzebroek, A.T.G. and Koop Wind (Eds.). (2008). *Guide to Cultivated Plants*. CABI. pp. 186–192. ISBN 978-1-84593-356-2.

[20] Rosengarten, Frederic, Jr. (2004). *The Book of Edible Nuts*. Dover Publications. pp. 65–93. ISBN 978-0-486-43499-5.

[21] Losada, Fernando Díez. (2004). *La tribuna del idioma*. Editorial Tecnologica de CR. p. 481. ISBN 978-9977-66-161-2. (Spanish)

[22] Figueiredo, Cândido. (1940). *Pequeno Dicionário da Lingua Portuguesa*. Livraria Bertrand. Lisboa. (Portuguese)

[23] "Coco". Merriam-Webster. Retrieved August 28, 2011.

[24] "Coco". Online Etymology Dictionary. Retrieved August 28, 2011.

[25] Werth, E. (1933). Distribution, Origin and Cultivation of the Coconut Palm. *Ber. Deutschen Bot. Ges.*, vol 51, pp. 301–304. (article translated into English by Dr. R. Child, Director, Coconut Research Scheme, Lunuwila, Sri Lanka).

[26] Grimwood, Brian E., F. Ashman, D.A.V. Dendy, C.G. Jarman, E.C.S. Little, and W.H. Timmins. (1975). *Coconut Palm Products – Their processing in developing countries*. Rome: FAO. pp. 3–4. ISBN 978-92-5-100853-9.

[27] Perera, Lalith, Suriya A.C.N. Perera, Champa K. Bandaranayake and Hugh C. Harries. (2009). "Chapter 12 – Coconut". In Johann Vollmann and Istvan Rajcan (Eds.). *Oil Crops*. Springer. pp. 370–372. ISBN 978-0-387-77593-7.

[28] Jackson, Eric. (August 20 – September 2, 2006). From whence come coconuts?. *The Panama News* (Volume 12, Number 16). Retrieved April 10, 2011.

[29] Cook, O.F. (1901) The Origin and Distribution of the Cocoa Palm. Washington: Government Printing Office. 37 p.

[30] Heyerdahl, Thor. (1950) Kon-Tiki: Across the Pacific by Raft. Mattituck: Amereon House. 240 p.

[31] Harries, H. C. (1978). "The evolution, dissemination and classification of *Cocos nucifera* L.". *The Botanical Review* **44** (3): 265–319. doi:10.1007/bf02957852.

[32] Harries, H (2012). "Germination rate is the significant characteristic determining coconut palm diversity". *AoB Plants*. doi:10.1093/aobpla/pls045.

[33] Harries, H.C.; Clement, C.R. (2013). "Long-distance dispersal of the coconut palm by migration within the coral atoll ecosystem". *Annals of Botany*. doi:10.1093/aob/mct293.

[34] Harries, H (2012). "Germination rate is the significant characteristic determining coconut palm diversity". *Annals of Botany*. doi:10.1093/aobpla/pls04.

[35] Lebrun, P.; Seguin, M.; Grivet, L.; Baudouin, L. (1998). "Genetic diversity in coconut (*Cocos nucifera* L.) revealed by restriction fragment length polymorphism (RFLP) markers". *Euphytica* **101**: 103–108.

[36] Shukla, A.; Mehrotra, R. C.; Guleria, J. S. (2012). "Cocos sahnii Kaul: A *Cocos nucifera* L.-like fruit from the Early Eocene rainforest of Rajasthan, western India". *Journal of Biosciences* **37** (4): 769–776. doi:10.1007/s12038-012-9233-3.

[37] Santos, G.A., Batugal, P.A., Othman, A., Baudouin, L., and Labouisse J.P. 1996. Manual on standardised techniques in coconut breeding. IPGRI–COGENT publication. Stamford Press, Singapore. Accessed at http://www.bioversityinternational.org/fileadmin/bioversity/publications/Web_version/108/ch02.htm#Chapter%201%20BOTANY%20OF%20THE%20COCONUT%20PALM

[38] Huang, Y.-Y.; Matzke, A. J. M.; Matzke, M. (2013). "Complete sequence and comparative analysis of the chloroplast genome of coconut palm (*Cocos nucifera*)". *PLOS ONE* **8** (8): e74736. doi:10.1371/journal.pone.0074736.

[39] Rivera, R.; Edwards, K. J.; Barker, J. H.; Arnold, G. M.; Ayad, G.; Hodgkin, T.; Karp, A. (1999). "Isolation and characterization of polymorphic microsatellites in *Cocos nucifera* L". *Genome / National Research Council Canada = Genome / Conseil national de recherches Canada* **42** (4): 668–675. doi:10.1139/gen-42-4-668. PMID 10464790.

[40] Edmondson, C.H. (1941). "Viability of coconut seeds after floating in sea". *Bernice P. Bishop Museum Occasional Papers* **16**: 293–304.

[41] Ward, R. G.; Brookfield, M. (1992). "Special Paper: the dispersal of the coconut: did it float or was it carried to Panama?". *Journal of Biogeography* **19** (5): 467–480. doi:10.2307/2845766.

[42] Gunn, Bee; Luc Baudouin; Kenneth M. Olsen (2011). "Independent Origins of Cultivated Coconut (*Cocos nucifera* L.) in the Old World Tropics". *PLoS ONE* **6** (6): e21143. doi:10.1371/journal.pone.0021143. PMC 3120816. PMID 21731660. Retrieved November 28, 2011.

[43] Foale, Mike. (2003). The Coconut Odyssey – the bounteous possibilities of the tree of life. Australian Centre for International Agricultural Research. Retrieved May 30, 2009.

[44] Ferguson, John. (1898). *All about the "coconut palm" (*Cocos nucifera*) (2nd edition).

[45] Chan, Edward and Craig R. Elevitch. (April 2006). *Cocos nucifera* (coconut) (version 2.1). In C.R. Elevitch (Ed.). *Species Profiles for Pacific Island Agroforestry*. Hōlualoa, Hawai ʻi: Permanent Agriculture Resources (PAR).

[46] Remo, Amy R. (September 27, 2007). Beetles infest coconuts in Manila, 26 provinces. *Philippine Daily Inquirer*.

[47] "Coconuts, Production/Crops". Food And Agriculture Organization of the United Nations: Statistical Division (FAOSTAT). 2013. Retrieved 17 October 2015.

[48] World Wildlife Fund. (December 17, 2010). "Petenes mangroves". In Mark McGinley, C. Michael Hogan & Cutler J. Cleveland *Encyclopedia of Earth*. Washington, D.C.: Environmental Information Coalition, National Council for Science and the Environment. Retrieved April 14, 2011.

[49] Bertrand, Mireille. (January 27, 1967). Training without Reward: Traditional Training of Pig-tailed Macaques as Coconut Harvesters. *Science* **155** (3761): 484–486.

[50] Department of Agriculture Karshika Keralam. Government of Kerala. India. (n.d.). "Coconut Cultivation". Retrieved December 6, 2009.

[51] Halsall, Paul. (Ed). (February 21, 2001). "Medieval Sourcebook: Ibn Battuta: Travels in Asia and Africa 1325–1354". Fordham University Center for Medieval Studies. Retrieved April 14, 2011.

[52] Kaakeh, Walid, Fouad El-Ezaby, Mahmoud M. Aboul-Nour, and Ahmed A. Khamis (2001). "Management of the red palm weevil, *Rhynchophorus ferrugineus* Oliv., by a pheromone/food-based trapping system" (PDF). Retrieved December 6, 2009.

[53] Kaunitz, H. (1986). "Medium chain triglycerides (MCT) in aging and arteriosclerosis". *Journal of Environmental Pathology, Toxicology and Oncology : official organ of the International Society for Environmental Toxicology and Cancer* **6** (3–4): 115–121. PMID 3519928.

[54] http://www.floridagardener.com/palms/coconutpalm.htm

[55] Margolis, Jason. (December 13, 2006). Coconut fuel. *PRI's The World*. Retrieved April 10, 2011.

[56] Grimwood 1975, p. 182.

[57] "Full Report (All Nutrients): 12117, Nuts, coconut milk, raw (liquid expressed from grated meat and water)". US Department of Agriculture, National Nutrient Database, version SR-28. 2015.

[58] Naik A, Raghavendra SN, Raghavarao KS (2012). "Production of coconut protein powder from coconut wet processing waste and its characterization". *Appl Biochem Biotechnol* **167** (5): 1290–302. doi:10.1007/s12010-012-9632-9. PMID 22434355.

[59] Porter, Jolene V. (2005). "Lambanog: A Philippine Drink". Washington, D.C.: American University. Retrieved April 10, 2011.

[60] Grimwood 1975, p. 20.

[61] FOOD AND AGRICULTURE ORGANIZATION OF THE UNITED NATIONS, FAO

[62] Grimwood 1975, p. 22.

[63] Grimwood 1975, p. 19.

[64] "Coconut Shell Lump Charcoal". Supreme Carbon Indonesia.

[65] "Hans van Amsterdam: Coconut Cup with Cover (17.190.622ab) - Heilbrunn Timeline of Art History - The Metropolitan Museum of Art". *metmuseum.org*.

[66] Somyos Kijkar. "Handbook: Coconut husk as a potting medium". ASEAN-Canada Forest Tree Seed Centre Project 1991, Muak-Lek, Saraburi, Thailand. ISBN 974-361-277-1.

[67] Herming, George. (March 6, 2006). Wagina whips offenders. *Solomon Star*.

[68] ongnaturalbodycare.co.uk. (April 1, 2012)

[69] Dallapiccola, Anna. *Dictionary of Hindu Lore and Legend*. ISBN 0-500-51088-1.

[70] "Hainuwele - Oxford Reference". *oxfordreference.com*.

[71] Romero-Frias, Xavier (2012) *Folk tales of the Maldives*, NIAS Press, ISBN 978-87-7694-104-8, ISBN 978-87-7694-105-5

[72] Yong, JW. Ge L. Ng YF. Tan SN. (2009). "The chemical composition and biological properties of coconut (*Cocos nucifera* L.) water". *Molecules* **14** (12): 5144–64. doi:10.3390/molecules14125144.

[73] "Data sheet about delta-decalactone and its properties". Thegoodscentscompany.com. July 18, 2000. Retrieved February 14, 2012.

[74] Finn, Julian K.; Tregenza, Tom; Norman, Mark D. (2009). "Defensive tool use in a coconut-carrying octopus". *Curr. Biol.* **19** (23): R1069–R1070. doi:10.1016/j.cub.2009.10.052. PMID 20064403.

[75] Gelineau, Kristen (December 15, 2009). "Aussie scientists find coconut-carrying octopus". Associated Press. Retrieved December 15, 2009.

[76] Harmon, Katherine (December 14, 2009). "A tool-wielding octopus? This invertebrate builds armor from coconut halves". *Scientific American*.

[77] Henderson, Mark (December 15, 2009). "Indonesia's veined octopus 'stilt walks' to collect coconut shells". *Times Online*.

[78] Living with food allergies; Venugopal P. (Sept–Dec. 2006). Food Allergy. *Pulmon – The Journal of Respiratory Sciences* **8** (3). ISSN: 0973-3809.

[79] Australasian Society of Clinical Immunology and Allergy. Coconut Allergy; National Health Service. United Kingdom. (January 12, 2010). Causes of a food allergy. *NHS Choices*.

[80] Australasian Society of Clinical Immunology and Allergy. Coconut Allergy.

[81] USFDA. (October 2009). Guidance for Industry: A Food Labeling Guide.

19.14   Further reading

- Adkins S.W., M. Foale and Y.M.S. Samosir (eds.) (2006). Coconut revival – new possibilities for the 'tree of life'. Proceedings of the International Coconut Forum held in Cairns, Australia, November 22–24, 2005. ACIAR Proceedings No. 125. ISBN 1-86320-515-2

- Batugal, P., V.R. Rao and J. Oliver (2005). *Coconut Genetic Resources*. Bioversity International. ISBN 978-92-9043-629-4.

- Frison, E.A.; Putter, C.A.J.; Diekmann, M. (eds.). (1993). *Coconut*. ISBN 978-92-9043-156-5.

- International Plant Genetic Resources Institute (IP-GRI). (1995). Descriptors for Coconut (*Cocos nucifera* L.). ISBN 978-92-9043-215-9.

- Mathur, P.N.; Muralidharan, K.; Parthasarathy, V.A.; Batugal, P.; Bonnot, F. (2008). *Data Analysis Manual for Coconut Researchers*. ISBN 978-92-9043-736-9.

- Salunkhe, D.K., J.K. Chavan, R.N. Adsule, and S.S. Kadam. (1992). *World Oilseeds – Chemistry, Technology, and Utilization*. Springer. ISBN 978-0-442-00112-4.

19.15   External links

Chapter 20

Cooking oil

Cooking oil is plant, animal, or synthetic fat used in frying, baking, and other types of cooking. It is also used in food preparation and flavouring not involving heat, such as salad dressings and bread dips, and in this sense might be more accurately termed edible oil.

Cooking oil is typically a liquid at room temperature, although some oils that contain saturated fat, such as coconut oil, palm oil and palm kernel oil are solid.*[1]

There is a wide variety of cooking oils from plant sources such as olive oil, palm oil, soybean oil, canola oil (rapeseed oil), corn oil, peanut oil and other vegetable oils, as well as animal-based oils like butter and lard.

Oil can be flavoured with aromatic foodstuffs such as herbs, chillies or garlic.

20.1 Health and nutrition

The appropriate amount of fat as a component of daily food consumption is established as a guideline by regulatory agencies such as the FDA which recommends that 10% or fewer of calories consumed daily should be from saturated fat and 20-35% of total daily calories come from polyunsaturated and monounsaturated fats.*[2]

While consumption of small amounts of saturated fats is common in diets,*[3] meta-analyses found a significant correlation between high consumption of saturated fats and blood LDL concentration,*[4] a risk factor for cardiovascular diseases.*[5] Other meta-analyses based on cohort studies and on controlled, randomized trials found a positive*[6] or neutral*[7] effect from shifting consumption from saturated fats to polyunsaturated fats (10% lower risk for 5% replacement).*[7]

Mayo Clinic has highlighted oils that are high in saturated fats, including coconut, palm oil and palm kernel oil. Those having lower amounts of saturated fats and higher levels of unsaturated (preferably monounsaturated) fats like olive oil, peanut oil, canola oil, soy and cottonseed oils are generally

Olive oil

Sunflower seed oil

Several large studies[15][16][17][18] indicate a link between consumption of high amounts of trans fat and coronary heart disease and possibly some other diseases. The United States Food and Drug Administration (FDA), the National Heart, Lung and Blood Institute and the American Heart Association (AHA) all have recommended limiting the intake of trans fats.

20.1.2   Cooking with oil

Heating an oil changes its characteristics. Oils that are healthy at room temperature can become unhealthy when heated above certain temperatures. When choosing a cooking oil, it is important to match the oil's heat tolerance with the cooking method.[19]

Palm oil contains more saturated fats than canola oil, corn oil, linseed oil, soybean oil, safflower oil, and sunflower oil. Therefore, palm oil can withstand the high heat of deep frying and is resistant to oxidation compared to highly unsaturated vegetable oils.[20] Since about 1900, palm oil has been increasingly incorporated into food by the global commercial food industry because it remains stable in deep frying or in baking at very high temperatures[21][22] and for its high levels of natural antioxidants.[23]

Oils that are suitable for high-temperature frying (above 230 °C or 446 °F) because of their high smoke point:

- Avocado oil

- Mustard oil

- Palm oil

- Peanut oil (marketed as "groundnut oil" in the UK and India)

- Rice bran oil

- Safflower oil

- Semirefined sesame oil

- Semirefined sunflower oil[24]

healthier.[8] The US National Heart, Lung and Blood Institute[9] urged saturated fats be replaced with polyunsaturated and monounsaturated fats, listing olive and canola oils as sources of healthier monounsaturated oils while soybean and sunflower oils as good sources of polyunsaturated fats. One study showed that consumption of non-hydrogenated unsaturated oils like soybean and sunflower are preferable to the consumption of palm oil for lowering the risk of heart disease.[10]

Peanut, cashew and other nut-based oils may present a hazard to persons with a nut allergy.

20.1.1   Trans fats

Unlike other dietary fats, trans fats are not essential, and they do not promote good health.[11] The consumption of trans fats increases one's risk of coronary heart disease[12] by raising levels of "bad" LDL cholesterol and lowering levels of "good" HDL cholesterol.[13] Trans fats from partially hydrogenated oils are more harmful than naturally occurring oils.[14]

20.1.3   Storing and keeping oil

Whether refined or not, all oils are sensitive to heat, light, and exposure to oxygen. Rancid oil has an unpleasant aroma and acrid taste, and its nutrient value is greatly diminished. To delay the development of rancid oil, a blanket of an inert gas, usually nitrogen, is applied to the vapor space in the storage container immediately after production. This is referred to as tank blanketing. Vitamin E oil is a natural antioxidant that can also be added to cooking oils to prevent rancidification.

All oils should be kept in a cool, dry place. Oils may thicken, but they will soon return to liquid if they stand at room temperature. To prevent negative effects of heat and light, oils should be removed from cold storage just long enough for use. Refined oils high in monounsaturated fats keep up to a year (olive oil will keep up to a few years), while those high in polyunsaturated fats keep about six months. Extra-virgin and virgin olive oils keep at least 9 months after opening. Other monounsaturated oils keep well up to eight months, while unrefined polyunsaturated oils will keep only about half as long.

In contrast, saturated oils, such as coconut oil and palm oil, have much longer shelf lives and can be safely stored at room temperature. Their lack of polyunsaturated content causes them to be more stable.*[25]

20.2 Types of oils and their characteristics

Lighter, more refined oils tend to have a higher smoke point. Experience using an oil is generally a sufficiently reliable guide. Although outcomes of empirical tests are sensitive to the qualities of particular samples (brand, composition, refinement, process), the data below should be helpful in comparing the properties of different oils.

Smoking oil indicates a risk of combustion, and left unchecked can also set off a fire alarm. When using any cooking oil, should it begin to smoke, *reduce the heat immediately*. The cook should be fully prepared to extinguish a burning oil fire *before* beginning to heat the oil, by having on hand the lid to place on the pan, or (for the worst case) having on hand the proper fire extinguisher.

20.3 Cooking oil extraction and refinement

Cooking oil extraction and refinement are separate processes. Extraction first removes the oil, typically from a seed, nut or fruit. Refinement then alters the appearance, texture, taste, smell, or stability of the oil to meet buyer expectations.

20.3.1 Extraction

There are three broad types of oil extraction:

Olive oil production in Croatia

- Chemical solvent extraction, most commonly using hexane.

- Pressing, using an expeller press or cold press (pressing at low temperatures to prevent oil heating).

- Decanter centrifuge.

In large-scale industrial oil extraction you will often see some combination of pressing, chemical extraction and/or centrifuging in order to extract the maximum amount of oil possible.*[38]

20.3.2 Refinement

Cooking oil can either be unrefined, or refined using one or more of the following refinement processes (in any combination):

- Distilling, which heats the oil to evaporate off chemical solvents from the extraction process.

- Degumming, by passing hot water through the oil to precipitate out gums and proteins that are soluble in oil but not in water, then discarding the water along with the impurities.

- Neutralization, or deacidification, which treats the oil with sodium hydroxide or sodium carbonate to pull out free fatty acids, phospholipids, pigments, and waxes.

- Bleaching, which removes "off-colored" components by treatment with fuller's earth, activated carbon, or activated clays, followed by heating, filtering, then drying to recoup the oil.

- Dewaxing, or winterizing, improves clarity of oils intended for refrigeration by dropping them to low temperatures and removing any solids that form.

- Deodorizing, by treating with high-heat pressurized steam to evaporate less stable compounds that might cause "unusual" odors or tastes.

- Preservative addition, such as BHA and BHT to help preserve oils that have been made less stable due to high-temperature processing.

Filtering, a non-chemical process which screens out larger particles, could be considered a step in refinement, although it doesn't alter the state of the oil.

Most large-scale commercial cooking oil refinement will involve all of these steps in order to achieve a product that's uniform in taste, smell and appearance, and has a longer shelf life.*[38] Cooking oil intended for the health food market will often be unrefined, which can result in a less stable product but minimizes exposure to high temperatures and chemical processing.

20.4   Waste cooking oil

A bin for spent cooking oil in Austin, Texas, USA, managed by a recycling company.

Proper disposal of used cooking oil is an important waste-management concern. Oil is lighter than water and tends to spread into thin and broad membranes which hinder the oxygenation of water. Because of this, a single litre of oil can contaminate as much as 1 million litres of water. Also, oil can congeal on pipes provoking blockages.*[39]

Because of this, cooking oil should never be dumped in the kitchen sink or in the toilet bowl. The proper way to dispose of oil is to put it in a sealed non-recyclable container and discard it with regular garbage.*[40] Placing the container of oil in the refrigerator to harden also makes disposal easier and less messy.

20.4.1   Recycling

Main article: Vegetable oil recycling

Cooking oil can be recycled. It can be used as animal feed, directly as fuel, and to produce biodiesel,*[41] soap, and other industrial products.

In the recycling industry, used cooking oil recovered from restaurants and food-processing industries (typically from deep fryers or griddles) is called recycled vegetable oil (RVO), used vegetable oil (UVO), waste vegetable oil (WVO), or **yellow grease**.*[42]

Yellow grease is used to feed livestock, and to make soap, make-up, clothes, rubber, detergents, and biodiesel fuel.*[43]*[44]

Used cooking oil, besides being converted to biodiesel, can be used directly in modified diesel engines and for heating.

Grease traps or interceptors collect fats and oils from kitchen sinks and floor drains which would othewise clog sewer lines and interfere with septic systems and sewage treatment. The collected product is called **brown grease** in the recycling industry.*[42] Brown grease is contaminated with rotted food solids and considered unsuitable for re-use in most applications.

Gutter oil or Trench Oil are terms used in Asia for recycled oil which is processed to resemble virgin oil but contains toxic contaminents and is illegally sold for cooking; its origin is frequently brown grease from garbage.*[45]

20.5   Notes

[1] The smoke point of an oil depends primarily on its free fatty acid content (FFA) and molecular weight. Through repeated use, as in a deep fryer, the oil accumulates food residues or by-products of the cooking process, that lower its smoke point further. The values shown in the table must therefore be taken as approximate, and are not suitable for accurate or scientific use.*[26]*[27]

[2] The smoke point of margarine varies depending on the types of oils used in its formulation, but can be generally assumed to be similar to that of butter.

20.6   References

[1] "Dietary fats explained" . Retrieved May 4, 2012.

[2] "Dietary Guidelines for Americans 2005; Key Recommendations for the General Population" . US Department of Health and Human Services and Department of Agriculture. 2005.

[3] Yanai H, Katsuyama H, Hamasaki H, Abe S, Tada N, Sako A (2015). "Effects of Dietary Fat Intake on HDL Metabolism". *J Clin Med Res* **7** (3): 145–9. doi:10.14740/jocmr2030w. PMID 25584098.

[4] Clarke, R; Frost, C; Collins, R; Appleby, P; Peto, R (1997). "Dietary lipids and blood cholesterol: quantitative meta-analysis of metabolic ward studies". *BMJ* **314** (7074): 112–7. doi:10.1136/bmj.314.7074.112. PMC 2125600. PMID 9006469.

[5] Mensink, RP; Zock, PL; Kester, AD; Katan, MB (2003). "Effects of dietary fatty acids and carbohydrates on the ratio of serum total to HDL cholesterol and on serum lipids and apolipoproteins: a meta-analysis of 60 controlled trials". *Am J Clin Nutr* **77** (5): 1146–55. PMID 12716665.

[6] Jakobsen, M. U; O'Reilly, E. J; Heitmann, B. L; Pereira, M. A; Balter, K.; Fraser, G. E; Goldbourt, U.; Hallmans, G.; et al. (2009). "Major types of dietary fat and risk of coronary heart disease: a pooled analysis of 11 cohort studies". *American Journal of Clinical Nutrition* **89** (5): 1425–32. doi:10.3945/ajcn.2008.27124. PMC 2676998. PMID 19211817.

[7] Katan, Martijn B.; Mozaffarian, Dariush; Micha, Renata; Wallace, Sarah (2010). Katan, Martijn B., ed. "Effects on Coronary Heart Disease of Increasing Polyunsaturated Fat in Place of Saturated Fat: A Systematic Review and Meta-Analysis of Randomized Controlled Trials". *PLoS Medicine* **7** (3): e1000252. doi:10.1371/journal.pmed.1000252. PMC 2843598. PMID 20351774.

[8] "Dietary fats: Know which types to choose". Mayo Clinic Staff. 2015.

[9] "Choose foods low in saturated fat". National Heart, Lung, and Blood Institute (NHLBI), NIH Publication No. 97-4064. 1997.

[10] Kabagambe, Baylin, Ascherio & Campos, EK; Baylin, A; Ascherio, A; Campos, H (November 2005). "The Type of Oil Used for Cooking Is Associated with the Risk of Nonfatal Acute Myocardial Infarction in Costa Rica". *Journal of Nutrition* (135 ed.) **135** (11): 2674–2679. PMID 16251629.

[11] Food and nutrition board, institute of medicine of the national academies (2005). *Dietary Reference Intakes for Energy, Carbohydrate, Fiber, Fat, Fatty Acids, Cholesterol, Protein, and Amino Acids (Macronutrients)*. National Academies Press. p. 423. ISBN 0-309-08537-3.

[12] Food and nutrition board, institute of medicine of the national academies (2005). *Dietary Reference Intakes for Energy, Carbohydrate, Fiber, Fat, Fatty Acids, Cholesterol, Protein, and Amino Acids (Macronutrients)*. National Academies Press. p. 504. ISBN 0-309-08537-3.

[13] "Trans fat: Avoid this cholesterol double whammy". Mayo Foundation for Medical Education and Research (MFMER). Retrieved 2007-12-10.

[14] Mozaffarian, Dariush; Katan, Martijn B.; Ascherio, Alberto; Stampfer, Meir J.; Willett, Walter C. (2006). "Trans Fatty Acids and Cardiovascular Disease". *New England Journal of Medicine* **354** (15): 1601–113. doi:10.1056/NEJMra054035. PMID 16611951.

[15] Willett, WC; Stampfer, MJ; Manson, JE; Colditz, GA; Speizer, FE; Rosner, BA; Sampson, LA; Hennekens, CH (1993). "Intake of trans fatty acids and risk of coronary heart disease among women". *Lancet* **341** (8845): 581–5. doi:10.1016/0140-6736(93)90350-P. PMID 8094827.

[16] Hu, Frank B.; Stampfer, Meir J.; Manson, Joann E.; Rimm, Eric; Colditz, Graham A.; Rosner, Bernard A.; Hennekens, Charles H.; Willett, Walter C. (1997). "Dietary Fat Intake and the Risk of Coronary Heart Disease in Women". *New England Journal of Medicine* **337** (21): 1491–9. doi:10.1056/NEJM199711203372102. PMID 9366580.

[17] Hayakawa, Kyoko; Linko, Yu-Yen; Linko, Pekka (2000). "The role of trans fatty acids in human nutrition". *Starch - Stärke* **52** (6–7): 229–35. doi:10.1002/1521-379X(200007)52:6/7<229::AID-STAR229>3.0.CO;2-G.

[18] The Nurses' Health Study (NHS)

[19] Orna Izakson. "Oil right: choose wisely for heart-healthy cooking - Eating Right". *E: the Environmental Magazine*.

[20] De Marco, Elena; Savarese, Maria; Parisini, Cristina; Battimo, Ilaria; Falco, Salvatore; Sacchi, Raffaele (2007). "Frying performance of a sunflower/palm oil blend in comparison with pure palm oil". *European Journal of Lipid Science and Technology* **109** (3): 237–246. doi:10.1002/ejlt.200600192.

[21] Che Man, YB; Liu, J.L.; Jamilah, B.; Rahman, R. Abdul (1999). "Quality changes of RBD palm olein, soybean oil and their blends during deep-fat frying". *Journal of Food Lipids* **6** (3): 181–193. doi:10.1111/j.1745-4522.1999.tb00142.x.

[22] Matthäus, Bertrand (2007). "Use of palm oil for frying in comparison with other high-stability oils". *European Journal of Lipid Science and Technology* **109** (4): 400–409. doi:10.1002/ejlt.200600294.

[23] Sundram, K; Sambanthamurthi, R; Tan, YA (2003). "Palm fruit chemistry and nutrition" (PDF). *Asia Pacific journal of clinical nutrition* **12** (3): 355–62. PMID 14506001.

[24] "Smoke Points of Various Fats - Kitchen Notes - Cooking For Engineers". *cookingforengineers.com*. 2012. Retrieved July 3, 2012.

[25] Articledashboard.com

[26] F. D. Gunstone; D. Rousseau (2004). *Rapeseed and canola oil: production, processing, properties and uses*. Oxford: Blackwell Publishing Ltd. p. 91. ISBN 0-8493-2364-9. Retrieved 2011-01-17.

[27] Brown, Amy L. (2010). *Understanding Food: Principles and Preparation*. Belmont, CA: Wadsworth Publishing. p. 468. ISBN 0-538-73498-1. Retrieved 2011-01-16.

[28] A. G. Vereshagin and G. V. Novitskaya (1965) The triglyceride composition of linseed oil. Journal of the American Oil Chemists' Society 42, 970-974.

[29] http://www.goodeatsfanpage.com/collectedinfo/ oilsmokepoints.htm

[30] Sunflowernsa.com

[31] "Triglyceride composition of tea seed oil". doi:10.1002/jsfa.2740271206.

[32] "Cooking Oil Smoke Points". Retrieved January 3, 2011.

[33] nutritiondata.com → Oil, vegetable, sunflower Retrieved on September 27, 2010

[34] USDA → Basic Report: 04042, Oil, peanut, salad or cooking Retrieved on January 16, 2015

[35] nutritiondata.com → Egg, yolk, raw, fresh Retrieved on August 24, 2009

[36] "09038, Avocados, raw, California". *National Nutrient Database for Standard Reference, Release 26*. United States Department of Agriculture, Agricultural Research Service. Retrieved 14 August 2014.

[37] "Feinberg School > Nutrition > Nutrition Fact Sheet: Lipids". Northwestern University. Archived from the original on 2011-07-20.

[38] "How cooking oil is made". Retrieved May 18, 2012.

[39] "Tips to avoid water waste and to require the preservation of hydro-resources". *Natureba - Educação Ambiental*. Retrieved 2007-09-05.

[40] "Grease Disposal Tips to Help the City's Environment". NYC Department of Environmental Protection. Retrieved 2007-08-05.

[41] "Production of biodiesel based on waste oils and/or waste fats from biogenic origin for use as fuel" (PDF). CDM - Executive Board. Retrieved 2007-09-05.

[42] Brown Grease Feedstocks for Biodiesel. K. Shaine Tyson, National Renewable Energy Laboratory. Available from Northeast Regional Biomass Program. Retrieved January 31, 2009

[43] Murphy, Denis J. *Plant lipids: biology, utilisation, and manipulation*. Wiley-Blackwell, 2005, p. 117.

[44] Radich, Anthony *Biodiesel Performance, Costs, and Use*

[45] Austin Ramzy, China Cracks Down on "Gutter Oil," a Substance Even Worse Than its Name, *Time*, 13 September 2011.

20.7 Further reading

- Warner, K. (1999). "Impact of high-temperature food processing on fats and oils". *Advances in experimental medicine and biology* **459**: 67–77. doi:10.1007/978-1-4615-4853-9_5. PMID 10335369.

- Fox, R. (2001). Frying oils. In Kaarin Goodburn (Ed.) *EU Food Law*. Woodhead. pp. 195–224. ISBN 978-1-85573-557-6.

20.8 External links

Chapter 21

Cork (material)

For other uses, see Cork.

Cork is an impermeable, buoyant material, a prime-subset

Untreated cork panel

of bark tissue that is harvested for commercial use primarily from *Quercus suber* (the Cork Oak), which is endemic to southwest Europe and northwest Africa. Cork is composed of suberin, a hydrophobic substance, and because of its impermeable, buoyant, elastic, and fire retardant properties, it is used in a variety of products, the most common of which is for wine stoppers. The montado landscape of Portugal produces approximately 50% of cork harvested annually worldwide, with Corticeira Amorim being the leading company in the industry.*[1] Cork was examined microscopically by Robert Hooke, which led to his discovery and naming of the cell.*[2]

Quercus suber *(Cork Oak) bark, Portugal*

21.1 Sources

There are about 2,200,000 hectares of cork forest worldwide; 34% in Portugal and 27% in Spain. Annual production is about 200,000 tons; 49.6% from Portugal, 30.5% from Spain, 5.8% from Morocco, 4.9% from Algeria, 3.5% from Tunisia, 3.1% Italy, and 2.6% from France.*[3]

Once the trees are about 25 years old the cork is traditionally

stripped from the trunks every nine years, with the first two harvests generally producing lower quality cork. The trees live for about 300 years.

The cork industry is generally regarded as environmentally friendly.*[4] Cork production is generally considered sustainable because the cork tree is not cut down to obtain cork; only the bark is stripped to harvest the cork.*[5] The tree continues to live and grow. The sustainability of production and the easy recycling of cork products and by-products are two of its most distinctive aspects. Cork Oak forests also prevent desertification and are a particular habitat in the Iberian Peninsula and the refuge of various endangered species.[*][6]

Carbon footprint studies committed by Corticeira Amorim, Oeneo Bouchage of France and the Cork Supply Group of Portugal concluded that cork is the most environmentally friendly wine stopper in comparison to other alternatives. The Corticeira Amorim's study, in particular ("Analysis of the life cycle of Cork, Aluminum and Plastic Wine Closures"), was developed by PricewaterhouseCoopers, according to ISO 14040.[*][7] Results concluded that, concerning the emission of greenhouse gases, each plastic stopper released 10 times more CO_2, whilst an aluminium stopper releases 26 times more CO_2 than does a cork stopper.

The Cork Oak is unrelated to the "cork trees" (*Phellodendron*), which have corky bark but are not used for cork production.

21.2 Harvesting

Cork extraction near Aracena, Spain

Cork is extracted only from early May to late August, when the cork can be separated from the tree without causing permanent damage. When the tree reaches 25–30 years of age and about 24in (60 cm) in circumference, the cork can be removed for the first time. However, this first harvest almost always produces poor quality or "male" cork (Portuguese *cortiça virgem*;[*][8] Spanish *corcho bornizo* or *corcho virgen*[*][9]). Bark from initial harvests can be used to make flooring, shoes, insulation and other industrial products. Subsequent extractions usually occur at intervals of 9 years, though it can take up to 13 for the cork to reach an acceptable size. If the product is of high quality it is known as "gentle" cork (Portuguese *cortiça amadia*,[*][10] but also *cortiça secundeira* only if it is the second time;[*][8] Spanish *corcho segundero*, also restricted to the "second time"[*][9]), and, ideally, is used to make stoppers for wine and champagne bottles.[*][11]

The workers who specialize in removing the cork are known as extractors. Extractors use a very sharp axe to make two types of cuts on the tree: one horizontal cut around the plant, called a crown or necklace, at a height of about 2-3 times the circumference of the tree, and several vertical cuts called rulers or openings. This is the most delicate phase of the work because, even though cutting the cork requires quite a bit of strength, the extractor must not damage the underlying phellogen or the tree will be harmed.

To free the cork from the tree, the extractor pushes the handle of the axe into the rulers. A good extractor needs to use a firm but precise touch in order to free a large amount of cork without damaging the product or tree.

These freed portions of the cork are called planks. The planks are usually carried off by hand since cork forests are rarely accessible to vehicles. The cork is stacked in piles in the forest or in yards at a factory, and traditionally, left to dry, after which it can be loaded onto a truck and shipped to a processor.

21.3 Properties and uses

Cork's elasticity combined with its near-impermeability makes it suitable as a material for bottle stoppers, especially for wine bottles. Cork stoppers represent about 60% of all cork based production.

Cork is also an essential element in the production of badminton shuttlecocks.

Cork's bubble-form structure and natural fire retardant make it suitable for acoustic and thermal insulation in house walls, floors, ceilings and facades. The by-product of more lucrative stopper production, corkboard is gaining popularity as a non-allergenic, easy-to-handle and safe alterna-

Varnished cork tiles can be used for flooring, as an alternative for linoleum, stone or ceramic tiles

A cork stopper for a wine bottle

tive to petrochemical-based insulation products which are flammable and emit highly toxic fumes when burned.

Sheets of cork, also often the by-product of stopper production, are used to make bulletin boards as well as floor and wall tiles.

Cork's low density makes it a suitable material for fishing floats and buoys, as well as handles for fishing rods (as an alternative to neoprene).

Granules of cork can also be mixed into concrete. The composites made by mixing cork granules and cement have lower thermal conductivity, lower density and good energy absorption. Some of the property ranges of the composites are density (400–1500 kg/m^3), compressive strength (1–26 MPa) and flexural strength (0.5–4.0 MPa).[12]

21.3.1 Use in wine bottling

As late as the mid-17th century, French vintners did not use cork stoppers, using oil-soaked rags stuffed into the necks of bottles instead.[13]

Wine corks can be made of either a single piece of cork, or composed of particles, as in champagne corks; corks made of granular particles are called "agglomerated corks".[14]

Natural cork closures are used for about 80% of the 20 billion bottles of wine produced each year. After a decline in use as wine-stoppers due to the increase in the use of cheaper synthetic alternatives, cork wine-stoppers are making a comeback and currently represent approximately 60% of wine-stoppers today.

Because of the cellular structure of cork, it is easily compressed upon insertion into a bottle and will expand to form a tight seal. The interior diameter of the neck of glass

High-speed air-gap flash image of a champagne bottle being uncorked

bottles tends to be inconsistent, making this ability to seal through variable contraction and expansion an important attribute. However, unavoidable natural flaws, channels, and cracks in the bark make the cork itself highly inconsistent. In a 2005 closure study, 45% of corks showed gas leakage during pressure testing both from the sides of the cork as well as through the cork body itself.[15]

Since the mid-1990s, a number of wine brands have switched to alternative wine closures such as synthetic plastic stoppers, screw caps, or other closures. In some countries, screw caps are often seen as a cheap alternative destined only for the low grade wines; however, in Australia, for example, much of the non-sparkling wine production now uses these caps as a cork alternative, although some have recently switched back to cork citing issues using screw caps.[16] These alternatives to real cork have both advantageous and controversial properties. For example, while screwtops are generally considered to offer a trichloroanisole (TCA) free seal, it is possible to find TCA contamination in a screw cap bottle.[17] Additionally, they reduce the oxygen transfer rate to almost zero, which can lead to reductive qualities in the wine. TCA is one of the primary causes of cork taint in wine. However, in recent years major cork producers (Amorim, Álvaro Coelho & Irmãos, Ganau, Cork Supply Group, and Oeneo) have developed methods that remove most TCA from natural wine corks. Natural cork stoppers are important because they allow oxygen to interact with wine for proper aging, and are best suited for wines purchased with the intent to age.[18] Stoppers which resemble natural cork very closely can be made by isolating the suberin component of the cork from the undesirable lignin, mixing it with the same substance used for contact lenses and an adhesive, and molding it into a standardized product, free of TCA or other undesirable substances.[19]

The study "Analysis of the life cycle of Cork, Aluminum and Plastic Wine Closures," commissioned by cork manufacturer Amorim and made public in December 2008, concluded that cork is the most environmentally responsible stopper, in a one-year life cycle analysis comparison with plastic stoppers and aluminum screw caps.[20]

21.3.2   Other uses

Cork is used in musical instruments, particularly woodwind instruments, where it is used to fasten together segments of the instrument, making the seams airtight. Conducting baton handles are also often made out of cork.

It is also used in shoes, especially those using Goodyear Welt Construction.

Cork can be used to make bricks for the outer walls of houses, as in Portugal's pavilion at Expo 2000.

On November 28, 2007, the Portuguese national postal ser-

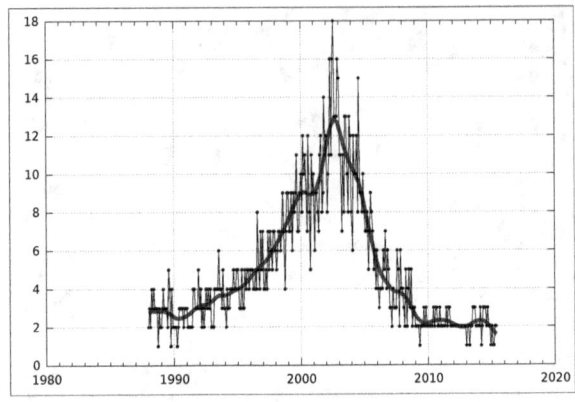

Import value of cork manufactures to Australia since 1988 ($million/month), showing a peak in 2002

vice CTT issued the world's first postage stamp made of cork.[21][22]

Cork is used as the core of both baseballs and cricket balls. A corked bat is made by replacing the interior of a baseball bat with cork a practice known as "corking". It was historically a method of cheating at baseball; the efficacy of the practice is now discredited.

Cork is often used, in various forms, in spacecraft heat shields and fairings.

Cork can be used in the paper pick-up mechanisms in inkjet and laser printers.

Cork is also used inside footwear to improve climate control and comfort.

Corks are also hung from hats to keep insects away. See cork hat.

Cork has been used as a core material in sandwich composite construction.

Cork can be used as the friction lining material of an automatic transmission clutch, as designed in certain mopeds.[23]

Cork can be used instead of wood or aluminium in automotive interiors.[24]

Cork can also be used to make watch bands and faces as seen with Sprout Watches.

Cork slabs are sometimes used by orchid growers as a natural mounting material.

21.4   See also

- APCOR, Portuguese Cork Association

- Bung

- Cork Boat (vessel)

- Cork borer

- Corkscrew

- Corky

21.5 Notes

[1] J. L. CALHEIROS E MENESES, President, Junta Nacional da Cortiça, Portugal. "The cork industry in Portugal"

[2] "Robert Hooke". Retrieved 2010-11-03.

[3] "Cork Production - Area of cork oak forest". *apcor.pt.* APCOR.

[4] Skidmore, Sarah, *USA Today* (August 26, 2007). "Stopper pulled on cork debate"

[5] McClellan, Keith. "Apples, Corks, and Age". Blanco County News. Retrieved 22 May 2014.

[6] Henley, Paul, BBC.com (September 18, 2008) "Urging vintners to put a cork in it"

[7] PricewaterhouseCoopers/ECOBILAN (October 2008). Analysis of the life cycle of Cork, Aluminium and Plastic Wine Closures.

[8] Portuguese wikipedia

[9] DRAE

[10] *Amadio* comes from and is synonym of *amavio*, "beberage or spell to seduce" (Dicionário Houaiss da Língua Portuguesa), from *amar*, "to love".

[11] "Harvesting Cork Is as Natural as Shearing Sheep". *newsusa.com.*

[12] Karade SR. 2003. An Investigation of Cork Cement Composites. PhD Thesis. BCUC. Brunel University, UK.

[13] Prlewe, J. Wine From Grape to Glass. New York: Abbeville Press, 1999, p. 110.

[14] http://www.brewerylane.com/corks.html

[15] Gibson, Richard, *Scorpex Wine Services* (2005). "variability in permeability of corks and closures" (PDF).

[16] "http://www.harpers.co.uk/news/rusden-wines-abandons-screwcap-for-cork/312560.article". *Harpers.co.uk.* Harpers.co.uk.

[17] Mobley, Esther. "Is it possible to find TCA contamination in a screw cap bottle?". *WineSpectator.com.* Wine Spectator.

[18] "Cork or screw cap – which is best for your wine?". *Corklink.com.*

[19] Diam Corks, The Wine Society

[20] Easton, Sally, *Decanter.com* (December 4, 2008). "Cork is the most sustainable form of closure, study finds".

[21] Publico.pt Cork stamp almost sold out (Portuguese)

[22] Wine Storage Guide Cork stamp debuts in Portugal

[23] Tomos A35 Clutch Repair

[24] Motor Trend Faurecia Takes to the Automotive Interior Fashion Runway

21.6 References

- Margarida Pi i Contallé. 2006. Laboratory head in Manuel Serra Hongos y micotoxinas en tapones de corcho. Propuesta de límites micológicos aceptables

- Cork production corkfacts.com

- Instituto de Promoción del Corcho, Extremadura iprocor.org (Spanish)

- Analysis of the life cycle of Cork, Aluminium and Plastic Wine Closures

- Henley, Paul, *BBC.com* (September 18, 2008). "Urging vintners to put a cork in it".

- PricewaterhouseCoopers/ECOBILAN (October 2008). Analysis of the life cycle of Cork, Aluminium and Plastic Wine Closures

- Cork - Forest in a Bottle. 2008.

21.7 External links

- Cork Quality Council

- Book review: To cork or not to cork

- Material Properties Data: Cork

Chapter 22

Cratoxylum formosum

Cratoxylum formosum or pink mempat is a species of flowering plant in the Hypericaceae family. Its commercial name in timber production is "**mampat**".*[1]

It is a tropical plant found in Brunei, Burma, Cambodia, China, Indonesia, Laos, Malaysia, the Philippines, Singapore, Thailand, and Vietnam. The trees flower when there is dry weather followed by wet weather than dry weather again. It has pink flowers and can be up to 20 meters tall, though they rarely achieve the size required for timber exploitation.*[2]

In Laos, *Cratoxylum fomosum* trees are used:

- for the production of charcoal *[3]

- for their edible young leaves, which can be differentiated as either sour (☐☐), smooth (☐☐☐) or blood-red (☐☐☐☐), possibly depending on subspecies (such as *sp. prunifolium*).

Local names:

- Laotian: ☐☐☐☐ [mâi tî:w]

- Malay: mampat

- Thai: ผักติ้ว Phak tiu

- Vietnamese: thành ngạnh đẹp (*sp. prunifolium* : thành ngạnh vàng)

22.1 References

- 2011. IUCN Red List of Threatened Species. Version 2011.1. "World Conservation Monitoring Centre 1998. Cratoxylum formosum." . Retrieved 23 June 2011.

[1] "'Mampat' entry in the Wood Explorer database" . The Wood Explorer database. Retrieved 23 June 2011.

[2] "'Mampat' entry in the Wood Explorer database" . The Wood Explorer database. Retrieved 23 June 2011.

[3] "Charcoal maker rewards villagers for growing mai tiew" . *Vientiane Times*. 2011-06-21.

Chapter 23

Creosote

For other uses, see Creosote (disambiguation).

Creosotes are a category of carbonaceous chemicals

Railcar-loads of wood railroad ties before and after impregnation with creosote, at a facility of the Santa Fe Railroad, in Albuquerque, New Mexico, in March 1943. This U.S. wartime governmental photo reports that "The steaming black ties in the [left of photo]··· have just come from the retort where they have been impregnated with creosote for eight hours." Ties are "made of pine and fir... seasoned for eight months" [as seen in the untreated railcar load at right].[1]

formed by the distillation of various tars, and by pyrolysis of plant-derived material, such as wood or fossil fuel. They are typically used as preservatives or antiseptics.*[2] Some creosote types were used historically as a treatment for components of seagoing and outdoor wood structures to prevent rot (e.g., railroad ties and bridgework, see image). Samples may be commonly found inside chimney flues where the wood or coal burns under variable conditions, producing soot and tarry smoke. Creosotes are the principal chemicals responsible for the stability, scent, and flavor which is characteristic of smoked meat; the name is derived from the Greek *kréas* (κρέας), meaning "meat", and *sōtēr* (σωτήρ), meaning "preserver" .*[3]

The two main kinds recognized in industry are **wood-tar creosote** and **coal-tar creosote**. The coal-tar variety, having stronger and more toxic properties, has chiefly been used as a preservative for wood, while the wood-tar variety has been used for meat preservation, ship treatment, and for medical purposes as an expectorant, antiseptic, astringent,

anaesthetic, and laxative, though these have mostly been replaced by modern medicines. Coal-tar creosote was formerly used as an escharotic to burn malignant skin tissue and in dentistry to prevent necrosis before its carcinogenic properties became known. Varieties of creosote have also been made from both petroleum and oil shale and are known as **oil-tar creosote** when derived from oil tar and **water-gas-tar creosote** when derived from the tar of water gas. Creosote also has been made from pre-coal formations such as lignite, yielding **lignite-tar creosote**, and peat, yielding **peat-tar creosote**.

23.1 Creosote oils

For some part of their history, wood-tar creosote, and coal-tar creosote were suggested to be the same substance only of distinct origins accounting for their common name; the two were determined only later to be chemically different substances. All types of creosote are composed of phenol derivatives and share some quantity of monosubstituted phenols,*[4] but these are not the only active element of creosote. For its useful effect, wood-tar creosote relies on the presence of methyl ethers of phenol, and coal-tar creosote on the presence of naphthalenes and anthracenes; otherwise either type of tar would dissolve in water.

Creosote was first discovered in its wood-tar form in 1832 by Carl Reichenbach, when he found it both in the tar and in pyroligneous acids obtained by a dry distillation of beechwood. Because pyroligneous acid was known as an antiseptic and meat preservative, Reichenbach did experiments with dipping meat in a dilute solution of distilled creosote. He found that the meat was dried without undergoing putrefaction and had attained a smoky flavor.*[5] This led him to reason that creosote was the antiseptic component contained in smoke, and he further argued that the creosote he had found in wood tar was also in coal tar, animal tar, and amber tar in the same abundance as in wood tar.*[3]

Soon after, in 1834, Friedrich Ferdinand Runge discovered carbolic acid in coal-tar, and Auguste Laurence obtained it from phenylhydrate, which was soon determined to be the same compound. There was no clear view on the relationship between carbolic acid and creosote; Runge described it as having similar caustic and antiseptic properties, but noted that it was different, in that it was an acid and formed salts. Nonetheless, Reichenbach argued that creosote was also the active element, as it was in pyroligneous acid. Despite evidence to the contrary, his view held sway with most chemists, and it became commonly accepted wisdom that creosote, carbolic acid, and phenylhydrate were identical substances, with different degrees of purity.[3]

Carbolic acid was soon commonly sold under the name "creosote", and the scarcity of wood-tar creosote in some places led chemists to believe that it was the same substance as described by Reichenbach. In the 1840s, Eugen Freiherr von Gorup-Besanez after realizing that two samples of substances labeled as creosote were different, started a series of investigations to determine the chemical nature of carbolic acid, leading to a conclusion that it more resembled chlorinated quinones and must have been a different, entirely unrelated substance. Independently, there were investigations into the chemical nature of creosote. A study by F.K. Völkel revealed that the smell of purified creosote resembled that of guaiacol, and later studies by Heinrich Hlasiwetz identified a substance common to guaiacum and creosote that he called creosol and determined that creosote contained a mixture of creosol and guaiacol. Later investigations by Gorup-Besanez, A.E. Hoffmann and Siegfried Marasse showed that wood-tar creosote also contained phenols, giving it a feature in common with coal-tar creosote.[6]

Historically, coal-tar creosote has been distinguished from what was thought of as creosote proper the original substance of Reichenbach's discovery and referred to specifically as "creosote oil". But because creosote from coal-tar and wood-tar are obtained from a similar process and have some common uses, they have also been placed in the same class of substances, with the terms "creosote" or "creosote oil" referring to either product.[2]

23.1.1 Wood-tar creosote

The term creosote has a broad range of definitions depending on the origin of the coal tar oil and end use of the material. With respect to wood preservatives the United States Environmental Protection Agency (EPA) considers the term creosote to mean that it is a pesticide for use as a wood preservative meeting the American Wood Protection Association (AWPA) Standards P1/P13 and P2.[10] The AWPA Standards require that creosote "shall be a pure coal tar product derived entirely from tar produced by the carbonization of bituminous coal."[11][12] Currently

all creosote treated wood products railroad crossties, utility poles, foundation and marine piling, posts, lumber, and timbers are manufactured using this type of wood preservative. The manufacturing process can only be a pressure process under the supervision of a licensed applicator certified by the State Departments of Agriculture. No brush-on, spray or non-pressure uses of creosote are allowed as specified by the EPA approved label for the use of creosote.[13] The use of creosote according to the AWPA Standards does not allow for mixing with other types of "creosote type" materials such as wood-tar creosote, lignite-tar creosote, peat-tar creosote, oil-tar creosote, and water-gas-tar creosote. The AWPA Standard P3 does however, allow blending of a high-boiling petroleum oil meeting the AWPA Standard P4.[12][14]

The information that follows describing the other various types of creosote materials and its uses should be considered as primarily being of only historical value. This history is important, because it traces the origin of these different material used during the 19th and early 20th centuries. Further it must be considered that these other types of creosotes -- lignite-tar, wood-tar, water-gas-tar, etc. -- are not currently being manufactured and have either been replaced with more economical materials, or replaced by products that are more efficacious.

Wood-tar creosote is a colourless to yellowish greasy liquid with a smoky odor, produces a sooty flame when burned, and has a burned taste. It is non-buoyant in water, with a specific gravity of 1.037 to 1.087, retains fluidity at a very low temperature, and boils at 205-225 °C. When transparent, it is in its purest form. Dissolution in water requires up to 200 times the amount of water as the base creosote.[15] The creosote is a combination of natural phenols: primarily guaiacol and creosol (4-methylguaiacol), which will typically constitute 50% of the oil; second in prevalence, cresol and xylenol; the rest being a combination of monophenols and polyphenols.

The simple phenols are not the only active element in wood-tar creosote. In solution, they coagulate albumin, which is a water-soluble protein found in meat; so they serve as a preserving agent, but also cause denaturation. Most of the phenols in the creosote are methoxy derivatives they contain the methoxy group linked to the benzene nucleus (O–CH$_3$). The high level of methyl derivates created from the action of heat on wood (also apparent in the methyl alcohol produced through distillation) make wood-tar creosote substantially different from coal-tar creosote. Guaiacol is a methyl ether of pyrocatechin, while creosol is a methyl ether of methylpyrocatechin, the next homolog of pyrocatechin. Methyl ethers differ from simple phenols in being less hydrophilic, caustic and poisonous.[17] This allows meat to successfully be preserved without tissue denaturation, and allows creosote to be used as a medical ointment.[18]

Because wood-tar creosote is used for its guaiacol and creosol content, it is generally derived from beechwood rather than other woods, since it distills with a higher proportion of those chemicals to other phenolics. The creosote can be obtained by distilling the wood tar and treating the fraction heavier than water with a sodium hydroxide solution. The alkaline solution is then separated from the insoluble oily layer, boiled in contact with air to reduce impurities, and decomposed by diluted sulphuric acid. This produces a crude creosote, which is purified by re-solution in alkali and re-precipitation with acid and then redistilled with the fraction passing over between 200° and 225° constituting the purified creosote.[20]

When ferric chloride is added to a dilute solution, it will turn green; a characteristic of ortho-oxy derivatives of benzene.[17] It dissolves in sulphuric acid to a red liquid, which slowly changes to purple-violet. Shaken with hydrochloric acid in the absence of air, it becomes red, the color changing in the presence of air to dark brown or black.[18]

In preparation of food by smoking, guaiacol contributes mainly to the smoky taste, while the dimethyl ether of pyrogallol, syringol, is the main chemical responsible for the smoky aroma.

Historical uses

Industrial   Soon after it was discovered and recognized as the principle of meat smoking, wood-tar creosote became used as a replacement for the process. Several methods were used to apply the creosote. One was to dip the meat in pyroligneous acid or a water of diluted creosote, as Reichenbach did, or brush it over with them, and within one hour the meat would have the same quality of that of traditionally smoked preparations.[21] Sometimes the creosote was diluted in vinegar rather than water, as vinegar was also used as a preservative.[22] Another was to place the meat in a closed box, and place with it a few drops of creosote in a small bottle. Because of the volatility of the creosote, the atmosphere was filled with a vapor containing it, and it would cover the flesh.[21]

The application of wood tar to seagoing vessels was practiced through the 18th century and early 19th century, before the creosote was isolated as a compound. Wood-tar creosote was found not to be as effective in wood treatments, because it was harder to impregnate the creosote into the wood cells, but still experiments[23] were done, including by many governments, because it proved to be less expensive on the market.[24]

Medical   Even before creosote as a chemical compound was discovered, it was the chief active component of medic-

inal remedies in different cultures around the world.

Larrea tridentata

Larrea tridentata, or the so-called creosote bush, as named after its distinct creosote smell, was used by Native Americans in the Southwest as a treatment for many maladies. The Coahuilla Indians used the plant for intestinal complaints and tuberculosis. The Pima drank a decoction of the leaves as an emetic, and applied the boiled leaves as poultices to wounds or sores.[25] Papago Indians prepared it medicinally for stiff limbs, snake bites, and menstrual cramps.[26] Guaiacum, after which the guaiacol in creosote was named, was used by native Caribbean islanders to treat tropical diseases and later for syphilis.[27][28]

In antiquity, pitches and resins were used commonly as medicines. Pliny mentions a variety of tar-like substances being used as medicine, including *cedria* and *pissinum*.[29] *Cedria* was the pitch and resin of the cedar tree, being equivalent to the oil of tar and pyroligneous acid which are used in the first stage of distilling creosote.[30][31] He recommends cedria to ease the pain in a toothache, as an injection in the ear in case of hardness of hearing, to kill parasitic worms, as a preventative for impregnation, as a treatment for phthiriasis and porrigo, as an antidote for the poison of the sea hare, as a liniment for elephantiasis, and as an ointment to treat ulcers both on the skin and in the lungs.[31] He further speaks of cedria being used as the embalming agent for preparing mummies.[29] *Pissinum* was a tar water that was made by boiling cedria, spreading wool fleeces over the vessels to catch the steam, and then wringing them out.[32][33]

The *Pharmacopeé of Lyons*, published in 1786, says that cedar tree oil can induce vomiting, and suggests it helps medicate tumors and ulcers.[34][35] Physicians contemporary to the discovery of creosote recommended ointments and pills made from tar or pitch to treat skin diseases.[34] Tar water had been used as a folk remedy since the Middle Ages to treat affections like dyspepsia. Bishop Berkeley wrote several works on the medical virtues of tar

Portrait of Bishop Berkeley by John Smybert, 1727

water, including a philosophical work in 1744 titled *Siris: a chain of philosophical reflexions and inquiries concerning the virtues of tar water, and divers other subjects connected together and arising one from another*, and a poem where he praised its virtues.[*][36] Pyroligneous acid was also used at the time in a medicinal water called *Aqua Binelli*.[*][34]

Given this history, and the antiseptic properties known to creosote, it became popular among physicians in the 19th century. A dilution of creosote in water was sold in pharmacies as *Aqua creosoti*, as suggested by the previous use of pyroligneous acid. It was prescribed to quell the irritability of the stomach and bowels and detoxify, treat ulcers and abscesses, neutralize bad odors, and stimulate the mucous tissues of the mouth and throat.[*][37][*][38] Creosote in general was listed as an irritant, styptic, antiseptic, narcotic, and diuretic, and in small doses when taken internally as a sedative and anaesthetic. It was used to treat ulcers, and as a way to sterilize the tooth and deaden the pain in case of a tooth-ache.[*][37]

Creosote was suggested as a treatment for tuberculosis by Reichenbach as soon as 1833. Following Reichenbach, it was argued for by John Elliotson and Sir John Rose Cormack.[*][37] Elliotson, inspired by the use of creosote to arrest vomiting during an outbreak of cholera, suggested its use for tuberculosis through inhalation. He also suggested it

for epilepsy, neuralgia, diabetes and chronic glanders.[*][39] The idea of using it for tuberculosis failed to take hold, and use of this purpose was dropped, until the idea was revived later in 1876 by the British doctor G. Anderson Imlay, who suggested it be applied locally in spray to the bronchial mucous membrane.[*][37][*][40][*][41] This was followed up in 1877 when it was argued for in a clinical paper by Charles Bouchard and Henri Gimbert.[*][42] Germ theory had been established by Pasteur in 1860, and Bouchard, arguing that a bacillus was responsible for the disease, sought to rehabilitate creosote for its use as an antiseptic to treat it. He began a series of trials with Gimbert to convince the scientific community, and claimed a promising cure rate.[*][43] A number of publications in Germany confirmed his results in the following years.[*][42]

Following that, that was a period of experimentation of different techniques and chemicals using creosote in tuberculosis, which lasted until about 1910, when radiation therapy looked to be a more promising treatment. Guaiacol, instead of a full creosote solution, was suggested by Hermann Sahli in 1887; he argued it had the active chemical of creosote and had the advantage of being of definite composition and of having a less unpleasant taste and odor.[*][44] A number of solutions of both creosote and guaiacol appeared on the market, such as *phosphotal* and *guaicophosphal*, phosphites of creosote and guaiacol; *eosot* and *geosot*, valerinates of creosote and guaicol; *phosot* and *taphosot*, phosphate and tannophospate of creosote; and *creosotal* and *tanosal*, tannates of creosote.[*][45] Creosote and eucalptus oil were also a remedy used together, administered through a vaporizor and inhaler. Since then, more effective and safer treatments for tuberculosis have been developed.

In the 1940s, Canadian-based Eldon Boyd experimented with guaiacol and a recent synthetic modification glycerol guaiacolate (guaifenesin) on animals. His data showed that both drugs were effective in increasing secretions into the airways in laboratory animals, when high enough doses were given.

Current uses

Industrial Wood-tar creosote is to some extent used for wood preservation, but it is generally mixed with coal-tar creosote, since the former is not as effective. Commercially available preparations of "liquid smoke", marketed to add a smoked flavor to meat and aid as a preservative, consist primarily of creosote and other constituents of smoke.[*][46] Creosote is the ingredient that gives liquid smoke its function; guaicol lends to the taste and the creosote oils help act as the preservative.

Medical The guaifenesin developed by Eldon Boyd is still commonly used today as an expectorant, sold over the counter, and usually taken by mouth to assist the bringing up of phlegm from the airways in acute respiratory tract infections. Guaifenesin is a component of Mucinex, Robitussin DAC, Cheratussin DAC, Robitussin AC, Cheratussin AC, Benylin, DayQuil Mucous Control, Meltus, and Bidex 400.

Seirogan is a popular Kampo medicine in Japan, used as an anti-diarrheal, and has 133 mg wood creosote from beech, pine, maple or oak wood per adult dose as its primary ingredient. Seirogan was first used as a gastrointestinal medication by the Imperial Japanese Army in Russia during the Russo-Japanese War of 1904-5.*[47] Creomulsion is a cough medicine in the United States, introduced in 1925, that is still sold and contains beechwood creosote.

Creosote, in the form of samples from the creosote bush, is often found as a herbal remedy and supplement under the name *chaparral*, and in the form of beechwood creosote under the name *kreosotum* or *kreosote*.

23.1.2   Coal-tar creosote

The term creosote has a broad range of definitions depending on the origin of the coal tar oil and end use of the material. With respect to wood preservatives the United States Environmental Protection Agency (EPA) considers the term creosote to mean that it is a pesticide for use as a wood preservative meeting the American Wood Protection Association (AWPA) Standards P1/P13 and P2. (1) The AWPA Standards require that creosote "shall be a pure coal tar product derived entirely from tar produced by the carbonization of bituminous coal." (2) (3) Currently all creosote treated wood products railroad crossties, utility poles, foundation and marine piling, posts, lumber, and timbers are manufactured using this type of wood preservative. The manufacturing process can only be a pressure process under the supervision of a licensed applicator certified by the State Departments of Agriculture. No brush-on, spray or non-pressure uses of creosote are allowed as specified by the EPA approved label for the use of creosote. (2) The use of creosote according to the AWPA Standards does not allow for mixing with other types of "creosote type" materials such as wood-tar creosote, lignite-tar creosote, peat-tar creosote, oil-tar creosote, and water-gas-tar creosote. The AWPA Standard P3 does however, allow blend-ing of a high boiling petroleum oil meeting the AWPA Standard P4. (3) (4)

The information that follows describing the other various types of creosote materials and its uses should be considered as primarily being of only historical value. This history is important, because it traces the origin of these different material used during the 19th and early 20th centuries.

Further it must be considered that these other types of creosotes lignite-tar, wood-tar, water-gas-tar, etc. are not currently being manufactured and have either been replaced with more economical materials, or replaced by products that are more efficacious.

(1) Communication between United States Environmental Protection Agency (EPA) and the Creosote Council. (2) Reregistration Eligibility Decision Document for Creosote, United States Environmental Protection Agency 2008. (3) American Wood Protection Association Book of Standards 2013 (4) Preservative Treatment of Wood by Pressure Methods, United States Department of Agriculture, Forest Service, Handbook No. 40, 1952.

Coal-tar creosote is greenish-brown liquid, with different degrees of darkness, viscosity, and fluorescence depending on how it's made. When freshly made, the creosote is a yellow oil with a greenish cast and highly fluorescent; the fluorescence increased by exposure to air and light. After settling, the oil is dark green by reflected light and dark red by transmitted light.*[50] To the naked eye, it will generally appear brown. The creosote (often called "creosote oil") consists almost wholly of aromatic hydrocarbons, with some amount of bases and acids and other neutral oils. The flash point is 70–75 °C and burning point is 90–100 °C,*[51] and when burned it releases a greenish smoke.*[52] The smell largely depends on the naptha content in the creosote; if there is a high amount, it will have a naptha-like smell; otherwise it will smell more of tar.

In the process of coal-tar distillation, the distillate is collected into four fractions; the "light oil", which remains lighter than water, the "middle oil" which passes over when the light oil is removed; the "heavy oil", which sinks; and the "anthracene oil", which when cold is mostly solid and greasy, of a buttery consistence. Creosote refers to the portion of coal tar which distills as "heavy oil", typically between 230–270 °C, also called "dead oil"; it sinks into water but still is fairly liquid. Carbolic acid is produced in the second fraction of distillation and is often distilled into what is referred to as "carbolic oil".*[53]*[54]*[55]*[56]

Commercial creosote will contain substances from six groups.*[48] The two groups occur in the greatest amounts and are the products of the distillation process the "tar acids", which distill below 205 °C and consist mainly of phenols, cresols, and xylenols, including carbolic acid and aromatic hydrocarbons, which divide into naphthalenes, which distill approximately between 205° and 255 °C, and constituents of an anthracene nature, which distill above 255 °C.*[58] The quantity of each varies based on the quality of tar and temperatures used, but generally, the tar acids won't exceed 5%, the naphthalenes will make up 15 to 50%, and the anthracenes will make up 45% to 70%.*[58] The hydrocarbons are mainly aromatic; derivatives of benzene and related cyclic compounds such as naphthalene, anthracene, phenanthrene, acenapthene, and fluorene. Creosotes from vertical-retort and low temperature tars contain, in addition, some paraffinic and olefinic hydrocarbons. The tar-acid content also depends on the source of the tar it may be less than 3% in creosote from coke-oven tar and as high as 32% in creosote from vertical retort tar.*[59] All of these have antiseptic properties. The tar acids are the strongest antiseptics but have the highest degree of solubility in water and are the most volatile; so, like with wood-tar creosote, phenols are not the most valued component, as by themselves they would lend to being poor preservatives.*[60] In addition, creosote will contain several products naturally occurring in coal nitrogen-containing heterocycles, such as acridines, carbazoles, and quinolines, referred to as the "tar bases" and generally make up about 3% of the creosote sulfur-containing heterocycles, generally benzothiophenes*[61] and oxygen-containing heterocycles, dibenzofurans.*[62] Lastly, creosote will contain a small number of aromatic amines produced by the other substances during the distillation process and likely resulting from a combination of thermolysis and hydrogenation.*[63]*[64] The tar bases are often extracted by washing the creosote with aqueous mineral acid,*[59] although they're also suggested to have antiseptic ability similar to the tar acids.

Commercially used creosote is often treated to extract the carbolic acid, naphthalene, or anthracene content. The carbolic acid or naphthalene is generally extracted to be used in other commercial products.*[65] American produced creosote oils typically will have low amounts of anthracene and high amounts of naphthalene, because when forcing the distillate at a temperature that produces anthracene the soft pitch will be ruined and only the hard pitch will remain; this ruins it for use in roofing purposes, and only leaves a product which isn't commercially useful.*[64]

Historical uses

Industrial   The use of coal-tar creosote on a commercial scale began in 1838, when a patent covering the use of creosote oil to treat timber was taken out by John Bethell. The "Bethell process" or as it later became known, the full-cell process involves placing wood to be treated in a sealed chamber and applying a vacuum to remove air and moisture from wood "cells". The wood is then pressure-treated to impregnate it with creosote or other preservative chemicals, after which vacuum is reapplied separate the excess treatment chemicals from the timber. Alongside the zinc chloride-based "Burnett process", use of creosoted wood prepared by the Bethell process became a principal way of preserving railway timbers (e.g., ties, sleepers) so wood rot and need for replacement could be avoided.*[66]

Besides treating wood, it was also used for lighting and fuel. In the beginning, it was only used for lighting needed in harbor and outdoor work, where the smoke that was produced from burning it was of little inconvenience. By 1879, lamps had been created that ensured a more complete combustion by using compressed air, removing the drawback of the smoke. Creosote was also processed into gas and used for lighting that way. As a fuel, it was used to power ships at sea and blast furnaces for different industrial needs, once it was discovered to be more efficient than unrefined coal or wood. It was also used industrially for the softening of hard pitch, and burned produce lamp black. By 1890, the production of creosote in the United Kingdom totaled approximately 29,900,000 gallons per year.*[52]

In 1854, Alexander McDougall and Angus Smith developed and patented a product called McDougall's Powder as a sewer deodorant; it was mainly composed from carbolic acid derived from creosote. McDougall, in 1864, experimented with his solution to remove entozoa parasites from cattle pasturing on a sewage farm.*[67] This later led to widespread use of creosote as a cattle wash and sheep dip. External parasites would be killed in a creosote diluted dip, and drenching tubes would be used to administer doses to the animals stomach to kill internal parasites.*[68]

Two later methods for creosoting wood were introduced after the turn of the century, referred to as empty-cell processes, because they involve compressing the air inside the wood so that the preservative can only coat the inner cell walls rather than saturating the interior cell voids. This is a less effective, though usually satisfactory, method of treating the wood, but is used because it requires less of the creosoting material. The first method, the "Rüping process" was patented in 1902, and the second, the "Lowry process" was patented in 1906. Later in 1906, the "Allardyce process" and "Card process" were patented to treat wood with a combination of both creosote and zinc chloride.*[66] In 1912, it was estimated that a total of 150,000,000 gallons were produced in the United States per year.

Medical   Coal-tar creosote, despite its toxicity, was used as a stimulant and escharotic, as a caustic agent used to treat ulcers and malignancies and cauterize wounds and prevent infection and decay. It was particularly used in dentistry to destroy tissues and arrest necrosis.*[69]*[70]*[71]

Current uses

Industrial   Coal-tar creosote is the most widely used wood treatment today; both industrially, processed into wood using pressure methods such as "full-cell process" or "empty-cell process", and more commonly applied to wood through brushing. In addition to toxicity to fungi, insects, and marine borers, it serves as a natural water repellant. It's commonly used to preserve and waterproof cross ties, pilings, telephone poles, power line poles, marine pilings, and fence posts. Although suitable for use in preserving the structural timbers of buildings, it is not generally used that way because it is difficult to apply.

Due to its carcinogenic character, the European Union has regulated the quality of creosote for the EU market *[72] and requires that the sale of creosote be limited to professional users.*[73]*[74] The United States Environmental Protection Agency regulates the use of coal tar creosote as a wood preservative under the provisions of the Federal Insecticide, Fungicide, and Rodenticide Act. Creosote is considered a restricted-use pesticide and is only available to licensed pesticide applicators*[75]*[76]

Health effects

According to the Agency for Toxic Substances and Disease Registry (ATSDR), eating food or drinking water contaminated with high levels of coal tar creosote may cause a burning in the mouth and throat, and stomach pains. ATSDR also states that brief direct contact with large amounts of coal tar creosote may result in a rash or severe irritation of the skin, chemical burns of the surfaces of the eyes, convulsions and mental confusion, kidney or liver problems, unconsciousness, and even death. Longer direct skin contact with low levels of creosote mixtures or their vapors can result in increased light sensitivity, damage to the cornea, and skin damage. Longer exposure to creosote vapors can cause irritation of the respiratory tract.

The International Agency for Research on Cancer (IARC) has determined that coal tar creosote is probably carcinogenic to humans, based on adequate animal evidence and limited human evidence. It is instructive to note that the animal testing relied upon by IARC involved the continuous application of creosote to the shaved skin of rodents. After weeks of creosote application, the animals developed cancerous skin lesions and in one test, lesions of the lung. The United States Environmental Protection Agency has stated that coal tar creosote is a probable human carcinogen based on both human and animal studies.*[77] As such, the Federal Occupational Safety and Health Administration (OSHA) has set a permissible exposure limit of 0.2 milligrams of coal tar creosote per cubic meter of air (0.2 mg/m3) in the workplace during an 8-hour day, and the Environmental Protection Agency (EPA) requires that spills or accidental releases into the environment of one pound (0.454 kg) or more of creosote be reported to them.*[78]

There is no unique exposure pathway of children to creosote. Children exposed to creosote will probably experience the same health effects seen in adults exposed to creosote. It is unknown whether children differ from adults in their susceptibility to health effects from creosote.

A 2005 mortality study of creosote workers found no evidence supporting an increased risk of cancer death, as a result of exposure to creosote. Based on the findings of the largest mortality study to date of workers employed in creosote wood treating plants, there is no evidence that employment at creosote wood-treating plants or exposure to creosote-based preservatives was associated with any significant mortality increase from either site-specific cancers or non-malignant diseases. The study consisted of 2,179 employees at eleven plants in the United States where wood was treated with creosote preservatives. Some workers began work in the 1940s to 1950s. The observation period of the study covered 1979- 2001. The average length of employment was 12.5 years. One third of the study subjects were employed for over 15 years.*[79]

The largest health effect of creosote is deaths caused by residential chimney fires due to chimney tar (creosote) build-up. This is entirely unconnected with its industrial production or use.*[80]

23.1.3   Oil-tar creosote

Oil-tar creosote is derived from the tar that forms when using petroleum or shale oil in the manufacturing of gas. The distillation of the tar from the oil occurs at very high temperatures; around 980 °C. The tar forms at the same time as the gas, and once processed for creosotes contains a high percentage of cyclic hydrocarbons, a very low amount of tar acids and tar bases, and no true anthracenes have been identified.*[81] Historically, this has mainly been produced in the United States in the Pacific coast, where petroleum has been more abundant than coal. Limited quantities have been used industrially, either alone, mixed with coal-tar creosote, or fortified with pentachlorophenol.*[82]

23.1.4   Water-gas-tar creosote

Water-gas-tar creosote is also derived from petroleum oil or shale oil, but by a different process; its distilled during the production of water-gas. The tar is a by-product resulting from enrichment of water gas with gases produced by thermal decomposition of petroleum. Of the creosotes derived from oil, its practically the only one used for wood preservation. It has the same degree of solubility as coal-tar creosote and is easy to impregnate into wood. Like standard oil-tar creosote, it has a low amount of tar acids and tar bases, and has less antiseptic qualities.*[57] Petri dish tests have shown that water-gas-tar creosote is one-sixth as anti-septically effective as that of coal-tar.*[83]

23.1.5 Lignite-tar creosote

Lignite-tar creosote is produced from lignite rather than bituminous coal, and varies considerably from coal-tar creosote. Also called "lignite oil", it has a very high content of tar acids, and has been used to increase the tar acids in normal creosote when necessary.[84] When it has been produced, its generally been applied in mixtures with coal-tar creosote or petroleum. Its effectiveness when used alone has not been established. In an experiment with southern yellow pine fence posts in Mississippi, straight lignite-tar creosote was giving good results after about 27 years exposure, although not as good as the standard coal-tar creosote used in the same situation.[85]

23.1.6 Peat-tar creosote

There have also been attempts to distill creosote from peat-tar, although mostly unsuccessful due to the problems with winning and drying peat on an industrial scale.[86] Peat tar by itself has in the past been used as a wood preservative.

23.2 Build-up in chimneys

Burning wood and fossil fuels at low temperature causes incomplete combustion of the oils in the wood, which are off-gassed as volatiles in the smoke. As the smoke rises through the chimney it cools, causing water, carbon, and volatiles to condense on the interior surfaces of the chimney flue. The black oily residue that builds up is referred to as creosote, which is similar in composition to the commercial products by the same name, but with a higher content of carbon black.

Over the course of a season creosote deposits can become several inches thick. This creates a compounding problem, because the creosote deposits reduce the draft (airflow through the chimney) which increases the probability that the wood fire is not getting enough air to burn at high temperature. Since creosote is highly combustible, a thick accumulation creates a fire hazard. If a hot fire is built in the stove or fireplace, and the air control left wide open, this may allow hot oxygen into the chimney where it comes in contact with the creosote which then ignites causing a chimney fire. Chimney fires often spread to the main building because the chimney gets so hot that it ignites any combustible material in direct contact with it, such as wood. The fire can also spread to the main building from sparks emitting from the chimney and landing on combustible roof surfaces. In order to properly maintain chimneys and heaters that burn wood or carbon-based fuels, the creosote buildup must be removed. Chimney sweeps perform this service for a fee.[80]

73% of heating fires and 25% of all residential fires in the United States are caused by failure to clean out creosote buildup. Since 1990, creosote buildup has caused 75% fewer fires..[80] This is partly due to the use of efficient wood-burning stoves that fully combust the carbon from fuel.

23.3 See also

- Pentachlorophenol

23.4 Notes

[1] See Jack Delnao, 1943, "At the Santa Fe R.R. tie plant, Albuquerque, N[ew] Mex[ico]···", Library of Congress, Prints & Photographs Online Catalog, accessed 16 February 2015.

[2] Price, Kelogg & Cox 1909, p. 7

[3] Schorlemmer 1885, p. 152

[4] Roscoe & Schorlemmer 1888, p. 37

[5] Roscoe & Schorlemmer 1888, p. 33

[6] Schorlemmer 1885, p. 153

[7] Allen 1910, p. 353

[8] American Pharmaceutical Association 1894, p. 1073

[9] Royal Chemical Society 1895, p. 294

[10] Communication between United States Environmental Protection Agency (EPA) and the Creosote Council.

[11] Reregistration Eligibility Decision Document for Creosote, United States Environmental Protection Agency 2008.

[12] American Wood Protection Association Book of Standards 2013.

[13] Reregistration Eligibility Decision Document for Creosote, United States Environmental Protection Agency 2008

[14] Preservative Treatment of Wood by Pressure Methods, United States Department of Agriculture, Forest Service, Handbook No. 40, 1952.

[15] Thorpe 1890, p. 614

[16] Lee et al. 2005, p. 1483

[17] Pharmaceutical Society of Great Britain 1898, p. 468

[18] Allen 1910, p. 348

[19] Price, Kelogg & Cox 1909, p. 13

[20] Allen 1910, p. 347

[21] Abel & Smith 1857, p. 23

[22] Letheby 1870, pp. 225–226

[23] Joerin 1909, p. 767

[24] Bradbury 1909, p. 107

[25] United States Herbarium 1890, p. 521

[26] Wignall & Bowers 1993, p. 104

[27] Foster & Johnson 2006, p. 190

[28] Bostock & Alison 1832, p. 553

[29] Cormack 1836, p. 58

[30] Parr 1809, p. 383

[31] Pliny 1856, p. 8

[32] Berkeley 1744, p. 9

[33] Pliny 1855, p. 290

[34] Cormack 1836, p. 50

[35] Vitet 1778, p. 427

[36] Chemist and Druggist 1889, p. 300

[37] King, Felter & Llyod 1905, p. 617

[38] Taylor 1902, p. 207

[39] Whittaker 1893, p. 77

[40] Imlay 1876, p. 514

[41] Dobbell 1878, p. 315

[42] Kinnicutt 1892, p. 514

[43] Contrepois 2002, p. 211

[44] Kinnicutt 1892, p. 515

[45] Coblentz 1908

[46] Chenoweth 1945, p. 206

[47] Seirogan 2011

[48] Melber, Kielhorn & Mangelsdorf 2004, p. 11

[49] Speight 1994, p. 456

[50] Allen 1910, p. 366

[51] Bateman 1922, p. 50

[52] Thorpe 1890, p. 615

[53] Philips 1891, p. 255

[54] Martin 1913, pp. 416–419

[55] Nelson 1907, p. 204

[56] Noller 1965, p. 185

[57] Price, Kelogg & Cox 1909, p. 12

[58] Engineering and Contracting 1912, p. 531

[59] Greenhow 1965, p. 58

[60] American Railway Bridge and Building Association 1914, p. 287

[61] Orr & White 2002, p. 39

[62] Speight 1994, p. 77

[63] Orr & White 2002, p. 255

[64] Bateman 1922, p. 47

[65] Mushrush & Speight 1995, p. 115

[66] Angier 1910, p. 408

[67] Brock 2008, p. 91

[68] Salmon 1901, pp. 7–14

[69] Farrar 1880, pp. 412–417

[70] Farrar 1893, pp. 1–25

[71] Pease 1862

[72] Commission of the European Communities 2001

[73] Commission of the European Communities 2007

[74] Health and Safety Executive 2011

[75] Creosote Council 2011

[76] Ibach, Miller & 2010 14-1–14-9

[77] EPA 1988

[78] LOSH 2003

[79] Wong 2005, pp. 683–697

[80] DHS 2006

[81] Voorhies 1940

[82] Hunt & Garratt 1967, p. 88

[83] Stimson 1915, p. 626

[84] Richardson 1993, p. 103

[85] Hunt & Garratt 1967, p. 97

[86] Encyclopaedia Britannica 1949, p. 821

23.5 References

- Schorlemmer, C. (1885). "The history of creosote, cedriret, and pittacal". *Journal of the Society of Chemical Industry* (Society of Chemical Industry) **4**: 152–157.

- Thorpe, Sir Thomas Edward (1890). "Creosote". *A dictionary of applied chemistry* (Longmans, Green, and Co) **1**: 614–620.

- Allen, Alfred Henry (1910). "Creosote and Creosote oils". *Allen's commercial organic analysis* (P. Blakiston's Son & Co) **3**: 346–391.

- Roscoe, Henry Enfield; Schorlemmer, Carl (1888). "Creosote and Creosote oils". *A Treatise on Chemistry: The hydrocarbons and their derivatives or organic chemistry* (Appleton) **3:4**: 32–37.

- American Pharmaceutical Association (1895). "Creosote and Creosote oils". *Proceedings of the American Pharmaceutical Association at the annual meeting* (The Association) **43**: 1073.

- Royal Chemical Society (1895). "Creosote and Creosote oils". *Journal of the Chemical Society* (Royal Society of Chemistry) **68:1**: 294.

- Lee, Kwang-Guen; Lee, Sung-Eun; Takeoka, Gary R.; Kim, Jeong-Han; Park, Byeoung-Soo (July 2005). "Antioxidant activity and characterization of volatile constituents of beechwood creosote". *Journal of the Science of Food and Agriculture* (USDA) **85:9**: 1580–1586. doi:10.1002/jsfa.2156.

- Pharmaceutical Society of Great Britain (1898). "Creosotum". *Pharmaceutical journal: A weekly record of pharmacy and allied sciences* (J. Churchill) **61**: 468.

- Abel, Ambrose; Smith, Elizur Goodrich (1857). *The preservation of food: From the "Aus der natur" of Abel.* Press of Case, Lockwood and company.

- Letheby, Henry (1870). *On food: its varieties, chemical composition, nutritive value, comparative digestibility, physiological functions and uses, preparation, culinary treatment, preservation, adulteration, etc.* Longmans, Green.

- United States Herbarium (1890). *Contributions from the United States National Herbarium* **23**. Smithsonian Institution Press. p. 521.

- Wignall, Brian; Bowers, Janice Emily (1993). *Shrubs and trees of the Southwest deserts.* Western National Parks Association. p. 104.

- Foster, Stephen; Johnson, Rebecca L. (2006). *Desk reference to nature's medicine.* National Geographic Books. p. 190.

- Bostock, John; Alison, William Pulteney (1832). *The Cyclopædia of practical medicine: comprising treatises on the nature and treatment of diseases, materia medica and therapeutics, medical jurisprudence, etc. etc* **1**. Sherwood, Gilbert, and Piper. p. 553.

- Cormack, Sir John Rose (1836). *A treatise on the chemical, medicinal, and physiological properties of creosote: illustrated by experiments on the lower animals: with some considerations on the embalment of the Egyptians. Being the Harveian prize dissertation for 1836.* J. Carfrae & Son.

- Parr, Bartholemew (1809). *The London Medical Dictionary, including under distinct heads every branch of medecine* **1**. J. Johnson.

- Berkeley, George (1744). *Siris: a chain of philosophical reflexions and inquiries concerning the virtues of tar water, and divers other subjects connected together and arising one from another.* Reprinted for W. Innys.

- Pliny (1855). *Pliny's Natural History* **3**. H. G. Bohn.

- Pliny (1856). *Pliny's Natural History* **5**. G. Bell and sons.

- Vitet, Louis (1778). *Pharmacopée de Lyon, ou exposition méthodique des médicaments simples et composés.* Chez les Freres Perisse.

- Chemist and Druggist (1889). "Tar Water". *Chemist and druggist: the newsweekly for pharmacy* (Benn Brothers) **35**: 300.

- King, John; Felter, Harvey Wickes; Lloyd, John Uri (1905). "Creosote". *King's American dispensatory* (Ohio Valley Co) **1**: 616–617.

- Taylor, C.F. (1902). "Creosote". *The Medical world* **20**: 207.

- Whittaker, J.T. (1893). "Creosote in Tuberculosis Pulmonum". *Transactions of the Association of American Physicians* (W.J. Dornan, inc) **8**: 77–90.

- Contrepois, Alain (2002). "The Clinician, Germs and Infectious Diseases: The Example of Charles Bouchard in Paris". *Medical History* **46** (2): 197–220. PMC 1044495. PMID 12024808.

- Kinnicutt, Sir Francis P. (1892). "New outlooks in the prophylaxis and treatment of tuberculosis". *Boston medical and surgical journal* **126**: 513–518. doi:10.1056/nejm189205261262101.

- Imlay, G. Anderson (1876). "New outlooks in the prophylaxis and treatment of tuberculosis". *The Medical times and gazette* (J. & A. Churchill) **2**: 514.

- Dobbell, Horace (1878). "Carbolic acid and creosote". *Annual reports on diseases of the chest* (Smith) **3**: 315.

- Bernheim, Samuel (1901). *La Tuberculose et la médication créosotée*. Paris: Maloine.

- Coblentz, Virgil (1908). *The newer remedies: including their synonyms, sources, tests, solubilities, incompatibilities, medicinal properties and doses as far as known, together with such proprietaries as have similar titles; a reference manual for physicians, pharmacists and students*. The Apothecary Pub. Co.

- Engineering and Contracting (1912). "Wood Preserving Creosotes: Methods of Production, Properties, Quality, Price and Quantity Consumed in the United States". *Engineering and contracting* (Chicago: Myron C. Clark Publishing Co) **38**: 350–353.

- American Railway Bridge and Building Association (1914). "Wood Preserving Creosotes: Methods of Production, Properties, Quality, Price and Quantity Consumed in the United States". *Proceedings of the annual convention of the American Railway, Bridge and Building Association* (Bretheren Publishing House) **23**: 287–288.

- Bateman, Ernest (1922). *Coal-tar and water-gas tar creosotes*. Govt. print. off.

- Angier, F.J. (1910). "The seasoning and preservative treatment of wood ties". *Railway age gazette* (Railway Age Gazette) **48**: 408–411.

- Brock, William Hodson (2008). *William Crookes and the commercialization of science*. Ashgate Publishing, Ltd.

- Price, Overton W.; Kellogg, R.S.; Cox, W.T. (1909). *Forests of the United States: Their Use*. Government printing office.

- Hodson, E.R. (1906). *Rules and Regulations for the Grading of Lumber*. Government printing office.

- Melber, Christine; Kielhorn, Janet; Mangelsdorf, Inge (2004). "COAL TAR CREOSOTE" (PDF). *who.int/*. World Health Organization.

- Mueller, J.G.; Chapman, P.J.; Pritchard, P.H. (December 1989). "Action of a Fluoranthene-Utilizing Bacterial Community on Polycyclic Aromatic Hydrocarbon Components of Creosote". *Applied and Environmental Microbiology* (American Society for Microbiology) **55** (12): 3085–90. PMC 203227. PMID 16348069.

- Orr, Wilson L.; White, Curt M. (1990). *Geochemistry of sulfur in fossil fuels*. American Chemical Society.

- Speight, J.G. (1994). *The chemistry and technology of coal*. CRC Press.

- Mushrush, George C.; Speight, J.G. (1995). *Petroleum products: instability and incompatibility*. CRC Press.

- Greenhow; E.J. (1965). *Wood* **30**. Tothill Press.

- Philips; H. Joshua (1891). *Engineering chemistry: a practical treatise for the use of analytical chemists, engineers, ironmasters, iron founders, students, and others*. C. Lockwood & son.

- Martin; Geoffrey (1913). *Industrial and manufacturing chemistry: a practical treatise* **1**. Appleton.

- Nelson; Thomas (1907). *Nelson's encyclopaedia: everybody's book of reference* **3**. Thomas Nelson.

- Noller; Carl Robert (1965). *Chemistry of organic compounds*. Saunders.

- Salmon, D.E. (1901). *Relationship of bovine tuberculosis to public health*. Government printing office.

- Seirogan (2011). "A Gift from the Forest". *seirogan.co.jp/*.

- Commission of the European Communities (2001). "COMMISSION DIRECTIVE 2001/90/EC". *eur-lex.europa.eu/*.

- Commission of the European Communities (2007). "COMMISSION DIRECTIVE 76/769/EEC". *eur-lex.europa.eu/*.

- Health and Safety Executive (2011). "Revocation of approvals for amateur creosote/coal tar creosote wood preservatives". *hse.gov.uk/*.

- Creosote Council (2011). "Regulation". *creosote-council.org/*.

- Ibach, Rebecca E.; Miller, Regis B. (2007). *The Encyclopedia of Wood*. Skyhorse Publishing Inc.

- Joerin, A.E. (December 1909). "The seasoning and preservative treatment of wood ties". *Popular Mechanics* (Popular Mechanics) **48**: 767.

- Bradbury, Robert H. (1909). "Increase in the use of wood preservatives indicates progress in wood preservation". *Journal of the Franklin Institute* (Pergamon Press) **168**: 107. doi:10.1016/s0016-0032(09)90070-9.

- Farrar, J.N. (1880). "On the comparative value of sulphuric acid and creosote in the treatment of alveolar cavities". *Annals of anatomy and surgery* **2**: 412–418.

- Farrar, J.N. (1893). "Pulpless teeth; abscess; treatment, especially surgical treatment". *Transactions of the New York Ondontological Society* (J.P. Lippincott Company): 1–25.

- Pease, William A. (1862). "Arsenic, its application and use". *British journal of dental science* (Oxford Press) **5**: 417–426.

- Martin, Stanlisas (1862). "Solidified Creosote". *British journal of dental science* (Oxford Press) **5**: 290.

- Hunt, George McMonies; Garratt, George Alfred (1967). *Wood preservation*. McGraw-Hill.

- Voorhies, Glenn (June 1940). "Oil tar creosote for wood preservation". *ir.library.oregonstate.edu*.

- Stimson, Earl (1914). "Report of the committee XVII on wood preservation". *Proceedings of the annual convention of the American Railway, Bridge and Building Association* (Bretheren Publishing House) **15**: 625–633.

- Richardson, Barry A. (1993). *Wood preservation*. Taylor & Francis.

- Encyclopaedia Britannica (1949). *Encyclopaedia britannica: a new survey of universal knowledge* **21**. Encyclopaedia Britannica.

- "Creosote: What you need to know" (PDF). *losh.ucla.edu/*. 2003.

- "Creosote (CASRN 8001-58-9)". *epa.gov/*. 1988.

- Wong O, Harris F (July 2005). "Retrospective cohort mortality study and nested case-control study of workers exposed to creosote at 11 wood-treating plants in the United States". *J. Occup. Environ. Med.* **47** (7): 683–97. doi:10.1097/01.jom.0000165016.71465.7a. PMID 16010195.

- "Heating Fires in Residential Buildings" (PDF). *usfa.dhs.gov/*. 2006.

- Chenoweth, Walter Winfred (1945). *How to preserve food*. Houghton Mifflin company.

23.6 External links

- Creosote Council

Chapter 24

Cycas circinalis

Cycas circinalis, also known as the **queen sago**, is a species of cycad known in the wild only from southern India.

24.1 Distribution

Cycas circinalis is the only gymnosperm species found among native Sri Lankan flora.

24.2 Cultivation

The plant is widely cultivated in Hawaii, both for its appearance in landscape and interiors, and for cut foliage.[*][1] In the Philippines, it is locally known as *patubo*, *pitogo* or *bitogo*.

- Male cone, new

- Male cone, old

- Young shoots

- Seed

- Collected seeds

24.3 Use as food

The seed is poisonous. The potent poison in the seeds is removed by soaking them in water. Water from the first seed-soaking will kill birds, goats, sheep and hogs. Water from the following soakings is said to be harmless.

After the final soaking, the seeds are dried and ground into flour. The flour is used to make tortillas, tamales, soup and porridge.

24.4 Lytico-bodig disease

The plant was thought to be linked with the degenerative disease lytico-bodig on the island of Guam; however, the cycad native to Guam has since been recognised as a separate species, *Cycas micronesica*, by K.D. Hill in 1994.

24.5 Chemistry

Leaflets of *C. circinalis* contain biflavonoids such as (2S, 2″S)−2,3,2″,3″-tetrahydro-4′,4‴-di-O-methylamentoflavone (tetrahydroisoginkgetin).[*][2]

24.6 References

[1] Iwata, Ruth Y.; Rauch, Fred D. (October 1988). "King and Queen Sago" . University of Hawaii. Retrieved September 24, 2012.

[2] Phytochemical Investigation of Cycas circinalis and Cycas revoluta Leaflets: Moderately Active Antibacterial Biflavonoids. Abeer Moawad, Mona Hetta, Jordan K. Zjawiony, Melissa R. Jacob, Mohamed Hifnawy, Jannie P. J. Marais and Daneel Ferreira, Planta Med., 2010, 76(8), pages 796-802, doi:10.1055/s-0029-1240743

24.7 External links

- Floridata: *Cycas circinalis*

- EDIS: Queen sago

- USDA Plants Profile

- University of Hawai'i at Manoa: King and Queen Sago

Chapter 25

Dehesa (pastoral management)

A dehesa in Bollullos Par del Condado, Huelva, southern Spain

A dehesa in the Sierra de Aracena.

Dehesa is a multifunctional agro-sylvo-pastoral system (a type of agroforestry) and cultural landscape of southern and central Spain and southern Portugal, where it is known as **montado**.*[1] Dehesas may be private or communal property (usually belonging to the municipality). Used primarily for grazing, they produce a variety of products including non-timber forest products such as wild game, mushrooms, honey, cork, and firewood. The tree component is oaks, usually holm (*Quercus ilex*) and cork (*Quercus suber*). Other oaks, including melojo (*Quercus pyrenaica*) and quejigo (*Quercus faginea*), may be used to form dehesa, the species depending on geographical location and elevation. Dehesa is an anthropogenic system that provides not only a variety of foods, but also wildlife habitat for endangered species such as the Iberian lynx and the Spanish imperial eagle.*[2]

25.1   Nature

The dehesa is derived from the Mediterranean forest ecosystem, consisting of pastureland featuring herbaceous species for grazing and tree species belonging to the genus *Quercus* (oak), such as the holm oak (*Quercus ilex* sp. *ballota*), although other tree species such as beech and pine

trees may also be present. Oaks are protected and pruned to produce acorns, which the famous black Iberian pigs feed on in the fall during the *montanera*.*[3] Ham produced from Iberian pigs fattened with acorns and air-dried at high elevations is known as *Jamón ibérico*, and sells for premium prices, especially if only acorns have been used for fattening.

There is debate about the origins and maintenance of the dehesa, and whether or not the oaks can reproduce adequately under the grazing densities now forced upon the dehesa or *montado*. Goats, cattle, and sheep also graze in dehesa. In a typical dehesa, oaks are managed to persist for about 250 years. If cork oaks are present, the cork is harvested about every 9 to 12 years, depending on the productivity of the site. The understory is usually cleared every 7 to 10 years, to prevent the takeover of the woodland by shrubs of the rock rose family (Cistaceae), often referred to as "jara", or by oak sprouts. Oaks are spaced to maximize light for the grasses in the understory, water use in the soils, and acorn production for pigs and game.*[4] Periodic hunts in the dehesa are known as the *monteria*. Groups attend a hunt at a private estate, and wait at hunting spots for game to be driven to them with dogs. They usually pay well for the privilege, and hunt wild boar, red deer and other species.

Dehesa in Extremadura, Spain

A dehesa in the Montes de Toledo.

25.2 Importance

The dehesa system has great economic and social importance on the Iberian peninsula because of both the large amount of land involved and its importance in maintaining rural population levels. The major source of income for dehesa owners is usually cork, a sustainable product that supports this ancient production system and old growth oaks.*[5] High end ibérico pigs and sale of hunting rights also represent significant income sources.

25.3 Economic context

The area of dehesa usually coincides with areas that could be termed "marginal" because of both their limited agricultural potential (due to the poor quality of the soil) and a lack of local industry, which results in isolated agro-industries and very low capitalization.

25.4 Extent

Dehesa covers nearly 20,000 square kilometers on the Iberian peninsula, mainly in:

Portugal (33% of total dehesa world's area)*[6]*[7]

- Alentejo
- The Algarve

Spain (23% of total dehesa world's area)*[8]*[9]

- Córdoba
- Extremadura
- Salamanca
- Sierra Morena
  - Sierra Norte de Sevilla
  - Sierra de Aracena

25.5 Other uses of the term

Dehesa also refers to the type of rangeland management of estates for private agro-livestock exploitation in Mediterranean-type forests from which multiple resources are obtained simultaneously.

25.6 See also

- Cabañeros National Park
- List of types of formally designated forests

25.7 References

25.7.1 Notes

[1] Fra. Paleo (2010)

[2] Joffre et al. (1999); Huntsinger et al. (2004); McGrath (2007)

[3] Parsons (1962)

[4] Joffre et al. (1999)

[5] McGrath (2007)

[6] http://ga2014.fsc.org/opinion-analysis-74.
the-dehesas-and-cork-production-today-and-its-alliance-with-fsc

[7] Francisco Manuel Parejo Moorish, 2010

[8] http://ga2014.fsc.org/opinion-analysis-74.
the-dehesas-and-cork-production-today-and-its-alliance-with-fsc

[9] Francisco Manuel Parejo Moorish, 2010

25.7.2   Bibliography

- Fra. Paleo, Urbano. (2010). "The *dehesa/montado* landscape". pp. 149–151 in *Sustainable Use of Biological Diversity in Socio-ecological Production Landscapes,* eds. Bélair, C., Ichikawa, K., Wong, B.Y.L. and Mulongoy, K.J. Montreal: Secretariat of the Convention on Biological Diversity. Technical Series no. 52.

- Huntsinger, Lynn; Adriana Sulak; Lauren Gwin; and Tobias Plieninger. (2004). "Oak woodland ranchers in California and Spain: Conservation and diversification". In *Advances in Geoecology*, ed. S. F. A. Schnabel.

- Joffre, R; Rambal, S; Ratte, JP. (1999). "The dehesa system of southern Spain and Portugal as a natural ecosystem mimic," *Journal of Agroforestry* 45(1-3): 57-79.

- McGrath, Susan. (2007). "Corkscrewed," *Audubon magazine*, January–February.

25.8   External links

Media related to Dehesas at Wikimedia Commons

- Plataforma integralDehesa - Página web de agentes del sector.

- Proyecto Biodehesa

- Foro encinal

- Acción por la dehesa

- Dehesas ibéricas

- Observatorio de la dehesa y el montado

Chapter 26

Diospyros melanoxylon

Coromandel Ebony or **East Indian Ebony** (*Diospyros melanoxylon*) is a species of flowering tree in the family *Ebenaceae* that is native to India and Sri Lanka and that has a hard, dry bark. Its common name derives from Coromandel, the coast of southeastern India. Locally it is known as *temburini* or by its Hindi name *tendu*. In Odisha and Jharkhand it is known as *kendu*. The leaves can be wrapped around tobacco to create the Indian *beedi*,[*][3] which has outsold conventional cigarettes in India.[*][4]

26.1 Common Names

- (Oriya) : Kendu

- (Bengali) : kend, kendu

- (Hindi) : abnus, kendu, tendu, timburni

- (Nepali) : abnush, tendu

- (Sanskrit) : dirghapatraka

- (Tamil) : karai, karundumbi, tumbi

- (Telugu) : tuniki, beedi aaku

- (Trade name) : ebony

26.2 Pharmacology

The leaf of the tree contains valuable flavones.[*][5] The pentacyclic triterpines found in the leaves possess antimicrobial properties,[*][6] while the bark shows antihyperglycemic activity.[*][7] The bark of four *Diospyros* species found in India has been determined to have significant antiplasmodial effects against *Plasmodium falciparum*, which causes malaria in humans.[*][8]

Tendu Patta (Leaf) Collection

26.3 References

[1] "*Diospyros melanoxylon* Roxb." . *Germplasm Resources Information Network*. United States Department of Agriculture. 2006-10-27. Retrieved 2009-04-09.

[2] "The Plant List: A Working List of All Plant Species" .

[3] Lal, Pranay (25 May 2009). "Bidi – A short history" (PDF). *Current Science* (Bangalore, India: Current Science Association) **96** (10): 1335–1337. Retrieved 5 May June 2013. Check date values in: |access-date= (help)

[4] "...bidis command 48 percent of the market while chewing tobacco commands 38 percent and cigarettes 14 percent..." , "The Tax Treatment of Bidis" , tobaccofreeunion.org

[5] NEW FLAVONOIDS FROM THE LEAVES OF DIOSPY-
ROS MELANOXYLON,Uppuluri V. Mallavadhani and
Anita Mahapatra

[6] Antimicrobial Activity of Some Pentacyclic Triterpenes
and Their Synthesized 3-O-Lipophilic Chains,Uppuluri
Venkata MALLAVADHANI,*,a Anita MAHAPATRA,a
Kaiser JAMIL,b and Peddi Srinivasa REDDYb , Biol.
Pharm. Bull. 27(10) 1576 1579 (2004) Vol. 27, No. 10

[7] Antihyperglycemic effect of *Diospyros melanoxylon* (Roxb.)
bark against Alloxan-induced diabetic rats Jadhav J.
K*.Masirkar V. J., Deshmukh V. N.International Journal of
PharmTech Research CODEN(USA): IJPRIF ISSN : 0974-
4304 ,Vol.1, No.2, pp 196-200 , April–June 2009

[8] Investigation of Indian *Diospyros* Species for Antiplasmodial
Properties,V. S. Satyanarayana Kantamreddi and Colin W.
Wright. eCAM 2008;5(2)187–190

Chapter 27

Durian

For other uses, see Durian (disambiguation).

The **durian** (/'djɔriən/)[*][4] is the fruit of several tree species belonging to the genus *Durio*. The name 'durian' is derived from the Malay-Indonesian languages word for duri or "spike", a reference to the numerous spike protuberances of the fruit, together with the noun-building suffix -an. There are 30 recognised *Durio* species, at least nine of which produce edible fruit. *Durio zibethinus* is the only species available in the international market: other species are sold in their local regions.

Regarded by many people in southeast Asia as the "king of fruits",[*][5] the durian is distinctive for its large size, strong odour, and formidable thorn-covered husk. The fruit can grow as large as 30 centimetres (12 in) long and 15 centimetres (6 in) in diameter, and it typically weighs one to three kilograms (2 to 7 lb). Its shape ranges from oblong to round, the colour of its husk green to brown, and its flesh pale yellow to red, depending on the species.

The edible flesh emits a distinctive odour that is strong and penetrating even when the husk is intact. Some people regard the durian as having a pleasantly sweet fragrance; others find the aroma overpowering and revolting. The smell evokes reactions from deep appreciation to intense disgust, and has been described variously as rotten onions, turpentine, and raw sewage. The persistence of its odour, which may linger for several days, has led to the fruit's banishment from certain hotels and public transportation in Southeast Asia.

The durian, native to Southeast Asia, has been known to the Western world for about 600 years. The nineteenth-century British naturalist Alfred Russel Wallace described its flesh as "a rich custard highly flavoured with almonds". The flesh can be consumed at various stages of ripeness, and it is used to flavour a wide variety of savoury and sweet edibles in Southeast Asian cuisines. The seeds can also be eaten when cooked.

There are hundreds of durian cultivars; many consumers express preferences for specific cultivars, which fetch higher prices in the market.

27.1 Taxonomy

For a complete list of known species of Durio, see List of Durio species.

The genus *Durio* is placed by some taxonomists in

Durian flowers are usually closed during the daytime.

the family Bombacaceae, or by others in a broadly defined Malvaceae that includes Bombacaceae, and

123

by others in a smaller family of just seven genera Durionaceae.*[3]*[6]*[7]

27.2 Description

Juvenile durian tree in Malaysia. Mature species can grow up to 50 metres (164 ft)

Durian trees are large, growing to 25–50 metres (80–164 ft) in height depending on the species.*[8] The leaves are evergreen, elliptic to oblong and 10–18 centimetres (4–7 in) long. The flowers are produced in three to thirty clusters together on large branches and directly on the trunk with each flower having a calyx (sepals) and five (rarely four or six) petals. Durian trees have one or two flowering and fruiting periods per year, although the timing varies depending on the species, cultivars, and localities. A typical durian tree can bear fruit after four or five years. The durian fruit can hang from any branch and matures roughly three months after pollination. The fruit can grow up to 30 centimetres (12 in) long and 15 centimetres (6 in) in diameter, and typically weighs one to three kilograms (2 to 7 lb).*[8] Its shape ranges from oblong to round, the colour of its husk green to brown, and its flesh pale-yellow to red, depending on the species.*[8] Among the thirty known species of *Durio*, nine of them have been identified as producing edible fruits: *D. zibethinus, D. dulcis, D. grandiflorus, D.*

graveolens, D. kutejensis, D. macrantha, D. oxleyanus, and *D. testudinarum.**[9] There are many species for which the fruit has never been collected or properly examined, however, so other species with edible fruit may exist.*[8] The durian is somewhat similar in appearance to the jackfruit, an unrelated species.

The name *durian* comes from the Malay word *duri* (thorn) together with the suffix *-an* (for building a noun in Malay).*[10]*[11] *D. zibethinus* is the only species commercially cultivated on a large scale and available outside of its native region. Since this species is open-pollinated, it shows considerable diversity in fruit colour and odour, size of flesh and seed, and tree phenology. In the species name, *zibethinus* refers to the Indian civet, *Viverra zibetha*. There is disagreement regarding whether this name, bestowed by Linnaeus, refers to civets being so fond of the durian that the fruit was used as bait to entrap them, or to the durian smelling like the civet.*[12]

Durian flowers are large and feathery with copious nectar, and give off a heavy, sour, and buttery odour. These features are typical of flowers pollinated by certain species of bats that eat nectar and pollen.*[13] According to research conducted in Malaysia in the 1970s, durians were pollinated almost exclusively by cave fruit bats (*Eonycteris spelaea*);*[8] however, a 1996 study indicated two species, *D. grandiflorus* and *D. oblongus*, were pollinated by spiderhunters (Nectariniidae) and another species, *D. kutejensis*, was pollinated by giant honey bees and birds as well as bats.*[14]

27.2.1 Cultivars

Different cultivars of durian often have distinct colours. D101 (right) has rich yellow flesh, clearly distinguishable from another variety (left).

Over the centuries, numerous durian cultivars, propagated by vegetative clones, have arisen in southeast Asia. They used to be grown with mixed results from seeds of trees bearing superior quality fruit, but now are propagated by layering, marcotting, or more commonly, by grafting, including bud, veneer, wedge, whip or U-grafting onto seedlings of randomly selected rootstocks. Different cultivars may be distinguished to some extent by variations in the fruit shape, such as the shape of the spines.[*][8] Durian consumers express preferences for specific cultivars, which fetch higher prices in the market.[*][15]

Most cultivars have a common name and a code number starting with "D". For example, some popular clones are Kop (D99 Thai: กบ ⬚ "frog" [kòp]), Chanee (D123, Thai: ชะนี ⬚ gibbon [tɕʰániː]), Berserah or Green Durian or Tuan Mek Hijau (D145 Thai: ทุเรียนเขียว ⬚ Green Durian [tʰúriːən kʰǐow]), Kan Yao (D158, Thai: ก้านยาว ⬚ Long Stem [kâːn jaːw]), Mon Thong (D159, Thai: หมอนทอง ⬚ Golden Pillow [mɔ̌ːn tʰɔːŋ]), Kradum Thong (Thai: กระ⬚ ดุมทอง ⬚ Golden Button [kràdum tʰɔːŋ]), and with no common name, D24 and D169. Each cultivar has a distinct taste and odour. More than 200 cultivars of *D. zibethinus* exist in Thailand.

Mon thong is the most commercially sought after for its thick, full-bodied creamy and mild sweet tasting flesh with relatively moderate smell emitted and smaller seeds, while Chanee is the best in terms of its resistance to infection by *Phytophthora palmivora*. Kan Yao is somewhat less common, but prized for its longer window of time when it is both sweet and odorless at the same time. Among all the cultivars in Thailand, five are currently in large-scale commercial cultivation: Chanee, Mon Thong, Kan Yao, Ruang, and Kradum.[*][16] There are more than 100 registered cultivars since 1920's in Malaysia[*][17] and up to 193 cultivar by 1992,[*][18] and many superior cultivars have been identified through competitions held at the annual Malaysian Agriculture, Horticulture, and Agrotourism Show. In Vietnam, the same process has been achieved through competitions held by the Southern Fruit Research Institute. A recently popular variety is, Cat Mountain King or Musang King.[*][19]

By 2007, Songpol Somsri, a Thai government scientist, had crossbred more than ninety varieties of durian to create Chantaburi No. 1, a cultivar without the characteristic odour.[*][20] Another hybrid, Chantaburi No. 3, develops the odour about three days after the fruit is picked, which enables an odourless transport yet satisfies consumers who prefer the pungent odour.[*][20] On 22 May 2012, two other cultivars from Thailand that also lack the usual odour, Long Laplae and Lin Laplae, were presented to the public by Yothin Samutkhiri, governor of Uttaradit province from where these cultivars were developed locally, while he announced the dates for the yearly durian fair of Laplae District, and the name giver to both cultivars.[*][21]

Popular cultivars in Malaysia and Singapore (Singapore im-

ports most of its durians from Malaysia hence the varieties are similar although there may be slight variation in the names) include "D24" which is a popular variety known for its bitter sweet taste; "XO" which has a pale color, thick flesh with a tinge of alcoholic fermentation; "Chook Kiok" (Cantonese meaning: bamboo leg) which has a distinctive yellowish core in the inner stem and "Musang King" (Musang is the Malay word for civet cat) which is usually the priciest of all cultivars. Musang King has bright yellow flesh and is almost like a more potent or enhanced version of the D24. This particular variety should be consumed last since it tends to make other durians taste bland in comparison.

27.3 Cultivation and availability

A durian stall in Malaysia

The durian is native to Brunei, Indonesia and Malaysia.[*][8] There is some debate as to whether the durian is native to the Philippines, particularly the Davao region in the island of Mindanao, or was introduced.[*][8] The durian is grown in other areas with a similar climate; it is strictly tropical and stops growing when mean daily temperatures drop below 22 °C (72 °F).[*][9]

The centre of ecological diversity for durians is the island of Borneo, where the fruit of the edible species of *Durio* including *D. zibethinus*, *D. dulcis*, *D. graveolens*, *D. kutejensis*, *D. oxleyanus* and *D. testudinarum* is sold in local markets. In Brunei, *D. zibethinus* is not grown because consumers prefer other species such as *D. graveolens*, *D. kutejensis* and *D. oxleyanus*. These species are commonly distributed in Brunei, and together with other species like *D. testudinarum* and *D. dulcis*, represent rich genetic diversity.[*][22]

Although the durian is not native to Thailand, the country is currently one of the major exporters of durians, growing 781,000 tonnes of the world's total harvest of

Durian on sale near Cirebon, Indonesia

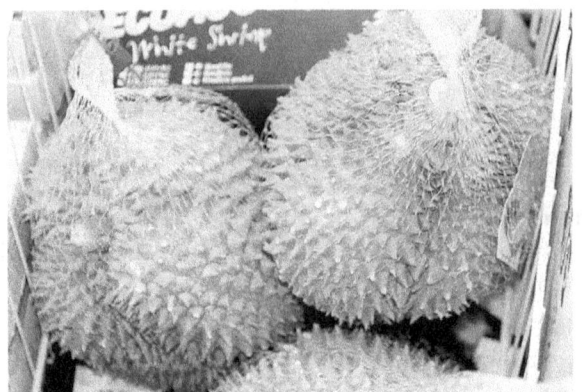

Durians being sold in mesh bags out of a freezer in a California market

ported 500 tonnes.*[23]

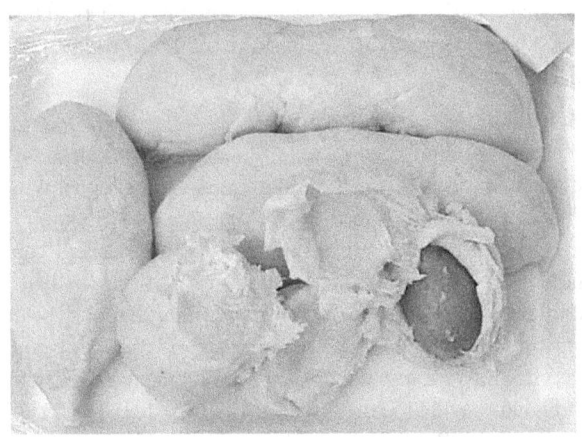

Durian flesh packed for sale, with an exposed seed

1,400,000 tonnes in 1999, 111,000 tonnes of which it exported to Taiwan, Hong Kong, Malaysia, Singapore and Canada.*[16]*[23] Malaysia and Indonesia follow, both producing about 265,000 tonnes each. Of this, Malaysia exported 35,000 tonnes in 1999.*[23] Chantaburi in Thailand each year holds the World Durian Festival in early May. This single province is responsible for half of the durian production of Thailand.*[24]*[25] In the Philippines, the centre of durian production is the Davao Region. The Kadayawan Festival is an annual celebration featuring the durian in Davao City. Other places where durian farms are located include Cambodia, Laos, Vietnam, Myanmar, Sri Lanka, India, the West Indies, Florida, Hawaii, Papua New Guinea, the Polynesian Islands, Madagascar, southern China (Hainan Island), northern Australia, and Singapore.

Durian was introduced into Australia in the early 1960s and clonal material was first introduced in 1975. Over thirty clones of *D. zibethinus* and six *Durio* species have been subsequently introduced into Australia.*[26] China is the major importer, purchasing 65,000 tonnes in 1999, followed by Singapore with 40,000 tonnes and Taiwan with 5,000 tonnes. In the same year, the United States imported 2,000 tonnes, mostly frozen, and the European Community im-

The durian is a seasonal fruit, unlike some other non-seasonal tropical fruits such as the papaya, which are available throughout the year. In Peninsular Malaysia and Singapore, the season for durians is typically from June to August, which coincides with that of the mangosteen.*[8] Prices of durians are relatively high as compared with other fruits. For example, in Singapore, the strong demand for high quality cultivars such as the D24, *Sultan*, and *Mao Shan Wang* has resulted in typical retail prices of between S$8 to S$15 (US$5 to US$10) per kilogram of whole fruit.*[15] With an average weight of about 1.5 kilograms (3.3 lb), a durian fruit would therefore cost about S$12 to S$22 (US$8 to US$15).*[15] The edible portion of the fruit, known as the aril and usually referred to as the "flesh" or "pulp", only accounts for about 15–30% of the mass of the entire fruit.*[27] Many consumers in Singapore are nevertheless quite willing to spend up to around S$75 (US$50) on a single purchase of about half a dozen of the favoured fruit to be shared by family members.*[15]

In-season durians can be found in mainstream Japanese supermarkets, while in the West they are sold mainly by Asian markets.

27.4 Flavour and odour

Sign forbidding durians on Singapore's Mass Rapid Transit

The unusual flavour and odour of the fruit have prompted many people to express diverse and passionate views ranging from deep appreciation to intense disgust. Writing in 1856, the British naturalist Alfred Russel Wallace provided a much-quoted description of the flavour of the durian:

> The five cells are silky-white within, and are filled with a mass of firm, cream-coloured pulp, containing about three seeds each. This pulp is the edible part, and its consistence and flavour are indescribable. A rich custard highly flavoured with almonds gives the best general idea of it, but there are occasional wafts of flavour that call to mind cream-cheese, onion-sauce, sherry-wine, and other incongruous dishes. Then there is a rich glutinous smoothness in the pulp which nothing else possesses, but which adds to its delicacy. It is neither acidic nor sweet nor juicy; yet it wants neither of these qualities, for it is in itself perfect. It produces no nausea or other bad effect, and the

more you eat of it the less you feel inclined to stop. In fact, to eat Durians is a new sensation worth a voyage to the East to experience. ... as producing a food of the most exquisite flavour it is unsurpassed.[28][a]

Wallace described himself as being at first reluctant to try it because of the aroma, "but in Borneo I found a ripe fruit on the ground, and, eating it out of doors, I at once became a confirmed Durian eater." [29] He cited one traveller from 1599:[b] "it is of such an excellent taste that it surpasses in flavour all other fruits of the world, according to those who have tasted it." [29] He cites another writer: "To those not used to it, it seems at first to smell like rotten onions, but immediately after they have tasted it they prefer it to all other food. The natives give it honourable titles, exalt it, and make verses on it." [29] Despite having tried many foods that are arguably more eccentric, Andrew Zimmern, host of Bizarre Foods, was unable to finish a durian upon sampling it, due to his intolerance of its strong taste.

While Wallace cautions that "the smell of the ripe fruit is certainly at first disagreeable", later descriptions by westerners are more graphic. Novelist Anthony Burgess writes that eating durian is "like eating sweet raspberry blancmange in the lavatory".[30] Chef Andrew Zimmern compares the taste to "completely rotten, mushy onions".[31] Anthony Bourdain, a lover of durian, relates his encounter with the fruit thus: "Its taste can only be described as...indescribable, something you will either love or despise. ...Your breath will smell as if you'd been French-kissing your dead grandmother."[32] Likewise, fellow chef Jamie Oliver has also expressed admiration for the fruit on his first sampling.[33] Travel and food writer Richard Sterling says:

> ... its odor is best described as pig-shit, turpentine and onions, garnished with a gym sock. It can be smelled from yards away. Despite its great local popularity, the raw fruit is forbidden from some establishments such as hotels, subways and airports, including public transportation in Southeast Asia.[34]

Other comparisons have been made with the civet, sewage, stale vomit, skunk spray and used surgical swabs.[35] The wide range of descriptions for the odour of durian may have a great deal to do with the variability of durian odour itself. Durians from different species or clones can have significantly different aromas; for example, red durian (*D. dulcis*) has a deep caramel flavour with a turpentine odour while red-fleshed durian (*D. graveolens*) emits a fragrance of roasted almonds.[36] Among the varieties of *D. zibethinus*, Thai varieties are sweeter in flavour and less odorous than Malay ones.[8] The degree of ripeness has an effect on the flavour as well.[8] Three scientific analyses of the composition of durian aroma – from 1972,[37] 1980, and 1995 – each found a mix of volatile compounds including esters, ketones, and different sulphur compounds, with no agreement on which may be primarily responsible for the distinctive odour.[8] People in South East Asia with frequent exposures to durian are able to easily distinguish its sweet-like ketones and esters scent from rotten or putrescine odours which are from volatile amines and fatty acids. Developmental or genetic differences in olfactory perception and mapping within the brain (for e.g. anterior piriform cortex to the orbitofrontal cortex) could possibly explain why some individuals are unable to differentiate these smells and find this fruit noxious.

This strong odour can be detected half a mile away by animals, thus luring them. In addition, the fruit is extremely appetising to a variety of animals, including squirrels, mouse deer, pigs, orangutan, elephants, and even carnivorous tigers.[38][39] While some of these animals eat the fruit and dispose of the seed under the parent plant, others swallow the seed with the fruit and then transport it some distance before excreting, with the seed being dispersed as a result.[40] The thorny, armoured covering of the fruit discourages smaller animals; larger animals are more likely to transport the seeds far from the parent tree.[41]

A customer sniffs a durian before purchasing it.

27.4.1   Ripeness and selection

According to *Larousse Gastronomique*, the durian fruit is ready to eat when its husk begins to crack.[42] However, the ideal stage of ripeness to be enjoyed varies from region to region in Southeast Asia and by species. Some species grow so tall that they can only be collected once they have fallen to the ground, whereas most cultivars of *D. zibethinus* are nearly always cut from the tree and allowed to ripen while waiting to be sold. Some people in southern Thailand prefer their durians relatively young when the clusters of fruit within the shell are still crisp in texture and mild in flavour. For some people in northern Thailand, the preference is for the fruit to be soft and aromatic. In Malaysia and Singapore, most consumers prefer the fruit to be as ripe and pungent in aroma as possible and may even risk allowing the fruit to continue ripening after its husk has already cracked open. In this state, the flesh becomes richly creamy, slightly alcoholic,[35] the aroma pronounced and the flavour highly complex.

The various preferences regarding ripeness among consumers make it hard to issue general statements about choosing a "good" durian. A durian that falls off the tree continues to ripen for two to four days, but after five or six days most would consider it overripe and unpalatable.[43]

The usual advice for a durian consumer choosing a whole fruit in the market is to examine the quality of the stem or stalk which loses moisture as it ages: a big, solid stem is a sign of freshness.[44] Reportedly, unscrupulous merchants wrap, paint, or remove the stalks altogether. Due to the popularity of Kan Yao, street vendors may sometimes sell a lesser variety with a long stem to unsuspecting customers. Another frequent piece of advice is to shake the fruit and listen for the sound of the seeds moving within, indicating the durian is very ripe and the pulp has dried out a bit.[44]

27.5   History

The durian has been known and consumed in Southeast Asia since prehistoric times, but has only been known to the western world for about 600 years. The earliest native reference to durian is the several bas relief panels of 9th-century Borobudur depicting durian as one of fruit offering for Javanese king, and also as one of the fruits sold in marketplace.[45]

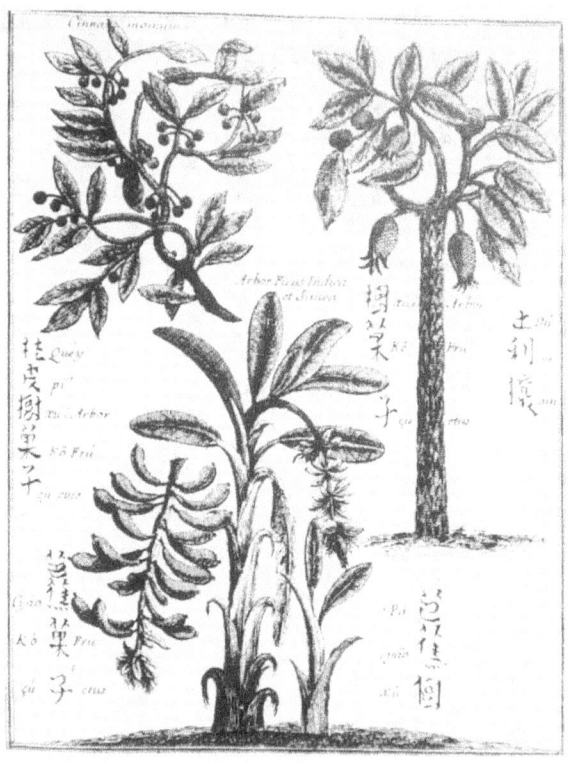

Michał Boym, a Jesuit missionary to China, provided one of the early (1655) reports on durian (upper right) to European scholars.

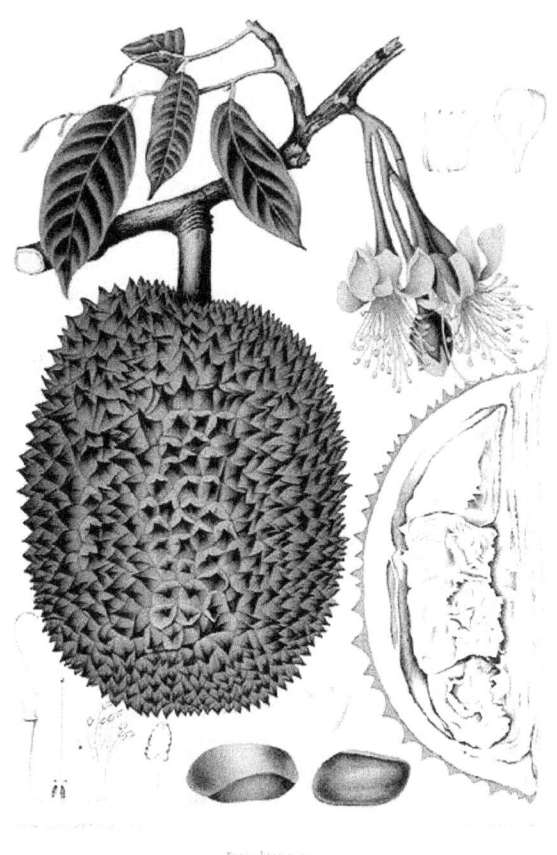

Durio zibethinus. *Chromolithograph by Hoola Van Nooten, circa 1863*

The earliest known European reference to the durian is the record of Niccolò Da Conti, who travelled to southeastern Asia in the 15th century.*[46] Translated from the Latin in which Poggio Bracciolini recorded Da Conti's travels: *"They* (people of Sumatra) *have a green fruit which they call durian, as big as a watermelon. Inside there are five things like elongated oranges, and resembling thick butter, with a combination of flavours."* *[47] The Portuguese physician Garcia de Orta described durians in *Colóquios dos simples e drogas da India* published in 1563. In 1741, *Herbarium Amboinense* by the German botanist Georg Eberhard Rumphius was published, providing the most detailed and accurate account of durians for over a century. The genus *Durio* has a complex taxonomy that has seen the subtraction and addition of many species since it was created by Rumphius.*[9] During the early stages of its taxonomical study, there was some confusion between durian and the soursop (*Annona muricata*), for both of these species had thorny green fruit.*[48] It is also interesting to note the Malay name for the soursop is *durian Belanda*, meaning *Dutch durian*.*[49] In the 18th century, Johann Anton Weinmann considered the durian to belong to Castaneae as its fruit was similar to the horse chestnut.*[48]

D. zibethinus was introduced into Ceylon by the Portuguese in the 16th century and was reintroduced many times

later. It has been planted in the Americas but confined to botanical gardens. The first seedlings were sent from the Royal Botanic Gardens, Kew, to Auguste Saint-Arroman of Dominica in 1884.*[50]

In southeastern Asia, the durian has been cultivated for centuries at the village level, probably since the late 18th century, and commercially since the mid-20th century.*[8] In *My Tropic Isle*, Australian author and naturalist Edmund James Banfield tells how, in the early 20th century, a friend in Singapore sent him a durian seed, which he planted and cared for on his tropical island off the north coast of Queensland.*[51]

In 1949, the British botanist E. J. H. Corner published *The Durian Theory, or the Origin of the Modern Tree*. His theory was that endozoochory (the enticement of animals to transport seeds in their stomach) arose before any other method of seed dispersal, and that primitive ancestors of *Durio* species were the earliest practitioners of that dispersal method, in particular red durian (*D. dulcis*) exemplifying the primitive fruit of flowering plants.

Since the early 1990s, the domestic and international demand for durian in the Association of Southeast Asian Nations (ASEAN) region has increased significantly, partly due to the increasing affluence of Asia.[*][8]

27.6 Uses

27.6.1 Culinary

Durian fruit is used to flavour a wide variety of sweet edibles such as traditional Malay candy, *ice kacang*, *dodol*, *lempuk*,[*][52] rose biscuits, and, with a touch of modern innovation, ice cream, milkshakes, mooncakes, Yule logs, and cappuccino. *Es durian* (durian ice cream) is a popular dessert in Indonesia, sold at street side stall in Indonesian cities, especially in Java. *Pulut Durian* or *ketan durian* is glutinous rice steamed with coconut milk and served with ripened durian. In Sabah, red durian is fried with onions and chilli and served as a side dish.[*][53] Red-fleshed durian is traditionally added to *sayur*, an Indonesian soup made from freshwater fish.[*][5] *Ikan brengkes* is fish cooked in a durian-based sauce, traditional in Sumatra.[*][54] Traditionally Bollen pastry, specialty of Bandung is filled with banana and cheese. Today Bollen durian is also available, it is pastry filled with durian. Dried durian flesh can be made into kripik durian (durian chips).

Tempoyak refers to fermented durian, usually made from lower quality durian that is unsuitable for direct consumption.[*][55] Tempoyak can be eaten either cooked or uncooked, is normally eaten with rice, and can also be used for making curry. Sambal Tempoyak is a Sumatran dish made from the fermented durian fruit, coconut milk, and a collection of spicy ingredients known as sambal.

In Thailand, durian is often eaten fresh with sweet sticky rice, and blocks of durian paste are sold in the markets, though much of the paste is adulterated with pumpkin.[*][43] Unripe durians may be cooked as a vegetable, except in the Philippines, where all uses are sweet rather than savoury. Malaysians make both sugared and salted preserves from durian. When durian is minced with salt, onions and vinegar, it is called *boder*. The durian seeds, which are the size of chestnuts, can be eaten whether they are boiled, roasted or fried in coconut oil, with a texture that is similar to taro or yam, but stickier. In Java, the seeds are sliced thin and cooked with sugar as a confection. Uncooked durian seeds are toxic due to cyclopropene fatty acids and should not be ingested.[*][56]

Young leaves and shoots of the durian are occasionally cooked as greens. Sometimes the ash of the burned rind is added to special cakes.[*][43] The petals of durian flowers are eaten in the North Sumatra province of Indonesia, while in the Moluccas islands the husk of the durian fruit is used as fuel to smoke fish. The nectar and pollen of the durian flower that honeybees collect is an important honey source, but the characteristics of the honey are unknown.[*][57]

- *Tempoyak*, made from fermented durian

- *Ketan durian*, glutinous rice with durian sauce

- Durian cake made of durian-flavoured *dodol*

- *Keripik durian* (durian chips)

- A street side durian ice cream in Bogor

- Durian gelato in Singapore

- Durian-flavoured Yule log

- *Durian Keju Bollen*, a pastry filled with cheese and durian cream

- Durian pancake

- Durian crêpe in Malaysia

- The vendors in Davao Region Philippines.

27.6.2 Nutritions and folk medicine

Durian fruit contains a high amount of sugar,[*][41] vitamin C, potassium, and the serotonergic amino acid tryptophan,[*][59] and is a good source of carbohydrates, proteins, and fats.[*][5][*][50] It is recommended as a good source of raw fats by several raw food advocates,[*][60][*][61] while others classify it as a high-glycemic food, recommending to minimise its consumption.[*][62][*][63]

In Malaysia, a decoction of the leaves and roots used to be prescribed as an antipyretic. The leaf juice is applied on the head of a fever patient.[*][43] The most complete description of the medicinal use of the durian as remedies for fevers is a Malay prescription, collected by Burkill and Haniff in 1930. It instructs the reader to boil the roots of *Hibiscus rosa-sinensis* with the roots of *Durio zibethinus*, *Nephelium longan*, *Nephelium mutabile* and *Artocarpus integrifolia*, and drink the decoction or use it as a poultice.[*][64]

In the 1920s, Durian Fruit Products, Inc., of New York City launched a product called "Dur-India" as a health food supplement, selling at US$9 for a dozen bottles, each containing 63 tablets. The tablets allegedly contained durian and a species of the genus *Allium* from India and vitamin E. The company promoted the supplement saying that it provides "more concentrated healthful energy in food form than any other product the world affords".[*][43]

27.7 Customs and beliefs

Southeast Asian traditional beliefs, as well as traditional Chinese medicine, consider the durian fruit to have warming properties liable to cause excessive sweating.[65] The traditional method to counteract this is to pour water into the empty shell of the fruit after the pulp has been consumed and drink it.[35] An alternative method is to eat the durian in accompaniment with mangosteen, which is considered to have cooling properties. Pregnant women or people with high blood pressure are traditionally advised not to consume durian.[20][66]

Another common local belief is that the durian is harmful when eaten with coffee[35] or alcoholic beverages.[8] The latter belief can be traced back at least to the 18th century when Rumphius stated that one should not drink alcohol after eating durians as it will cause indigestion and bad breath. In 1929, J. D. Gimlette wrote in his *Malay Poisons and Charm Cures* that the durian fruit must not be eaten with brandy. In 1981, J. R. Croft wrote in his *Bombacaceae: In Handbooks of the Flora of Papua New Guinea* that "a feeling of morbidity" often follows the consumption of alcohol too soon after eating durian. Several medical investigations on the validity of this belief have been conducted with varying conclusions,[8] though a study by the University of Tsukuba finds the fruit's high sulphur content inhibits the activity of aldehyde dehydrogenase, causing a 70% reduction of the ability to clear toxins from the body.[67]

The Javanese believe durian to have aphrodisiac qualities, and impose a set of rules on what may or may not be consumed with it or shortly thereafter.[35] A saying in Indonesian, *durian jatuh sarung naik*, meaning "the durian falls and the sarong comes up", refers to this belief.[68] The warnings against the supposed lecherous quality of this fruit soon spread to the West – the Swedenborgian philosopher Herman Vetterling commented on so-called "erotic properties" of the durian in the early 20th century.[69]

Durian fruit is armed with sharp thorns, capable of drawing blood.

A durian falling on a person's head can cause serious injuries because it is heavy, armed with sharp thorns, and can fall from a significant height. Wearing a hardhat is recommended when collecting the fruit. Alfred Russel Wallace writes that death rarely ensues from it, because the copious effusion of blood prevents the inflammation which might otherwise take place.[28] A common saying is that a durian has eyes and can see where it is falling because the fruit allegedly never falls during daylight hours when people may be hurt.[70] A saying in Indonesian, *ketiban durian runtuh*, which translates to "getting a durian avalanche", is the equivalent of the English phrase "windfall gain".[71] Nevertheless, signs warning people not to linger under durian trees are found in Indonesia.[72] Strong nylon or woven rope nettings are often strung between durian trees in orchards, serving a threefold purpose: the nets aid in the collection of the mature fruits, deter ground-level scavengers, and most importantly, prevent the durians from falling onto people.

A naturally spineless variety of durian growing wild in Davao, Philippines, was discovered in the 1960s; fruits borne from these seeds also lacked spines.[8] Since the bases of the scales develop into spines as the fruit matures, sometimes spineless durians are produced artificially by scraping scales off immature fruits.[8] In Malaysia, a spineless durian clone D172 is registered by Agriculture Department on 17 June 1989. It was called "Durian Botak" (Bald Durian).[18] In Indonesia, Ir Sumeru Ashari, head of Durian Research Centre, Universitas Brawijaya reported spineless durian from Kasembon, Malang. Another cultivar is from Lombok, Nusa Tenggara Barat, Indonesia.[73][74]

Animals such as Sumatran elephants are known to consume durians. Curiously, the carnivorous Sumatran tiger is also known to consume durian occasionally.[75] The strong odour of the fallen fruits in the jungle probably attracts the tiger to inspect the fruit and lick it.[39]

27.8 Cultural influence

The durian is commonly known as the "King of the Fruits",[5] a label that can be attributed to its formidable look and overpowering odour.[76] In its native southeastern Asia, the durian is an everyday food and portrayed in the local media in accordance with the cultural perception it has in the region. The durian symbolised the subjective nature of ugliness and beauty in Hong Kong director Fruit Chan's 2000 film *Durian Durian* (榴 飄飄, *lau lin piu piu*), and was a nickname for the reckless but lovable protagonist of the eponymous Singaporean TV comedy *Durian King* played by Adrian Pang.[77] Likewise, the oddly shaped Esplanade building in Singapore is often called "The Durian" by lo-

Singapore's Esplanade building, nicknamed "The Durian"

cals,[77] and "The Big Durian" is the nickname of Jakarta, Indonesia.[78]

One of the names Thailand contributed to the list of storm names for Western North Pacific tropical cyclones was 'Durian',[79] which was retired after the second storm of this name in 2006. Being a fruit much loved by a variety of wild beasts, the durian sometimes signifies the long-forgotten animalistic aspect of humans, as in the legend of Orang Mawas, the Malaysian version of Bigfoot, and Orang Pendek, its Sumatran version, both of which have been claimed to feast on durians.[80][81]

Frozen whole durians are shipped from Thailand to Asian markets and Chinatowns in Western countries.

27.9   See also

- Breadfruit

- Jackfruit

- Limburger

- List of delicacies

27.10   Notes

a. [*][^] Wallace makes an almost identical comment in his 1866 publication *The Malay Archipelago: The land of the orang-utang and the bird of paradise.*[82]

b. [*][^] The traveller Wallace cites is Linschott (Wallace's spelling for Jan Huyghen van Linschoten), whose name appears repeatedly in Internet searches on durian, with such

citations themselves tracing back to Wallace. In translations of Linschoten's writings, the fruit is spelled as *duryoen.*[83]

27.11   References

[1] *Durio zibethinus* at **worldagroforestry.org**

[2] *A traveler's guide to Durian Season* at **yearofthedurian.com**

[3] "*Durio* L." . *Germplasm Resources Information Network.* United States Department of Agriculture. 2007-03-12. Retrieved 2010-02-16.

[4] Pronunciation common to *Oxford English Dictionary* (2 ed.). Oxford University Press. and "Random House Dictionary" . dictionary.com. 2008-10-09.

[5] Heaton, Donald D. (2006). *A Consumers Guide on World Fruit.* BookSurge Publishing. pp. 54–56. ISBN 1-4196-3955-2.

[6] "USDA GRIN Taxonomy, Durionaceae" . Retrieved 2014-06-22.

[7] "Angiosperm Phylogeny Website – Malvales" . Missouri Botanical Garden.

[8] Brown, Michael J. (1997). *Durio – A Bibliographic Review.* International Plant Genetic Resources Institute (IPGRI). ISBN 92-9043-318-3. Retrieved 2008-11-20.

[9] O'Gara, E., Guest, D. I. and Hassan, N. M. (2004). "Botany and Production of Durian (*Durio zibethinus*) in Southeast Asia" . In Drenth, A. and Guest, D. I. *Diversity and management of* Phytophthora *in Southeast Asia. ACIAR Monograph No. 114* (PDF). Australian Centre for International Agricultural Research (ACIAR). pp. 180–186. ISBN 1-86320-405-9. Retrieved 2008-11-20.

[10] *Oxford English Dictionary.* Oxford University Press. 1897. Via *durion*, the Indonesia name for the plant.

[11] Huxley, A. (Ed.) (1992). *New RHS Dictionary of Gardening.* Macmillan. ISBN 1-56159-001-0.

[12] Brown, Michael J. (1997). *Durio – A Bibliographic Review.* International Plant Genetic Resources Institute (IPGRI). p. 2. ISBN 92-9043-318-3. Retrieved 2012-09-04. See also pp. 5–6 regarding whether Linnaeus or Murray is the correct authority for the binomial name

[13] Whitten, Tony (2001). *The Ecology of Sumatra.* Periplus. p. 329. ISBN 962-593-074-4.

[14] Yumoto, Takakazu (2000). "Bird-pollination of Three *Durio* Species (Bombacaceae) in a Tropical Rainforest in Sarawak, Malaysia" . *American Journal of Botany* (American Journal of Botany, Vol. 87, No. 8) **87** (8): 1181–1188. doi:10.2307/2656655. JSTOR 2656655. PMID 10948003.

[15] "ST Foodies Club – Durian King". The Straits Times. 2006. Archived from the original on 2007-12-15. Retrieved 2007-07-25.

[16] "Durian Exporting Strategy, National Durian Database (กลยุทธการส่งออกทุเรียน)" (in Thai). Department of Agriculture, Thailand. Archived from the original on 2010-07-26. Retrieved 2010-07-26.

[17] "Comprehensive List of Durian Clones Registered by the Agriculture Department (of Malaysia)". Durian OnLine. Archived from the original on 2008-11-20. Retrieved 2006-03-05.

[18] Boosting Durian Productivity

[19] Teo, Wan Gek (2009-06-23). "Durian lovers head north on day tours". *The Straits Times*. Retrieved 2009-09-19.

[20] Fuller, Thomas (2007-04-08). "Fans Sour on Sweeter Version of Asia's Smelliest Fruit". *New York Times*. Retrieved 2008-11-20.

[21] "Odourless durians to hit the market". *The Nation*. 23 May 2012.

[22] M.B. Osman, Z.A. Mohamed, S. Idris and R. Aman (1995). *Tropical fruit production and genetic resources in Southeast Asia: Identifying the priority fruit species*. International Plant Genetic Resources Institute (IPGRI). ISBN 92-9043-249-7. Retrieved 2008-11-20.

[23] "Committee on Commodity Problems – VI. Overview of Minor Tropical Fruits". FAO. December 2001. Retrieved 2008-11-20.

[24] "World Durian Festival 2005". *Thailand News – Thailand official news and information*. Foreign Office, The Government Public Relations Department. 2005-06-05. Retrieved 2008-11-20.

[25] "Thailand's Durian growing areas". *Food Market Exchange*. 2003. Retrieved 2008-11-20.

[26] Watson, B. J (1983). "Durian". *Fact Sheet No. 6.: Rare Fruits Council of Australia*.

[27] Brown, Michael J. (1997). *Durio – A Bibliographic Review* (PDF). International Plant Genetic Resources Institute (IPGRI). p. 35. ISBN 92-9043-318-3. Retrieved 2011-06-12.

[28] Wallace, Alfred Russel (1856). "On the Bamboo and Durian of Borneo". Retrieved 2008-11-20.

[29] Wallace, Alfred Russel (1886). "The Malay Archipeligo: The land of the orang-utang and the bird of paradise". London: Macmillan & Co. p. 74. Retrieved 2010-06-04.

[30] Burgess, Anthony (1993) [1956]. *The Long Day Wanes: A Malayan Trilogy*. W. W. Norton & Company. p. 68. ISBN 0-393-30943-6.

[31] "Bizarre Foods: Asia". *Bizarre Foods with Andrew Zimmern*. Season 1. Episode 0 (Pilot). 2006-11-01. Travel Channel. Video from YouTube. Retrieved on 2008-03-22

[32] "Anthony Bourdain tries out durian in Indonesia". *Anthony Bourdain: No Reservations*. Season 2. Episode 12. 2006-06-19. Travel Channel. Video from YouTube. Retrieved on 2008-06-29

[33] Jamie Oliver Tries Durian

[34] Winokur, Jon (Ed.) (2003). *The Traveling Curmudgeon: Irreverent Notes, Quotes, and Anecdotes on Dismal Destinations, Excess Baggage, the Full Upright Position, and Other Reasons Not to Go There*. Sasquatch Books. p. 102. ISBN 1-57061-389-3.

[35] Davidson, Alan (1999). *The Oxford Companion to Food*. Oxford University Press. p. 263. ISBN 0-19-211579-0.

[36] O'Gara, E., Guest, D. I. and Hassan, N. M. (2004). "Occurrence, Distribution and Utilisation of Durian Germplasm". In Drenth, A. and Guest, D. I. *Diversity and Management of* Phytophthora *in Southeast Asia ACIAR Monograph No. 114* (PDF). Australian Centre for International Agricultural Research (ACIAR). pp. 187–193. ISBN 1-86320-405-9. Retrieved 2008-11-20.

[37] PII: S0031-9422(00)90176-6

[38] "In praise of the delectable durian". *Telegraph.co.uk*. 18 October 2004.

[39] "Sumatran tiger inspects durian fruit on forest floor". *BBC*. ARKive. Retrieved 2012-07-02.

[40] Marinelli, Janet (Ed.) (1998). *Brooklyn Botanic Garden Gardener's Desk Reference*. Henry Holt and Co. p. 691. ISBN 0-8050-5095-7.

[41] McGee, Harold (2004). *On Food and Cooking (Revised Edition)*. Scribner. p. 379. ISBN 0-684-80001-2.

[42] Montagne, Prosper (Ed.) (2001). *Larousse Gastronomique*. Clarkson Potter. p. 439. ISBN 0-609-60971-8.

[43] Morton, J. F. (1987). *Fruits of Warm Climates*. Florida Flair Books. ISBN 0-9610184-1-0.

[44] "Durian & Mangosteens". Prositech.com. Retrieved 2008-11-20.

[45] Akhyari Hananto. "Riwayat Durian di Nusantara" (in Indonesian). Good News From Indonesia. Retrieved 28 June 2015.

[46] Brown, Michael J. (1997). *Durio – A Bibliographic Review* (PDF). International Plant Genetic Resources Institute (IPGRI). p. 3. ISBN 92-9043-318-3. Retrieved 2011-06-12.

[47] "Hobson-Jobson". *google.com*.

[48] Brown, Michael J. (1997). *Durio – A Bibliographic Review* (PDF). International Plant Genetic Resources Institute (IPGRI). p. 6. ISBN 92-9043-318-3. Retrieved 2011-06-12.

[49] Davidson, Alan (1999). *The Oxford Companion to Food.* Oxford University Press. p. 737. ISBN 0-19-211579-0.

[50] "Agroforestry Tree Database – *Durio zibethinus*". International Center for Research in Agroforestry. Retrieved 2008-11-20.

[51] Banfield, E. J., (1911). *My Tropic Isle.* T. Fisher Unwin. Retrieved 2008-11-20.

[52] Mardi – Lempuk technology.

[53] "Traditional Cuisine". Sabah Tourism Promotion Corporation. Retrieved 2008-11-20.

[54] Vaisutis, Justine; Neal Bedford; Mark Elliott; Nick Ray; Ryan Ver Berkmoes (2007). *Indonesia (Lonely Planet Travel Guides).* Lonely Planet Publications. p. 83. ISBN 1-74104-435-9.

[55] "Durian Recipe Gallery". Durian Online. Archived from the original on 2007-08-22. Retrieved 2008-11-20.

[56] "Question No. 18085: Is it true that durian seeds are poisonous?". Singapore Science Centre. 2006. Archived from the original on 2008-02-20. Retrieved 2008-11-20.

[57] Crane, E. (Ed.) (1976). *Honey: A Comprehensive Survey.* Bee Research Association. ISBN 0-434-90270-5.

[58] "USDA National Nutrient Database". U.S. Department of Agriculture. Retrieved 2013-02-22.

[59] Wolfe, David (2002). *Eating For Beauty.* Maul Brothers Publishing. ISBN 0-9653533-7-0.

[60] Boutenko, Victoria (2001). *12 Steps to Raw Foods: How to End Your Addiction to Cooked Food.* Raw Family. p. 6. ISBN 0-9704819-3-4.

[61] Mars, Brigitte (2004). *Rawsome!: Maximizing Health, Energy, and Culinary Delight With the Raw Foods Diet.* Basic Health Publications. p. 103. ISBN 1-59120-060-1.

[62] Cousens, Gabriel (2003). *Rainbow Green Live-Food Cuisine.* North Atlantic Books. p. 34. ISBN 1-55643-465-0.

[63] Klein, David (2005). "Vegan Healing Diet Guidelines". *Self Healing Colitis & Crohn's.* Living Nutrition Publications. ISBN 0-9717526-1-3.

[64] Burkill, I.H. and Haniff, M. (1930). "Malay village medicine, prescriptions collected". *Gardens Bulletin Straits Settlements* (6): 176–177.

[65] Huang, Kee C. (1998). *The Pharmacology of Chinese Herbs (Second Edition).* CRC Press. p. 2. ISBN 0-8493-1665-0.

[66] McElroy, Anne and Townsend, Patricia K. (2003). *Medical Anthropology in Ecological Perspective.* Westview Press. p. 253. ISBN 0-8133-3821-2.

[67] "Durians and booze: worse than a stinking hangover". *New Scientist.* 2009-09-16. Retrieved 2009-10-15.

[68] Stevens, Alan M. (2000). Schmidgall-Tellings, A., ed. *A Comprehensive Indonesian-English Dictionary.* Ohio University Press. p. 255. ISBN 0-8214-1584-0.

[69] Vetterling, Herman (2003) [1923]. *Illuminate of Gorlitz or Jakob Bohme's Life and Philosophy, Part 3.* Kessinger Publishing. p. 1380. ISBN 0-7661-4788-6.

[70] Solomon, Charmaine (1998). "Encyclopedia of Asian Food". Periplus. Retrieved 2008-11-20.

[71] Echols, John M.; Hassan Shadily (1989). *An Indonesian-English Dictionary.* Cornell University Press. p. 292. ISBN 0-8014-2127-6.

[72] Vaisutis, Justine; Neal Bedford; Mark Elliott; Nick Ray; Ryan Ver Berkmoes (2007). *Indonesia (Lonely Planet Travel Guides).* Lonely Planet Publications. pp. 393–394. ISBN 1-74104-435-9.

[73] "Multiply.com". *multiply.com.*

[74] "! BUJANG SUSAH !: DURIAN BOTAK". *rubbertapperz.blogspot.com.*

[75] Bogor Zoological Museum

[76] The mangosteen, called the "queen of fruits", is petite and mild in comparison. The mangosteen season coincides with that of the durian and is seen as a complement, which is probably how the mangosteen received the complementary title.

[77] "Uniquely Singapore – July 2006 Issue". Singapore Tourism Board. 2006. Archived from the original on 2007-08-23. Retrieved 2007-07-31.

[78] Suryodiningrat, Meidyatama (2007-06-22). "Jakarta: A city we learn to love but never to like". *The Jakarta Post.* Archived from the original on 2008-02-21.

[79] "Tropical Cyclone Names". Japan Meteorological Agency. Retrieved 2007-03-10.

[80] Lian, Hah Foong (2000-01-02). "Village abuzz over sighting of 'mawas'". Star Publications, Malaysia. Retrieved 2008-11-20.

[81] "Do 'orang pendek' really exist?". Jambiexplorer.com. Archived from the original on 2008-01-22. Retrieved 2006-03-19.

[82] "The Malay Archipelago: The Land of the Orang-utan, and the Bird of Paradise : a Narrative of ...". *archive.org.*

[83] Burnell, Arthur Coke & Tiele, P.A. (1885). "The voyage of John Huyghen van Linschoten to the East Indies". from the old English translation of 1598: the first book, containing his description of the East. London: The Hakluyt Society. p. 51 (n72 in electronic page field).. Full text at Internet Archive

27.12   External links

- Germplasm Resources Information Network: *Durio*

- Durio zibethinus (Bombacaceae)

- the controversial durian

- The durian: stinky fruit, killing fruit

- Nutritional value of durian

- Year of the Durian: a durian lover's travelogue and guide.

- A Durian How To.

- Media related to Durian at Wikimedia Commons

Chapter 28

Durio zibethinus

Durio zibethinus, is the most commonly available of those tree species in the genus *Durio* that are known as **durian**, and have edible fruit also known as durian.

As with other durian species, the edible flesh emits a distinctive odour that is strong and penetrating even when the husk is intact. Some people regard the durian as having a pleasantly sweet fragrance; others find the aroma overpowering and revolting. The smell evokes reactions from deep appreciation to intense disgust, and has been described variously as rotten onions, turpentine, and raw sewage. The persistence of its odour has led to the fruit's banishment from certain hotels and public transportation in Southeast Asia.

There are 30 recognised *Durio* species, at least nine of which produce edible fruit. *Durio zibethinus* is the only species available in the international market: other species are sold in their local regions. There are hundreds of cultivars of *Durio zibethinus*; many consumers express preferences for specific cultivars, which fetch higher prices in the market.

28.1 Description

The wood of *D. zibethinus* is reddish brown.[*][4]

28.2 Ecology

D. zibethinus flowers are visited by bats, which eat the pollen and pollinate the flowers.[*][4] The flowers open in the afternoon and shed pollen in the evening. By the following morning, the calyx, petals, and stamens have fallen off to leave only the gynoecium of the flower.[*][4]

Different cultivars of durian often have distinct colours. D101 (right) has rich yellow flesh, clearly distinguishable from another variety (left).

28.3 Cultivars

Over the centuries, numerous durian cultivars, propagated by vegetative clones, have arisen in southeast Asia. They used to be grown with mixed results from seeds of trees bearing superior quality fruit, but now are propagated by layering, marcotting, or more commonly, by grafting, including bud, veneer, wedge, whip or U-grafting onto seedlings of randomly selected rootstocks. Different cultivars may be distinguished to some extent by variations in the fruit shape, such as the shape of the spines.[*][4] Durian consumers express preferences for specific cultivars, which fetch higher prices in the market.[*][5]

Most cultivars have a common name and a code number starting with "D". For example, some popular clones are Kop (D99 Thai: กบ ⬚ "frog" [kòp]), Chanee (D123, Thai: ชะนี ⬚ gibbon [tɕʰáni:]), Berserah or Green Durian or Tuan Mek Hijau (D145 Thai: ทุเรียนเขียว ⬚ Green Durian [tʰúri:ən kʰǐow]), Kan Yao (D158, Thai: ก้านยาว ⬚ Long Stem [kâ:n ja:w]), Mon Thong (D159, Thai: หมอนทอง ⬚ Golden Pillow [mɔ̌:n tʰɔ:ŋ]), Kradum Thong (Thai: กระ⬚

ดุมทอง ⬚ Golden Button [kràdum tʰɔːŋ]), and with no common name, D24 and D169. Each cultivar has a distinct taste and odour. More than 200 cultivars of *D. zibethinus* exist in Thailand.

Mon thong is the most commercially sought after for its thick, full-bodied creamy and mild sweet tasting flesh with relatively moderate smell emitted and smaller seeds, while Chanee is the best in terms of its resistance to infection by *Phytophthora palmivora*. Kan Yao is somewhat less common, but prized for its longer window of time when it is both sweet and odorless at the same time. Among all the cultivars in Thailand, five are currently in large-scale commercial cultivation: Chanee, Mon Thong, Kan Yao, Ruang, and Kradum.*[6] There are more than 100 registered cultivars since 1920's in Malaysia*[7] and up to 193 cultivar by 1992,*[8] and many superior cultivars have been identified through competitions held at the annual Malaysian Agriculture, Horticulture, and Agrotourism Show. In Vietnam, the same process has been achieved through competitions held by the Southern Fruit Research Institute. A recently popular variety is, Cat Mountain King.*[9]

By 2007, Songpol Somsri, a Thai government scientist, had crossbred more than ninety varieties of durian to create Chantaburi No. 1, a cultivar without the characteristic odour.*[10] Another hybrid, Chantaburi No. 3, develops the odour about three days after the fruit is picked, which enables an odourless transport yet satisfies consumers who prefer the pungent odour.*[10] On 22 May 2012, two other cultivars from Thailand that also lack the usual odour, Long Laplae and Lin Laplae, were presented to the public by Yothin Samutkhiri, governor of Uttaradit province from where these cultivars were developed locally, while he announced the dates for the yearly durian fair of Laplae District, and the name giver to both cultivars.*[11]

28.3.1 Nutritional information

28.4 See also

- Breadfruit, an unrelated fruit that looks similar

- Jackfruit, an unrelated fruit that looks similar

28.5 References

[1] *Durio zibethinus* at **worldagroforestry.org**

[2] *A traveler´s guide to Durian Season* at **yearofthedurian.com**

[3] "The Plant List: A Working List of All Plant Species". Retrieved July 3, 2014.

[4] Brown, Michael J. (1997). *Durio – A Bibliographic Review*. International Plant Genetic Resources Institute (IPGRI). ISBN 92-9043-318-3. Retrieved 2014-06-03.

[5] "ST Foodies Club – Durian King". The Straits Times. 2006. Archived from the original on 2007-12-15. Retrieved 2007-07-25.

[6] "Durian Exporting Strategy, National Durian Database (กลยุทธ์การส่งออกทุเรียน)" (in Thai). Department of Agriculture, Thailand. Archived from the original on 2010-07-26. Retrieved 2010-07-26.

[7] "Comprehensive List of Durian Clones Registered by the Agriculture Department (of Malaysia)". Durian OnLine. Archived from the original on 2008-11-20. Retrieved 2006-03-05.

[8] Boosting Durian Productivity

[9] Teo, Wan Gek (2009-06-23). "Durian lovers head north on day tours". *The Straits Times.* Retrieved 2009-09-19.

[10] Fuller, Thomas (2007-04-08). "Fans Sour on Sweeter Version of Asia's Smelliest Fruit". *New York Times.* Retrieved 2008-11-20.

[11] Odourless durians to hit the market – The Nation

[12] "USDA National Nutrient Database". U.S. Department of Agriculture. Retrieved 2013-02-22.

28.6 External links

- Germplasm Resources Information Network: *Durio*

- *Durio zibethinus* (Bombacaceae)

- the controversial durian

- The durian: stinky fruit, killing fruit

- Nutritional value of durian

- Year of the Durian: a durian lover's travelogue and guide.

- A Durian How To.

Chapter 29

Eucalyptus oil

Eucalyptus oil is the generic name for distilled oil from the leaf of *Eucalyptus*, a genus of the plant family Myrtaceae native to Australia and cultivated worldwide. Eucalyptus oil has a history of wide application, as a pharmaceutical, antiseptic, repellent, flavouring, fragrance and industrial uses. The leaves of selected Eucalyptus species are steam distilled to extract eucalyptus oil.

29.1 Types and production

Eucalyptus oils in the trade are categorized into three broad types according to their composition and main end-use: medicinal, perfumery and industrial.[1] The most prevalent is the standard cineole-based "oil of eucalyptus", a colourless mobile liquid (yellow with age) with a penetrating, camphoraceous, woody-sweet scent.[2]

China produces about 75% of the world trade, but most of this is derived from camphor oil fractions rather than being true eucalyptus oil.[3] Significant producers of true eucalyptus oil include South Africa, Portugal, Spain, Brazil, Australia, Chile and Swaziland.

Global production is dominated by *Eucalyptus globulus*. However, *Eucalyptus kochii* and *Eucalyptus polybractea* have the highest cineole content, ranging from 80-95%. The British Pharmacopoeia states that the oil must have a minimum cineole content of 70% if it is pharmaceutical grade. Rectification is used to bring lower grade oils up to the high cineole standard required. Global annual production of eucalyptus oil is estimated at 3,000 tonnes.[4] The eucalyptus genus also produces non-cineole oils, including piperitone, phellandrene, citral, methyl cinnamate and geranyl acetate.

Eucalyptus oil should not be confused with the term "eucalyptol", another name for cineole.

29.2 Uses

29.2.1 Medicinal and antiseptic

The cineole-based oil is used as component in pharmaceutical preparations to relieve the symptoms of influenza and colds, in products like cough sweets, lozenges, ointments and inhalants. Eucalyptus oil has antibacterial effects on pathogenic bacteria in the respiratory tract.[5] Inhaled eucalyptus oil vapor is a decongestant and treatment for bronchitis.[6] Cineole controls airway mucus hypersecretion and asthma via anti-inflammatory cytokine inhibition.[7][8] Eucalyptus oil also stimulates immune system response by effects on the phagocytic ability of human monocyte derived macrophages.[9]

Eucalyptus oil also has anti-inflammatory and analgesic qualities as a topically applied liniment ingredient.[10][11]

Eucalyptus oil is also used in personal hygiene products for antimicrobial properties in dental care[12] and soaps. It can also be applied to wounds to prevent infection.

29.2.2 Repellent and biopesticide

Cineole-based eucalyptus oil is used as an insect repellent and biopesticide. In the U.S., eucalyptus oil was first registered in 1948 as an insecticide and miticide.[13]

29.2.3 Flavouring

Eucalyptus oil is used in flavouring. Cineole-based eucalyptus oil is used as a flavouring at low levels (0.002%) in various products, including baked goods, confectionery, meat products and beverages.[14] Eucalyptus oil has antimicrobial activity against a broad range of foodborne human pathogens and food spoilage microorganisms.[15] Non-cineole peppermint gum, strawberry gum and lemon ironbark are also used as flavouring.

Eucalyptus oil for pharmaceutical use.

Eucalyptus polybractea *or Blue-leaf Mallee, a species yielding high quality eucalyptus oil*

29.2.5 Industrial

Research shows that cineole-based eucalyptus oil (5% of mixture) prevents the separation problem with ethanol and petrol fuel blends. Eucalyptus oil also has a respectable octane rating and can be used as a fuel in its own right. However, production costs are currently too high for the oil to be economically viable as a fuel.[*][16]

Phellandrene- and piperitone-based eucalyptus oils have been used in mining to separate sulfide minerals via flotation.

29.3 Safety and toxicity

If consumed internally at low dosage as a flavouring component or in pharmaceutical products at the recommended rate, cineole-based 'oil of eucalyptus' is safe for adults. However, systemic toxicity can result from ingestion or topical application at higher than recommended doses.[*][17]

The probable lethal dose of pure eucalyptus oil for an adult is in the range of 0.05 mL to 0.5 mL/per kg of body weight.[*][18] Because of their high body surface area to mass ratio, children are more vulnerable to poisons absorbed transdermally. Severe poisoning has occurred in children after ingestion of 4 mL to 5 mL of eucalyptus oil.[*][19]

29.4 History

Australian Aboriginals use eucalyptus leaf infusions (which contain eucalyptus oil) as a traditional medicine for treating body pains, sinus congestion, fever, and colds.[*][20][*][21]

29.2.4 Fragrance

Eucalyptus oil is also used as a fragrance component to impart a fresh and clean aroma in soaps, detergents, lotions and perfumes. It is known for its pungent, intoxicating scent.

Dennis Considen and John White, surgeons on the First Fleet, distilled eucalyptus oil from *Eucalyptus piperita* found growing on the shores of Port Jackson in 1788 to treat convicts and marines.[*][22][*][23][*][24][*][25] Eucalyptus oil was subsequently extracted by early colonists, but was not commercially exploited for some time.

Baron Ferdinand von Mueller, Victorian botanist, promoted the qualities of Eucalyptus as a disinfectant in "fever districts" , and also encouraged Joseph Bosisto, a Melbourne pharmacist, to investigate the commercial potential of the oil.[*][26] Bosisto started the commercial eucalyptus oil industry in 1852 near Dandenong, Victoria, Australia, when he set up a distillation plant and extracted the essential oil from the cineole chemotype of *Eucalyptus radiata*. This resulted in the cineole chemotype becoming the generic 'oil of eucalyptus', and "Bosisto's Eucalyptus Oil" still survives as a brand.

French chemist, F.S. Cloez, identified and ascribed the name *eucalyptol* also known as cineole to the dominant portion of *E. globulus* oil.[*][27] By the 1870s oil from *Eucalyptus globulus*, Tasmanian blue gum, was being exported worldwide and eventually dominated world trade, while other higher quality species were also being distilled to a lesser extent. Surgeons were using eucalyptus oil as an antiseptic during surgery by the 1880s.[*][28]

The Australian eucalyptus oil industry peaked in the 1940s, the main area of production being the central goldfields region of Victoria, particularly Inglewood; then the global establishment of eucalyptus plantations for timber resulted in increased volumes of eucalyptus oil as a plantation by-product. By the 1950s the cost of producing eucalyptus oil in Australia had increased so much that it could not compete against cheaper Spanish and Portuguese oils (closer to European Market therefore less costs). Non-Australian sources now dominate commercial eucalyptus oil supply, although Australia continues to produce high grade oils, mainly from blue mallee (*E. polybractea*) stands.

29.5   Species utilised

Commercial cineole-based eucalyptus oils are produced from several species of *Eucalyptus*:

- *Eucalyptus cneorifolia*
- *Eucalyptus dives*
- *Eucalyptus dumosa*
- *Eucalyptus globulus*
- *Eucalyptus goniocalyx*

- *Eucalyptus horistes*
- *Eucalyptus kochii*
- *Eucalyptus leucoxylon*
- *Eucalyptus oleosa*
- *Eucalyptus polybractea*
- *Eucalyptus radiata*
- *Eucalyptus sideroxylon*
- *Eucalyptus smithii*
- *Eucalyptus tereticornis*
- *Eucalyptus viridis*

Non-cineole oil producing species:

- *Eucalyptus dives* - phellandrene variant
- *Eucalyptus dives* - piperitone variant
- *Eucalyptus elata* - piperitone variant
- *Eucalyptus macarthurii* - geranyl acetate
- *Eucalyptus olida* - methyl cinnamate
- *Eucalyptus radiata* - phellandrene variant
- *Eucalyptus staigeriana* - citral, limonene

The former lemon eucalyptus species *Eucalyptus citriodora* is now classified as *Corymbia citriodora*, which produces a citronellal-based oil.

29.6   Compendial status

- British Pharmacopoeia[*][29]

29.7   See also

- Essential oil
- Eucalypts, woody plants belonging to three closely related genera: *Eucalyptus*, *Corymbia* and *Angophora*

29.8 References

[1] William M. Ciesla. "Types of oil and uses". *Non-wood Forest Products from Temperate Broad-leaved Trees*. Food & Agriculture Org (2002). p. 30.

[2] Lawless, J., *The Illustrated Encyclopedia of Essential Oils*, Element Books 1995 ISBN 1-85230-661-0

[3] Ashurst, P. R (1999-07-31). *Food Flavorings*. ISBN 9780834216211.

[4] FOA

[5] Salari, M. H.; Amine, G.; Shirazi, M. H.; Hafezi, R.; Mohammadypour, M. (2006). "Antibacterial effects of Eucalyptus globulus leaf extract on pathogenic bacteria isolated from specimens of patients with respiratory tract disorders". *Clinical Microbiology and Infection* 12 (2): 194–6. doi:10.1111/j.1469-0691.2005.01284.x. PMID 16441463.

[6] Lu, XQ; Tang, FD; Wang, Y; Zhao, T; Bian, RL (2004). "Effect of Eucalyptus globulus oil on lipopolysaccharide-induced chronic bronchitis and mucin hypersecretion in rats". *Zhongguo Zhong yao za zhi = Zhongguo zhongyao zazhi = China journal of Chinese materia medica* 29 (2): 168–71. PMID 15719688.

[7] Juergens, U.R; Dethlefsen, U; Steinkamp, G; Gillissen, A; Repges, R; Vetter, H (2003). "Anti-inflammatory activity of 1.8-cineol (eucalyptol) in bronchial asthma: A double-blind placebo-controlled trial". *Respiratory Medicine* 97 (3): 250–6. doi:10.1053/rmed.2003.1432. PMID 12645832.

[8] Juergens, Uwe R.; Engelen, Tanja; Racké, Kurt; Stöber, Meinolf; Gillissen, Adrian; Vetter, Hans (2004). "Inhibitory activity of 1,8-cineol (eucalyptol) on cytokine production in cultured human lymphocytes and monocytes". *Pulmonary Pharmacology & Therapeutics* 17 (5): 281–7. doi:10.1016/j.pupt.2004.06.002. PMID 15477123.

[9] Serafino, A; Sinibaldi Vallebona, PS; Andreola, F; Zonfrillo, M; Mercuri, L; Federici, M; Rasi, G; Garaci, E; Pierimarchi, P (2008). "Stimulatory effect of Eucalyptus essential oil on innate cell-mediated immune response". *BMC Immunology* 9: 17. doi:10.1186/1471-2172-9-17. PMC 2374764. PMID 18423004.

[10] Göbel, H; Schmidt, G; Soyka, D (1994). "Effect of peppermint and eucalyptus oil preparations on neurophysiological and experimental algesimetric headache parameters". *Cephalalgia : an international journal of headache* 14 (3): 228–34; discussion 182. doi:10.1046/j.1468-2982.1994.014003228.x. PMID 7954745.

[11] Hong, CZ; Shellock, FG (1991). "Effects of a topically applied counterirritant (Eucalyptamint) on cutaneous blood flow and on skin and muscle temperatures. A placebo-controlled study". *American journal of physical medicine & rehabilitation / Association of Academic Physiatrists* 70 (1): 29–33. doi:10.1097/00002060-199102000-00006. PMID 1994967.

[12] Nagata, Hideki; Inagaki, Yoshika; Tanaka, Muneo; Ojima, Miki; Kataoka, Kosuke; Kuboniwa, Masae; Nishida, Nobuko; Shimizu, Katsumasa; Osawa, Kenji; Shizukuishi, Satoshi (2008). "Effect of Eucalyptus Extract Chewing Gum on Periodontal Health: A Double-Masked, Randomized Trial". *Journal of Periodontology* 79 (8): 1378–85. doi:10.1902/jop.2008.070622. PMID 18672986.

[13] Flower and Vegetable Oils, R.E.D. Facts, EPA

[14] Harborne, J.B., Baxter, H., *Chemical Dictionary of Economic Plants*, ISBN 0-471-49226-4

[15] Zhao, J., Agboola, S., Functional Properties of Australian Bushfoods - A Report for the Rural Industries Research and Development Corporation, *2007, RIRDC Publication No 07/030*

[16] Boland, D.J., Brophy, J.J., and A.P.N. House, *Eucalyptus Leaf Oils*, 1991, p. 8 ISBN 0-909605-69-6

[17] Darben, T; Cominos, B; Lee, CT (1998). "Topical eucalyptus oil poisoning". *The Australasian journal of dermatology* 39 (4): 265–7. doi:10.1111/j.1440-0960.1998.tb01488.x. PMID 9838728.

[18] Hindle, R.C. (1994). "Eucalyptus oil ingestion". *New Zealand Medical Journal*: 185–186.

[19] Foggie, WE (1911). "Eucalyptus Oil Poisoning". *British Medical Journal* 1 (2616): 359–360. doi:10.1136/bmj.1.2616.359. PMC 2332914. PMID 20765463.

[20] Low, T., *Bush Medicine, A Pharmacopeia of Natural Remedies*, Angus & Robertson, p. 85, 1990.

[21] Barr, A., Chapman, J., Smith, N., Beveridge, M., *Traditional Bush Medicines, An Aboriginal Pharmacopoeia*, Greenhouse Publications, pp. 116–117, 1988, ISBN 086436167X

[22] Maiden, J.H., *The Forest Flora of New South Wales*, vol. 4, Government Printer, Sydney, 1922.

[23] Copy of letter received by Dr Anthony Hamiltion, from Dennis Considen, 18 November 1788, and sent onto Joseph Banks.

[24] Lassak, E.V., & McCarthy, T., *Australian Medicinal Plants*, Methuen Australia, 1983, p. 15, ISBN 0-454-00438-9.

[25] White, J., *Journal of a Voyage to New South Wales*, 1790

[26] Grieve, M.,(author) & Leyel, C.F., (ed), *A Modern Herbal*, Jonathon Cape, 1931, p. 287.

[27] Boland, D.J., Brophy, J.J., and A.P.N. House, *Eucalyptus Leaf Oils*, 1991, p. 6 ISBN 0-909605-69-6

[28] Maiden, J.H., *The Useful Native Plants of Australia*, pp. 255, 1889

[29] The British Pharmacopoeia Secretariat (2009). "Index, BP 2009" (PDF). Retrieved 10 September 2009.

29.9 Further reading

- Boland, D.J., Brophy, J.J., and A.P.N. House, *Eucalyptus Leaf Oils*, 1991, ISBN 0-909605-69-6

- FAO Corporate Document Repository, Flavours and fragrances of plant origin

29.10 External links

- Toxicity Eucalyptus oil profile, Chemical Safety Information from Intergovernmental Organizations

- Eucalyptus oil (E. globulus Labillardiere, E. fructicetorum F. Von Mueller, E. smithii R.T. Baker) MedlinePlus, U.S. National Library of Medicine, U.S. National Institutes of Health evidence-based monograph prepared by the Natural Standard Research Collaboration

Chapter 30

Fern

This article is about the group of pteridophyte plants. For other uses, see Fern (disambiguation).

"Ferns" redirects here. For the city in Ireland, see Ferns, County Wexford. For other uses, see Ferns (disambiguation).

A **fern** is a member of a group of approximately 12,000 species[3] of vascular plants that reproduce via spores and have neither seeds nor flowers. They differ from mosses by being vascular (i.e. having water-conducting vessels). They have stems and leaves, like other vascular plants. Most ferns have what are called fiddleheads that expand into fronds, which are each delicately divided.[4]

Leptosporangiate ferns (sometimes called true ferns) are by far the largest group, but ferns as defined here (ferns *sensu lato*) include horsetails, whisk ferns, marattioid ferns, and ophioglossoid ferns. This group may be referred to as **monilophytes**. The term **pteridophyte** traditionally refers to ferns plus a few other seedless vascular plants (see the classification section below), although some recent authors have used the term to refer strictly to the monilophytes.

Ferns first appear in the fossil record 360 million years ago in the late Devonian period[5] but many of the current families and species did not appear until roughly 145 million years ago in the early Cretaceous, after flowering plants came to dominate many environments. The fern *Osmunda claytoniana* is a paramount example of evolutionary stasis. Paleontological evidence indicates it has remained unchanged, even at the level of fossilized nuclei and chromosomes, for at least 180 million years.[6]

Ferns are not of major economic importance, but some are grown or gathered for food, as ornamental plants, for remediating contaminated soils, and have been the subject of research for their ability to remove some chemical pollutants from the air. Some are significant weeds. They also play a role in mythology, medicine, and art.

30.1 Description

30.1.1 Life cycle

Gametophyte (thalloid green mass) and sporophyte (ascendent frond) of Onoclea sensibilis

Ferns are vascular plants differing from lycophytes by having true leaves (megaphylls), which are often pinnate. They differ from seed plants (gymnosperms and angiosperms) in their mode of reproduction lacking flowers and seeds. Like all other vascular plants, they have a life cycle referred to as alternation of generations, characterized by alternating diploid sporophytic and haploid gametophytic phases. The diploid sporophyte has $2n$ paired chromosomes, where n varies from species to species. The haploid gametophyte has n unpaired chromosomes, i.e. half the number of the sporophyte. The gametophyte of ferns is a free-living organism, whereas the gametophyte of the gymnosperms and angiosperms is dependent on the sporophyte.

Life cycle of a typical fern:

1. (a) i.

2. A diploid sporophyte phase produces haploid spores by meiosis (a process of cell division which reduces the number of chromosomes by a half).

3. A spore grows into a haploid gametophyte by mitosis (a process of cell division which maintains the number of chromosomes). The gametophyte typically consists of a photosynthetic prothallus.

4. The gametophyte produces gametes (often both sperm and eggs on the same prothallus) by mitosis.

5. A mobile, flagellate sperm fertilizes an egg that remains attached to the prothallus.

6. The fertilized egg is now a diploid zygote and grows by mitosis into a diploid sporophyte (the typical "fern" plant).

30.1.2 Morphology

Ferns at the Royal Melbourne Botanical Gardens

Like the sporophytes of seed plants, those of ferns consist of stems, leaves and roots.

Stems: Fern stems are often referred to as rhizomes, even though they grow underground only in a some of the species. Epiphytic species and many of the terrestrial ones have above-ground creeping stolons (e.g., Polypodiaceae), and many groups have above-ground erect semi-woody trunks (e.g., Cyatheaceae). These can reach up to 20 metres (66 ft) tall in a few species (e.g., *Cyathea brownii* on Norfolk Island and *Cyathea medullaris* in New Zealand).

Leaf: The green, photosynthetic part of the plant is technically a megaphyll and in ferns, it is often referred to as a frond. New leaves typically expand by the unrolling of a tight spiral called a crozier or fiddlehead fern. This uncurling of the leaf is termed circinate vernation. Leaves are divided into two types a trophophyll and a sporophyll. A trophophyll frond is a vegetative leaf analogous to the

Tree ferns, probably Dicksonia antarctica, *growing in Nunniong, Australia*

typical green leaves of seed plants that does not produce spores, instead only producing sugars by photosynthesis. A sporophyll frond is a fertile leaf that produces spores borne in sporangia that are usually clustered to form sori. In most ferns, fertile leaves are morphologically very similar to the sterile ones, and they photosynthesize in the same way. In some groups, the fertile leaves are much narrower than the sterile leaves, and may even have no green tissue at all (e.g., Blechnaceae, Lomariopsidaceae). The anatomy of fern leaves can either be simple or highly divided. In tree ferns, the main stalk that connects the leaf to the stem (known as the stipe), often have multiple leafy. The leafy structures that grow from the stipe are known as pinnae and are often again divided into smaller pinnules.[7]

Roots: The underground non-photosynthetic structures that take up water and nutrients from soil. They are always fibrous and are structurally very similar to the roots of seed plants.

The gametophytes of ferns, however, are very different from those of seed plants. Instead, they resemble liverworts. A fern gametophyte typically consists of:

- Prothallus: A green, photosynthetic structure that is one cell thick, usually heart or kidney shaped, 3–10 mm long and 2–8 mm broad. The prothallus produces gametes by means of:

  - Antheridia: Small spherical structures that produce flagellate sperm.

- Archegonia: A flask-shaped structure that produces a single egg at the bottom, reached by the sperm by swimming down the neck.

- Rhizoids: root-like structures (not true roots) that consist of single greatly elongated cells, water and mineral salts are absorbed over the whole structure. Rhizoids anchor the prothallus to the soil.

One difference between sporophytes and gametophytes might be summed up by the saying that "Nothing eats ferns, but everything eats gametophytes." This is an oversimplification, but it is true that gametophytes are often difficult to find in the field because they are far more likely to be food than are the sporophytes.

30.2 Evolution, phylogeny and classification

See also: List of fern families

Ferns first appear in the fossil record in the early-Carboniferous period. By the Triassic, the first evidence of ferns related to several modern families appeared. The "great fern radiation" occurred in the late-Cretaceous, when many modern families of ferns first appeared.

One problem with fern classification is the problem of cryptic species. A cryptic species is a species that is morphologically similar to another species, but differs genetically in ways that prevent fertile interbreeding. A good example of this is the currently designated species *Asplenium trichomanes*, the maidenhair spleenwort. This is actually a species complex that includes distinct diploid and tetraploid races. There are minor but unclear morphological differences between the two groups, which prefer distinctly differing habitats. In many cases such as this, the species complexes have been separated into separate species, thus raising the number of overall fern species. Possibly many more cryptic species are yet to be discovered and designated.

Ferns have traditionally been grouped in the Class Filices, but modern classifications assign them their own phylum or division in the plant kingdom, called Pteridophyta, also known as Filicophyta. The group is also referred to as Polypodiophyta, (or Polypodiopsida when treated as a subdivision of tracheophyta (vascular plants), although Polypodiopsida sometimes refers to only the leptosporangiate ferns). The term "pteridophyte" has traditionally been used to describe all seedless vascular plants, making it synonymous with "ferns and fern allies". This can be confusing since members of the fern phylum Pteridophyta are also sometimes referred to as pteridophytes.

Traditionally, three discrete groups of plants have been considered ferns: two groups of eusporangiate ferns families Ophioglossaceae (adders-tongues, moonworts, and grape-ferns) and Marattiaceae and the leptosporangiate ferns. The Marattiaceae are a primitive group of tropical ferns with a large, fleshy rhizome, and are now thought to be a sibling taxon to the main group of ferns, the leptosporangiate ferns. Several other groups of plants were considered "fern allies": the clubmosses, spikemosses, and quillworts in the Lycopodiophyta, the whisk ferns in Psilotaceae, and the horsetails in the Equisetaceae. More recent genetic studies have shown that the Lycopodiophyta are more distantly related to other vascular plants, having radiated evolutionarily at the base of the vascular plant clade, while both the whisk ferns and horsetails are as much "true" ferns as are the Ophioglossoids and Marattiaceae. In fact, the whisk ferns and Ophioglossoids are demonstrably a clade, and the horsetails and Marattiaceae are arguably another clade. Molecular data which remain poorly constrained for many parts of the plants' phylogeny have been supplemented by recent morphological observations supporting the inclusion of *Equisetaceae* within the ferns, notably relating to the construction of their sperm, and peculiarities of their roots.*[2] However, there are still differences of opinion about the placement of the Equisetum species (see Equisetopsida for further discussion). One possible means of treating this situation is to consider only the leptosporangiate ferns as "true" ferns, while considering the other three groups as "fern allies" . In practice, numerous classification schemes have been proposed for ferns and fern allies, and there has been little consensus among them.

A 2006 classification by Smith *et al.* is based on recent molecular systematic studies, in addition to morphological data. Their phylogeny is a consensus of a number of studies. This phylogeny has been refined in more recent years,*[8]*[9] and is shown below (to the level of orders).*[2]*[10]

Their classification based on this phylogeny divides extant ferns into four classes:

- Psilotopsida (whisk ferns and ophioglossoid ferns), about 92 species*[11]

- Equisetopsida (horsetails), about 15 species*[11]

- Marattiopsida, about 150 species*[11]

- Polypodiopsida (leptosporangiate ferns), over 9000 species*[11]

Others have divided extant ferns into five groups (Psilotales, Ophioglossales, Equisetales, Marattiales and Polypodiopsida), separating the whisk ferns and ophioglossoid ferns.*[8]

The leptosporangiate ferns are sometimes called "true ferns".[12] This group includes most plants familiarly known as ferns. Modern research supports older ideas based on morphology that the Osmundaceae diverged early in the evolutionary history of the leptosporangiate ferns; in certain ways this family is intermediate between the eusporangiate ferns and the leptosporangiate ferns. Research by Rai and Graham (2010) since this 2006 classification broadly supports the main groups, but queries their relationships, concluding that "at present perhaps the best that can be said about all relationships among the major lineages of monilophytes in current studies is that we do not understand them very well".[13] Grewe et al. (2013) confirmed the inclusion of horsetails within ferns *sensu lato*, but also suggested that uncertainties remained in their precise placement.[8] For the most recent classification of ferns and lycopods, see: Christenhusz & Chase (2014) [14]

30.3   Ecology

Ferns at Muir Woods, California

The stereotypical image of ferns growing in moist shady woodland nooks is far from a complete picture of the habitats where ferns can be found growing. Fern species live in a wide variety of habitats, from remote mountain elevations, to dry desert rock faces, to bodies of water or in open fields. Ferns in general may be thought of as largely being specialists in marginal habitats, often succeeding in places where various environmental factors limit the success of flowering plants. Some ferns are among the world's most serious weed species, including the bracken fern growing in the Scottish highlands, or the mosquito fern (*Azolla*) growing in tropical lakes, both species forming large aggressively spreading colonies. There are four particular types of habitats that ferns are found in: moist, shady forests; crevices in rock faces, especially when sheltered from the full sun; acid wetlands including bogs and swamps; and tropical trees, where many species are epiphytes (something like a quarter to a

third of all fern species[15]).

Many ferns depend on associations with mycorrhizal fungi. Many ferns only grow within specific pH ranges; for instance, the climbing fern (*Lygodium palmatum*) of eastern North America will only grow in moist, intensely acid soils, while the bulblet bladder fern (*Cystopteris bulbifera*), with an overlapping range, is only found on limestone.

The spores are rich in lipids, protein and calories, so some vertebrates eat these. The European woodmouse (*Apodemus sylvaticus*) has been found to eat the spores of *Culcita macrocarpa* and the bullfinch (*Pyrrhula murina*) and the New Zealand lesser short-tailed bat (*Mystacina tuberculata*) also eat fern spores.[16]

30.4   Uses

Ferns are not as important economically as seed plants but have considerable importance in some societies. Some ferns are used for food, including the fiddleheads of bracken, *Pteridium aquilinum*, ostrich fern, *Matteuccia struthiopteris*, and cinnamon fern, *Osmundastrum cinnamomeum*. *Diplazium esculentum* is also used by some tropical people as food. Tubers from the King Fern or *para* (*Ptisana salicina*) are a traditional food in New Zealand and the South Pacific. Fern tubers were used for food 30,000 years ago in Europe.[17][18] Fern tubers were used by the Guanches to make gofio in the Canary Islands. Ferns are generally not known to be poisonous to humans.[19] Licorice fern rhizomes were chewed by the natives of the Pacific Northwest for their flavor.

Ferns of the genus *Azolla* are very small, floating plants that do not resemble ferns. Called mosquito fern, they are used as a biological fertilizer in the rice paddies of southeast Asia, taking advantage of their ability to fix nitrogen from the air into compounds that can then be used by other plants.

Many ferns are grown in horticulture as landscape plants, for cut foliage and as houseplants, especially the Boston fern (*Nephrolepis exaltata*) and other members of the genus *Nephrolepis*. The Bird's Nest Fern (*Asplenium nidus*) is also popular, as are the staghorn ferns (genus *Platycerium*). Perennial (also known as hardy) ferns planted in gardens in the northern hemisphere also have a considerable following.

Several ferns are noxious weeds or invasive species, including Japanese climbing fern (*Lygodium japonicum*), mosquito fern and sensitive fern (*Onoclea sensibilis*). Giant water fern (*Salvinia molesta*) is one of the world's worst aquatic weeds. The important fossil fuel coal consists of the remains of primitive plants, including ferns.

Ferns have been studied and found to be useful in the removal of heavy metals, especially arsenic, from the soil. Other ferns with some economic significance include:

- *Dryopteris filix-mas* (male fern), used as a vermifuge, and formerly in the US Pharmacopeia; also, this fern accidentally sprouting in a bottle resulted in Nathaniel Bagshaw Ward's 1829 invention of the terrarium or Wardian case

- *Rumohra adiantiformis* (floral fern), extensively used in the florist trade

- *Microsorum pteropus* (Java fern), one of the most popular freshwater aquarium plants.

- *Osmunda regalis* (royal fern) and *Osmunda cinnamomea* (cinnamon fern), the root fiber being used horticulturally; the fiddleheads of *O. cinnamomea* are also used as a cooked vegetable

- *Matteuccia struthiopteris* (ostrich fern), the fiddleheads used as a cooked vegetable in North America

- *Pteridium aquilinum* or *Pteridium esculentum* (bracken), the fiddleheads used as a cooked vegetable in Japan and are believed to be responsible for the high rate of stomach cancer in Japan. It is also one of the world's most important agricultural weeds, especially in the British highlands, and often poisons cattle and horses.

- *Diplazium esculentum* (vegetable fern), a source of food for some native societies

- *Pteris vittata* (brake fern), used to absorb arsenic from the soil

- *Polypodium glycyrrhlza* (licorice fern), roots chewed for their pleasant flavor

- Tree ferns, used as building material in some tropical areas

- *Cyathea cooperi* (Australian tree fern), an important invasive species in Hawaii

- *Ceratopteris richardii*, a model plant for teaching and research, often called C-fern

30.5 Culture

30.5.1 Pteridologist

The study of ferns and other pteridophytes is called **pteridology**. A **pteridologist** is a specialist in the study of pteridophytes in a broader sense that includes the more distantly related lycophytes.

Blätter des Manns Walfarn. *by Alois Auer, Vienna: Imperial Printing Office, 1853*

30.5.2 Pteridomania

"Pteridomania"' is a term for the Victorian era craze of fern collecting and fern motifs in decorative art including pottery, glass, metals, textiles, wood, printed paper, and sculpture "appearing on everything from christening presents to gravestones and memorials." The fashion for growing ferns indoors led to the development of the Wardian case, a glazed cabinet that would exclude air pollutants and maintain the necessary humidity.[20]

The dried form of ferns was also used in other arts, being used as a stencil or directly inked for use in a design. The botanical work, *The Ferns of Great Britain and Ireland*, is a notable example of this type of nature printing. The process, patented by the artist and publisher Henry Bradbury, impressed a specimen on to a soft lead plate. The first publication to demonstrate this was Alois Auer's *The Discovery of the Nature Printing-Process*.

30.5.3 Folklore

Ferns figure in folklore, for example in legends about mythical flowers or seeds.[21] In Slavic folklore, ferns are believed to bloom once a year, during the Ivan Kupala night.

Barnsley fern created using chaos game, through an Iterated function system (IFS).

Although alleged to be exceedingly difficult to find, anyone who sees a "fern flower" is thought to be guaranteed to be happy and rich for the rest of their life. Similarly, Finnish tradition holds that one who finds the "seed" of a fern in bloom on Midsummer night will, by possession of it, be guided and be able to travel invisibly to the locations where eternally blazing Will o' the wisps called aarnivalkea mark the spot of hidden treasure. These spots are protected by a spell that prevents anyone but the fern-seed holder from ever knowing their locations.*[22]

30.6   Organisms confused with ferns

30.6.1   Misunderstood names

Several non-fern plants (and even animals) are called "ferns" and are sometimes confused with true ferns. These include:

- "Asparagus fern" This may apply to one of several species of the monocot genus *Asparagus*, which are flowering plants.

- "Sweetfern" A flowering shrub of the genus *Comptonia*.

- "Air fern" A group of animals called hydrozoan that are distantly related to jellyfish and corals. They are harvested, dried, dyed green, and then sold as a "plant" that can "live on air". While it may look like a fern, it is merely the skeleton of this colonial animal.

- "Fern bush" *Chamaebatiaria millefolium* a rose family shrub with fern-like leaves.

30.6.2   Fern-like flowering plants

Some flowering plants such as palms and members of the carrot family have pinnate leaves that somewhat resemble fern fronds. However, these plants have fully developed seeds contained in fruits, rather than the microscopic spores of ferns.

30.7   Gallery

- *Adiantum lunulatum*

- Fern leaf, probably *Blechnum nudum*

- A tree fern unrolling a new frond

- Tree fern, probably *Dicksonia antarctica*

- Tree ferns, probably *Dicksonia antarctica*

- "Filicinae" from Ernst Haeckel's *Kunstformen der Natur*, 1904

- Unidentified tree fern in Oaxaca

- Tree Fern Spores San Diego, CA

- Leaf of fern

- Unidentified fern with spores showing in Rotorua, NZ.

- Ferns in one of many natural Coast Redwood undergrowth settings Santa Cruz, CA.

- Nature prints in The Ferns of Great Britain and Ireland used fronds to produce the plates

- A young, newly formed fern frond

- Fern bed under a forest canopy in woods near Franklin, Virginia

- *Pyrrosia piloselloides*, Dragon's Scale, in Malaysia

- Fern growing on a wall

- Spores of Dryopteris filix-mas

30.8   See also

- British Pteridological Society

- Chirosia betuleti - Fern gall

- Fern spike

- Fern sports

30.9 References

[1] *Wattieza*, Stein, W. E.; Mannolini, F.; Hernick, L. V.; Landling, E.; Berry, C. M. (2007). "Giant cladoxylopsid trees resolve the enigma of the Earth's earliest forest stumps at Gilboa". *Nature* **446**: 904–907. doi:10.1038/nature05705.

[2] Smith, A.R.; Pryer, K.M.; Schuettpelz, E.; Korall, P.; Schneider, H.; Wolf, P.G. (2006). "A classification for extant ferns" (PDF). *Taxon* **55** (3): 705–731. doi:10.2307/25065646. JSTOR 25065646. Retrieved 2008-02-12.

[3] Chapman, Arthur (2010-08-26). "Numbers of Living Species in Australia and the World. Report for the Australian Biological Resources Study. Canberra, Australia. September 2009.". Environment.gov.au. Retrieved 2013-09-07.

[4] McCausland, Jim (2009-09-20). "Rediscover ferns". *Sunset.com*. Retrieved 2013-09-07.

[5] "Pteridopsida: Fossil Record". University of California Museum of Paleontology. Retrieved 2014-03-11.

[6] Bomfleur B, McLoughlin S, Vajda V (March 2014). "Fossilized nuclei and chromosomes reveal 180 million years of genomic stasis in royal ferns". *Science* **343** (6177): 1376–7. doi:10.1126/science.1249884. PMID 24653037.

[7] "Fern Fronds". Basic Biology. Retrieved December 6, 2014.

[8] Grewe, Felix; et al. (2013). "Complete plastid genomes from Ophioglossum californicum, Psilotum nudum, and Equisetum hyemale reveal an ancestral land plant genome structure and resolve the position of Equisetales among monilophytes". *BMC Evolutionary Biology* **13** (1): 1–16. doi:10.1186/1471-2148-13-8. ISSN 1471-2148. Retrieved 21 May 2013.

[9] Karol, Kenneth G; et al. (2010). "Complete plastome sequences of Equisetum arvense and Isoetes flaccida: implications for phylogeny and plastid genome evolution of early land plant lineages.". *BMC Evolutionary Biology* **10** (1): 321–336. doi:10.1186/1471-2148-10-321. ISSN 1471-2148. PMID 20969798. Retrieved 21 May 2013.

[10] Li, F-W; Kuo, L-Y; Rothfels, CJ; Ebihara, A; Chiou, W-L; et al. (2011). "rbcL and matK Earn Two Thumbs Up as the Core DNA Barcode for Ferns". *PLoS ONE* **6** (10): e26597. doi:10.1371/journal.pone.0026597.

[11] Eric Schuettpelz (2007), "table 1", *The evolution and diversification of epiphytic ferns* (PDF), Duke University PhD thesis

[12] Stace, Clive (2010b). *New Flora of the British Isles* (3rd ed.). Cambridge, UK: Cambridge University Press. p. xxviii. ISBN 978-0-521-70772-5.

[13] Rai, Hardeep S. & Graham, Sean W. (2010). "Utility of a large, multigene plastid data set in inferring higher-order relationships in ferns and relatives (monilophytes)". *American Journal of Botany* **97** (9): 1444–1456. doi:10.3732/ajb.0900305., p. 1450

[14] Christenhusz, Maarten J.M. & Chase, Mark W. (2014). "Trends and concepts in fern classification". *Annals of Botany* **113** (9): 571–594. doi:10.1093/aob/mct299.

[15] Schuettpelz, Eric. "Fern Phylogeny Inferred from 400 Leptosporangiate Species and Three Plastid Genes," contained in "The Evolution and Diversification of Epiphytic Ferns." Doctoral dissertation, Duke University. 2007. http://dukespace.lib.duke.edu/dspace/bitstream/10161/181/1/D_Schuettpelz_Eric_a_052007.pdf

[16] Walker, Matt (19 February 2010). "A mouse that eats ferns like a dinosaur". BBC Earth News. Retrieved 20 February 2010.

[17] "Stone Age humans liked their burgers in a bun", Sonia Van Gilder Cooke, *New Scientist*, 23 Oct. 2010, p. 18.

[18] "Thirty thousand-year-old evidence of plant food processing" by Anna Revedin et al., PNAS, published online Oct. 18, 2010.

[19] Pelton, Robert (2011). *The Official Pocket Edible Plant Survival Manual*. Freedom and Liberty Foundation Press. p. 25. BNID 2940013382145.

[20] • Boyd, Peter D. A. (2002-01-02). "Pteridomania - the Victorian passion for ferns". Revised: web version. Antique Collecting 28, 6, 9–12. Retrieved 2007-10-02.

[21] May, Lenore Wile (1978). "The economic uses and associated folklore of ferns and fern allies". *The Botanical Review* **44** (4): 491–528. doi:10.1007/BF02860848.

[22] "Traditional Finnish Midsummer celebration". Saunalahti.fi. Retrieved 2013-09-07.

30.10 Further reading

• Christenhusz, Maarten J.M.; Chase, Mark W. (2014). "Trends and concepts in fern classification". *Annals of Botany* **113** (4): 571–594. doi:10.1093/aob/mct299.

• Lord, Thomas R. (2006). *Ferns and Fern Allies of Pennsylvania*. Indiana, PA: Pinelands Press.

• Melan, M. A.; Whittier, D. P. (1990). "Effects of Inorganic Nitrogen Sources on Spore Germination and Gametophyte Growth in Botrychium Dissectum". *Plant, Cell and Environment* **13** (5): 477–82. doi:10.1111/j.1365-3040.1990.tb01325.x.

- Moran, Robbin C. (2004). *A Natural History of Ferns*. Portland, OR: Timber Press. ISBN 0-88192-667-1.

- Pryer, Kathleen M.; Schneider, Harald; Smith, Alan R.; Cranfill, Raymond; Wolf, Paul G.; Hunt, Jeffrey S.; Sipes, Sedonia D. (2001). "Horsetails and ferns are a monophyletic group and the closest living relatives to seed plants". *Nature* **409** (6820): 618–622. doi:10.1038/35054555. PMID 11214320.

- Pryer, Kathleen M.; Schuettpelz, Eric; Wolf, Paul G.; Schneider, Harald; Smith, Alan R.; Cranfill, Raymond (2004). "Phylogeny and evolution of ferns (monilophytes) with a focus on the early leptosporangiate divergences". *American Journal of Botany* **91** (10): 1582–1598. doi:10.3732/ajb.91.10.1582. PMID 21652310.

30.11   External links

- Tree of Life Web Project: Filicopsida

- A classification of the ferns and their allies

- A fern book bibliography

- Register of fossil Pteridophyta

- L. Watson and M.J. Dallwitz (2004 onwards). The Ferns (Filicopsida) of the British Isles.

- Ferns and Pteridomania in Victorian Scotland

- Non-seed plant images at bioimages.vanderbilt.edu

- "American Fern Society"

- "British Pteridological Society"

- Checklist of Ferns and Lycophytes of the World

- Images of ferns of Hawaii

Chapter 31

Forage

For other uses, see Forage (disambiguation).

Forage is plant material (mainly plant leaves and stems) eaten by grazing livestock.[*][1] Historically, the term *forage* has meant only plants eaten by the animals directly as pasture, crop residue, or immature cereal crops, but it is also used more loosely to include similar plants cut for fodder and carried to the animals, especially as hay or silage.[*][2] The term **forage fish** refers to small schooling fish that are preyed on by larger aquatic animals.[*][3]

While the term *forage* has a broad definition, the term **forage crop** is used to define crops, annual or biennial, which are grown to be utilized by grazing or harvesting as a whole crop.[*][4]

31.1 Common forages

Bull feeding on grass

31.1.1 Grasses

Grass forages include:[*][5][*][6]

Horse-drawn transport of fodder in Romania

Meadow of perennial ryegrass (Lolium perenne)

- *Agrostis* spp. – bentgrasses
  - *Agrostis capillaris* – common bentgrass
  - *Agrostis stolonifera* – creeping bentgrass
- *Andropogon hallii* – sand bluestem
- *Arrhenatherum elatius* – false oat-grass
- *Bothriochloa bladhii* – Australian bluestem

- *Bothriochloa pertusa* – hurricane grass
- *Brachiaria decumbens* – Surinam grass
- *Brachiaria humidicola* – koronivia grass
- *Bromus* spp. – bromegrasses
- *Cenchrus ciliaris* – buffelgrass
- *Chloris gayana* – Rhodes grass
- *Cynodon dactylon* – bermudagrass
- *Dactylis glomerata* – orchard grass
- *Echinochloa pyramidalis* – antelope grass
- *Entolasia imbricata* – bungoma grass
- *Festuca* spp. – fescues
  - *Festuca arundinacea* – tall fescue
  - *Festuca pratensis* – meadow fescue
  - *Festuca rubra* – red fescue
- *Heteropogon contortus* – black spear grass
- *Hymenachne amplexicaulis* – West Indian marsh grass
- *Hyparrhenia rufa* – jaragua
- *Leersia hexandra* – southern cutgrass
- *Lolium* spp. – ryegrasses
  - *Lolium multiflorum* – Italian ryegrass
  - *Lolium perenne* – perennial ryegrass
- *Megathyrsus maximus* – Guinea grass
- *Melinis minutiflora* – molasses grass
- *Paspalum dilatatum* – dallisgrass
- *Phalaris arundinacea* – reed canarygrass
- *Phleum pratense* – timothy
- *Poa* spp. – bluegrasses, meadow-grasses
  - *Poa arachnifera* – Texas bluegrass
  - *Poa pratensis* – Kentucky bluegrass
  - *Poa trivialis* – rough bluegrass
- *Setaria sphacelata* – African bristlegrass
- *Themeda triandra* – kangaroo grass
- *Thinopyrum intermedium* – intermediate wheatgrass

Alfalfa

*White clover (*Trifolium repens*)*

31.1.2 Herbaceous legumes

Herbaceous legume forages include:[*][7]

- *Arachis pintoi* – pinto peanut
- *Chamaecrista rotundifolia* – roundleaf sensitive pea
- *Clitoria ternatea* – butterfly-pea
- *Lotus corniculatus* – bird's-foot trefoil
- *Macroptilium atropurpureum* – purple bush-bean
- *Macroptilium bracteatum* – burgundy bean
- *Medicago* spp. – medics
  - *Medicago sativa* – alfalfa, lucerne
  - *Medicago truncatula* – barrel medic
- *Melilotus* spp. – sweetclovers

- *Neonotonia wightii* – perennial soybean

- *Onobrychis viciifolia* – common sainfoin

- *Stylosanthes* spp. – stylo

  - *Stylosanthes humilis* – Townsville stylo

  - *Stylosanthes scabra* – shrubby stylo

- *Trifolium* spp. – clovers

  - *Trifolium hybridum* – alsike clover

  - *Trifolium incarnatum* – crimson clover

  - *Trifolium pratense* – red clover

  - *Trifolium repens* – white clover

- *Vicia* spp. – vetches

  - *Vicia articulata* – oneflower vetch

  - *Vicia ervilia* – bitter vetch

  - *Vicia narbonensis* – narbon vetch

  - *Vicia sativa* – common vetch, tare

  - *Vicia villosa* – hairy vetch

- *Vigna parkeri* – creeping vigna

31.1.3   Tree legumes

Tree legume forages include:

- *Acacia aneura* – mulga

- *Albizia* spp. – silk trees

- *Albizia canescens* – Belmont siris

- *Albizia lebbeck* – lebbeck

- *Enterolobium cyclocarpum* – earpodtree

- *Leucaena leucocephala* – leadtree

31.1.4   Silage

Silage may be composed by the following:*[8]

- Alfalfa

- Maize (corn)

- Grass-legume mix

- Sorghums

- Oats

Sheep with silage

31.1.5   Crop residue

Crop residues used as forage include:

- Sorghum

- Corn or soybean stover

31.2   See also

- Grass-fed beef

31.3   References

[1] Fageria, N.K. (1997). *Growth and Mineral Nutrition of Field Crops.* NY,NY: Marcel Dekker. p. 595.

[2] Fageria, N.K. (1997). *Growth and Mineral Nutrition of Field Crops.* NY,NY: Marcel Dekker. p. 583.

[3] Karpouzi V, R Watson and D Pauly (2006) "Forage fish consumption by marine mammals and seabirds" *Fisheries Centre Research Reports,* **14** (3): 33–46.

[4] Givens, D. Ian (2000). *Forage evaluation in ruminant nutrition.* CABI. p. 1. ISBN 978-0-85199-344-7.

[5] Murphy, B. (1998). *Greener Pastures On Your Side of the Fence.* Colchester, Vermont: Arriba Publishing. pp. 19–20.

[6] "Pasture". *New International Encyclopedia.* 1905.

[7] Murphy, B. (1998). *Greener Pastures On Your Side of the Fence.* Colchester, Vermont: Arriba Publishing. p. 20.

[8] George, J. R. (1994). *Extension Publications: Forage and Grain Crops.* Dubuque,Iowa: Kendall/Hunt. p. 152.

Chapter 32

Forest farming

Forest farming is the cultivation of high-value specialty crops under a forest canopy that is intentionally modified or maintained to provide shade levels and habitat that favor growth and enhance production levels. Forest farming encompasses a range of cultivated systems from introducing plants into the understory of a timber stand to modifying forest stands to enhance the marketability and sustainable production of existing plants.[*][1]

Forest farming is a type of agroforestry practice characterized by the "four I's": intentional, integrated, intensive and interactive.[*][2] Agroforestry is a land management system that combines trees with crops or livestock, or both, on the same piece of land. It focuses on increasing benefits to the landowner as well as maintaining forest integrity and environmental health. The practice involves cultivating non-timber forest products or niche crops, some of which, such as ginseng or shiitake mushrooms, can have high market value.

Non-timber forest products (NTFPs) are plants, parts of plants, fungi, and other biological materials harvested from within and on the edges of natural, manipulated, or disturbed forests.[*][3] Examples of crops are ginseng, shiitake mushrooms, decorative ferns, and pine straw.[*][4] Products typically fit into the following categories: edible, medicinal and dietary supplements, floral or decorative, or specialty wood-based products.

32.1 History

Forest farming, though not always by that name, is practiced around the world. For centuries, humans have relied on fruits, nuts, seeds, parts of foliage and pods from trees and shrubs in the forests to feed themselves and their livestock.[*][5] Over time, certain species have been selected for cultivation near homes or livestock to provide food or medicine. For example, in the southern United States, mulberry trees are used as a feedstock for pigs and often cultivated near pig quarters.

Toyohiko Kagawa, forest farming pioneer.

In 1929, J. Russell Smith, Emeritus Professor of Economic Geography at Columbia University, published "Tree Crops – A Permanent Agriculture" which stated that crop-yielding trees could provide useful substitutes for cereals in animal feeding programs, as well as conserve environmental health.[*][6] Toyohiko Kagawa read and was heavily influenced by Smith's publication and began experimental cultivation under trees in Japan during the 1930s. Through forest farming, or three-dimensional forestry, Kagawa addressed problems of soil erosion by persuading many of Japan's upland farmers to plant fodder trees to conserve soil, supply food and feed animals. He combined extensive plantings of walnut trees, harvested the nuts and fed them to the pigs, then sold the pigs as a source of income. When the walnut trees matured, they were sold for timber and more trees were planted so that there was a continuous cycle of economic cropping that provided both short-term and long-term income to the small landowner.[*][7] The success of these trials prompted similar research in other countries.

154

Unfortunately, World War II disrupted communication and slowed advances in forest farming.*[8] In the mid-1950s research resumed in places such as southern Africa. Kagawa was also an inspiration to Robert Hart pioneered forest gardening in temperate climates in the sixties in Shropshire, England.*[9]

In earlier years, livestock were often considered part of the forest farming system. Now they are typically excluded and agroforestry systems that integrate trees, forages and livestock are referred to as silvopastures.*[10] Because forest farming combines ecological stability of natural forests and productive agriculture systems, it is considered to have great potential for regenerating soils, restoring ground water supplies, controlling floods and droughts and cultivating marginal lands.*[11] In addition to these benefits for re-establishing productive forests on marginal lands, forest farming is way to add financial value while conserving land that is currently forested, as discussed in the methods section.

In more recent years, there has been growing interest in locally grown and organic foods throughout the United States. There has been an increase in farmer's markets and community-supported agriculture small enterprises. These have also become outlets for NTFPs. In order to stay competitive, many farmers look to add unique crops to their product line. With the quantity and quality of resources developing online that offer tutorials and educational information on how to create and maintain forest farms, forest gardens, how to cultivate specific crops such as shiitake mushrooms and how to successfully market these items, forest farming is expanding as a viable land management practice. Good places to look for research-based resources are the USDA National Agroforestry Center's publication section, the Center for Agroforestry at the University of Missouri, the Cornell Cooperative Extension, the Non Timber Forest Products website by The Virginia Tech Department of Wood Science and Forest Products, the USDA Forest Service Southern Research Station and the Top of the Ozarks RC&D in Missouri and the collaborative Forest Farming community of practice on eXtension.org, the online presence of the Cooperative Extension System of the US Land Grant Universities.

32.2 Principles

Forest farming principles constitute an ecological approach to forest management. Forest resources are judiciously used while biodiversity and wildlife habitat are conserved. Forest farms have the potential to restore ecological balance to fragmented second growth forests through intentional manipulation to create the desired forest ecosystem.

In some instances, the intentional introduction of species for botanicals, medicinals, food or decorative products is accomplished using existing forests. The tree cover, soil type, water supply, land form and other site characteristics determine what species will thrive. Developing an understanding of species/site relationships as well as understanding the site limitations is necessary to utilize these resources for production needs, while conserving adequate resources for the long-term health of the forest.

Apart from the environmental benefits, forest farming can increase the economic value of forest property and provide short- and long-term benefits to the landowner. Forest farming provides economic return from intact forest ecosystems, but timber sales can remain part of the long-term management strategy.

32.3 Methods

Forest farming methods may include: Intensive, yet careful thinning of overstocked, suppressed tree stands; multiple integrated entries to accomplish thinning so that systemic shock is minimized; and interactive management to maintain a cross-section of healthy trees and shrubs of all ages and species. Physical disturbance to the surrounding area should be minimized. The following are forest farming techniques described in the Agroforestry Training Manual produced by the Center for Agroforestry at the University of Missouri.*[12]

32.3.1 Level of management required

(from most intense to least intense)

1. Forest gardening is the most intensive of forest farming methods. In addition to thinning the overstory, this method involves clearing the understory of undesirable vegetation and other practices that are closely related to agronomy (tillage, fertilization, weeding, and control of disease and insects and wildlife management). Due to input levels, this method often produces lower valued products compared to other methods. Forest gardens take advantage of the vertical levels of light availability and space under the forest canopy so that more than one crop can be grown at once if desired.

2. Wild-simulated seeks to maintain a natural growing environment, yet enriches local NTFP populations to create an abundant renewable supply of the products. Minimal disturbance and natural growing conditions ensure products will be similar in appearance and quality of those harvested from the wild. Rather than till, practitioners often rake leaves to expose soil, sow seed directly onto the ground, and then cover with leaves again. Since this method produces NTFPs that closely resemble wild plants; they often command a higher price than NTFPs produced using the forest gardening method.

3. Forest tending involves adjusting tree crown density to manipulate light levels that favor natural reproduction of desirable NTFPs. This low intensity management approach does not involve supplemental planting to increase populations of desired NTFPs.

4. Wildcrafting is the harvesting of naturally growing NTFPs. It is not considered a forest farming practice since there is no human involvement in the plant's establishment and maintenance. However, wildcrafters often take steps to protect NTFPs with future harvests in mind.*[13] It becomes agroforestry once forest thinnings, or other inputs, are applied to sustain or maintain plant populations that might otherwise succumb to successional changes in the forest. The most important difference between forest farming and wildcrafting is that forest farming intentionally produces NTFPS, whereas wildcrafting seeks and gathers from naturally growing NTFPs.

32.4   Production considerations

Forest farming can be a small business opportunity for landowners and requires careful planning, including a business and marketing plan. Learning how to market the NTFPs on the Internet is an option, but may entail higher shipping costs. Landowners should consider all options for selling their products including, farmer's markets or restaurants that focus on locally grown ingredients. The development phase should include a forest management plan that states the landowner's objectives and a resource inventory. Start-up costs should be analyzed as specific equipment may be necessary to harvest or process the product, whereas other crops require minimal initial investment. Local incentives for sustainable forest management, as well as regulations and policies should be explored. The Convention on International Trade in Endangered Species of Wild Fauna and Flora (CITES) regulates international trade of certain plant (American ginseng and goldenseal) and animal species. To be legally exported, regulated plants must be harvested and records kept according to CITES rules and restrictions. Many states also have harvesting regulations for certain native plants that are searchable online. Another good source to start with on information is the Medicinal Plants at Risk 2008 report by the Center for Biological Diversity in the U.S.

32.5   Examples of crops

(from the National Agroforestry Center)

Medicinal herbs:

- Ginseng (*Panax quinquefolius*)
- Black Cohosh (*Actaea racemosa*)
- Goldenseal (*Hydrastis canadensis*)
- Bloodroot (*Sanguinaria canadensis*)
- Pacific yew (*Taxus brevifolia*)
- Mayapple (*Podophyllum peltatum*)
- Saw palmetto (*Serenoa repens*)
- American Pokeweeed (*Phytolacca americana*)

Nuts:

- Black walnut (*Juglans nigra*)
- Hazelnut (*Corylus avellana*)
- Shagbark hickory (*Carya ovata*)
- Beechnut (*Fagus sylvatica*)

Fruit:

- Pawpaw (*Asimina triloba*)
- Currants (*Ribes spp*)
- Elderberry (*Sambucas spp*)
- Serviceberry (*Amelanchier spp*)
- Blackberry (*Rubus spp*)
- Huckleberry (*Gaylussacia brachycera*)

Other food crops:

- Ramps (wild leeks) (*Allium tricoccum*)
- Syrups (maple)
- Honey
- Mushrooms
- Other edible roots

Other products: (mulch, decoratives, crafts, dyes)

- Pine straw

- Willow twigs

- Vines

- Beargrass (*Xerophyllum tenax*)

- Ferns

- Pine cones

- Moss

Native ornamentals:

- Rhododendron (*Rhododendron catawbiense*)

- Highbush cranberry (*Viburnum trilobum*)

- Flowering dogwood (*Cornus florida*)

32.6   References

[1] Chamberlain, J.L.; D. Mitchell; T. Brigham; T. Hobby (2009). "Forest Farming Practices" . *North American Agroforestry: an integrated science and practice* (2nd ed.). Madison, Wisconsin: American Society of Agronomy. pp. 219–254.

[2] Gold, M.A., W.J. Rietveld, H.E. Garrett and R.F. Fisher, 2000. In: H.E. Garrett, W.J. Rietveld, and R.F. Fisher (eds.) *North American Agroforestry: An Integrated Science and Practice*. Madison, WI: American Society of Agronomy. As cited at "Forest Farming," Cornell University.

[3] Chamberlain 2009

[4] Vaughan, R. C.; J. F. Munsell; J. L. Chamberlain (2013). "Opportunities for Enhancing Nontimber Forest Products Management in the United States." . *Journal of Forestry* **111** (1): 26–33.

[5] Douglas, J.S.; R. A. de J. Hart (1984). *Forest farming: towards a solution to problems of world hunger and conservation*. London: Intermediate Technology Publications.

[6] Douglas 1984

[7] Douglas 1984

[8] Douglas 1984

[9] Hart, Robert A. de J. (1996). *Forest gardening: cultivating an edible landscape*. White River Junction, VT: Chelsea Green Pub. Co.

[10] Garrett, H.E. (2009). *North American agroforestry: an integrated science and practice.* (2nd ed.). Madison, WI: American Society of Agronomy.

[11] Garrett 2009

[12] University of Missouri Center for Agroforestry (2006). *Training manual for applied agroforestry practices.* Columbia, Mo.: University of Missouri Center for Agroforestry.

[13] Vaughn et al. 2013

32.7   External links

- National Agroforestry Center (USDA)

- Agroforestry Practices by The Center for Agroforestry, University of Missouri.

- http://hwwff.cce.cornell.edu/

- http://www.centerforagroforestry.org/pubs/training/sec7.pdf

- http://www.ces.ncsu.edu/fletcher/programs/herbs/crops/medicinal/forest-farming-2007.html

- http://www.ntfpinfo.us/

- http://www.dcnr.state.pa.us/Forestry/wildplant/forest_farming_proceedings.pdf

Chapter 33

Forest gardening

"Home gardens" redirects here. For other uses, see Home garden (disambiguation).

Forest gardening is a low-maintenance sustainable plant-

Robert Hart's forest garden in Shropshire

based food production and agroforestry system based on woodland ecosystems, incorporating fruit and nut trees, shrubs, herbs, vines and perennial vegetables which have yields directly useful to humans. Making use of companion planting, these can be intermixed to grow in a succession of layers, to build a woodland habitat.

Forest gardening is a prehistoric method of securing food in tropical areas. In the 1980s, Robert Hart coined the term "forest gardening" after adapting the principles and applying

them to temperate climates.[1]

33.1 History

Forest gardens are probably the world's oldest form of land use and most resilient agroecosystem.[2][3] They originated in prehistoric times along jungle-clad river banks and in the wet foothills of monsoon regions. In the gradual process of families improving their immediate environment, useful tree and vine species were identified, protected and improved whilst undesirable species were eliminated. Eventually superior foreign species were selected and incorporated into the gardens.[4]

Forest gardens are still common in the tropics and known by various names such as: *home gardens* in Kerala in South India, Nepal, Zambia, Zimbabwe and Tanzania; *Kandyan forest gardens* in Sri Lanka;[5] *huertos familiares*, the "family orchards" of Mexico; and *pekarangan*, the gardens of "complete design", in Java.[6] These are also called agroforests and, where the wood components are short-statured, the term shrub garden is employed. Forest gardens have been shown to be a significant source of income and food security for local populations.[7]

Robert Hart adapted forest gardening for the United Kingdom's temperate climate during the 1980s.[1] His theories were later developed by Martin Crawford from the Agroforestry Research Trust and various permaculturalists such as Graham Bell, Patrick Whitefield, Dave Jacke and Geoff Lawton.

33.2 In tropical climates

Forest gardens, or home gardens, are common in the tropics, using intercropping to cultivate trees, crops, and livestock on the same land. In Kerala in south India as well as in northeastern India, the home garden is the most common form of land use and is also found in Indonesia.

One example combines coconut, black pepper, cocoa and pineapple. These gardens exemplify polyculture, and conserve much crop genetic diversity and heirloom plants that are not found in monocultures. Forest gardens have been loosely compared to the religious concept of the Garden of Eden.[*][8]

33.2.1 Americas

The BBC's *Unnatural Histories* claimed that the Amazon rainforest, rather than being a pristine wilderness, has been shaped by humans for at least 11,000 years through practices such as forest gardening and *terra preta*.[*][9] This was also explored in the bestselling book *1491* by author Charles C. Mann. Since the 1970s, numerous geoglyphs have also been discovered on deforested land in the Amazon rainforest, furthering the evidence about Pre-Columbian civilizations.[*][10][*][11]

On the Yucatán Peninsula, much of the Maya food supply was grown in "orchard-gardens", known as *pet kot*.[*][12] The system takes its name from the low wall of stones (*pet* meaning circular and *kot* wall of loose stones) that characteristically surrounds the gardens.[*][13]

33.2.2 Africa

In many African countries, for example Zambia, Zimbabwe, Ethiopia and Tanzania, gardens are widespread in rural, periurban and urban areas and they play an essential role in establishing food security. Most well known are the Chaga or Chagga gardens on the slopes of Mt. Kilimanjaro in Tanzania. These are an excellent example of an agroforestry system. In many countries, women are the main actors in home gardening and food is mainly produced for subsistence. In North-Africa, oasis layered gardening with palm trees, fruit trees and vegetables is a traditional type of forest garden.

33.2.3 Nepal

In Nepal, the *Ghar Bagaincha*, literally "home garden", refers to the traditional land use system around a homestead, where several species of plants are grown and maintained by household members and their products are primarily intended for the family consumption (Shrestha et al., 2002). The term "home garden" is often considered synonymous to the kitchen garden. However, they differ in terms of function, size, diversity, composition and features (Sthapit et al., 2006). In Nepal, 72% of households have home gardens of an area 2–11% of the total land holdings (Gautam et al., 2004). Because of their small size, the government

has never identified home gardens as an important unit of food production and they thereby remain neglected from research and development. However, at the household level the system is very important as it is an important source of quality food and nutrition for the rural poor and, therefore, are important contributors to the household food security and livelihoods of farming communities in Nepal. The gardens are typically cultivated with a mixture of annual and perennial plants that can be harvested on a daily or seasonal basis. Biodiversity that has an immediate value is maintained in home gardens as women and children have easy access to preferred food. Home gardens, with their intensive and multiple uses, provide a safety net for households when food is scarce. These gardens are not only important sources of food, fodder, fuel, medicines, spices, herbs, flowers, construction materials and income in many countries, they are also important for the in situ conservation of a wide range of unique genetic resources for food and agriculture (Subedi et al., 2004). Many uncultivated, as well as neglected and underutilised species could make an important contribution to the dietary diversity of local communities (Gautam et al., 2004).

In addition to supplementing diet in times of difficulty, home gardens promote whole-family and whole-community involvement in the process of providing food. Children, the elderly, and those caring for them can participate in this infield agriculture, incorporating it with other household tasks and scheduling. This tradition has existed in many cultures around the world for thousands of years.[*][14][*][15]

33.3 In mediteranean climates

The mediterranean climate has long, hot, rainless summers and relatively short, cool, rainy winters (Köppen climate classification *Csa*).[*][16] Its climate conditions are highly variable within an area and modified locally by altitude, latitude, and the proximity to the Mediterranean.[*][16] In the 1950s the Forest Research Department of the Ministry of Agriculture founded a botanical forest garden in the Sharon region in Israel, the Ilanot Forest.[*][17] As the only one of its kind in Israel, it harbours more than seven hundred and fifty species of trees from locations all over the world, including the Japanese sago palm cycas revoluta, fig trees (ficus glomerata), stone pine trees (pinus pinea) that produce tasty pine nuts and adds to the biodiversity of Israel.

33.4 In temperate climates

Robert Hart coined the term "forest gardening" during the 1980s. Hart began farming at Wenlock Edge in Shropshire

Flash flood at Ein Avdat

Robert Hart, forest gardening pioneer

with the intention of providing a healthy and therapeutic environment for himself and his brother Lacon.[18] Starting as relatively conventional smallholders, Hart soon discovered that maintaining large annual vegetable beds, rearing livestock and taking care of an orchard were tasks beyond their strength. However, a small bed of perennial vegetables and herbs he planted was looking after itself with little intervention.

Following Hart's adoption of a raw vegan diet for health and personal reasons, he replaced his farm animals with plants. The three main products from a forest garden are fruit, nuts and green leafy vegetables.[19] He created a model forest garden from a 0.12 acre (500 m²) orchard on his farm and intended naming his gardening method *ecological horticulture* or *ecocultivation*.[20] Hart later dropped these terms once he became aware that *agroforestry* and *forest gardens* were already being used to describe similar systems in other parts of the world.[21] He was inspired by the forest farming methods of Toyohiko Kagawa and James Sholto Douglas, and the productivity of the Keralan home gardens as Hart explains:[22]

> From the agroforestry point of view, perhaps the world's most advanced country is the Indian state of Kerala, which boasts no fewer than three and a half million forest gardens...As an example of the extraordinary intensivity of cultivation of some forest gardens, one plot of only 0.12 hectares (0.30 acres) was found by a study group to have twenty-three young coconut palms, twelve cloves, fifty-six bananas, and forty-nine pineapples, with thirty pepper vines trained up its trees. In addition, the small holder grew fodder for his house-cow.[23]

33.4.1   Seven-layer system

The seven layers of the forest garden

Robert Hart pioneered a system based on the observation that the natural forest can be divided into distinct levels. He used intercropping to develop an existing small orchard of apples and pears into an edible polyculture landscape consisting of the following layers:

1. 'Canopy layer' consisting of the original mature fruit trees.

2. 'Low-tree layer' of smaller nut and fruit trees on dwarfing root stocks.

3. 'Shrub layer' of fruit bushes such as currants and berries.

4. 'Herbaceous layer' of perennial vegetables and herbs.

5. 'Rhizosphere' or 'underground' dimension of plants grown for their roots and tubers.

6. 'Ground cover layer' of edible plants that spread horizontally.

7. 'Vertical layer' of vines and climbers.

A key component of the seven-layer system was the plants he selected. Most of the traditional vegetable crops grown today, such as carrots, are sun loving plants not well selected for the more shady forest garden system. Hart favoured shade tolerant perennial vegetables.

33.4.2 Further development

The Agroforestry Research Trust (ART), managed by Martin Crawford, runs experimental forest gardening projects on a number of plots in Devon, United Kingdom.*[24] Crawford describes a forest garden as a low-maintenance way of sustainably producing food and other household products.*[25]

Ken Fern had the idea that for a successful temperate forest garden a wider range of edible shade tolerant plants would need to be used. To this end, Fern created the organisation Plants for a Future (PFAF) which compiled a plant database suitable for such a system. Fern used the term *woodland gardening*, rather than forest gardening, in his book *Plants for a Future*.*[26]*[27]

The Movement for Compassionate Living (MCL) promote forest gardening and other types of vegan organic gardening to meet society's needs for food and natural resources. Kathleen Jannaway, the founder of MCL, wrote a book outlining a sustainable vegan future called *Abundant Living in the Coming Age of the Tree* in 1991. In 2009, the MCL provided a grant of £1,000 to the Bangor Forest Garden project in Gwynedd, North West Wales.*[28]

In 2005, Dave Jacke and Eric Toensmeier's two-volume book *Edible Forest Gardens* provided a deeply researched reference focused on North American forest gardening climates, habitats, and species. The book attempts to ground forest gardening deeply in ecological science. The Apios Institute wiki grew out of their work, and seeks to document and share the experience of people around the world working with the species in polycultures.

33.4.3 Permaculture

Bill Mollison, who coined the term *permaculture*, visited Robert Hart at his forest garden in Wenlock Edge in October 1990.*[29] Hart's seven-layer system has since been adopted as a common permaculture design element.

Numerous permaculturalists are proponents of forest gardens, or food forests, such as Graham Bell, Patrick Whitefield, Dave Jacke, Eric Toensmeier and Geoff Lawton. Bell started building his forest garden in 1991 and wrote the book *The Permaculture Garden* in 1995, Whitefield wrote the book *How to Make a Forest Garden* in 2002, Jacke and Toensmeier co-authored the two volume book set *Edible Forest Gardening* in 2005, and Lawton presented the film *Establishing a Food Forest* in 2008.*[30]*[31]*[32]

Austrian Sepp Holzer practices "Holzer Permaculture" on his *Krameterhof* farm, at varying altitudes ranging from 1,100 to 1,500 metres above sea level. His designs create micro-climates with rocks, ponds and living wind barriers, enabling the cultivation of a variety of fruit trees, vegetables and flowers in a region that averages 4°C, and with temperatures as low as −20°C in the winter.

33.5 Projects

El Pilar on the Belize-Guatemala border features a forest garden to demonstrate traditional Maya agricultural practices.*[33]*[34] A further 1-acre model forest garden, called Känan K' aax (meaning well-tended garden in Mayan), is being funded by the National Geographic Society and developed at Santa Familia Primary School in Cayo.*[35]

In the United States the largest known food forest on public land is believed to be the 7-acre Beacon Food Forest in Seattle, WA.*[36] Other forest garden projects include those at the Central Rocky Mountain Permaculture Institute in Basalt, Colorado and Montview Neighborhood farm in Northampton, Massachusetts.*[37]*[38]

In Canada food forester Richard Walker has been developing and maintaining food forests in the province of British Columbia for over 30 years. He developed a 3-acre food forest that when at maturity provided raw materials for a nursery and herbalism business as well as food for his family.*[39] The Living Centre have developed various forest garden projects in Ontario.*[40]

In the United Kingdom, other than those run by the Agroforestry Research Trust (ART), there are numerous forest garden projects such as the Bangor Forest Garden in Gwynedd, North West Wales.*[41] Martin Crawford from ART administers the Forest Garden Network, an informal network of people and organisations around the world who are cultivating their own forest gardens.*[42]*[43]

33.6   See also

- Agroecology
- Analog forestry
- Climate-friendly gardening
- Deep ecology
- Forest farming
- List of companion plants
- Mycoforestry
- Multiple cropping
- Natural farming
- Nutrient cycle
- Orchard
- Permaculture
- Polyculture
- Vegan organic gardening

33.7   Notes

[1] Crawford, Martin (2010). *Creating a Forest Garden*. Green Books. p. 18.

[2] Hart, Robert A. de J. (1996a), p.124: "Forest gardening, in the sense of finding uses for and attempting to control the growth of wild plants, is undoubtedly the oldest form of land use in the world."

[3] Douglas John McConnell (2003). *The Forest Farms of Kandy: And Other Gardens of Complete Design*, p.1, "Forest garden farms are probably the world's oldest and most resilient agroecosystem."

[4] Douglas John McConnell (1992). *The Forest-Garden Farms of Kandy, Sri Lanka*. p. 1. ISBN 9789251028988.

[5] Jacob, V. J.; Alles, W. S. (1987). "Kandyan gardens of Sri Lanka". *Agroforestry Systems* **5** (2): 123. doi:10.1007/BF00047517.

[6] timeshighereducation.co.uk

[7] Douglas John McConnell (1973). *The economic structure of Kandyan forest-garden farms.*

[8] Graham Bell (2004). *The Permaculture Garden*, p.129, "The Forest Garden...This is the original garden of Eden. It could be your garden too."

- Also see Rob Hopkins (foreword), Martin Crawford (2010). *Creating a Forest Garden: Working with Nature to Grow Edible Crops*, p.10 "Perhaps what Hart created was the closest to what we imagine the Garden of Eden as being."
- Helmut Lieth (1989). *Tropical Rain Forest Ecosystems: Biogeographical and Ecological Studies*, p.611 "Important food plants, such as sago-producing palms, fruit-producing trees and medicinal plants were purposefully aggregated and tended in convenient places. Eventually, the forest garden, a kind of Garden of Eden, emerged. These jungle gardens on good soils of easy access required little maintenance and hardly any hard work."
- Dave Jacke and Eric Toensmeier (2005). *Edible Forest Gardens - Volume One*, p.1
- Robert Hart (1996a), p.80
- Deborha d'Arms (2011). *Jardin D'Or: A Treatise on Forest Gardening, Recreating Sustainable Gardens of Eden*

[9] "Unnatural Histories - Amazon". *BBC Four.*

[10] Simon Romero (January 14, 2012). "Once Hidden by Forest, Carvings in Land Attest to Amazon's Lost World". *The New York Times.*

[11] Martti Pärssinen, Denise Schaan and Alceu Ranzi (2009). "Pre-Columbian geometric earthworks in the upper Purús: a complex society in western Amazonia". *Antiquity* **83** (322): 1084–1095.

[12] Michael Ernest Smith and Marilyn A. Masson (2000). *The Ancient Civilizations of Mesoamerica*. p. 127. ISBN 9780631211167.

[13] David L. Lentz, ed. (2000). *Imperfect Balance: Landscape Transformations in the Precolumbian Americas*. p. 212. ISBN 9780231111577.

[14] Killion, Thomas W., *Gardens of Prehistory: The Archaeology of Settlement Agriculture in Greater Mesoamerica*, University of Alabama Press, 1992

[15] Heidelberg, Kurt, "Ethnographic Analogy and Its Problems in the Northern Maya Lowlands". In *Lifeways in the Northern Maya Lowlands: New Approaches to Archaeology in the Yucatan Peninsula*. Edited by Jennifer Mathews. University of Arizona Press. 2006

[16] "Climate". U.S. Library of Congress. Retrieved 8 April 2008.

[17] "Ilanot Forest – A Botanical Forest Garden". *kkl.org.il.* KKL JNF. Retrieved 22 September 2015.

[18] Graham Burnett. "Seven Storeys of Abundance; A visit to Robert Hart's Forest Garden".

[19] Patrick Whitefield (2002). *How to Make a Forest Garden.* p. 5. ISBN 9781856230087.

[20] Hart, Robert A. de J. (1996a), p. 45

[21] Hart, Robert A. de J. (1996a), pages 28 and 43

[22] Hart, Robert A. de J. (1996a), p. 41

[23] Hart, Robert A. de J. (1996a), pages 4–5

[24] "Agroforestry Research Trust".

[25] "Forest gardening". Agroforestry Research Trust. Retrieved 13 Feb 2013.

[26] "Woodland Gardening".

[27] "Plants for a Future - The book".

[28] "Bangor Forest Garden" (PDF). The Movement for Compassionate Living - New Leaves (issue no.93). 2009. pp. 6–8.

[29] Hart, Robert A. de J. (1996a), p. 149

[30] "Graham Bell's Forest Garden".

[31] "Edible Forest Gardening".

[32] "*Establishing a Food Forest* review".

[33] Ford, Anabel (May 2, 2009). "El Pilar Archaeological Reserve for Maya Flora and Fauna". *The Guatemala Times.* Retrieved 2009-07-26.

[34] Ford, Anabel (December 15, 2010). "Legacy of the Ancient Maya: The Maya Forest Garden". Popular Archaeology.

[35] "National Geographic Society Funds Mayan Garden".

[36] Mellinger, Robert (16 February 2012). "Nations Largest Food Forest takes root on Beacon Hill". *Crosscut.* Retrieved 14 March 2012.

[37] "The Central Rocky Mountain Permaculture Institute".

[38] "Montview Neighborhood farm".

[39] "Richard Walker".

[40] "Forest Gardening".

[41] "Bangor Forest Garden".

[42] "The Agroforestry and Forest Garden Network".

[43] Martin Crawford (2014). "List of visitable forest garden and agroforestry projects in the UK, Europe and North America". Agroforestry Research Trust.

33.8 References

- Crawford, Martin 2010. *Creating a Forest Garden: Working with Nature to Grow Edible Crops.* Totnes: Green Books. ISBN 1-900322-62-5.

- d'Arms, Deborha 2011. *Jardin d' Or (Garden of Gold): A Treatise on Forest Gardening, Recreating Sustainable Gardens of Eden.* Los Gatos, CA: Robertson Publishing. ISBN 978-1611700299.

- Douglas, J. Sholto and Hart, Robert A. de J. 1985. *Forest Farming.* Intermediate Technology. ISBN 0-946688-30-3.

- Fern, Ken 1997. *Plants for a Future: Edible and Useful Plants for a Healthier World.* Hampshire: Permanent Publications. ISBN 1-85623-011-2.

- Hart, Robert A. de J. (1996a). *Forest Gardening: Cultivating an Edible Landscape.* White River Junction, VT: Chelsea Green. ISBN 0-930031-84-9.

- Hart, Robert A. de J. 1996b. *Beyond the Forest Garden.* Gaia Books. ISBN 1-85675-037-X.

- Jacke, Dave, and Toensmeier, Eric 2005. *Edible Forest Gardens.* Two volume set. Volume One: *Ecological Vision and Theory for Temperate Climate Permaculture,* ISBN 1-931498-79-2. Volume Two: *Ecological Design and Practice for Temperate Climate Permaculture,* ISBN 1-931498-80-6. White River Junction, VT: Chelsea Green.

- Jannaway, Kathleen 1991. *Abundant Living in the Coming Age of the Tree.* Movement for Compassionate Living. ISBN 0-9517328-0-3.

- Smith, Joseph Russell 1988 (first published in 1929). *Tree Crops: A Permanent Agriculture.* Island Press. ISBN 0-933280-44-0

- Whitefield, P. 2002. *How to Make a Forest Garden.* Hampshire: Permanent Publications. ISBN 1-85623-008-2.

33.9 External links

- Why Food Forests?, Permaculture Research Institute

- Plant an Edible Forest Garden, *Mother Earth News*

- The garden of the future?, *The Guardian*

- Edible Forest Gardens: an Invitation to Adventure, *The Natural Farmer*

- Forest gardens, Permaculture Association

- El Pilar Forest Garden Network, information on traditional Maya forest gardening

Chapter 34

Forest produce (India)

Forest produce is defined under section 2(4) of the Indian Forest Act, 1927. Its legal definition includes timber, charcoal, caoutchouc, catechu, wood-oil, resin, natural varnish, bark, lac, myrobalans, mahua flowers (whether found inside or brought from a forest or not), trees and leaves, flowers and fruit, plants (including grass, creepers, reeds and moss), wild animals, skins, tusks, horns, bones, cocoons, silk, honey, wax, other parts or produce of animals, and also includes peat, surface soil, rocks and minerals etc. when found inside or brought from a forest, among other things.

Mahua

Forest produce can be divided into several categories. From the point of view of usage, forest produce can be categorized into three types: Timber, Non Timber and Minor Minerals. Non-timber forest products [NTFPs] are known also as *minor forest produce* (MFP) or *non-wood forest produce* (NWFP). The NTFP can be further categorized into medicinal and aromatic plants (MAP), oil seeds, fiber & floss, resins, edible plants, bamboo, reeds and grasses.

34.1   Wood products

Timber The Odisha Forest Development Corporation (OFDC) trades timber in round and sawn forms, in different dimensions, from several depots. Round timbers are sold monthly from each depot, through general auction.

Sandalwood The sandalwood tree is found in southern Indian forests, i.e. in Kerala, Tamil Nadu, Karnataka, etc.

Plywood The plywood industry at Kuikeda near Saintala of Bolangir district, Odisha, was incorporated during 1983, and started commercial production during the year, 1986-87; it operated until 1992-93.

34.2   Non-wood forest produce

Non-timber forest product (NTFP) refers to all biological materials other than timber extracted from natural forests for human and animal use.

Kendu leaves Orissa is the third largest producer of Kendu leaf in India. The uniqueness of kendu leaf in Orissa is because of its specification of Color, Texture, Size and Body condition of the leaf.

Bamboo The collection and marketing of Bamboo from the natural forest is done either by OFDC or through the RMP (Raw Material Procurer) as per the decision of the Government to regulate the collection and trade of Bamboo.

Sal seed Sal seed is a nationalized product since 1973 and is one of the important Produce obtained from Sal (*Shorea robursta*) tree, which is predominantly available in Orissa.

Tendu Patta (Leaf) Collection

Honey OFDC is involved in collection, processing and trading of honey from natural forest with an assurance of pure and genuine in quality.

Medicinal plants With the financial aid of National Medicinal Plants Board, Government of India, 16 projects for promotional activities are currently running in Orissa.

Medicinal Plant Karra

Rubber OFDC is having rubber plantation and processing unit in Baripada & Bhubaneswar zone, since 2003.

OFDC is extracting the rubber from the matured trees and marketing it.

Pickle & squash OFDC is manufacturing and marketing high quality, delicious pickles such as Mango Pickle, Mixed Pickle – free from preservatives.

Cashew & spice OFDC Ltd. have raised cashew plantation over an area of 18704.99 ha. from 1978-79 to 1992-93 in Bhubaneswar and Berhampur Division. Out of which pure cashew plantation over an area of 11,053.99 ha.

34.3 Minerals

Biodiesel plant Biodiesel is the name of a clean burning alternative fuel, produced from domestic, renewable resources. Biodiesel contains no petroleum, but it can be blended at any level with petroleum diesel to create a biodiesel blend.

34.4 See also

- Forestry in India

- Indian Council of Forestry Research and Education

- Van Vigyan Kendra (VVK) Forest Science Centres

34.5 References

Chapter 35

Fur

For other uses, see Fur (disambiguation) and Furs (disambiguation).

"Furriness" redirects here. For the act of being a "furry", see Furry fandom.

Fur is used in reference to the hair of animals, usually

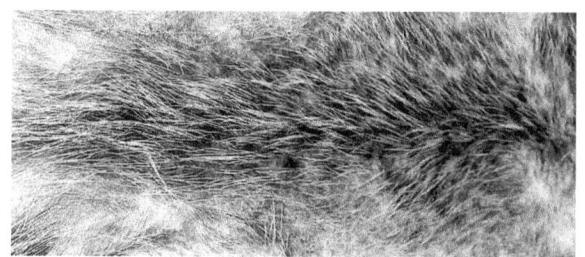

Opossum fur

Fur mosaic with portrait of Emperor Franz Joseph

mammals, particularly those with extensive body hair coverage. The term "**pelage**" (French, from Middle French, from *poil* hair, from Old French *peilss*, from Latin *pilus*; first known use in English circa 1828.[*][1]) is sometimes used to refer to the body hair of an animal as a complete coat. *Fur* is also used to refer to animal pelts which have been processed into leather with the hair still attached. The words *fur* or *furry* are also used, more casually, to refer to hair-like growths or formations, particularly when the subject being referred to exhibits a dense coat of fine, soft "hairs."

Animal *fur*, if layered, rather than grown as a single coat, may consist of short down hairs, long guard hairs, and, in some cases, medium awn hairs. Mammals with reduced amounts of fur are often called "naked", such as naked mole-rat and naked dogs.

An animal with commercially valuable fur is known within the fur industry as a furbearer. The use of fur as clothing and/or decoration is considered controversial by some people: most animal welfare advocates object to the trapping and killing of wildlife, and to the confinement and killing of animals on fur farms.

Fur has been a major challenge for 3D computer graphics artists due to its visual complexity and physical properties. The first movie which made extensive use of CGI fur was Pixar's 2001 film *Monsters, Inc.*

35.1 Composition

Fur usually consists of two main layers:

- Down hair (known also as undercoat or ground hair) the bottom layer consisting of wool hairs, usually wavy or curly without straight portions or sharp points; down hairs tend to be shorter, flat, curly, and more numerous than the top layer. Its principal function is thermoregulation; it maintains a layer of dry air next to the skin and repels water, thus providing thermal insulation.

- Guard hair the top layer consisting of longer, generally coarser, nearly straight shafts of hair that protrude through the down hair layer. The distal

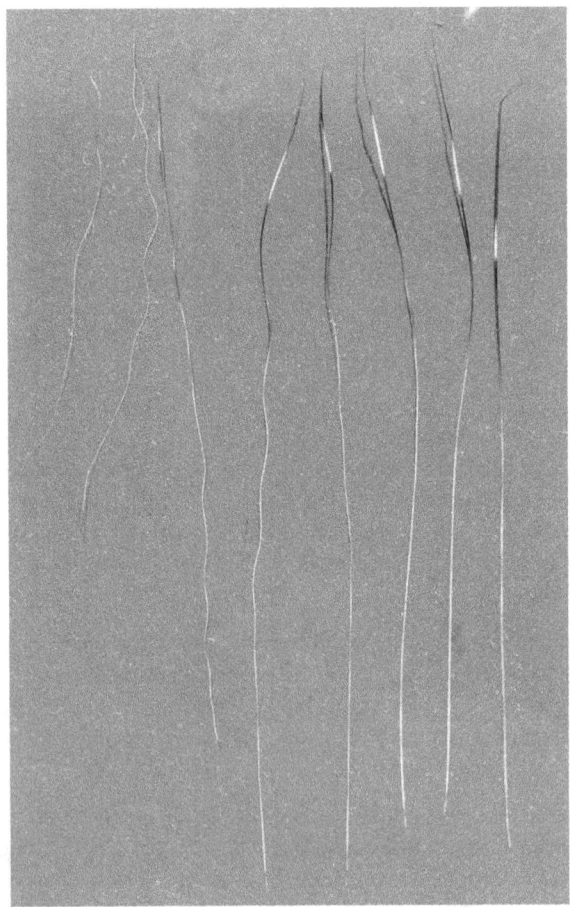

Down, awn and guard hairs of a domestic tabby cat

guard hairs.

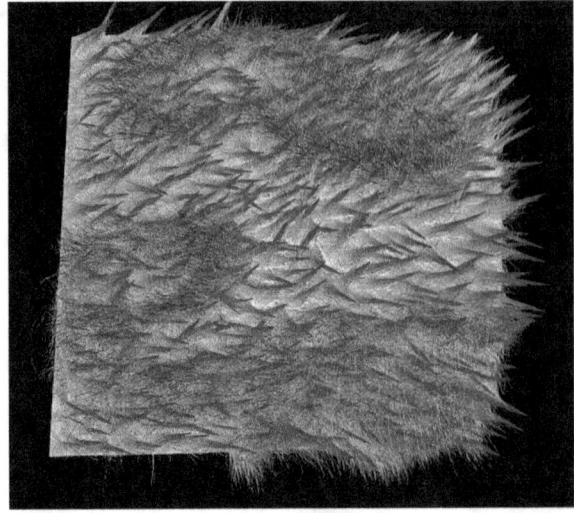

Computer-generated wet fur

ends of the guard hairs provide the externally visible layer of the coat of most mammals with well-developed fur. This layer of the coat displays the most marked pigmentation and gloss, including coat patterns adapted to display or camouflage. It is also adapted to shedding water and blocking sunlight, protecting the undercoat and skin from external factors such as rain and ultraviolet radiation. Many animals, such as domestic cats, erect their guard hairs as part of their threat display when agitated.

Mammals with well-developed down and guard hairs also usually have large numbers of awn hairs. These begin their growth much as guard hairs do, but change their mode of growth, usually when less than half the length of the hair has emerged. This portion of the hair is called awn. The rest of the growth is thin and wavy, much like down hair. In many species of mammals, the awn hairs comprise the bulk of the visible coat. The proximal part of the awn hair shares the function of the down hairs, whereas the distal part aids the water-shedding function of the guard hairs, though their thin basal portion prevents their being erected like true

35.2 Mammals without fur

Hair is one of the defining characteristics of mammals, however, several species or breeds have considerably reduced amounts of fur. These are often called "naked" or "hairless".

35.2.1 Natural selection

Some mammals naturally have reduced amounts of fur. Some semiaquatic or aquatic mammals such as cetaceans, pinnipeds and hippopotamuses have evolved hairlessness, presumably to reduce resistance through water. The naked mole-rat has evolved hairlessness, perhaps as an adaptation to their sub-terranian life-style. Two of the largest extant mammals, the elephant and the rhinoceros, are largely hairless. The hairless bat is mostly hairless but does have short bristly hairs around its neck, on its front toes, and around the throat sac, along with fine hairs on the head and tail membrane.*[2]

Humans are the only primate species that have undergone significant hair loss. The hairlessness of humans compared to related species may be due to loss of functionality in the pseudogene KRTHAP1 (which helps produce keratin) in the human lineage about 240,000 years ago.*[3] Mutations in the gene HR can lead to complete hair loss, though this is not typical in humans.*[4]

Sheep have not become hairless, however, their pelage is usually referred to as "wool" rather than fur.

35.2.2 Artificial selection

Humans have artificially selected some domesticated mammalian species to have breeds that are hairless. There are several breeds of hairless cats, perhaps the most commonly known being the Sphynx cat. Similarly, there are several breeds of hairless dogs. Other examples of artificially selected hairless animals include the hairless guinea-pig, nude mouse and the hairless rat.

35.3 Use in clothing

Carl Ben Eielson, US Pilot and Arctic explorer wearing a seal fur coat

Main article: Fur clothing

In clothing, fur is usually leather with the hair retained for its aesthetic and insulating properties. Fur has long served as a source of clothing for hominoids including the Neanderthal. Animal furs used in garments and trim may be dyed bright colors or to mimic exotic animal patterns, or shorn down to imitate the feel of a soft velvet fabric. The term "a fur" is often used to refer to a fur coat, wrap, or shawl.

Usual animal sources for fur clothing and fur trimmed accessories include fox, rabbit, mink, beavers, ermine, otters, sable, seals, coyotes, chinchilla, raccoon, and possum. The import and sale of seal products was banned in the U.S. in 1972 over conservation concerns about Canadian seals. The import and sale is still banned even though the Marine Animal Response Society estimates the harp seal population is thriving at approximately 8 million.[5] The import, export and sales of domesticated cat and dog fur were also banned in the U.S. under the Dog and Cat Protection Act of 2000.[6]

The manufacturing of fur clothing involves obtaining animal pelts where the hair is left on the animal's processed skin. In contrast, making leather involves removing the hair from the hide or pelt and using only the skin. The use of wool involves shearing the animal's fleece from the living animal, so that the wool can be regrown but sheepskin shearling is made by retaining the fleece to the leather and shearing it.[7] Shearling is used for boots, jackets and coats and is probably the most common type of skin worn.

Fur is also used to make felt. A common felt is made from beaver fur and is used in high-end cowboy hats.[8]

35.3.1 Controversy

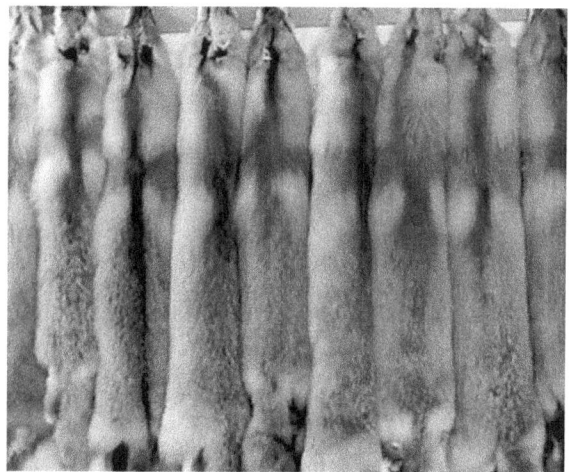

Red fox furs

Main article: Fur farming
Further information: Fur trade

Most animal rights activists are opposed to the trapping and killing of wildlife, and the confinement and killing of animals on fur farms. According to Humane Society International, over 8 million animals are trapped yearly for fur, while more than 30 million were raised in fur farms.[9]

According to Statistics Canada, 2.6 million fur-bearing animals raised on farms were killed in 2010. Another 700,000 were killed for fur by traps.[10][11] Based on a controversial animal rights video, the People for the Ethical Treatment of Animals (PETA) exposed fur farming in China as particularly inhumane.[12]

35.4 See also

- Animal coloration

- Cat coat genetics

- Coat (animal)

Human activities

- Fur farming

- Tanning

35.5 References

[1] "Pelage". Merriam-Webster. Retrieved January 9, 2013.

[2] Thomson, Paul (2002). "Cheiromeles torquatus". Animal Diversity Web. Retrieved 29 October 2013.

[3] Winter, H.; Langbein, L.; Krawczak, M.; Cooper, D. N.; Jave-Suarez, L. F.; Rogers, M. A.; Praetzel, S.; Heidt, P. J.; Schweizer, J. (2001). "Human type I hair keratin pseudogene phihHaA has functional orthologs in the chimpanzee and gorilla: Evidence for recent inactivation of the human gene after the Pan-Homo divergence". *Human genetics* **108** (1): 37–42. doi:10.1007/s004390000439. PMID 11214905.

[4] "Molecular evolution of HR, a gene that regulates the postnatal cycle of the hair follicle". Retrieved 2012-03-13.

[5] "Harp Seal", Marine Animal Response Society.

[6] Rules and Regulations Under the Fur Products Labeling Act.

[7] Australian Wool Corporation, Australian Wool Classing, Raw Wool Services, 1990.

[8] Chamber's journal, Published by Orr and Smith, 1952, pg 200, Original from the University of Michigan.

[9] Humane Society International Fur Trade.

[10] "Fur production, by province and territory".

[11] The fur industry. Peta, n.d.

[12] China's Shocking Dog and Cat Fur Trade

35.6 External links

- "Fur-Bearing Animals". *Encyclopedia Americana*. 1920.

Chapter 36

Game (hunting)

Common pheasant, widely introduced and hunted as game

Game or **quarry** is any animal hunted for sport or for food.

The type and range of animals hunted for food varies in different parts of the world. This is influenced by climate, animal diversity, local taste and locally accepted views about what can or cannot be legitimately hunted. Sometimes a distinction is also made between varieties and species of a particular animal, such as wild turkey and domestic turkey. Fish caught for sport are referred to as game fish.

The term *game* arises in medieval hunting terminology by the late 13th century and is particular to English, from the generic meaning of Old English *gamen* (Germanic **gamanan*) "joy, amusement, sport, merriment". *Quarry* in the generic meaning is early modern (first recorded 1610), in the more specific sense "bird targeted in falconry" late 14th and 15th centuries as *quirre* "entrails of deer placed on the hide and given to the hunting-dogs as a reward", from Old French *cuiriee* "spoil, quarry" (ultimately Latin *corium* "hide"), but influenced by *corée* "viscera, entrails" (Late Latin **corata* "entrails", from *cor* "heart").

In some countries, game is classified, including legal classification with respect to licences required, as either "small game" or "large game". Small game includes small an-

imals, such as rabbits, pheasants, geese or ducks. A single small game licence may cover all small game species and be subject to yearly bag limits. Large game includes animals like deer and bear and are often subject to individual licensing where a separate licence is required for each individual animal taken (tags).

Big game is a term sometimes used interchangeably with large game although in other contexts it refers to large, typically African, mammals (specifically "big five game" or "dangerous game") which are hunted mainly for trophies in safaris.

36.1 By region

36.1.1 Africa

See also: Elephant meat

In some parts of Africa, wild animals hunted for their meat are called bushmeat; see that article for more detailed information on how this operates within the economy (for personal consumption and for money) and the law (including overexploitation and illegal imports). Animals hunted for bushmeat include, but are not limited to:

- Various species of antelope, including duikers

- Various species of primates like mandrills or gorillas

- Rodents like porcupines or cane rats

Some of these animals are endangered or otherwise protected, and thus it is illegal to hunt them.

In Africa, animals hunted for their pelts or ivory are sometimes referred to as *big game*.

Also see the legal definition of game in Swaziland.[*][1]

An African buffalo bull

South Africa

South Africa has 62 species of gamebirds, including guineafowl, francolin, partridge, quail, sandgrouse, duck, geese, snipe, bustard and korhaan. Some of these species are no longer hunted, and of the 44 indigenous gamebirds that can potentially be utilised in South Africa, only three, namely the yellow-throated sandgrouse, Delegorgue's pigeon and the African pygmy goose warrant special protection. Of the remaining 41 species, 24 have shown an increase in numbers and distribution range in the last 25 years or so. The status of 14 species appears unchanged, with insufficient information being available for the remaining three species. The gamebirds of South Africa where the population status in 2005 was secure or growing are listed below:

- Helmeted guineafowl
- Crested partridge
- Greywing partridge
- Redwing partridge
- Orange River partridge
- Cape francolin
- Natal francolin
- Swainson's francolin
- Common quail
- Harlequin quail
- Namaqua sandgrouse
- Double-banded sandgrouse

- Burchell's sandgrouse
- Rock pigeon
- Rameron pigeon
- Red-eyed dove
- Cape turtle dove
- Laughing dove
- White-faced duck
- Egyptian goose
- Yellow-billed duck
- Red-billed teal
- Cape shoveller
- Southern pochard
- Knob-billed duck
- Spur-winged goose

36.1.2 Australia

In Australia, game includes:

- Deer
- Duck
- Magpie geese
- European rabbit
- Feral cat
- Red fox
- Wild pig
- Wild goat
- Kangaroo
- Emu
- Crocodile
- Feral buffalo
- Banteng ("Scrub bull")
- Feral camel
- Australian feral horse
- Quail

Species of deer include:

- Red deer

- Sambar deer

- Rusa deer

- Chital deer

- Hog deer

- Fallow deer

36.1.3 Canada and the United States

Bobwhite quail, an important North American gamebird

See also: Upland game bird

In the U.S. and Canada, deer are the most commonly hunted big game. Other game species include:

Reptiles and amphibians

See also: Alligator meat

- American alligator

- American bullfrog

- Common snapping turtle

Birds (predator)

- Crow

- Raven

Birds (galliforms)

- Dove (non-galliform)

- Chukar partridge

- Grouse

- Partridge

- Pheasant

- Ptarmigan

- Quail

- Turkey

Birds (waterfowl)

- Duck

- Goose

Birds (waders)

- Snipe

- Woodcock

Ungulates

- American Bison

- Bighorn Sheep

- Dall sheep

- Deer

- Pronghorn

White-tailed deer

Carnivores

- American black bear
- Bobcat
- Coyote
- Gray wolf
- Fox
- Grizzly bear
- Cougar
- Raccoon

Rodents

- Beaver
- River rat
- Squirrel

Misc. mammals

- Hare
- Opossum
- Rabbit
- Wild boar

"The Hunters at Rest" by Vasily Perov, 1871

36.1.4 Russia

See also: Hunting in Russia

- Anser
- Beaver
- Black grouse
- Brown bear
- Common quail
- Deer
- Duck
- European hare
- Fox
- Ground squirrel
- Goose
- Gray wolf
- Hazel grouse
- Eurasian lynx
- Mountain hare
- Perdix
- Pheasant
- Rabbit
- Raven
- Siberian ibex
- Squirrel
- Wild boar
- Woodcock

36.1.5 People's Republic of China

In the PRC there is a special cuisine category called ye wei, which includes animals in the wild.

36.1.6 United Kingdom

Game birds at Borough Market in London

In the UK game is defined in law by the Game Act 1831. It is illegal to shoot game on Sundays or at night. Other (non-game birds) that are hunted for food in the UK are specified under the Wildlife and Countryside Act 1981. UK law defines game as including:

- Black grouse (No longer hunted due to decline in numbers)
- Red grouse
- Brown hare
- Ptarmigan
- Grey partridge and red-legged partridge
- Common pheasant

Deer are not included in the definition, but similar controls provided to those in the Game Act apply to deer (from the Deer Act 1991). Deer hunted in the UK are:

- Red deer
- Roe deer
- Fallow deer
- Sika deer
- Muntjac deer
- Chinese water deer
- and hybrids of these deer

Other animals which are hunted in the UK include:

- Duck, including mallard, tufted duck, teal, pintail and pochard
- Goose, including greylag goose, Canada goose, pink-footed goose and in England and Wales white-fronted goose
- Woodpigeon
- Woodcock
- Snipe
- Rabbit
- Golden plover

Capercaillie are not currently hunted in the UK because of a recent decline in numbers and conservation projects towards their recovery. The ban is generally considered voluntary on private lands, and few birds live away from RSPB or Forestry Commission land allegedly.

See also: Hunting and shooting in the United Kingdom

36.1.7 Iceland

In Iceland game includes:

- Reindeer
- Ptarmigan, a popular Christmas dish in Iceland
- Auk
- Goose
- Mallard

36.1.8 Nordic countries

Game in Norway, Sweden, Denmark and Finland includes:

- Moose, *Alces alces*. Moose hunting season in October is close to a national pastime.
- Red deer
- Roe deer
- Wild reindeer
- Mountain hare
- Rock ptarmigan

Roe deer

- Willow ptarmigan

- Mallard

- Auk in Norway

- Black grouse

- Boar in Denmark and southern Sweden. Once hunted to extinction, boars were re-introduced in the late 20th century and are now considered a pest by farmers, but an asset by hunters.

- Woodcock

- Common pheasant

- Common wood pigeon

- Goose

36.2 Preparation

A kitchen interior with a maid and a lady preparing game, c. 1600

Once obtained, game meat must be processed. The method of processing varies by game species and size. Small game and fowl may simply be carried home to be butchered. Large game such as deer is quickly field-dressed by removing the viscera in the field, while very large animals like moose may be partially butchered in the field because of the difficulty of removing them intact from their habitat. Commercial processors often handle deer taken during deer seasons, sometimes even at supermarket meat counters. Otherwise the hunter handles butchering. The carcass is kept cool to minimize spoilage.

Traditionally, game meat used to be hung until "high", i.e. approaching a state of decomposition. The term 'gamey'/'gamy' refers to this usually desirable taste (*haut goût*). However, this adds to the risk of contamination. Small game can be processed essentially intact; after gutting and skinning or defeathering (by species), small animals are ready for cooking although they may be disjointed first. Large game must be processed by techniques commonly practised by commercial butchers.

36.3 Cooking

Generally game is cooked in the same ways as farmed meat.[2] Because some game meat is leaner than store-bought beef, overcooking is a common mishap which can be avoided if properly prepared.[3][4] It is sometimes grilled or cooked longer or by slow cooking or moist-heat methods to make it more tender, since some game tends to be tougher than farm-raised meat. Other methods of tenderizing include marinating as in the dish Hasenpfeffer, cooking in a game pie or as a stew such as Burgoo.

36.4 Safety

The Norwegian Food Safety Authority considers that children, pregnant women, fertile-aged women, and people with high blood pressure should not consume game shot with lead-based ammunition more than once a month. Children who often eat such game might develop a slightly lower IQ, as lead influences the development of the central nervous system.[5]

36.5 See also

- Big game hunting

- British Association for Shooting and Conservation

- Bushfood, something quite different

- Bushmeat

- Waterfowl hunting

- Endangered species

- Fishing

- Legislation on Hunting with Dogs

- Game & Wildlife Conservation Trust

- Hunter-gatherer

- Hunting horn

- Hunting and shooting in the United Kingdom

- Hunting

- Ornithology

- Overfishing

- Persistence hunting

- Animal trapping

- Wildlife

- World Hunting Association

36.6 References

[1] The Game Act Swaziland Legislation

[2] "Game-to-Eat". 2007-05-02.

[3] "About Game Meat". 2007-05-19. Archived from the original on 2007-05-19. Retrieved 2011-10-17.

[4] Venison Direct to Your Door Highland Game

[5] "Mattilsynet: – Barn kan få lavere IQ av storvilt" (in Norwegian). NRK.no. Retrieved August 30, 2013.

36.7 External links

- Media related to Game birds at Wikimedia Commons

Chapter 37

Ginseng

This article is about the plant species. For the town, see Ginseng, Kentucky.

Ginseng (/ˈdʒɪnsɛŋ/*[1]) is any one of the 11 species of slow-growing perennial plants with fleshy roots, belonging to the genus *Panax* of the family Araliaceae.

Ginseng is found in North America and in eastern Asia (mostly Korea, northeast China, Bhutan, eastern Siberia), typically in cooler climates. *Panax vietnamensis*, discovered in Vietnam, is the southernmost ginseng known. This article focuses on the series *Panax* ginsengs, which are the adaptogenic herbs, principally *Panax ginseng* and *P. quinquefolius*. Ginseng is characterized by the presence of ginsenosides and gintonin.

Siberian ginseng (*Eleutherococcus senticosus*) is in the same family, but not genus, as true ginseng. Like ginseng, it is considered to be an adaptogenic herb. The active compounds in Siberian ginseng are eleutherosides, not ginsenosides. Instead of a fleshy root, Siberian ginseng has a woody root.

37.1 Etymology

The English word ginseng derives from the Chinese term *rénshēn* (simplified: 人 ; traditional: 人蔘). *Rén* means "Person" and *shēn* means "plant root"; this refers to the root's characteristic forked shape, which resembles the legs of a person.*[2] The English pronunciation derives from a southern Chinese reading, similar to Cantonese *yun sum* (Jyutping: jan4sam1) and the Hokkien pronunciation "jîn-sim".

The botanical/genus name *Panax* means "all-heal" in Greek, sharing the same origin as "panacea" was applied to this genus because Linnaeus was aware of its wide use in Chinese medicine as a muscle relaxant.

Besides *P. ginseng*, many other plants are also known as or mistaken for the ginseng root. The most commonly known examples are *xiyangshen*, also known as American ginseng 西洋 (*P. quinquefolius*), Japanese ginseng 東洋 (*P. japonicus*), crown prince ginseng 太子參 (*Pseudostellaria heterophylla*), and Siberian ginseng 刺五加 (*Eleutherococcus senticosus*). Although all have the name ginseng, each plant has distinctively different functions. However, true ginseng plants belong only to the *Panax* genus.*[3]

37.2 Economics

In 2010, nearly all of the world's 80,000 tons of ginseng in international commerce was produced in four countries: South Korea, China, Canada,*[4] and the United States. The product was marketed in over 35 countries. Sales exceeded $2.1 billion, of which half came from South Korea.*[5] Historically, Korea has been the largest provider, and China the largest consumer. Control over the ginseng fields was an issue in the 16th century.*[6]

37.3 Medicinal uses

The root is most often available in dried form, either whole or sliced. Ginseng leaf, although not as highly prized, is sometimes also used. Folk medicine attributes various benefits to oral use of American ginseng and Asian ginseng (*P. ginseng*) roots, including roles as an aphrodisiac, stimulant, type II diabetes treatment, or cure for sexual dysfunction in men.*[7] There are not enough studies to understand the effectiveness of ginseng for treating diabetes*[8]*[9]*[10] nor to draw conclusions about whether it affects erectile dysfunction.*[11]

Ginseng may be included in small doses in energy drinks*[12] or herbal teas, such as ginseng coffee.*[13] It may be found in hair tonics and cosmetic preparations, as well, but those uses have not been shown to be clinically effective.

37.4   Research

Ginsenosides, unique compounds of the *Panax* species, are under basic and clinical research to investigate their potential for use in medicine*[14] or when taken as a dietary supplement.*[15]

37.5   Safety

37.5.1   Considerations

Ginseng is known to contain phytoestrogens.*[16]*[17]*[18]

37.5.2   Side effects

A common side effect of *P. ginseng* may be insomnia,*[19] but this effect is disputed.*[20] Other side effects can include nausea, diarrhea, headaches, nose bleeds,*[21] high blood pressure, low blood pressure, and breast pains.*[22]

37.5.3   Interactions

Ginseng has been shown to have adverse drug reactions with phenelzine and warfarin; it has been shown to decrease blood alcohol levels.*[23] A potential interaction has also been reported with imatinib*[24] resulting in hepatotoxicity, and with lamotrigine*[25] causing DRESS syndrome.

Ginseng may also lead to induction of mania in depressed patients who mix it with antidepressants.*[26]

37.5.4   Overdose

The common adaptogen ginsengs (*P. ginseng* and *P. quinquefolia*) are generally considered to be relatively safe even in large amounts. One of the most common and characteristic symptoms of acute overdose of *Panax ginseng* is bleeding. Symptoms of mild overdose may include dry mouth and lips, excitation, fidgeting, irritability, tremor, palpitations, blurred vision, headache, insomnia, increased body temperature, increased blood pressure, edema, decreased appetite, dizziness, itching, eczema, early morning diarrhea, bleeding, and fatigue.*[3]

Symptoms of gross overdose with *Panax ginseng* may include nausea, vomiting, irritability, restlessness, urinary and bowel incontinence, fever, increased blood pressure, increased respiration, decreased sensitivity and reaction to light, decreased heart rate, cyanotic (blue) facial complexion, red facial complexion, seizures, convulsions, and delirium.*[3]

Patients experiencing any of the above symptoms are advised to discontinue the herbs and seek any necessary symptomatic treatment, as well as medical advice in severe cases.*[3]

37.6   Common classification

Ginseng roots in a market in Seoul, 2003

37.6.1   Asian ginseng (root)

Ginseng and reishi mushrooms in bottles being sold in Seoul, Korea

Panax ginseng is available commercially as fresh, red, and white ginsengs; wild ginseng is used where available.

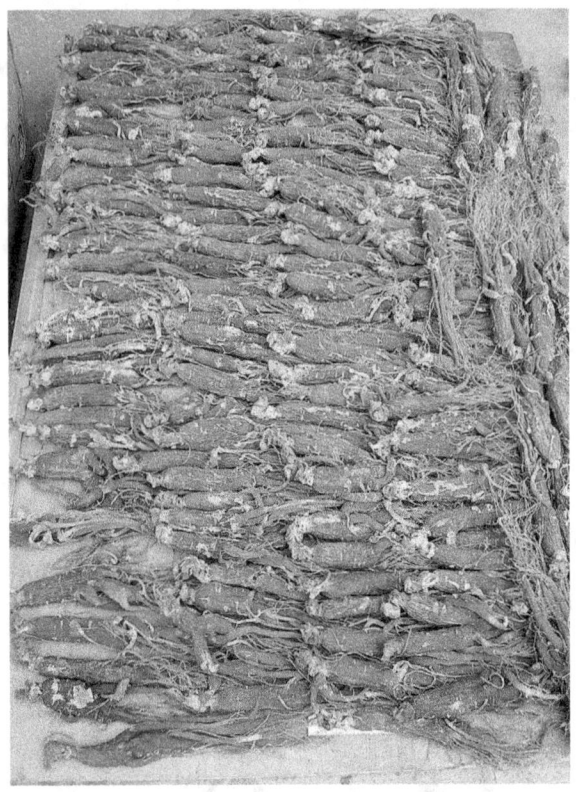

Red ginseng

Red ginseng

Red ginseng (Hangul: 홍삼; hanja: 紅蔘; RR: *hong-sam*; traditional Chinese: 紅蔘; simplified Chinese: 红参; pinyin: *hóng shēn*), *P. ginseng*, has been peeled, heated through steaming at standard boiling temperatures of 100 °C (212 °F), and then dried or sun-dried. It is frequently marinated in an herbal brew which results in the root becoming extremely brittle.

Fresh ginseng

Fresh ginseng is the raw product. Its use is limited by availability.

White ginseng

White ginseng, native to America, is fresh ginseng which has been dried without being heated. It is peeled and dried to reduce the water content to 12% or less. White ginseng air-dried in the sun may contain less of the therapeutic constituents. It is thought by some that enzymes contained in the root break down these constituents in the process of drying. Drying in the sun bleaches the root to a yellowish-white color.

Wild ginseng

Harvested ginseng in Germany

Wild ginseng grows naturally and is harvested from wherever it is found. It is relatively rare, and even increasingly endangered, due in large part to high demand for the product in recent years, which has led to the wild plants being sought out and harvested faster than new ones can grow (it requires years for a root to reach maturity). Wild ginseng can be either Asian or American, and can be processed to be red ginseng.

Woods-grown American ginseng programs in Vermont, Maine, Tennessee, Virginia, North Carolina, Colorado, West Virginia and Kentucky,*[27]*[28] and United Plant Savers have been encouraging the planting of ginseng both to restore natural habitats and to remove pressure from any remaining wild ginseng, and they offer both advice and sources of rootlets. Woods-grown plants have a value comparable to wild-grown ginseng of similar age.

Partially germinated ginseng seeds harvested the previous Fall can be planted from early Spring until late Fall, and will sprout the following Spring. If planted in a wild setting and left to their own devices, they will develop into mature plants which cannot be distinguished from native wild plants. Both Asian and American partially germinated ginseng seeds can be bought from May through December on various eBay sales. Some seed sales come with planting and growing instructions.

37.6.2 *P. quinquefolius* American ginseng (root)

According to traditional Chinese medicine, American ginseng promotes yin energy, cleans excess yang and calms the body. The reason it has been claimed that American ginseng promotes yin (shadow, cold, negative, female) while

Asian ginseng promotes yang (sunshine, hot, positive, male) is that, according to traditional Chinese medicine, things living in cold places or northern side of mountains or southern side of rivers are strong in yang and the converse, so the two are balanced. Chinese/Korean ginseng grows in Manchuria and Korea, the coldest area known to many Koreans in ancient times. Thus, ginseng from there is supposed to be very *yang*.

Originally, American ginseng was imported into China via subtropical Guangzhou, the seaport next to Hong Kong, so Chinese doctors believed American ginseng must be good for *yang*, because it came from a hot area. They did not know, however, that American ginseng can only grow in temperate regions. Nonetheless, the root is legitimately classified as more *yin* because it generates fluids.*[29]

Most North American ginseng is produced in the Canadian provinces of Ontario and British Columbia and the American state of Wisconsin.*[30] *P. quinquefolius* is now also grown in northern China.

The aromatic root resembles a small parsnip that forks as it matures. The plant grows 6″ to 18″ tall, usually bearing three leaves, each with three to five leaflets two to five inches long.

37.7 Other plants sometimes called ginseng

Several other plants are sometimes referred to as ginsengs, but they are either from a different family or genus.

- *Angelica sinensis* (female ginseng, *dong quai*)
- *Codonopsis pilosula* (poor man's ginseng)
- *Eleutherococcus senticosus* (Siberian ginseng)
- *Gynostemma pentaphyllum* (southern ginseng, *jiaogulan*)
- *Lepidium meyenii* (Peruvian ginseng, *maca*)
- *Oplopanax horridus* (Alaskan ginseng)
- *Panax notoginseng* (known as *san qi*, *tian qi* or *tien chi*; ingredient in *yunnan bai yao*)
- *Pfaffia paniculata* (Brazilian ginseng, *suma*)
- *Pseudostellaria heterophylla* (prince ginseng)
- *Schisandra chinensis* (five-flavoured berry)
- *Withania somnifera* (Indian ginseng, *ashwagandha*)

37.8 See also

- *Codonopsis pilosula* "poor man's ginseng"
- Food therapy
- Herbalism
- List of herbs with known adverse effects
- List of ineffective cancer treatments
- *Salvia miltiorrhiza*
- Gintonin

37.9 References

[1] "ginseng". *Cambridge Dictionaries Online*. Retrieved 2011-06-04.

[2] *Oxford Dictionaries Online*, s.v. "ginseng".

[3] Chinese Medical Herbology and Pharmacology, by John K. Chen, Tina T. Chen

[4] Brian L. Evans, "Ginseng: Root of Chinese-Canadian Relations," *Canadian Historical Review* (1985) 66#1 pp 1-26

[5] Baeg, In-Ho, and Seung-Ho So. "The world ginseng market and the ginseng." *Journal of ginseng research* 37.1 (2013): 1. online

[6] Seonmin Kim, "Ginseng and Border Trespassing Between Qing China and Choson Korea," *Late Imperial China* (2007) 28#1 pp 33-61

[7] "As ginseng prices soar, diggers take to the backcountry". *Fox News*. 2012-09-28. Retrieved 28 September 2012.

[8] Kim, Sina; Shin, Byung-Cheul; Lee, Myeong Soo; Lee, Hyangsook; Ernst, Edzard (3 December 2011). "Red ginseng for type 2 diabetes mellitus: A systematic review of randomized controlled trials". *Chinese Journal of Integrative Medicine* **17** (12): 937–944. doi:10.1007/s11655-011-0937-2. PMID 22139546.

[9] Yeh, GY; Eisenberg, DM; Kaptchuk, TJ; Phillips, RS (April 2003). "Systematic review of herbs and dietary supplements for glycemic control in diabetes.". *Diabetes care* **26** (4): 1277–94. doi:10.2337/diacare.26.4.1277. PMID 12663610.

[10] Shishtar, E; Sievenpiper, JL; Djedovic, V; Cozma, AI; Ha, V; Jayalath, VH; Jenkins, DJ; Meija, SB; de Souza, RJ; Jovanovski, E; Vuksan, V (2014). "The effect of ginseng (the genus panax) on glycemic control: a systematic review and meta-analysis of randomized controlled clinical trials.". *PloS one* **9** (9): e107391. doi:10.1371/journal.pone.0107391. PMID 25265315.

[11] Jang, DJ; Lee, MS; Shin, BC; Lee, YC; Ernst, E (October 2008). "Red ginseng for treating erectile dysfunction: a systematic review". *British journal of clinical pharmacology* **66** (4): 444–50. doi:10.1111/j.1365-2125.2008.03236.x. PMID 18754850.

[12] "Do You Know What's in Your Favorite Energy Drink?". Retrieved 28 October 2013.

[13] Clauson KA, Shields KM, McQueen CE, Persad N (2008). "Safety issues associated with commercially available energy drinks". *J Am Pharm Assoc (2003)* **48** (3): e55–63; quiz e64–7. doi:10.1331/JAPhA.2008.07055. PMID 18595815.

[14] Qi LW, Wang CZ, Yuan CS (June 2011). "Ginsenosides from American ginseng: chemical and pharmacological diversity". *Phytochemistry* **72** (8): 689–99. doi:10.1016/j.phytochem.2011.02.012. PMC 3103855. PMID 21396670.

[15] "Ginseng". *American Cancer Society.* Retrieved 5 May 2015.

[16] Lee YJ, Jin YR, Lim WC, et al. (January 2003). "Ginsenoside-Rb1 acts as a weak phytoestrogen in MCF-7 human breast cancer cells". *Arch. Pharm. Res.* **26** (1): 58–63. doi:10.1007/BF03179933. PMID 12568360.

[17] Chan RY, Chen WF, Dong A, Guo D, Wong MS (August 2002). "Estrogen-like activity of ginsenoside Rg1 derived from Panax notoginseng". *J. Clin. Endocrinol. Metab.* **87** (8): 3691–5. doi:10.1210/jc.87.8.3691. PMID 12161497.

[18] Lee Y, Jin Y, Lim W, et al. (March 2003). "A ginsenoside-Rh1, a component of ginseng saponin, activates estrogen receptor in human breast carcinoma MCF-7 cells". *J. Steroid Biochem. Mol. Biol.* **84** (4): 463–8. doi:10.1016/S0960-0760(03)00067-0. PMID 12732291.

[19] http://www.umass.edu/cnshp/faq.html

[20] "The Ginseng Book." Stephen Fulder, PhD

[21] "Ginseng definition - Medical Dictionary definitions of some medical terms defined on MedTerms". Medterms.com. 2012-09-20. Retrieved 2013-03-26.

[22] Kiefer D, Pantuso T (October 2003). "Panax ginseng". *Am Fam Physician* **68** (8): 1539–42. PMID 14596440.

[23] Izzo AA, Ernst E (2001). "Interactions between herbal medicines and prescribed drugs: a systematic review". *Drugs* **61** (15): 2163–75. doi:10.2165/00003495-200161150-00002. PMID 11772128.

[24] Bilgi N, Bell K, Ananthakrishnan AN, Atallah E (2010). "Imatinib and Panax ginseng: a potential interaction resulting in liver toxicity". *The Annals of Pharmacotherapy* **44** (5): 926–8. doi:10.1345/aph.1M715. PMID 20332334.

[25] Myers AP, Watson TA, Strock SB (2015). "Drug Reaction with Eosinophilia and Systemic Symptoms Syndrome Probably Induced by a Lamotrigine-Ginseng Drug Interaction". *Pharmacotherapy.* doi:10.1002/phar.1550. PMID 25756365. Retrieved 2015-03-16.

[26] Fugh-Berman A (January 2000). "Herb-drug interactions". *Lancet* **355** (9198): 134–8. doi:10.1016/S0140-6736(99)06457-0. PMID 10675182.

[27] state.tn.us TDEC: DNH: Ginseng Program

[28] "Care and Planting of Ginseng Seed and Roots". Ces.ncsu.edu. 1914-06-30. Retrieved 2013-03-26.

[29] Chinese Herbal Medicine: Materia Medica, Third Edition by Dan Bensky, Steven Clavey, Erich Stonger, and Andrew Gamble 2004

[30] Agri-food Canada

37.10 Further reading

- Baeg, In-Ho, and Seung-Ho So. "The world ginseng market and the ginseng." *Journal of ginseng research* 37.1 (2013): 1. online

- Evans, Brian L. "Ginseng: Root of Chinese-Canadian Relations," *Canadian Historical Review* (1985) 66#1 pp 1–26

- Johannsen, Kristin (2006). *Ginseng Dreams: The Secret World of America's Most Valuable Plant.* University Press of Kentucky.

- Kim, Seonmin. "Ginseng and Border Trespassing Between Qing China and Choson Korea," *Late Imperial China* (2007) 28#1 pp 33–61

- Pritts, K.D. (2010). *Ginseng: How to Find, Grow, and Use America's Forest Gold.* Stackpole Books. ISBN 978-0-8117-3634-3

- Taylor, D.A. (2006). *Ginseng, the Divine Root: The Curious History of the Plant That Captivated the World.* Algonquin Books. ISBN 978-1-56512-401-1

37.11 External links

- MedlinePlus-Ginseng - National Institutes of Health
- Asian Ginseng - NCCIH - National Institutes of Health
- Ginseng Abuse Syndrome disputed
- Panax ginseng - American Family Physician

Chapter 38

Henna

For other uses, see Henna (disambiguation).

Henna (*Lawsonia inermis*, also known as **hina**, the **henna tree**, the **mignonette tree**, and the **Egyptian privet**)[1][2] is a flowering plant and the sole species of the *Lawsonia* genus. The English name "henna" comes from the Arabic حِنَّاء (ALA-LC: *ḥinnāʾ*; pronounced [ħɪnˈnæːʔ]) or, colloquially حِنا, loosely pronounced as /ˈhɪnna/.

Hand with mehndi design.

The name *henna* also refers to the dye prepared from the plant and the art of temporary body art (staining) based on those dyes (see also mehndi). Henna has been used since antiquity to dye skin, hair and fingernails, as well as fabrics including silk, wool and leather. The name is used in other skin and hair dyes, such as *black henna* and *neutral henna*, neither of which is derived from the henna plant.[3][4]

Historically, henna was used for cosmetic purposes primarily in Ancient India. It was also found to be used in the Arabian Peninsula, South Asia, Carthage and other parts of North Africa, and the Horn of Africa. Bridal henna nights remain an important custom in many of these areas, particularly among traditional families.

38.1 Description

Henna is a tall shrub or small tree, standing 1.8 to 7.6 m tall (6 to 25 ft). It is glabrous and multi-branched, with spine-tipped branchlets. The leaves grow opposite each other on the stem. They are glabrous, sub-sessile, elliptical, and lanceolate (long and wider in the middle; average dimensions are 1.5–5.0 cm x 0.5–2 cm or 0.6–2 in x 0.2–0.8 in), acuminate (tapering to a long point), and have depressed veins on the dorsal surface.[2] Henna flowers have four sepals and a 2 mm (0.079 in) calyx tube, with 3 mm (0.12 in) spread lobes. Its petals are obvate, with white or red stamens found in pairs on the rim of the calyx tube. The ovary is four-celled, 5 mm (0.20 in) long, and erect. Henna fruits are small, brownish capsules, 4–8 mm (0.16–0.31 in) in diameter, with 32–49 seeds per fruit, and open irregularly into four splits.[5]

38.2 Cultivation

The henna plant is native to northern Africa, western and southern Asia, and northern Australasia, in semi-arid zones and tropical areas.[2][6] It produces the most dye when grown in temperatures between 35 and 45 °C (95 and 113 °F).[7] During the onset of precipitation intervals, the plant grows rapidly, putting out new shoots. Growth subsequently slows. The leaves gradually yellow and fall during prolonged dry or cool intervals. It does not thrive where minimum temperatures are below 11 °C (52 °F). Temperatures below 5 °C (41 °F) will kill the henna plant.

38.3 Preparation and application

38.3.1 Body art

Whole, unbroken henna leaves will not stain the skin. Henna will not stain skin until the lawsone molecules are made available (released) from the henna leaf. Dried henna

Henna powder

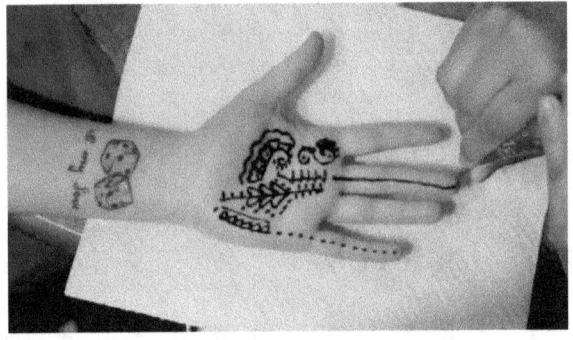

Lawsone, an active compound in Henna

leaves will stain the skin if they are mashed into a paste. The lawsone will gradually migrate from the henna paste into the outer layer of the skin and bind to the proteins in it, creating a fast stain.

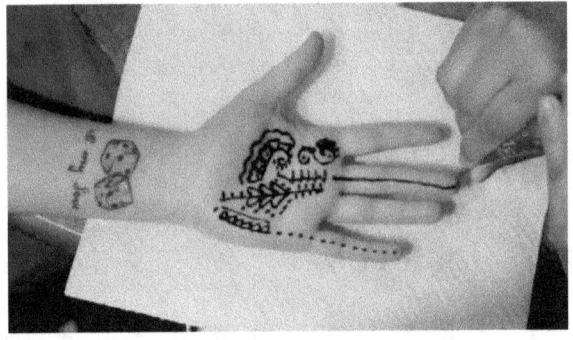

Video of Henna being applied

Since it is difficult to form intricate patterns from coarse crushed leaves, henna is commonly traded as a powder made by drying, milling and sifting the leaves. The dry powder is mixed with one of a number of liquids, including water, lemon juice, or strong tea, and other ingredients, depending on the tradition. Many artists use sugar or molasses in the paste to improve consistency and keep it stuck to the skin better. The henna mix must rest for 1 to 48 hours before use, to release the lawsone from the leaf matter. The timing depends on the crop of henna being used. Essential oils with high levels of monoterpene alcohols, such as tea tree, cajeput, or lavender, will improve skin stain characteristics. Other essential oils, such as eucalyptus and clove, are also useful but are too irritating and should not be used on skin.

The paste can be applied with many traditional and innova-

tive tools, starting with a basic stick or twig. In Morocco, a syringe is common. In India, a plastic cone similar to those used to pipe icing onto cakes. In the Western world, a cone is common, as is a Jacquard bottle, which is otherwise used to paint silk fabric. A light stain may be achieved within minutes, but the longer the paste is left on the skin, the darker and longer lasting the stain will be, so it needs to be left on as long as possible. To prevent it from drying or falling off the skin, the paste is often sealed down by dabbing a sugar/lemon mix over the dried paste, or simply adding some form of sugar to the paste. After time the dry paste is simply brushed or scraped away.

Henna stains are orange when the paste is first removed, but darkens over the following three days to a deep reddish brown. Soles and palms have the thickest layer of skin and so take up the most lawsone, and take it to the greatest depth, so that hands and feet will have the darkest and most long-lasting stains. Some also believe that steaming or warming the henna pattern will darken the stain, either during the time the paste is still on the skin, or after the paste has been removed. It is debatable whether this adds to the color of the end result as well. Chlorinated water and soaps may spoil the darkening process; alkaline products may hasten the darkening process. After the stain reaches its peak color, it holds for a few days, then gradually wears off by way of exfoliation.

38.3.2   Hair dye

History

Henna has been used as a cosmetic hair dye for 6,000 years. In Ancient Egypt, it is known to have been worn. Henna has also traditionally been used for centuries in other parts of North Africa, the Horn of Africa, the Arabian Peninsula, the Near East and South Asia.

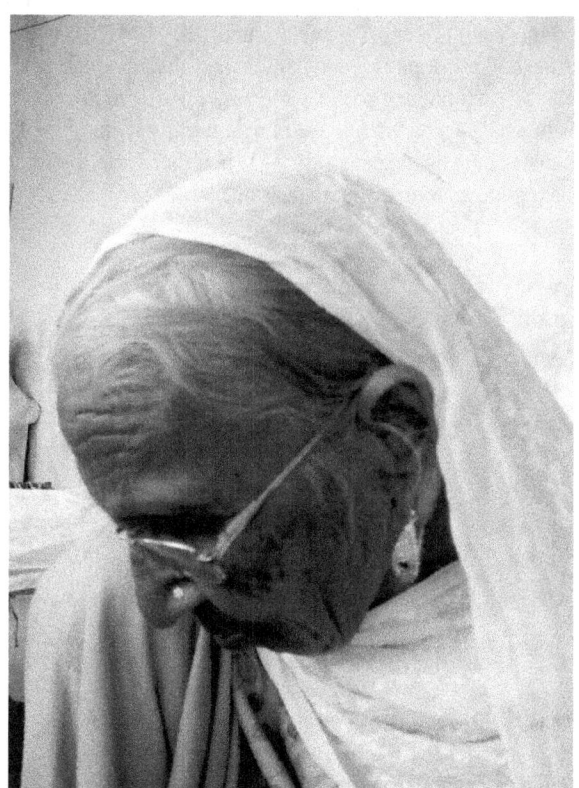

Elderly Punjabi woman whose hair is dyed with henna.

In Ancient Egypt, Ahmose-Henuttamehu (17th Dynasty, 1574 BCE): Henuttamehu was probably a daughter of Seqenenre Tao and Ahmose Inhapy. Smith reports that the mummy of Henuttamehu's own hair had been dyed a bright red at the sides, probably with henna.*[8]

In Europe, henna was popular among women connected to the aesthetic movement and the Pre-Raphaelite artists of England in the 1800s. Dante Gabriel Rossetti's wife and muse, Elizabeth Siddal, had naturally bright red hair. Contrary to the cultural tradition in Britain that considered red hair unattractive, the Pre-Raphaelites fetishized red hair. Siddal was portrayed by Rossetti in many paintings that emphasized her flowing red hair.*[9] The other Pre-Raphaelites, including Evelyn De Morgan and Frederick Sandys, academic classicists such as Frederic Leighton, and French painters such as Gaston Bussière and the Impressionists further popularized the association of henna-dyed hair and young bohemian women.

Opera singer Adelina Patti is sometimes credited with popularizing the use of henna in Europe in the late 1800s. Parisian courtesan Cora Pearl was often referred to as La Lune Rousse (the red-haired moon) for dying her hair red. In her memoirs, she relates an incident when she dyed her pet dog's fur to match her own hair.*[10] By the 1950s, Lucille Ball popularized "henna rinse" as her character,

Lucy Ricardo, called it on the television show I Love Lucy. It gained popularity among young people in the 1960s through growing interest in Eastern cultures.*[11]

Muslim men may use henna as a dye for hair and most particularly their beards. This is considered *sunnah*, a commendable tradition of the Prophet Muhammad. Furthermore, a *hadith* (narration of the Prophet) holds that he encouraged Muslim women to dye their nails with henna to demonstrate femininity and distinguish their hands from those of men. Thus, some Muslim women in the Middle East apply henna to their finger and toenails as well as their hands.

Today

Cosmetic henna for colouring hair.

Commercially packaged henna, intended for use as a cosmetic hair dye, is available in many countries, and is now popular in India, as well as the Middle East, Europe, Australia, Canada and the United States. The color that results from dying with henna depends on the original color of the hair, as well as the quality of the henna, and can range from orange to auburn to burgundy. Henna can be mixed with other natural hair dyes including Cassia Obovata for lighter shades of red or even blond, or with indigo to achieve brown and black shades. Some products sold as "henna" include these other natural dyes. Others may include metal salts that can interact with other chemical treatments, or oils and waxes that may inhibit the dye, or even chemical dyes which are common allergens. Any product that comes in a cream, block, or paste form has some sort of additives.

As with henna in body art, the dried leaf powder should be mixed with a mild acid such as lemon juice, orange juice, or vinegar and left to stand. The resulting paste is then applied to the hair, and covered with plastic wrap to keep it from drying out. This paste should be left in the hair for several hours in order for the dye to permanently bind to the hair strands. The paste is then washed away leaving hair that is permanently dyed. Sometimes henna is mixed with hot or boiling water and used immediately. This gives a color that may fade, and which is not as rich or deep.*[12]

38.4   Traditions of henna as body art

The different words for henna in ancient languages imply that it had more than one point of discovery and origin, as well as different pathways of daily and ceremonial use.

Mehndi on a hand.

Henna has been used to adorn young women's bodies as part of social and holiday celebrations since the late Bronze Age in the eastern Mediterranean. The earliest text mentioning henna in the context of marriage and fertility celebrations comes from the Ugaritic legend of Baal and Anath,*[13] which has references to women marking themselves with henna in preparation to meet their husbands, and Anath adorning herself with henna to celebrate a victory over the enemies of Baal. Wall paintings excavated at Akrotiri (dating prior to the eruption of Thera in 1680 BCE) show women with markings consistent with henna on their nails, palms and soles, in a tableau consistent with the henna bridal description from Ugarit.*[14] Many statuettes of young women dating between 1500 and 500 BCE along the Mediterranean coastline have raised hands with markings consistent with henna. This early connection between young, fertile women and henna seems to be the origin of the Night of the Henna, which is now celebrated worldwide.

The Night of the Henna was celebrated by most groups in the areas where henna grew naturally: Jews,*[15] Muslims,*[16] Sikhs, Hindus, Christians and Zoroastrians, among others, all celebrated marriages and weddings by adorning the bride, and often the groom, with henna.

Across the henna-growing region, Purim,*[15] Eid,*[17] Diwali,*[18] Karva Chauth, Passover, Nowruz, Mawlid,

and most saints' days were celebrated with some henna. Favorite horses, donkeys, and salukis had their hooves, paws, and tails hennaed. Battle victories, births, circumcision, birthdays, Zār, as well as weddings, usually included some henna as part of the celebration. When there was joy, there was henna, as long as henna was available.*[16]

Henna pattern on foot in Morocco.

Henna was regarded as having Barakah ("blessings"), and was applied for luck as well as joy and beauty.*[19] Brides typically had the most henna, and the most complex patterns, to support their greatest joy, and wishes for luck. Some bridal traditions were very complex, such as those in Yemen, where the Jewish bridal henna process took four or five days to complete, with multiple applications and resist work.

The fashion of "Bridal Mehndi" in Pakistan, Northern Libya and in North Indian diasporas is currently growing in complexity and elaboration, with new innovations in glitter, gilding, and fine-line work. Recent technological innovations in grinding, sifting, temperature control, and packaging henna, as well as government encouragement for henna cultivation, have improved dye content and artistic potential for henna.

Though traditional henna artists were Nai caste in India, and barbering castes in other countries (lower social classes), talented contemporary henna artists can command high fees for their work. Women in countries where women are discouraged from working outside the home can find socially acceptable, lucrative work doing henna.*[20] Morocco, Mauritania,*[21] Yemen, Libya, Somalia, Sudan, as well as India and many other countries have thriving women's henna businesses. These businesses are often open all night for Eid, Diwali and Karva Chauth. Many women may work together during a large wedding, wherein hundreds of guests have henna applied to their body parts. This particular event at a marriage is known as the Mehndi Celebration or Mehndi Night, and is mainly held for the bride and groom.

38.4.1   Regions

Henna being sold at the Egyptian Bazaar in Istanbul, Turkey.

Bridal henna nights are a popular tradition in North Africa, the Horn of Africa, the Arabian Peninsula, the Near East and South Asia.

Algeria

In Algeria, the bride's mother-in-law traditionally presents her with jewelry and paints the henna on her hands.

India

In India, As a part of Hindu and Sikh weddings, henna is applied during wedding ceremonies. Traditionally it is thought that the darker the henna on the bride's hand, the more intensely her husband will love her. It is an important part of many Hindu festivals (such as Karva Chauth and Diwali), as well as during Eid. It is a common practice among Indians, particularly elderly ones, to dye their hair using Henna.

Saudi Arabia

In Saudi Arabia, prenuptial Henna nights are common. Traditionally the bride's hands are painted with henna by one of her female relatives, the belief being that the relative must be happily married or else she will bring bad luck to the bride.*[22]

Pakistan and Bangladesh

In Pakistan, henna is used on hands and feet by brides before their wedding, and by many women for Eid al-Fitr and Eid al-Adha. Often the (women) friends and relatives of groom design or choose the design for the hands and feet of bride. It is also used by men to color their hair. As well as Bangladesh.

Israel

In Israel, Mizrahi and Sephardic Jewish women sometimes choose to have a henna party about a week before the wedding. The henna party is smaller than the wedding, as only closer friends and family members are invited. The bride and groom wear traditional costumes as do some of the main guests. There is much dancing and music, especially when the henna is brought out. The henna is usually presented in a deep dish with lit candles in it and carried by the grandmother. She applies the henna onto the palms of the bride and groom and they are blessed. Subsequently, guests stain their palms with henna as well. There are variations in customs and dress between the different Jewish communities (Yemenite, Moroccan, Indian, etc.).

Somalia

Somali singer Fartuun Birimo wearing henna hand and arm designs.

In Somalia, henna is worn by Somali women on their hands, arms, feet and neck during weddings, Eid ul-Fitr, Ramadan, and other festive occasions. Somali henna designs are similar to those in the Arabian peninsula, often featuring flower motifs and triangular shapes. The palm is also frequently decorated with a dot of henna and the fingertips are dipped in the dye. Henna parties are usually held before the wedding takes place.

Tunisia

In Tunisia, prenuptial henna celebrations last for seven days. On the 3rd day, the bride wears a traditional dress and has henna painted on her hands and feet. As for the groom, his pinky finger is painted with henna on the 6th day.

Turkey

In Turkey, henna is sold in convenience stores and markets. Among these are the Spice Bazaar in Istanbul.

38.5   As a medicine

Henna is known as a traditional Ayurveda medicine. It shows various health benefits such as hypoglycaemic and hypolipidemic activities, inhibits the tuberculosis bacteria, and useful in skin diseases. Moreover, henna extract prevents the liver damage occurred from exposure of carbon tetrachloride.[23]

According to *Ayurveda for All* by Murli Manohar, boiled aqueous extract of henna is effective remedy for the urinary stones.[24]

38.6   Health effects

Henna is known to be dangerous to people with glucose-6-phosphate dehydrogenase deficiency (G6PD deficiency), which is more common in males than females. Infants and children of particular ethnic groups, mainly from the Middle East and North Africa, are especially vulnerable.[25] Though user accounts cite few other negative effects of natural henna paste, save for occasional allergic reactions, pre-mixed henna body art pastes may have ingredients added to darken stain, or to alter stain color. The health risks involved in pre-mixed paste can be significant. The United States Food and Drug Administration (FDA) does consider these risks to be adulterants and therefore illegal for use on skin.[26] Some pastes have been noted to include: silver nitrate, carmine, pyrogallol, disperse orange dye, and chromium.[27] These have been found to cause allergic reactions, chronic inflammatory reactions, or late-onset allergic reactions to hairdressing products and textile dyes.[28][29]

38.6.1   Regulation

The U.S. FDA has not approved henna for direct application to the skin. It is however grandfathered in as a hair dye, and can only be imported for that purpose.[26][30] Henna imported into the U.S. that appears to be for use as body art is subject to seizure,[31] but prosecution is rare.

38.7   Varieties

38.7.1   Natural henna

Natural henna stains only a rich red brown. Products sold as "black henna" or "neutral henna" do not contain henna, but are instead made from other plants, or from other substances altogether.

38.7.2   Neutral henna

Neutral henna does not change the color of hair. This is not henna powder; it is usually the powder of the plant *Senna italica* (often referred to by the synonym *Cassia obovata*) or closely related *Cassia* and *Senna* species.

38.7.3   Black henna

Woman with henna stained hands in Khartoum, Sudan.

Black henna powder may be derived from indigo (from the plant *Indigofera tinctoria*). It may also contain unlisted dyes and chemicals[32] such as *p*-phenylenediamine (PPD), which can stain skin black quickly, but can cause severe allergic reactions and permanent scarring if left on for more than 2–3 days. The FDA specifically forbids PPD to be used for this purpose, and may prosecute those who produce black henna.[33] Artists who injure clients with black henna in the U.S. may be sued for damages.[34]

The name arose from imports of plant-based hair dyes into the West in the late 19th century. Partly fermented, dried indigo was called black henna because it could be used in combination with henna to dye hair black. This gave rise to the belief that there was such a thing as black henna which could dye skin black. Indigo will not dye skin black. Pictures of indigenous people with black body art (either alkalized henna or from some other source) also fed the belief that there was such a thing as black henna.

38.7.4 Para-phenylenediamine

In the 1990s, henna artists in Africa, India, Bali, the Arabian Peninsula and the West began to experiment with para-phenylenediamine (PPD) based black hair dye, applying it as a thick paste as they would apply henna, in an effort to find something that would quickly make jet black temporary body art. PPD can cause severe allergic reactions, with blistering, intense itching, permanent scarring, and permanent chemical sensitivities.[35][36] Estimates of allergic reactions range between 3% and 15%. Henna does not cause these injuries.[37] Black henna made with PPD can cause lifelong sensitization to coal tar derivatives.[38] while black henna made with gasoline, kerosene, lighter fluid, paint thinner, and benzene has been linked to adult leukemia.[39]

The most frequent serious health consequence of having a black henna temporary tattoo is sensitization to hair dye and related chemicals. If a person has had a black henna tattoo, and later dyes their hair with chemical hair dye, the allergic reaction may be life-threatening and require hospitalization.[40] Because of the epidemic of para-phenylenediamine allergic reactions, chemical hair dye products now post warnings on the labels: "Temporary black henna tattoos may increase your risk of allergy. Do not colour your hair if: ... – you have experienced a reaction to a temporary black henna tattoo in the past." [41]

Para-phenylenediamine is illegal for use on skin in western countries, though enforcement is difficult. Physicians have urged governments to legislate against black henna because of the frequency and severity of injuries, especially to children.[42] To assist prosecution of vendors, government agencies encourage citizens to report injuries and illegal use of PPD black henna.[43][44] When used in hair dye, the PPD amount must be below 6%, and application instructions warn that the dye not touch the scalp and the dye must be quickly rinsed away. Black henna pastes have PPD percentages from 10% to 80%, and are left on the skin for half an hour.[27][45]

Para-phenylenediamine black henna use is widespread, particularly in tourist areas.[46] Because the blistering reaction appears 3 to 12 days after the application, most tourists have left and do not return to show how much damage the artist has done. This permits the artists to continue injuring others, unaware they are causing severe injuries. The high profit margins of black henna and the demand for body art that emulates "tribal tattoos" further encourage artists to deny the dangers.[47][48]

It is not difficult to recognize and avoid para-phenylenediamine black henna:[49]

- if a paste stains skin on the torso black in less than ½ hour, it has PPD in it.

- if the paste is mixed with peroxide, or if peroxide is wiped over the design to bring out the color, it has PPD in it.

Anyone who has an itching and blistering reaction to a black body stain should go to a doctor, and report that they have had an application of para-phenylenediamine to their skin.

PPD sensitivity is lifelong. A person who has become sensitized through black henna tattoos may have future allergic reactions to perfumes, printer ink, chemical hair dyes, textile dye, photographic developer, sunscreen and some medications. A person who has had a black henna tattoo should consult their physician about health consequences of para-phenylenediamine sensitization.[50]

38.8 Gallery

- in Hyderabad, India.

- in Hyderabad, India.

- in Hyderabad, India.

- in Hyderabad, India.

- A branch of henna tree in Malaysia

38.9 See also

- Achiote (urucum, annatto), another plant that stains skin orange-red

- Genipapo, a plant that stains the skin blue-black

- Mehndi

38.10 References

[1] Bailey, L.H.; Bailey, E.Z. (1976). *Hortus Third: A concise dictionary of plants cultivated in the United States and Canada*. New York: Macmillan. ISBN 978-0025054707.

[2] "Henna" . *HowStuffWorks*. Retrieved 5 May 2013.

[3] Cartwright-Jones, Catherine (2004). "Cassia Obovata" . *Henna for Hair*. Retrieved 5 May 2013.

[4] Dennis, Brady (26 March 2013). "FDA: Beware of "black henna" tattoos" . *The Style Blog*. The Washington Post. Retrieved 5 May 2013.

[5] Kumar S., Singh Y. V., & Singh, M. (2005). "Agro-History, Uses, Ecology and Distribution of Henna (Lawsonia inermis L. syn. Alba Lam)". *Henna: Cultivation, Improvement, and Trade*. Jodhpur: Central Arid Zone Research Institute. pp. 11–12. OCLC 124036118.

[6] "henna (plant)". *Encyclopedia Britannica*. Retrieved 5 May 2013.

[7] Bechtold, Thomas; Mussak, Rita (6 April 2009). *Handbook of Natural Colorants*. John Wiley & Sons. p. 155. ISBN 9780470744963.

[8] G. Elliott Smith, The Royal Mummies, Duckworth Publishing; (September, 2000)

[9] "Aesthetics". Retrieved 15 August 2011.

[10] Pearl, Cora (2009). *The Memoirs of Cora Pearl* **13**. General Books LLC. ISBN 9781151590527.

[11] Sherrow, Victoria (2006). *Encyclopedia of Hair: A Cultural History*. Greenwood. pp. 206–207. ISBN 0313331456. ISBN 978-0-313-33145-9.

[12] Cartwright-Jones, Catherine. *Henna for Hair*. Free E-Book.

[13] de Moor, Johannes C. (1971). *The seasonal pattern in the Ugaritic myth of Balu, according to the version of Ilimilku, (Alter Orient und Altes Testament)*. Kevelaer: Butzon & Bercker. ISBN 978-3-7887-0293-9. OCLC 201316.

[14] Doumas, Christos (1992). *The wall-paintings of Thera*. Athens: Thera Foundation. ISBN 978-960-220-274-6. OCLC 30069766.

[15] Brauer, Erich; Raphael Patai (1993). *The Jews of Kurdistan*. Detroit: Wayne State University Press. ISBN 978-0-8143-2392-2. OCLC 27266639.

[16] Westermarck, Edward (1972) [1914]. *Marriage ceremonies in Morocco*. London: Curzon Press. ISBN 978-0-87471-089-2. OCLC 633323.

[17] Hammoudi, Abdellah (1993). *The victim and its masks: an essay on sacrifice and masquerade in the Maghreb*. Chicago: University of Chicago Press. ISBN 978-0-226-31525-6. OCLC 27265476.

[18] Saksena, Jogendra (1979). *Art of Rajasthan: Henna and Floor Decorations*. Delhi: Sundeep. OCLC 7219114.

[19] Westermarck, E. (1926). Ritual and Belief in Morocco Vols 1 & 2. London, UK: Macmillan and Company, Limited

[20] "Easy Mehndi Design Tutorial". 4 December 2014.

[21] Tauzin, Aline (1998). *Le henné, art des femmes de Mauritanie*. Paris: UNESCO. ISBN 978-92-3-203487-8.

[22] "Extraordinary Bridal Mehndi Design For Wedding". 4 December 2014.

[23] "Benefits of Henna (Mehndi)". 6 July 2013. Retrieved 13 November 2014.

[24] Manohar, Murli (2011). *Ayurveda for All*. India: V&S Publishers. p. 138. ISBN 9381384908.

[25] "Henna and Glucose-6-phosphate dehydrogenase deficiency". The Henna Page.

[26] "Temporary Tattoos & Henna/Mehndi". Food and Drug Administration.

[27] Kang IJ, Lee MH (July 2006). "Quantification of para-phenylenediamine and heavy metals in henna dye". *Contact Dermatitis* **55** (1): 26–9. doi:10.1111/j.0105-1873.2006.00845.x. PMID 16842550.

[28] Dron P, Lafourcade MP, Leprince F, et al. (June 2007). "Allergies associated with body piercing and tattoos: a report of the Allergy Vigilance Network". *European Annals of Allergy and Clinical Immunology* **39** (6): 189–92. PMID 17713170.

[29] Raupp P, Hassan JA, Varughese M, Kristiansson B (November 2001). "Henna causes life threatening haemolysis in glucose-6-phosphate dehydrogenase deficiency". *Archives of Disease in Childhood* **85** (5): 411–2. doi:10.1136/adc.85.5.411. PMC 1718961. PMID 11668106.

[30] "§ 73.2190 Henna". *Listing of Color Additives Exempt from Certification*. Federal Register. 30 July 2009. Retrieved 3 August 2009.

[31] Accessdate.fda.gov

[32] Singh, M., Jindal, S. K., Kavia, Z. D., Jangid, B. L., & Khem Chand (2005). "Traditional Methods of Cultivation and Processing of Henna. Henna, Cultivation, Improvement and Trade". *Henna: Cultivation, Improvement, and Trade*. Jodhpur: Central Arid Zone Research Institute. pp. 21–24. OCLC 124036118.

[33] FDA.gov

[34] Rosemariearnold.com

[35] Van den Keybus C, Morren M.A., Goossens A. (September 2005). "Walking difficulties due to an allergic reaction to a temporary tattoo". *Contact Dermatitis* **53** (3): 180–1. doi:10.1111/j.0105-1873.2005.0407m.x. PMID 16128770.

[36] Stante M, Giorgini S, Lotti T (April 2006). "Allergic contact dermatitis from henna temporary tattoo". *Journal of the European Academy of Dermatology and Venereology* **20** (4): 484–6. doi:10.1111/j.1468-3083.2006.01483.x. PMID 16643167.

[37] Jung P., Sesztak-Greinecker G., Wantke F., Götz M., Jarisch R., Hemmer W. (April 2006). "A painful experience: black henna tattoo causing severe, bullous contact dermatitis". *Contact Dermatitis* **54** (4): 219–20. doi:10.1111/j.0105-1873.2006.0775g.x. PMID 16650103.

[38] Lifelong damage from black henna, Hennapage.com

[39] Acute leukemia among the adult population of United Arab Emirates: an epidemiological study. 2009, vol. 50, no. 7, pp. 1138–1147. Inaam Bashir Hassan, Sherief I. A. M. Islam, Hussain Alizadeh, Jorgen Kristensen, Amr Kamba, Shanaaz Sonday and Roos M. D. Bernseen.

[40] Severe allergic hair dye reactions in 8 children. Heidi Sosted1,Jeanne Duus Johansen, Klaus Ejner Andersen, Torkil Menné, Contact Dermatitis, Volume 54, Issue 2, pages 87–91, February 2006

[41] Commission Directive 2009/134/EC of 28 October 2009 amending Council Directive 76/768/EEC concerning cosmetic products for the purposes of adapting Annex III thereto to technical progress

[42] "p-Phenylenediamine in Black Henna Tattoos A Practice in Need of Policy" in Children Sharon E. Jacob, MD; Tamar Zapolanski, BA; Pamela Chayavichitsilp, BA; Elizabeth Alvarez Connelly, MD; Lawrence F. Eichenfield, MD Arch Pediatr Adolesc Med. 2008;162(8):790–792.

[43] DOH.state.fl.us

[44] HC-SC-GC.ca

[45] "Acute fingertip dermatitis from a temporary tattoo and quantitative chemical analysis of the product" Avnstorp, C., Rastogi, S., and Menne, T. Contact Point, 2002, p. 119

[46] Marcoux, D.; Couture-Trudel, P.; Riboulet-Delmas, G.; Sasseville, D. 2002. Sensitization to Para-Phenylenediamine from a Streetside Temporary Tattoo. Pediatric Dermatology 19, 6:498–502.

[47] Önder, M. 2003. Temporary holiday tattoos may cause life-long allergic contact dermatitis when henna is mixed with PPD. Journal of Cosmetic Dermatology 2, 3–4: 126–130.

[48] Önder, Meltem, Çiğdem Asena Atahan, Pinar Öztaş, and Murat Orhan Öztaş. 2001. Temporary henna tattoo reactions in children. International Journal of Dermatology 40, 9: 577–579.

[49] HC-SC.GC.ca

[50] *CaresforHairs.com: P-Phenylenediamine and the dangers of henna.*

38.11 Further reading

Semwal, Ruchi Badoni; Semwal, Deepak Kumar; Combrinck, Sandra (August 2014). "Lawsonia inermis L. (henna): Ethnobotanical, phytochemical and pharmacological aspects". *Journal of Ethnopharmacology* **155** (1): 80-103. doi:10.1016/j.jep.2014.05.042.

38.12 External links

- The Henna Page

Chapter 39

Honey

For other uses, see Honey (disambiguation).

A jar of honey with a honey dipper and an American biscuit

Honey in honeycomb

Honey /'hʌni/ is a sweet food made by bees using nectar from flowers. The variety produced by honey bees (the genus *Apis*) is the one most commonly referred to, as it is the type of honey collected by most beekeepers and consumed by people. Honeys are also produced by bumblebees, stingless bees, and other hymenopteran insects such as honey wasps, though the quantity is generally lower and they have slightly different properties compared to honey from the genus *Apis*. Honey bees convert nectar into honey by a process of regurgitation and evaporation. They store it as a primary food source in wax honeycombs inside the beehive.

Honey gets its sweetness from the monosaccharides fructose and glucose, and has about the same relative sweetness as granulated sugar.[1][2] It has attractive chemical properties for baking and a distinctive flavor that leads some people to prefer it over sugar and other sweeteners.[1] Most microorganisms do not grow in honey because of its low water activity of 0.6.[3] However, honey sometimes contains dormant endospores of the bacterium *Clostridium botulinum*, which can be dangerous to infants, as the endospores can transform into toxin-producing bacteria in infants' immature intestinal tracts, leading to illness and even death.[4]

People who have a weakened immune system should not eat honey because of the risk of bacterial or fungal infection.[5] No evidence shows the benefit of using honey to treat diseases.[5] Honey is a source of empty calories and it is recommended that it be replaced with fruits and vegetables.[6] One tablespoon of honey provides 64 calories.[7]

Honey use and production has a long and varied history.[8] Honey collection is an ancient activity.[9] Humans apparently began hunting for honey at least 8,000 years ago, as evidenced by a cave painting in Valencia, Spain.[9]

A honey bee on calyx of goldenrod

39.1   Formation

Honey's sugars are dehydrated, which prevents fermentation, with added enzymes to modify and transform their chemical composition and pH. Invertases and digestive acids hydrolyze sucrose to give the monosaccharides glucose and fructose. Invertase is one of these enzymes synthesized by the body of the insect.

Honey bees transform saccharides into honey by a process of regurgitation, a number of times, until it is partially digested. The bees do the regurgitation and digestion as a group. After the last regurgitation, the aqueous solution is still high in water, so the process continues by evaporation of much of the water and enzymatic transformation.

Honey is produced by bees as a food source. To produce about 500 g of honey, foraging honey bees have to travel the equivalent of three times around the world.[10] In cold weather or when fresh food sources are scarce, bees use their stored honey as their source of energy.[11] By contriving for bee swarms to nest in artificial hives, people have been able to semidomesticate the insects and harvest excess honey. In the hive or in a wild nest, the three types of bees are:

- a single female queen bee

- a seasonally variable number of male drone bees to fertilize new queens

- 20,000 to 40,000 female worker bees[12]

The worker bees raise larvae and collect the nectar that will become honey in the hive. Leaving the hive, they collect sugar-rich flower nectar and return.

In the hive, the bees use their "honey stomachs" to ingest and regurgitate the nectar a number of times until it is partially digested.[13] Invertase synthesized by the bees and digestive acids hydrolyze sucrose to give the same mixture of glucose and fructose. The bees work together as a group with the regurgitation and digestion until the product reaches a desired quality. It is then stored in honeycomb cells. After the final regurgitation, the honeycomb is left unsealed. However, the nectar is still high in both water content and natural yeasts, which, unchecked, would cause the sugars in the nectar to ferment.[11] The process continues as bees inside the hive fan their wings, creating a strong draft across the honeycomb, which enhances evaporation of much of the water from the nectar.[11] This reduction in water content raises the sugar concentration and prevents fermentation. Ripe honey, as removed from the hive by a beekeeper, has a long shelf life, and will not ferment if properly sealed.[11]

Another source of honey is from a number of wasp species, such as the wasps *Brachygastra lecheguana* and *Brachygastra mellifica*, which are found in South and Central America. These species are known to feed on nectar and produce honey.[14]

Some wasps, such as the *Polistes versicolor*, even consume honey themselves, switching from feeding on pollen in the middle of their lifecycles to feeding on honey, which can better provide for their energy needs.[15]

39.2   Collection

Honey is collected from wild bee colonies, or from domesticated beehives. Wild bee nests are sometimes located by following a honeyguide bird. The bees may first be pacified by using smoke from a bee smoker. The smoke triggers a feeding instinct (an attempt to save the resources of the hive from a possible fire), making them less aggressive and the smoke obscures the pheromones the bees use to communicate.

The honeycomb is removed from the hive and the honey may be extracted from that, either by crushing or by using a honey extractor. The honey is then usually filtered to remove beeswax and other debris.

Before the invention of removable frames, bee colonies were often sacrificed in order to conduct the harvest. The harvester would take all the available honey and replace the entire colony the next spring. Since the invention of removable frames, the principles of husbandry lead most beekeepers to ensure that their bees will have enough stores to survive the winter, either by leaving some honey in the beehive or by providing the colony with a honey substitute such as sugar water or crystalline sugar (often in the form of a "candyboard"). The amount of food necessary to survive the winter depends on the race of bees and on the length and severity of local winters.

39.3   Modern uses

39.3.1   Food

Over its history as a food,[8] the main uses of honey are in cooking, baking, desserts, such as *mel i mató*, as a spread on bread, and as an addition to various beverages, such as tea, and as a sweetener in some commercial beverages. Honey barbecue and honey mustard are other common flavors used in sauces.

39.3.2   Fermentable

Honey is the main ingredient in the alcoholic beverage mead, which is also known as "honey wine" or "honey beer". Historically, the ferment for mead was honey's naturally occurring yeast. Honey is also used as an adjunct in some beers.

Honey wine, or mead, is typically (modern era) made with a honey and water mixture with yeast added for fermentation. Primary fermentation usually takes 40 days, after which the must needs to be racked into a secondary fermentation vessel and left to sit about 35–40 more days. If done properly, fermentation will be finished by this point (though if a sparkling mead is desired, fermentation can be restarted after bottling by the addition of a small amount of sugar), but most meads require aging for 6–9 months or more in order to be palatable.

39.4   Honey-producing and consuming countries

As of 2012, China, Turkey, and Argentina were the top producers of honey, followed by Ukraine and the United States.[16]

Mexico is also an important producer of honey, providing more than 4% of the world's supply.[17] Much of this (about one-third) comes from the Yucatán Peninsula. Honey production began there when the *Apis mellifera* and the *A. mellifera ligustica* were introduced there early in the 20th century. Most of Mexico's Yucatán producers are small, family operations who use original traditional techniques, moving hives to take advantage of the various tropical and subtropical flowers.[18]

Honey is also one of the gourmet products of the French island of Corsica. Corsican honey is certified as to its origin

(*Appellation d'origine contrôlée*) just as are French wines, like Champagne.[19]

Honey consumption per capita per year exceeds one kilogram in some countries like Austria, Germany and Switzerland.[20]

39.5   Physical and chemical properties

Crystallized honey. The inset shows a close-up of the honey, showing the individual glucose grains in the fructose mixture.

The physical properties of honey vary, depending on water content, the type of flora used to produce it (pasturage), temperature, and the proportion of the specific sugars it contains. Fresh honey is a supersaturated liquid, containing more sugar than the water can typically dissolve at ambient temperatures. At room temperature, honey is a supercooled liquid, in which the glucose will precipitate into solid granules. This forms a semisolid solution of precipitated glucose crystals in a solution of fructose and other ingredients.

39.5.1   Phase transitions

The melting point of crystallized honey is between 40 and 50 °C (104 and 122 °F), depending on its composition. Below this temperature, honey can be either in a metastable state, meaning that it will not crystallize until a seed crystal is added, or, more often, it is in a "labile" state, being saturated with enough sugars to crystallize spontaneously.[21] The rate of crystallization is affected by many factors, but the primary factor is the ratio of the main sugars: fructose to glucose. Honeys that are supersaturated with a very high percentage of glucose, such as brassica honey, will crystallize almost immediately after harvesting, while honeys with a low percentage of glucose, such as chestnut or tupelo honey, do not crystallize. Some types of honey may produce very large but few crystals, while others will produce many small crystals.[*][22]

Crystallization is also affected by water content, because a high percentage of water will inhibit crystallization, as will a high dextrin content. Temperature also affects the rate of crystallization, with the fastest growth occurring between 13 and 17 °C (55 and 63 °F). Crystal nuclei (seeds) tend to form more readily if the honey is disturbed, by stirring, shaking or agitating, rather than if left at rest. However, the nucleation of microscopic seed-crystals is greatest between 5 and 8 °C (41 and 46 °F). Therefore, larger but fewer crystals tend to form at higher temperatures, while smaller but more-numerous crystals usually form at lower temperatures. Below 5 °C, the honey will not crystallize and, thus, the original texture and flavor can be preserved indefinitely.[*][22]

Since honey normally exists below its melting point, it is a supercooled liquid. At very low temperatures, honey will not freeze solid. Instead, as the temperatures become lower, the viscosity of honey increases. Like most viscous liquids, the honey will become thick and sluggish with decreasing temperature. At −20 °C (−4 °F), honey may appear or even feel solid, but it will continue to flow at very low rates. Honey has a glass transition between −42 and −51 °C (−44 and −60 °F). Below this temperature, honey enters a glassy state and will become an amorphous solid (noncrystalline).[*][23][*][24]

39.5.2   Viscosity

The viscosity of honey is affected greatly by both temperature and water content. The higher the water percentage, the easier honey flows. Above its melting point, however, water has little effect on viscosity. Aside from water content, the composition of honey also has little effect on viscosity, with the exception of a few types. At 25 °C (77 °F), honey with 14% water content generally has a viscosity around 400 poise, while a honey containing 20% water has a viscosity around 20 poise. Viscosity increase due to temperature occurs very slowly at first. A honey containing 16% water, at 70 °C (158 °F), will have a viscosity around 2 poise, while at 30 °C (86 °F), the viscosity is around 70 poise. As cooling progresses, honey becomes more viscous at an increasingly rapid rate, reaching 600 poise around 14 °C (57 °F). However, while honey is very viscous, it has rather low surface tension.[*][25][*][26]

A few types of honey have unusual viscous properties. Honeys from heather or manuka display thixotropic properties. These types of honey enter a gel-like state when motionless, but then liquify when stirred.[*][27]

39.5.3   Electrical and optical properties

Because honey contains electrolytes, in the form of acids and minerals, it exhibits varying degrees of electrical conductivity. Measurements of the electrical conductivity are used to determine the quality of honey in terms of ash content.[*][26]

The effect honey has on light is useful for determining the type and quality. Variations in the water content alter the refractive index of honey. Water content can easily be measured with a refractometer. Typically, the refractive index for honey will range from 1.504 at 13% water content to 1.474 at 25%. Honey also has an effect on polarized light, in that it will rotate the polarization plane. The fructose will give a negative rotation, while the glucose will give a positive one. The overall rotation can be used to measure the ratio of the mixture.[*][26][*][28] Honey may vary in color between pale yellow and dark brown, but other bright colors may occasionally be found, depending on the source of the sugar harvested by the bees.[*][29]

39.5.4   Hygroscopy and fermentation

Honey has the ability to absorb moisture directly from the air, a phenomenon called hygroscopy. The amount of water the honey will absorb is dependent on the relative humidity of the air. Because honey contains yeast, this hygroscopic nature requires that honey be stored in sealed containers to prevent fermentation, which usually begins if the honey's water content rises much above 25%. Honey will tend to absorb more water in this manner than the individual sugars would allow on their own, which may be due to other ingredients it contains.[*][28]

Fermentation of honey will usually occur after crystallization because, without the glucose, the liquid portion of the honey primarily consists of a concentrated mixture of the fructose, acids, and water, providing the yeast with enough of an increase in the water percentage for growth. Honey that is to be stored at room temperature for long periods of time is often pasteurized, to kill any yeast, by heating it above 70 °C (158 °F).[*][28]

39.5.5   Thermal characteristics

Like all sugar compounds, honey will caramelize if heated sufficiently, becoming darker in color, and eventually burn. However, honey contains fructose, which caramelizes at lower temperatures than the glucose.[*][30] The temperature at which caramelization begins varies, depending on the composition, but is typically between 70 and 110 °C (158 and 230 °F). Honey also contains acids, which act as catalysts, decreasing the caramelization temperature even more.[31] Of these acids, the amino acids, which occur in very small amounts, play an important role in the darkening of honey. The amino acids form darkened compounds called melanoidins, during a Maillard reaction. The Maillard reaction will occur slowly at room temperature, taking from a few to several months to show visible darkening, but will speed-up dramatically with increasing temperatures. However, the reaction can also be slowed by storing the honey at colder temperatures.[32]

Unlike many other liquids, honey has very poor thermal conductivity, taking a long time to reach thermal equilibrium. Melting crystallized honey can easily result in localized caramelization if the heat source is too hot, or if it is not evenly distributed. However, honey will take substantially longer to liquify when just above the melting point than it will at elevated temperatures.[26] Melting 20 kilograms of crystallized honey, at 40 °C (104 °F), can take up to 24 hours, while 50 kilograms may take twice as long. These times can be cut nearly in half by heating at 50 °C (122 °F). However, many of the minor substances in honey can be affected greatly by heating, changing the flavor, aroma, or other properties, so heating is usually done at the lowest temperature possible for the shortest amount of time.[33]

39.6 Classification

Honey is classified by its floral source, and there are also divisions according to the packaging and processing used. There are also regional honeys. In the USA honey is also graded on its color and optical density by USDA standards, graded on the Pfund scale, which ranges from 0 for "water white" honey to more than 114 for "dark amber" honey.[34]

39.6.1 Floral source

Generally, honey is classified by the floral source of the nectar from which it was made. Honeys can be from specific types of flower nectars or can be blended after collection. The pollen in honey is traceable to floral source and therefore region of origin. The rheological and melissopalynological properties of honey can be used to identify the major plant nectar source used in its production.[35]

Blended

Most commercially available honey is blended, meaning it is a mixture of two or more honeys differing in floral source, color, flavor, density or geographic origin.[36]

Polyfloral

Polyfloral honey, also known as wildflower honey,[37] is derived from the nectar of many types of flowers.[38]

The taste may vary from year to year, and the aroma and the flavor can be more or less intense, depending on which bloomings are prevalent.[39]

Monofloral

Monofloral honey is made primarily from the nectar of one type of flower. Different monofloral honeys have a distinctive flavor and color because of differences between their principal nectar sources.[40] To produce monofloral honey, beekeepers keep beehives in an area where the bees have access to only one type of flower. In practice, because of the difficulties in containing bees, a small proportion of any honey will be from additional nectar from other flower types. Typical examples of North American monofloral honeys are clover, orange blossom, blueberry, sage, tupelo, buckwheat, fireweed, mesquite and sourwood. Some typical European examples include thyme, thistle, heather, acacia, dandelion, sunflower, honeysuckle, and varieties from lime and chestnut trees. In North Africa (e.g. Egypt) examples include clover, cotton, and citrus (mainly orange blossoms). The unique flora of Australia yields a number of distinctive honeys, with some of the most popular being yellow box, blue gum, ironbark, bush mallee, Tasmanian leatherwood, and macadamia.

Honeydew honey

Instead of taking nectar, bees can take honeydew, the sweet secretions of aphids or other plant sap-sucking insects. Honeydew honey is very dark brown in color, with a rich fragrance of stewed fruit or fig jam, and is not as sweet as nectar honeys.[40] Germany's Black Forest is a well known source of honeydew-based honeys, as well as some regions in Bulgaria, Tara (mountain) in Serbia and Northern California in the United States. In Greece, pine honey (a type of honeydew honey) constitutes 60–65% of the annual honey production.[41] Honeydew honey is popular in some areas, but in other areas beekeepers have difficulty selling the stronger flavored product.

The production of honeydew honey has some complications and dangers. The honey has a much larger proportion of indigestibles than light floral honeys, thus causing dysentery to the bees, resulting in the death of colonies in areas with cold winters. Good beekeeping management requires the removal of honeydew prior to winter in colder areas. Bees collecting this resource also have to be fed protein supplements, as honeydew lacks the protein-rich pollen accompaniment gathered from flowers.

39.6.2 Classification by packaging and processing

Generally, honey is bottled in its familiar liquid form. However, honey is sold in other forms, and can be subjected to a variety of processing methods.

Honeycomb

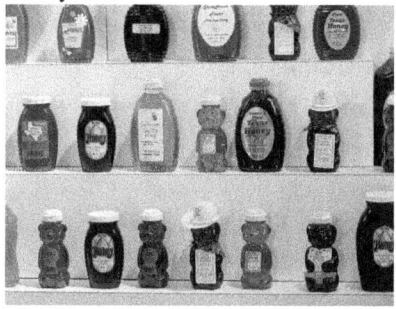

A variety of honey flavors and container sizes and styles from the 2008 Texas State Fair

- **Crystallized honey** is honey in which some of the glucose content has spontaneously crystallized from solution as the monohydrate. Also called "granulated honey" or "candied honey." Honey that has crystallized (or commercially purchased crystallized) can be returned to a liquid state by warming.*[42]

- **Pasteurized honey** is honey that has been heated in a pasteurization process which requires temperatures of 161 °F (72 °C) or higher. Pasteurization destroys yeast cells. It also liquefies any microcrystals in the honey, which delays the onset of visible crystallization. However, excessive heat exposure also results in product deterioration, as it increases the level of hydroxymethylfurfural (HMF) and reduces enzyme (e.g. diastase) activity. Heat also affects appearance (darkens the natural honey color), taste, and fragrance.*[43]

- **Raw honey** is honey as it exists in the beehive or as obtained by extraction, settling or straining, without

adding heat (although some honey that has been "minimally processed" is often labeled as raw honey).*[44] Raw honey contains some pollen and may contain small particles of wax.

- **Strained honey** has been passed through a mesh material to remove particulate material (pieces of wax, propolis, other defects) without removing pollen, minerals or enzymes.

- **Filtered honey** is honey of any type that has been filtered to the extent that all or most of the fine particles, pollen grains, air bubbles, or other materials normally found in suspension, have been removed.*[45] The process typically heats honey to 150–170 °F (66–77 °C) to more easily pass through the filter.*[46] Filtered honey is very clear and will not crystallize as quickly,*[46] making it preferred by the supermarket trade.*[47]

- **Ultrasonicated honey** has been processed by ultrasonication, a non-thermal processing alternative for honey. When honey is exposed to ultrasonication, most of the yeast cells are destroyed. Those cells that survive sonication generally lose their ability to grow, which reduces the rate of honey fermentation substantially. Ultrasonication also eliminates existing crystals and inhibits further crystallization in honey. Ultrasonically aided liquefaction can work at substantially lower temperatures of approximately 95 °F (35 °C) and can reduce liquefaction time to less than 30 seconds.*[48]

- **Creamed honey**, also called whipped honey, spun honey, churned honey, honey fondant, and (in the UK) set honey, has been processed to control crystallization. Creamed honey contains a large number of small crystals, which prevent the formation of larger crystals that can occur in unprocessed honey. The processing also produces a honey with a smooth, spreadable consistency.*[49]

- **Dried honey** has the moisture extracted from liquid honey to create completely solid, nonsticky granules. This process may or may not include the use of drying and anticaking agents.*[50] Dried honey is used in baked goods,*[50] and to garnish desserts.

- **Comb honey** is honey still in the honeybees' wax comb. It traditionally is collected by using standard wooden frames in honey supers. The frames are collected and the comb is cut out in chunks before packaging. As an alternative to this labor-intensive method, plastic rings or cartridges can be used that do not require manual cutting of the comb, and speed packaging. Comb honey harvested in the traditional manner is also referred to as "cut-comb honey".*[42]*:13*[51]

- **Chunk honey** is packed in widemouth containers consisting of one or more pieces of comb honey immersed in extracted liquid honey.[42][*]:13

- **Honey decoctions** are made from honey or honey by-products which have been dissolved in water, then reduced (usually by means of boiling). Other ingredients may then be added. (For example, abbamele has added citrus.) The resulting product may be similar to molasses.

39.7 Preservation

Sealed frame of honey

Because of its unique composition and chemical properties, honey is suitable for long-term storage, and is easily assimilated even after long preservation. Honey, and objects immersed in honey, have been preserved for centuries.[52][53] The key to preservation is limiting access to humidity. In its cured state, honey has a sufficiently high sugar content to inhibit fermentation. If exposed to moist air, its hydrophilic properties will pull moisture into the honey, eventually diluting it to the point that fermentation can begin.

Regardless of preservation, honey may crystallize over time. The crystals can be dissolved by heating the honey.[54][55]

39.8 Distinguishing

39.8.1 Grading

See also: Food grading

In the US, honey grading is performed voluntarily (USDA does offer inspection and grading "as on-line (in-plant) or lot inspection...upon application, on a fee-for-service basis.") based upon USDA standards. Honey is graded based upon a number of factors, including water content, flavor and aroma, absence of defects and clarity. Honey is also classified by color though it is not a factor in the grading scale.[56] The honey grade scale is:

Other countries may have differing standards on the grading of honey. India, for example, certifies honey grades based on additional factors, such as the Fiehe's test, and other empirical measurements.[57]

39.8.2 Indicators of quality

High-quality honey can be distinguished by fragrance, taste, and consistency. Ripe, freshly collected, high-quality honey at 20 °C (68 °F) should flow from a knife in a straight stream, without breaking into separate drops.[58] After falling down, the honey should form a bead. The honey, when poured, should form small, temporary layers that disappear fairly quickly, indicating high viscosity. If not, it indicates excessive water content (over 20%)[58] of the product. Honey with excessive water content is not suitable for long-term preservation.[59]

In jars, fresh honey should appear as a pure, consistent fluid, and should not set in layers. Within a few weeks to a few months of extraction, many varieties of honey crystallize into a cream-colored solid. Some varieties of honey, including tupelo, acacia, and sage, crystallize less regularly. Honey may be heated during bottling at temperatures of 40–49 °C (104–120 °F) to delay or inhibit crystallization. Overheating is indicated by change in enzyme levels, for instance, diastase activity, which can be determined with the Schade or the Phadebas methods. A fluffy film on the surface of the honey (like a white foam), or marble-colored or white-spotted crystallization on a containers sides, is formed by air bubbles trapped during the bottling process.

A 2008 Italian study determined nuclear magnetic resonance spectroscopy can be used to distinguish between different honey types, and can be used to pinpoint the area where it was produced. Researchers were able to identify differences in acacia and polyfloral honeys by the differing proportions of fructose and sucrose, as well as differing levels of aromatic amino acids phenylalanine and tyrosine. This ability allows greater ease of selecting compatible stocks.[60]

39.8.3 Acid content and flavor effects

The average pH of honey is 3.9, but can range from 3.4 to 6.1.*[61] Honey contains many kinds of acids, both organic and amino. However, the different types and their amounts vary considerably, depending on the type of honey. These acids may be aromatic or aliphatic (non-aromatic). The aliphatic acids contribute greatly to the flavor of honey by interacting with the flavors of other ingredients.*[61]

Organic acids comprise most of the acids in honey, accounting for 0.17–1.17% of the mixture, with gluconic acid formed by the actions of an enzyme called glucose oxidase as the most prevalent.*[61] Other organic acids are minor, consisting of formic, acetic, butyric, citric, lactic, malic, pyroglutamic, propionic, valeric, capronic, palmitic, and succinic, among many others.*[61]*[62]

39.9 Nutrition and composition

Honey is mainly devoid of essential nutrients, containing only trace amounts of protein, dietary fiber, vitamins or minerals (table).*[63]

A mixture of sugars and other carbohydrates, honey is mainly fructose (about 38-55%) and glucose (about 31%),*[1] with remaining sugars including maltose, sucrose, and other complex carbohydrates.*[1] Its glycemic index ranges from 31 to 78, depending on the variety.*[64] The specific composition, color, aroma and flavor of any batch of honey depend on the flowers foraged by bees that produced the honey.*[8]

Typical honey analysis:*[65]

- Fructose: 38.2%

- Glucose: 31.3%

- Maltose: 7.1%

- Sucrose: 1.3%

- Water: 17.2%

- Higher sugars: 1.5%

- Ash: 0.2%

- Other/undetermined: 3.2%

Honey has a density of about 1.36 kilograms per litre (36% denser than water).*[66]

39.10 Adulteration

Adulteration of honey is the addition of other sugars, syrups or compounds into honey to change its flavor, viscosity, make it cheaper to produce, or to increase the fructose content in order to stave off crystallization. According to the Codex Alimentarius of the United Nations, any product labeled as honey or pure honey must be a wholly natural product, although different nations have their own laws concerning labeling.*[67] Adulteration of honey is sometimes used as a method of deception when buyers are led to believe that the honey is pure. The practice was common dating back to ancient times, when crystallized honey was often mixed with flour or other fillers, hiding the adulteration from buyers until the honey was liquefied. In modern times the most common adulteration-ingredient became clear, almost-flavorless corn syrup, which, when mixed with honey, is often very difficult to distinguish from unadulterated honey.*[68]

Isotope ratio mass spectrometry can be used to detect addition of corn syrup and cane sugar by the carbon isotopic signature. Addition of sugars originating from corn or sugar cane (C4 plants, unlike the plants used by bees, and also sugar beet, which are predominantly C3 plants) skews the isotopic ratio of sugars present in honey,*[69] but does not influence the isotopic ratio of proteins. In an unadulterated honey, the carbon isotopic ratios of sugars and proteins should match. Levels as low as 7% of addition can be detected.*[70]

In one country, the USA, according to The National Honey Board (a USDA-overseen organization), "honey stipulates a pure product that does not allow for the addition of any other substance...this includes, but is not limited to, water or other sweeteners".*[71]

39.11 Health applications

Due mainly to low numbers and quality of human studies,*[72] little to no efficacy has been identified about the potential health benefits of honey.

39.11.1 Wounds and burns

There is some evidence that honey may help healing in skin wounds after surgery and mild (partial thickness) burns when used in a dressing.*[72]*[73]

Evidence does not support the use of honey-based products in the treatment of other burns, chronic wounds, pressure ulcers, Fournier's gangrene, venous stasis ulcers, minor acute wounds, diabetic foot ulcers,

Leishmaniasis,[*][72][*][74] or ingrown toenails.[*][75]

39.11.2 Cough

Little evidence supports honey as a treatment of coughs in children. For chronic cough and acute cough, a Cochrane review found no strong evidence for or against the use of honey.[*][76][*][77] For treating children, the study concluded that honey possibly helps more than no treatment.[*][77]

Whereas regulatory authorities recommend avoiding giving over the counter common cold medication to children,[*][78][*][79] UK authorities suggest "a homemade remedy containing honey and lemon is likely to be just as useful and safer to take" , but warn that honey should not be given to babies because of the risk of infant botulism.[*][78] The World Health Organization recommends honey as a treatment for coughs and sore throats, including for children, stating that there is no reason to believe it is less effective than a commercial remedy.[*][80] Honey is recommended by Canadian physicians for children over the age of 1 for the treatment of coughs as it is deemed as effective as dextromethorphan and more effective than diphenhydramine.[*][81]

39.11.3 Other

People who have a weakened immune system should not eat honey because of the risk of bacterial or fungal infection.[*][5]

No evidence shows the benefit of using honey to treat cancer,[*][5] although honey may be useful for controlling side effects of radiation therapy or chemotherapy applied in cancer treatment.[*][82]

Consumption is sometimes advocated as a treatment for seasonal allergies due to pollen, but there is inconclusive scientific evidence to support the claim.[*][5][*][83] Honey is generally considered ineffective for the treatment of allergic conjunctivitis.[*][5][*][84]

Honey is a source of empty calories and it is recommended that it be replaced with fruits and vegetables.[*][6] One tablespoon of honey provides 64 calories, while one tablespoon of sugar provides 48 calories.[*][7]

39.12 Health hazards

39.12.1 Botulism

Infants can develop botulism after consuming honey contaminated with Clostridium botulinum endospores.[*][85]

Infantile botulism shows geographical variation. In the UK, only six cases have been reported between 1976 and 2006,[*][86] yet the U.S. has much higher rates: 1.9 per 100,000 live births, 47.2% of which are in California.[*][87] While the risk honey poses to infant health is small, it is recommended not to take the risk until after one year of age, and then giving honey is considered safe.[*][88]

39.12.2 Toxic honey

Main article: Bees and toxic chemicals § Toxic honey

Mad honey intoxication is a result of eating honey containing grayanotoxins.[*][89] Honey produced from flowers of rhododendrons, mountain laurels, sheep laurel, and azaleas may cause honey intoxication. Symptoms include dizziness, weakness, excessive perspiration, nausea, and vomiting. Less commonly, low blood pressure, shock, heart rhythm irregularities, and convulsions may occur, with rare cases resulting in death. Honey intoxication is more likely when using "natural" unprocessed honey and honey from farmers who may have a small number of hives. Commercial processing, with pooling of honey from numerous sources, is thought to dilute any toxins.[*][90]

Toxic honey may also result when bees are proximate to tutu bushes (Coriaria arborea) and the vine hopper insect (Scolypopa australis). Both are found throughout New Zealand. Bees gather honeydew produced by the vine hopper insects feeding on the tutu plant. This introduces the poison tutin into honey.[*][91] Only a few areas in New Zealand (Coromandel Peninsula, Eastern Bay of Plenty and the Marlborough Sound) frequently produce toxic honey. Symptoms of tutin poisoning include vomiting, delirium, giddiness, increased excitability, stupor, coma, and violent convulsions. To reduce the risk of tutin poisoning, humans should not eat honey taken from feral hives in the risk areas of New Zealand. Since December 2001, New Zealand beekeepers have been required to reduce the risk of producing toxic honey by closely monitoring tutu, vine hopper, and foraging conditions within 3 kilometres (1.9 mi) of their apiary. Intoxication is rarely dangerous.[*][89]

39.13 In history, culture, and folklore

Honey use and production has a long and varied history.[*][8] In many cultures, honey has associations that go beyond its use as a food. Honey is frequently used as a talisman and symbol of sweetness.

39.13.1 Ancient times

Honey seeker depicted in an 8000-year-old cave painting at Araña Caves in Spain

Honey collection is an ancient activity.[9] Humans apparently began hunting for honey at least 8,000 years ago, as evidenced by a cave painting in Valencia, Spain.[9] The painting is a Mesolithic rock painting, showing two honey-hunters collecting honey and honeycomb from a wild bee nest. The figures are depicted carrying baskets or gourds, and using a ladder or series of ropes to reach the wild nest.

The greater honeyguide bird guides humans to wild bee hives[92] and this behavior may have evolved with early hominids.[93][94]

So far, the oldest remains of honey have been found in the country of Georgia. Archaeologists have found honey remains on the inner surface of clay vessels unearthed in an ancient tomb, dating back some 4,700–5,500 years.[95][96][97] In ancient Georgia, honey was packed for people's journeys into the afterlife, and more than one type, too – along for the trip were linden, berry, and a meadow-flower variety.[98]

In ancient Egypt, honey was used to sweeten cakes and biscuits, and was used in many other dishes. Ancient Egyptian and Middle Eastern peoples also used honey for embalming the dead.[99] The fertility god of Egypt, Min, was offered honey.

The spiritual and therapeutic use of honey in ancient India is documented in both the Vedas and the Ayurveda texts, which were both composed at least 4,000 years ago.[100]

Pliny the Elder devotes considerable space in his book *Naturalis Historia* to the bee and honey, and its many uses. In the absence of sugar, honey was an integral sweetening ingredient in Roman recipes, and references to its use in food can be found in the work of many Roman authors, including Athenaeus, Cato, and Bassus.

The art of beekeeping in ancient China has existed since time immemorial and appears to be untraceable to its origin. In the book *Golden Rules of Business Success* written by Fan Li (or Tao Zhu Gong) during the Spring and Autumn Period, some parts mention the art of beekeeping and the importance of the quality of the wooden box for beekeeping that can affect the quality of its honey.

Honey was also cultivated in ancient Mesoamerica. The Maya used honey from the stingless bee for culinary purposes, and continue to do so today. The Maya also regard the bee as sacred (see Mayan stingless bees of Central America).

Some cultures believed honey had many practical health uses. It was used as an ointment for rashes and burns, and to help soothe sore throats when no other practices were available.

39.13.2 Folk medicine and wound research

In myths and folk medicine, honey has been used both orally and topically to treat various ailments including gastric disturbances, ulcers, skin wounds, and skin burns by ancient Greeks, Egyptians and in Ayurveda and traditional Chinese medicine.[100]

Proposed for treating wounds and burns, honey may have antimicrobial properties as first reported in 1892 and be useful as a safe, improvisational wound treatment.[101][102] Though its supposed antimicrobial properties may be due to high osmolarity even when diluted with water, it is more effective than plain sugar water of a similar viscosity.[101][102] Definitive clinical conclusions about the efficacy and safety of treating wounds, however, are not possible from this limited research.[72]

The flora that bees use to make the honey may have a role in its properties, particularly by bees foraging from the manuka myrtle, *Leptospermum scoparium*, as proposed in one study.*[101]

39.13.3   Religious significance

In Hinduism, honey (Madhu) is one of the five elixirs of immortality (Panchamrita). In temples, honey is poured over the deities in a ritual called Madhu abhisheka. The Vedas and other ancient literature mention the use of honey as a great medicinal and health food.*[103]

In Jewish tradition, honey is a symbol for the new year, Rosh Hashanah. At the traditional meal for that holiday, apple slices are dipped in honey and eaten to bring a sweet new year. Some Rosh Hashanah greetings show honey and an apple, symbolizing the feast. In some congregations, small straws of honey are given out to usher in the new year.

The Hebrew Bible contains many references to honey. In the Book of Judges, Samson found a swarm of bees and honey in the carcass of a lion (14:8). In Old Testament law, offerings were made in the temple to God. The Book of Leviticus says that "Every grain offering you bring to the Lord must be made without yeast, for you are not to burn any yeast or honey in a food offering presented to the Lord" (2:11). In the Books of Samuel Jonathan is forced into a confrontation with his father King Saul after eating honey in violation of a rash oath Saul made (14:24–47). Proverbs 16:24 in the JPS Tanakh 1917 version says "Pleasant words are as a honeycomb, Sweet to the soul, and health to the bones." Book of Exodus famously describes the Promised Land as a "land flowing with milk and honey" (33:3). However, most Biblical commentators write that the original Hebrew in the Bible (דבש *devash*) refers to the sweet syrup produced from the juice of dates (silan).*[104] In 2005 an apiary dating from the 10th century B.C. was found in Tel Rehov, Israel that contained 100 hives and is estimated to produce half a ton of honey annually.*[105]*[106] Pure honey is considered kosher even though it is produced by a flying insect, a nonkosher creature; other products of nonkosher animals are not kosher.*[107]

In Buddhism, honey plays an important role in the festival of Madhu Purnima, celebrated in India and Bangladesh. The day commemorates Buddha's making peace among his disciples by retreating into the wilderness. The legend has it that while he was there, a monkey brought him honey to eat. On Madhu Purnima, Buddhists remember this act by giving honey to monks. The monkey's gift is frequently depicted in Buddhist art.*[103]

In the Christian New Testament, Matthew 3:4, John the Baptist is said to have lived for a long period of time in the wilderness on a diet consisting of locusts and wild honey.

In Islam, there is an entire chapter (Surah) in the Qur'an called an-Nahl (the Bee). According to his teachings (hadith), Muhammad strongly recommended honey for healing purposes.*[108] The Qur'an promotes honey as a nutritious and healthy food. Below is the English translation of those specific verses:

> And thy Lord taught the Bee to build its cells in hills, on trees, and in (men's) habitations; Then to eat of all the produce (of the earth), and find with skill the spacious paths of its Lord: there issues from within their bodies a drink of varying colours, wherein is healing for men: verily in this is a Sign for those who give thought [Al-Quran 16:68–69].*[109]

39.14   Gallery

- Extraction from a honeycomb
- Filtering from a honeycomb
- Pouring raw honey

39.15   See also

- Bee bread
- Honey hunting
- *More than Honey* a 2012 Swiss documentary film on the current state of honey bees and beekeeping
- National Honey Show
- Royal jelly

39.16   References

[1] National Honey Board. "Carbohydrates and the Sweetness of Honey". Last accessed 1 June 2012.

[2] Oregon State University. "What is the relative sweetness of different sugars and sugar substitutes?". Retrieved 1 June 2012.

[3] Prescott, Lansing; Harley, John P. and Klein, Donald A. (1999). *Microbiology*. Boston: WCB/McGraw-Hill. ISBN 0-697-35439-3.

[4] Shapiro RL, Hatheway C, Swerdlow DL (1998). "Botulism in the United States: A Clinical and Epidemiologic Review". *Annals of Internal Medicine* **129** (3): 221–8. doi:10.1059/0003-4819-129-3-199808010-00011 (inactive 13 January 2015). PMID 9696731.

[5] "Honey". Mayo Clinic. 1 November 2013. Retrieved 24 September 2015.

[6] Constance U. Battle (6 October 2009). *Essentials of Public Health Biology: A Guide for the Study of Pathophysiology.* Jones & Bartlett Publishers. pp. 153–. ISBN 978-0-7637-4464-9.

[7] Marilyn P. Shieh; Michelle Shieh (April 2012). *Getting Healthy with the Devip System.* Trafford Publishing. pp. 68–. ISBN 978-1-4669-1902-0.

[8] Hunt CL, Atwater HW (7 April 1915). *Honey and Its Uses in the Home.* US Department of Agriculture, Farmers' Bulletin, No. 653. Retrieved 2 April 2015.

[9] Crane, Eva (1983) *The Archaeology of Beekeeping*, Cornell University Press, ISBN 0-8014-1609-4

[10] International Bee Research Association.

[11] "Honey and Bees." at the Wayback Machine (archived 5 March 2010) National Honey Board.

[12] Whitmyre, Val. "The Plight of the Honeybees". University of California. Retrieved 14 April 2007.

[13] Standifer LN. "Honey Bee Nutrition And Supplemental Feeding". *Excerpted from "Beekeeping in the United States"*. Retrieved 14 April 2007.

[14] Bequaert, J.Q. (1932). "The Nearctic social wasps of the subfamily polybiinae (Hymenoptera; Vespidae)". *Entomologica Americana.*

[15] Britto, Fábio Barros, and Flávio Henrique Caetano. "Morphological Features and Occurrence of Degenerative Characteristics in the Hypopharyngeal Glands of the Paper Wasp Polistes versicolor (Olivier) (Hymenoptera: Vespidae)." Micron 37.8 (2006): 742-47. Web.

[16] FAOstat Browse data. FAOSTAT Domains >>Production>>Livestock Primary; Item: Honey, natural; Area: World; Year: as needed

[17] Where Honey Comes From. Wherefoodcomesfrom.com (9 November 2012). Retrieved on 2 August 2013.

[18] Lavin Tierra, Mariely (February 2008). "Yucatán y su miel". *México Desconocido* **372**: 78–83.

[19] Miel de Corse mele di Corsica, la gamme variétale AOC AOP. Miel-corse.eu. Retrieved on 6 February 2011.

[20] 14. HONEY MARKETING AND INTERNATIONAL TRADE. fao.org

[21] Root, p. 355

[22] Tomasik, Piotr (2004) *Chemical and functional properties of food saccharides*, CRC Press, p. 74, ISBN 0-8493-1486-0

[23] Kántor Z, Pitsi G, Thoen J (1999). "Glass Transition Temperature of Honey as a Function of Water Content As Determined by Differential Scanning Calorimetry". *Journal of Agricultural and Food Chemistry* **47** (6): 2327–2330. doi:10.1021/jf981070g. PMID 10794630.

[24] Russell EV, Israeloff NE (2000). "Direct observation of molecular cooperativity near the glass transition". *Nature* **408** (6813): 695–698. doi:10.1038/35047037. PMID 11130066.

[25] *Value-added products from beekeeping.* Food and Agriculture Organization of the United Nations. 1996. pp. 7–8. ISBN 978-92-5-103819-2.

[26] Bogdanov, Stefan (2009) "Physical Properties of Honey" at the Wayback Machine (archived 20 September 2009), Chapter 4 in *Book of Honey*, Bee Product Science.

[27] Krell, pp. 5–6

[28] Root, p. 348

[29] "Bees 'producing M&M's coloured honey'". *Telegraph.co.uk.* 4 October 2012. Retrieved 30 December 2014.

[30] Hans-Dieter Belitz, Werner Grosch, Peter Schieberle *Food chemistry* Springer Verlag, Berlin-Heidelberg 2004 p. 884 ISBN 3-540-69933-3

[31] Zdzisław E. Sikorski *Chemical and functional properties of food components* CRC Press 2007 p. 121 ISBN 0-8493-9675-1

[32] Root, p. 350

[33] Krell, pp. 40–43

[34] Value-added products from beekeeping. Chapter 2. Fao.org. Retrieved on 14 April 2011.

[35] "The Rheological & Mellisopalynological Properties of Honey" (PDF). Minerva Scientific. Retrieved 10 December 2012. If however, rheological measurements are made on a given sample it can be deduced that the sample is predominantly Manuka (Graph 2) or Kanuka (Graph 3) or a mixture of the two plant species

[36] "Definition of Honey and Honey Products" (PDF). National Honey Issac Board. Retrieved 3 February 2011. Blended Honey: A homogeneous mixture of two or more honeys differing in floral source, color, flavor, density or geographic origin.

[37] "Honey Color and Flavor". National Honey Board. Retrieved 3 February 2011. Wildflower honey is often used to describe honey from miscellaneous and undefined flower sources.

[38] "Varieties of honey: Polyfloral honey". The Honey Book. Retrieved 10 November 2007. Honey that is from wild or commercialized honeybees that is derived from many types of flowers is a resulting Polyfloral honey.

[39] Mountain Wildflower Honey. Mieliditalia.it. Retrieved on 6 February 2011.

[40] The Colours Of Honey. Mieliditalia.it. Retrieved on 6 February 2011.

[41] Gounari, Sofia (2006). "Studies on the phenology of Marchalina hellenica (gen.) (Hemiptera: coccoidea, margarodidae) in relation to honeydew flow". *Journal of apicultural research* **45** (1): 8–12. doi:10.3896/IBRA.1.45.1.03.

[42] Flottum, Kim (2010). *The Backyard Beekeeper: An Absolute Beginner's Guide to Keeping Bees in Your Yard and Garden.* Quarry Books. pp. 170–. ISBN 978-1-61673-860-0.

[43] Subramanian, R.; Hebbar, H. Umesh; Rastogi, N. K. (2007). "Processing of Honey: A Review". *International Journal of Food Properties* **10**: 127. doi:10.1080/10942910600981708.

[44] Definition of Honey and Honey Products at the Wayback Machine (archived 3 December 2007). honey.com. Approved by the National Honey Board 15 June 1996; Updated 27 September 2003

[45] "United States Standards for Grades of Extracted Honey". USDA / Agricultural Marketing Service. Retrieved 20 January 2012.

[46] Damerow, Gail (2011). *The Backyard Homestead Guide to Raising Farm Animals: Choose the Best Breeds for Small-Space Farming, Produce Your Own Grass-Fed Meat, Gather Fresh Eggs, Collect Fresh Milk, Make Your Own Cheese, Keep Chickens, Turkeys, Ducks, Rabbits, Goats, Sheep, Pigs, Cattle, & Bees.* Storey Publishing, LLC. pp. 167–. ISBN 978-1-60342-697-8.

[47] *First Regional Training Workshop for Beekeepers.* Bib. Orton IICA / CATIE. 1992. pp. 55–.

[48] Ultrasonic Honey Processing. Hielscher.com. Retrieved on 6 February 2011.

[49] Sharma, Rajeev (2005). *Improve your Health! with Honey.* Diamond Pocket Books. pp. 33–. ISBN 978-81-288-0920-0.

[50] Krell, Rainer (1996). *Value-added Products Froom Beekeeping.* Food & Agriculture Org. pp. 25–. ISBN 978-92-5-103819-2.

[51] Honey Processing. Beeworks.com. Retrieved on 6 February 2011.

[52] "The History of the Origin and Development of Museums". Dr. M. A. Hagen. The American naturalist, Volume 10. 1876.

[53] 1894. *The Mummy: A Handbook of Egyptian Funerary Archaeology.* 2nd ed. Cambridge: Cambridge University Press. (Reprinted New York: Dover Publications, 1989)

[54] What to Do About Crystallized Honey. Heavenly Homemakers. Retrieved on 2 August 2013.

[55] Decrystallizing Honey. Lynnskitchenadventures.com. Retrieved on 2 August 2013.

[56] "United States Standards for Grades of Extracted Honey". USDA. Retrieved 11 May 2010.

[57] NOTIFICATION, MINISTRY OF AGRICULTURE (Department of Agriculture and Co-operation) New Delhi, 24 December 2008

[58] Bogdanov, Stefan (2008) "Honey production" at the Wayback Machine (archived 5 March 2009). Bee Product Science

[59] Allan, Matthew. "Basic Honey Processing". *Beekeeping in a Nutshell* **5**.

[60] "Keeping Tabs on Honey". *Chemical & Engineering News* **86** (35): 43. 2008. doi:10.1021/cen-v086n035.p043.

[61] "pH and acids in honey" (PDF). National Honey Board Food Technology/Product Research Program. April 2006.

[62] Wilkins, Alistair L. and Lu, Yinrong (1995). "Extractives from New Zealand Honeys. 5. Aliphatic Dicarboxylic Acids in New Zealand Rewarewa (Knightea excelsa) Honey". *J. Agric. Food Chem.* **43** (12): 3021–3025. doi:10.1021/jf00060a006.

[63] "Nutrient data for 19296, Honey". USDA Nutrient Data Laboratory, version SR-27. 2014.

[64] Arcot, Jayashree and Brand-Miller, Jennie (March 2005) A Preliminary Assessment of the Glycemic Index of Honey. A report for the Rural Industries Research and Development Corporation. RIRDC Publication No 05/027. rirdc.infoservices.com.au

[65] "Beesource Beekeeping: Honey Composition and Properties". Beesource.com. October 1980. Retrieved 6 February 2011.

[66] Krell

[67] "Authenticity of honey". doi:10.1007/978-1-4613-1119-5_8#page-1 (inactive 2015-01-13). Retrieved 30 December 2014.

[68] *The Hive: The Story of the Honeybee and Us* By Bee Wilson --St. Martins Press 2004 Page 167

[69] Edwards, G and Walker, D A (1983). *C3,C4: Mechanisms, and Cellular and Environmental Regulation, of Photosynthesis.* University of California Press. pp. 469–. GGKEY: 05LA62Q2TQJ. Sucrose synthesized by a C4 plant (e.g. sugar beet) can be distinguished from sucrose synthesized by a C3 plant (e.g. sugar-cane) due to differences in δ values.

[70] Barry, Carla (1999). "The detection of C4 sugars in honey". *Hivelights* (Canadian Honey Council) **12** (1). Archived from the original on 17 June 2008.

[71] Definition of honey and honey products. (PDF). Retrieved on 9 January 2012.

[72] Jull AB, Cullum N, Dumville JC, Westby MJ, Deshpande S, Walker N (2015). "Honey as a topical treatment for wounds". *Cochrane Database Syst Rev* **3** (CD005083). doi:10.1002/14651858.CD005083.pub4. PMID 25742878. Honey appears to heal partial thickness burns more quickly than conventional treatment (which included polyurethane film, paraffin gauze, soframycin-impregnated gauze, sterile linen and leaving the burns exposed) and infected post-operative wounds more quickly than antiseptics and gauze.

[73] Majtan, J (2014). "Honey: an immunomodulator in wound healing". *Wound Repair Regen.* **22** (2 Mar-Apr): 187–192. doi:10.1111/wrr.12117. PMID 24612472.

[74] O'Meara S, Al-Kurdi D, Ologun Y, Ovington LG, Martyn-St James M, Richardson R (2014). "Antibiotics and antiseptics for venous leg ulcers". *Cochrane Database Syst Rev* (Systematic review) **1**: CD003557. doi:10.1002/14651858.CD003557.pub5. PMID 24408354.

[75] Eekhof JA, Van Wijk B, Knuistingh Neven A, van der Wouden JC (2012). "Interventions for ingrowing toenails". *Cochrane Database Syst Rev* (Systematic review) **4**: CD001541. doi:10.1002/14651858.CD001541.pub3. PMID 22513901.

[76] Mulholland S, Chang AB (2009). "Honey and lozenges for children with non-specific cough". *Cochrane Database Syst Rev* (Systematic review) (2): CD007523. doi:10.1002/14651858.CD007523.pub2. PMID 19370690.

[77] Oduwole O, Meremikwu MM, Oyo-Ita A, Udoh EE (2014). "Honey for acute cough in children". *Cochrane Database Syst Rev* (Systematic review) **3** (12): CD007094. doi:10.1002/14651858.CD007094.pub4. PMID 25536086.

[78] "Cough". NHS Choices. 20 June 2013. Retrieved June 2014.

[79] "Using Over-the-Counter Cough and Cold Products in Children". US Food and Drug Administration. 29 April 2015. Retrieved 24 September 2015.

[80] "Cough and cold remedies for the treatment of acute respiratory infections in young children". *World Health Organization*. Retrieved 15 October 2015.

[81] Goldman, Ran D. (December 2014). "Honey for treatment of cough in children.". *Canadian Family Physician* (Systematic review) **60** (12): 1107–1110. PMID 25642485. Retrieved 15 October 2015.

[82] Bardy J, Slevin NJ, Mais KL, Molassiotis A (2008). "A systematic review of honey uses and its potential value within oncology care". *J Clin Nurs* **17** (19): 2604–23. doi:10.1111/j.1365-2702.2008.02304.x. PMID 18808626.

[83] Dale Kiefer (4 May 2012). "Honey for Allergies".

[84] Rudmik L, Hoy M, Schlosser RJ, Harvey RJ, Welch KC, Lund V, Smith TL (April 2013). "Topical therapies in the management of chronic rhinosinusitis: an evidence-based review with recommendations". *Int Forum Allergy Rhinol* (Review) **3** (4): 281–98. doi:10.1002/alr.21096. PMID 23044832.

[85] Asked Questions The National Honey Board at the Wayback Machine (archived 1 February 2010). Honey.com. Retrieved on 6 February 2011.

[86] Report on Minimally Processed Infant Weaning Foods and the Risk of Infant Botulism. (PDF). Advisory Committee on the Microbiological Safety of Food (July 2006). food.gov.uk. Retrieved on 9 January 2012.

[87] Botulism in the United States, 1899–1996, Handbook for Epidemiologists, Clinicians, and Laboratory Workers, Atlanta, GA. Centers for Disease Control and Prevention (1998)

[88] Infant Botulism and Honey. Edis.ifas.ufl.edu. Retrieved on 9 January 2012.

[89] Jansen, Suze A.; Kleerekooper, Iris; Hofman, Zonne L. M.; Kappen, Isabelle F. P. M.; Stary-Weinzinger, Anna; van der Heyden, Marcel A. G. (2012). "Grayanotoxin Poisoning: 'Mad Honey Disease' and Beyond". *Cardiovascular Toxicology* **12** (3): 208–215. doi:10.1007/s12012-012-9162-2. ISSN 1530-7905. PMID 22528814.

[90] "Grayanotoxin" at the Wayback Machine (archived 14 March 2010) in the *Foodborne Pathogenic Microorganisms and Natural Toxins Handbook*, FDA Center for Food Safety and Applied Nutrition.

[91] Tutu Bush and Toxic Honey. National Beekeepers Association, New Zealand

[92] Isack HA, Reyer HU (1989). "Honeyguides and honey gatherers: interspecific communication in a symbiotic relationship". *Science* **243** (4896): 1343–6. Bibcode:1989Sci...243.1343I. doi:10.1126/science.243.4896.1343. PMID 17808267.

[93] Short, Lester, Horne, Jennifer and Diamond, A. W. (2003). "Honeyguides". In Christopher Perrins (Ed.). Firefly Encyclopedia of Birds. Firefly Books. pp. 396–397. ISBN 1-55297-777-3.

[94] Dean, W. R. J.; MacDonald, I. A. W. (1981). "A Review of African Birds Feeding in Association with Mammals". *Ostrich* **52** (3): 135. doi:10.1080/00306525.1981.9633599.

[95] Kvavadze, Eliso; Gambashidze, Irina; Mindiashvili, Giorgi; Gogochuri, Giorgi (2006). "The first find in southern Georgia of fossil honey from the Bronze Age, based on palynological data". *Vegetation History and Archaeobotany* **16** (5): 399. doi:10.1007/s00334-006-0067-5.

[96] Georgian ancient honey. cncworld.tv (31 March 2012). Retrieved on 10 July 2012.

[97] Report: Georgia Unearths the World's Oldest Honey. EurasiaNet (30 March 2012). Retrieved on 3 July 2015.

[98] The world's first winemakers were the world's first beekeepers. guildofscientifictroubadours.com (2 April 2012). Retrieved on 10 July 2012.

[99] Larry Gonick The Cartoon History of the Universe Vol.2

[100] Pećanac M, Janjić Z, Komarcević A, Pajić M, Dobanovacki D, Misković SS (2013). "Burns treatment in ancient times". Med Pregl 66 (5–6): 263–7. doi:10.1016/s0264-410x(02)00603-5. PMID 23888738.

[101] Maddocks, Sarah E; Jenkins, Rowena E (2013). "Honey: a sweet solution to the growing problem of antimicrobial resistance?". Future Microbiology 8 (11): 1419–1429. doi:10.2217/fmb.13.105. PMID 24199801.

[102] Stewart, JA; McGrane, OL; Wedmore, IS (2014). "Wound care in the wilderness: is there evidence for honey?". Wilderness Environ Med. 25 (1 (Mar)): 103–110. doi:10.1016/j.wem.2013.08.006. PMID 24393701.

[103] A Meaningful Story of Buddha, Elephant and Monkey by Marguerite Theophil, United Press International, 16 November 2006, accessed 9 August 2008

[104] Berel, Rabbi. (24 September 2005) Apples and Honey. Aish.com. Retrieved on 6 February 2011.

[105] Mazar, Amihai; Panitz-Cohen, Nava (2007). "It Is the Land of Honey: beekeeping at Tel Rehov". Near Eastern Archeology 70 (4): 202–219.

[106] The Hebrew University of Jerusalem. "First Beehives In Ancient Near East Discovered". ScienceDaily. ScienceDaily. Retrieved 6 October 2015.

[107] "Why is honey kosher?" Chabad.org. Retrieved 30 November 2010.

[108] Sahih Bukhari vol. 7, book 71, number 584, 585, 588 and 603.

[109] Yusuf 'Ali, 'Abdullah. An Nahl, Al-Quran Chapter 16 (The Bee) quoted from "The Holy Qur'an: Original Arabic Text with English Translation & Selected Commentaries". Saba Islamic Media. Retrieved 20 May 2013.

39.17   Bibliography

- Krell, R. (1996). Value-added products from beekeeping. Food and Agriculture Organization of the United Nations. ISBN 978-92-5-103819-2.

- Root, A. I. and Root, E. R. (2005). The ABC and Xyz of Bee Culture. Kessinger Publishing. ISBN 978-1-4179-2427-1.

39.18   External links

- Beekeeping and Sustainable Livelihoods (2004), Food and Agriculture Organization of the United Nations

- "Honey". The New Student's Reference Work. 1914.

Chapter 40

Honey hunting

Honey hunting or **Honey harvesting** is the gathering of honey from wild bee colonies and is one of the most ancient human activities and is still practiced by aboriginal societies in parts of Africa, Asia, Australia and South America. Some of the earliest evidence of gathering honey from wild colonies is from rock painting, dating to around 8,000 BC. In the Middle Ages in Europe, the gathering of honey from wild or semi-wild bee colonies was carried out on a commercial scale.

Gathering honey from wild bee colonies is usually done by subduing the bees with smoke and breaking open the tree or rocks where the colony is located, often resulting in the physical destruction of the colony.

40.1 Africa

Honey hunting in Africa is a part of the indigenous culture in many parts and hunters have hunted for thousands of years.

40.2 Asia

A documentary by freelance photo journalists Diane Summers and Eric Valli on the **Honey hunters of Nepal** documents Gurung tribesmen of west-central Nepal entering the jungle in search of wild honey where they use indigenous tools under precarious conditions to collect honey.

Twice a year high in the Himalayan foothills of central Nepal teams of men gather around cliffs that are home to the world's largest honeybee, Apis laboriosa. As they have for generations, the men come to harvest the Himalayan cliff bee's honey.

This was also documented in a BBC2 documentary in August 2008 entitled *Jimmy and the Wild Honey Hunters-Sun*. An English farmer travelled into the Himalayan foothills on a honey hunting expedition. The world's largest honeybee,

Honey seeker depicted on 8000 year old cave painting near Valencia, Spain at Cuevas de la Araña en Bicorp

Apis laboriosa is over twice the size of those in the UK where their larger bodies have adapted to the colder climate for insulation. The documentary involved ascending a 200-foot rope ladder and balancing a basket and a long pole to chisel away at a giant honey comb of up to 2 million bees and catch it in the basket.

40.2.1   India and Bangladesh

In the Sunderban forest, shared by West Bengal and Bangladesh, estuarine forests are the area of operation of honey hunters.[*][1] They are known as "Mawals". This is a dangerous occupation as many honeyhunters die in tiger attacks which are common in this area. The harvest ritual, which varies slightly from community to community, begins with a prayer and sacrifice of flowers, fruits, and rice. Then a fire is lit at the base of the cliff to smoke the bees from their honeycombs.

40.3   Europe

40.3.1   Function

As early as the Stone Age, people collected the honey of wild bees, but this was not done commercially. From the Early Middle Ages it became a trade, known in German-speaking central Europe, for example, as a *Zeidler* or *Zeitler*, whose job it was to collect the honey of wild, semi-wild or domestic bees in the forests. Unlike modern beekeepers, they did not keep the bees in man-made wooden beehives. Instead, they cut holes as hives in old trees at a height of about six meters and fitted a board over the entrance. Whether a colony of bees nested there or not, depended entirely on the natural environment and that could change every year. The tree tops were also cut off in order to prevent wind damage.

40.3.2   Distribution

Extremely valuable, if not a prerequisite for tree beekeeping, were conifer stands. Important locations for honey hunting in the Middle Ages were in the regions of the Fichtel Mountains and the Nuremberg Imperial Forest. In Bavaria forest beekeeping is recorded as early as the year 959 in the vicinity of Grabenstätt. But even in the area of today's Berlin, there was extensive honey gathering, especially in the then much larger Grunewald.

In the area around Nuremberg there are still numerous references to an earlier flourishing honey hunting tradition such as the castle of Zeidlerschloss in Feucht. Honey was important for Nuremberg's gingerbread production; the Nuremberg *Reichswald* ("The bee garden of the Holy Roman Empire") provided plenty of it.

A mannequin dressed as a honey hunter

40.4   References

[1] http://news.bbc.co.uk/2/hi/programmes/from_our_own_correspondent/2969234.stm

40.5   Literature

- Eva Crane: *The world history of beekeeping and honey hunting.* Duckworth, London, 2000. ISBN 0-7156-2827-5.

- Karl Hasel, Ekkehard Schwartz: *Forstgeschichte. Ein Grundriss für Studium und Praxis.* 2nd, updated edition. Kessel, Remagen, 2002, ISBN 3-935638-26-4.

- Richard B. Hilf: *Der Wald. Wald und Weidwerk in Geschichte und Gegenwart – Erster Teil* [reprint]. Aula, Wiebelsheim, 2003, ISBN 3-494-01331-4.

- Klaus Baake: *Das Zeidelprivileg von 1350*. Munich, 1990.

Chapter 41

Huckleberry

For other uses, see Huckleberry (disambiguation).
Huckleberry is a name used in North America for several plants in the family Ericaceae, in two closely related genera: *Vaccinium* and *Gaylussacia*.

The huckleberry is the state fruit of Idaho.

Bog Huckleberry at Polly's Cove, Nova Scotia

41.1 Nomenclature

The name 'huckleberry' is a North American variation of the English dialectal name variously called 'hurtleberry' or 'whortleberry' /'wɜrtəlˌbɛrɪ/ for the bilberry.[*][1] In North America the name was applied to numerous plant variations all bearing small berries with colors that may be red, blue or black.[*][2] It is the common name for various *Gaylussacia* species, and some *Vaccinium* species, such as *Vaccinium parvifolium*, the *red huckleberry*, and is also applied to other *Vaccinium* species which may also be called blueberries depending upon local custom, as in New England and parts of Appalachia.[*][2]

41.2 Edibility

The fruit of the various species of plant called huckleberries is generally edible and tasty. The berries are small and round, 5–10 mm in diameter and look like dark blueberries. In taste, the berries range from tart to sweet, with a flavor similar to that of a blueberry, especially in blue- and purple-colored varieties. However, many kinds of huckleberries have a noticeable, distinct taste different from blueberries, and some have noticeably larger seeds. Huckleberries are consumed by many animals including bears, birds, and humans.

The 'garden huckleberry' (*Solanum scabrum*) is not a true huckleberry but is instead a member of the nightshade family.

41.3 Taxonomy

Wild huckleberry in the Mount Hood National Forest in Oregon.

41.3.1 *Gaylussacia*

Four species of huckleberries in the genus *Gaylussacia* are common in eastern North America, especially *G. baccata*, also known as the black huckleberry.*[2]

41.3.2 *Vaccinium*

From coastal Central California to southern Washington and British Columbia, the red huckleberry (*Vaccinium parvifolium*) is found in the maritime-influenced plant community. In the Pacific Northwest and mountains of Montana and Idaho, this huckleberry species and several others, such as the black *Vaccinium* huckleberry (*V. membranaceum*) and blue (Cascade) huckleberry (*V. deliciosum*), grow in various habitats, such as mid-alpine regions up to 11,500 feet elevation, mountain slopes, forests or lake basins.*[2] The plant grows best in damp, acidic soil having volcanic origin, attaining under optimal conditions heights of 1.5 to 2 m (4.9 to 6.6 ft), usually ripening in mid-to-late summer or later at high elevations.*[2]

Where the climate is favorable, certain species of huckleberry, such as *V. membranaceum*, *V. parvifolium* and *V. deliciosum*, are used in ornamental plantings.*[2]

41.4 Nutrients and phytochemicals

Only limited research has been applied to define the content of essential nutrients in huckleberries, showing none with high content.*[3]

Two huckleberry species, *V. membranaceum* and *V. ovatum*, were studied for phytochemical content, showing that *V. ovatum* had greater total anthocyanin and polyphenols than did *V. membranaceum*.*[4] Each species contained 15 anthocyanins (galactoside, glucoside, and arabinoside of delphinidin, cyanidin, petunidin, peonidin, and malvidin) but in different proportions.*[4]

41.5 Use as food or traditional medicine

Huckleberries were traditionally collected by Native American and First Nations people along the Pacific coast, interior British Columbia, and Montana for use as food or traditional medicine.*[2]*[5]*[6]

Huckleberries can be processed into numerous food products including juice, tea, soup, syrup, jam, pudding, candy, pie, muffins, pancakes, and salad dressings.*[2]*[6] Tradi-

tional medical applications included treating pain, heart ailments, and infections.*[6]

41.6 Use in slang

Huckleberries hold a place in archaic American English slang. The tiny size of the berries led to their use as a way of referring to something small, often affectionately as in the lyrics of Moon River. The phrase "a huckleberry over my persimmon" was used to mean "a bit beyond my abilities". "I'm your huckleberry" is a way of saying that one is just the right person for a given job.*[7] The range of slang meanings of huckleberry in the 19th century was fairly large, also referring to significant persons or nice persons.*[8]*[9]

41.7 See also

- *Vaccinium ovatum* (known by the common names evergreen huckleberry, winter huckleberry and California huckleberry)

41.8 References

[1] Cited as "U.S. 1670" in Onions, CT (1933). *Shorter Oxford English Dictionary* **1** (3rd ed.). Oxford: Oxford University Press. p. 930.

[2] Barney DL (1999). "Growing Western Huckleberries" (PDF). University of Idaho. Retrieved August 12, 2014.

[3] "Nutrition facts for Huckleberries, raw (Alaska Native) per 100 g, from US Department of Agriculture Nutrient Tables, version SR-21". Conde Nast. 2014. Retrieved 2014-08-14.

[4] Lee, J; Finn, C. E.; Wrolstad, R. E. (2004). "Comparison of anthocyanin pigment and other phenolic compounds of Vaccinium membranaceum and Vaccinium ovatum native to the Pacific Northwest of North America". *Journal of agricultural and food chemistry* **52** (23): 7039–44. doi:10.1021/jf049108e. PMID 15537315.

[5] Foster, Steven; Hobbs, Christopher (April 2002). *A Field Guide to Western Medicinal Plants and Herbs*. Houghton Mifflin Harcourt. ISBN 039583807X.

[6] Strass K (2010). "Huckleberry Harvesting of the Salish and Kootenai of the Flathead Reservation" (PDF). Retrieved 2014-08-14.

[7] "World Wide Words: Huckleberry". *World Wide Words*.

[8] Gullible Gulls, Huckleberry, Jumbi, Wooden Nickels, Realtors, and Calling a Spade a Spade, *The Word Detective*, apparently based on the *Dictionary of American Regional English*

[9] Huckleberry, Douglas Harper, *Online Etymology Dictionary*, 2001

Chapter 42

Illicium verum

"Star anise" redirects here. For other uses, see Star anise (disambiguation).

Illicium verum is a medium-sized native evergreen tree of northeast Vietnam and southwest China. A spice commonly called **star anise**, **star anise seed**, or **Chinese star anise** that closely resembles anise in flavor is obtained from the star-shaped pericarp of the fruit of *Illicium velum* which are harvested just before ripening. Star anise oil is a highly fragrant oil used in cooking, perfumery, soaps, toothpastes, mouthwashes, and skin creams. 90% of the world's star anise crop is used for extraction of shikimic acid, a chemical intermediate used in the synthesis of oseltamivir.

Reverse side of fruit

42.1 Nomenclature

'Illicium' comes from the from Latin *illicio* meaning "entice". In Persian, star anise is called بادیان *bādiyān*, hence its French name *badiane*. In India it is called *badian* or *phoolchakri* and in Pakistan, it is called *badian*.

42.2 Use

42.2.1 Culinary use

Star anise contains anethole, the same ingredient that gives the unrelated anise its flavor. Recently, star anise has come into use in the West as a less expensive substitute for anise in baking as well as in liquor production, most distinctively in the production of the liquor Galliano. It is also used in the production of *sambuca*, *pastis*, and many types of absinthe. Star anise enhances the flavour of meat.[2] It is used as a spice in preparation of *biryani* and *masala chai* all over the Indian subcontinent. It is widely used in Chinese cuisine, and in Indian cuisine where it is a major component of *garam masala*, and in Malay and Indonesian cuisines. It is widely grown for commercial use in China, India, and most other countries in Asia. Star anise is an ingredient of the traditional five-spice powder of Chinese cooking. It is also a major ingredient in the making of *phở*, a Vietnamese noodle soup.It is also used in the French recipe of mulled wine : called vin chaud (hot wine).

42.2.2 Medicinal use

Star anise is the major source of the chemical compound shikimic acid, a primary precursor in the pharmaceutical synthesis of anti-influenza drug oseltamivir (Tamiflu).[3] Shikimic acid is produced by most autotrophic organisms, and whilst it can be obtained in commercial quantities elsewhere, star anise remains the usual industrial source. In 2005, a temporary shortage of star anise was caused by its use in the production of Tamiflu. Later that year, a method for the production of shikimic acid using bacteria was discovered.[4][5][6] Roche now derives some of the raw material it needs from the fermentation of *E. coli* bacteria. The 2009 swine flu outbreak led to another series of shortages as stocks of Tamiflu were built up around the world, sending prices soaring.[7]

Plate from François-Pierre Chaumeton's 1833 Flore Medicale

Star anise is grown in four provinces in China and harvested between March and May. It is also found in the south of New South Wales. The shikimic acid is extracted from the seeds in a 10-stage manufacturing process which takes a year.

In traditional Chinese medicine, star anise is considered a warm and moving herb, and used to assist in relieving cold-stagnation in the middle jiao.

Japanese star anise (*Illicium anisatum*), a similar tree, is highly toxic and inedible; in Japan, it has instead been burned as incense. Cases of illness, including "serious neurological effects, such as seizures", reported after using star anise tea, may be a result of using this species. Japanese star anise contains anisatin, which causes severe inflammation of the kidneys, urinary tract, and digestive organs. The toxicity of *I. anisatum*, also known as shikimi, is caused by its potent neurotoxins anisatin, neoanisatin,

and pseudoanisatin which are noncompetitive antagonists of GABA receptors.*[8]

42.3   Standardization of its products and services

- ISO 676:1995 - contains the information about the nomenclature of the variety and cultivars*[9]

42.3.1   Identification

- Refer to the 4th edition of the *European Pharmacopoeia* [1153].

42.3.2   Differentiation with other species

Joshi *et al.* have used fluorescent microscopy and gas chromatography*[10] to distinguish the species, while Lederer *et al.* employed TLC with HPLC-MS/MS.*[11]

42.3.3   Specifications

- ISO 11178:1995 - a specification for its dried fruits*[12]
- GB/T 7652:2006 - a Chinese standard of the product*[13]

42.4   See also

- *Pimpinella anisum*

42.5   References

[1]   "The Plant List: A Working List of All Plant Species". Retrieved 3 September 2015.

[2]   "Spaghetti Bolognese". *In Search of Perfection*. BBC Two.

[3]   Wang, G. W.; Hu, W. T.; Huang, B. K.; Qin, L. P. (2011). "*Illicium verum*: A review on its botany, traditional use, chemistry and pharmacology". *Journal of Ethnopharmacology* **136** (1): 10–20. doi:10.1016/j.jep.2011.04.051. PMID 21549817.

[4]   Bradley, D. . (Dec 2005). "Star role for bacteria in controlling flu pandemic?". *Nature reviews. Drug discovery* **4** (12): 945–946. doi:10.1038/nrd1917. ISSN 1474-1776. PMID 16370070.

[5] Krämer, M.; Bongaerts, J.; Bovenberg, R.; Kremer, S.; Müller, U.; Orf, S.; Wubbolts, M.; Raeven, L. (2003). "Metabolic engineering for microbial production of shikimic acid". *Metabolic Engineering* **5** (4): 277–283. doi:10.1016/j.ymben.2003.09.001. PMID 14642355.

[6] Johansson, L.; Lindskog, A.; Silfversparre, G.; Cimander, C.; Nielsen, K. F.; Lidén, G. (Dec 2005). "Shikimic acid production by a modified strain of E. Coli (W3110.shik1) under phosphate-limited and carbon-limited conditions". *Biotechnology and Bioengineering* **92** (5): 541–552. doi:10.1002/bit.20546. ISSN 0006-3592. PMID 16240440.

[7] Louisa Lim (18 May 2009). "Swine Flu Bumps Up Price Of Chinese Spice". NPR.

[8] Perret, C.; Tabin, R.; Marcoz, J. -P.; Llor, J.; Cheseaux, J. -J. (2011). "Malaise du nourrisson pensez à une intoxication à l'anis étoilé". *Archives de Pédiatrie* **18** (7): 750–753. doi:10.1016/j.arcped.2011.03.024. PMID 21652187. ("Apparent life-threatening event in infants: think about star anise intoxication!")

[9] International Organization for Standardization. "ISO 676:1995 Spices and condiments -- Botanical nomenclature". Retrieved 8 June 2009.

[10] Joshi, Vaishali C.; Ragone, S; Bruck, IS; Bernstein, JN; Duchowny, M; Peña, BM (2005). "Rapid and easy identification of *Illicium verum* Hook. f. and its adulterant *Illicium anisatum* Linn. by fluorescent microscopy and gas chromatography". *Journal of AOAC International* (AOAC International) **88** (3): 703–706. PMID 16001842. Retrieved 10 November 2007.

[11] Lederer, Ines; Schulzki, G; Gross, J; Steffen, JP (2006). "Combination of TLC and HPLC-MS/MS methods. Approach to a rational quality control of Chinese star anise". *Journal of Agricultural and Food Chemistry* (American Chemical Society) **54** (6): 1970–1974. doi:10.1021/jf058156b. PMID 16536563.

[12] International Organization for Standardization. "ISO 11178:1995 Star anise (Illicium verum Hook. f.) -- Specification". Retrieved 8 June 2009.

[13] 供 杜南京野生植物 合利用 究院. "GB/T 7652-2006 八角". Retrieved 8 June 2009.

42.5.1 Bibliography

- ITIS 505892

- US FDA Advisory on star anise "teas"

- Fooducation:Star Anise

Chapter 43

Jackfruit

The **jackfruit** (*Artocarpus heterophyllus*), also known as jack tree, jakfruit, or sometimes simply jack or jak[*][6] is a species of tree in the mulberry and fig family (Moraceae).

It is native to parts of South and Southeast Asia, and is believed to have originated in the southwestern rain forests of India, in present-day Andhra Pradesh,Goa, Kerala, Tamil Nadu,[*][7] coastal Karnataka, and Maharashtra.[*][8] The jackfruit tree is well suited to tropical lowlands, and its fruit is the largest tree-borne fruit,[*][9] reaching as much as 35 kg (80 lb) in weight, 90 cm (35 in) in length, and 50 cm (20 in) in diameter.[*][10]

The jackfruit tree is a widely cultivated and popular food item throughout the tropical regions of the world. Jackfruit is the national fruit of Bangladesh, by name Kanthal (কাঁঠাল) in Bengali language.[*][11] The Jackfruit tree can produce about 100 to 200 fruits in a year.

43.1 Etymology

Jackfruit hanging from the trunk

The word "jackfruit" comes from Portuguese *jaca*, which in turn, is derived from the Malayalam language term, *chakka* (Malayalam *chakka pazham* : ചക്കപ്പഴം).[*][12] When

The jackfruit illustrated by Michael Boym in the 1656 book Flora Sinensis.

the Portuguese arrived in India at Kozhikode (Calicut) on the Malabar Coast (Kerala) in 1498, the Malayalam name *chakka* was recorded by Hendrik van Rheede (1678–1703) in the *Hortus Malabaricus*, vol. iii in Latin. Henry Yule translated the book in Jordanus Catalani's (f. 1321–1330) *Mirabilia descripta: the wonders of the East.*[*][13]

The common English name "jackfruit" was used by the physician and naturalist Garcia de Orta in his 1563 book *Colóquios dos simples e drogas da India.*[*][14][*][15] Centuries later, botanist Ralph Randles Stewart suggested it was named after William Jack (1795–1822), a Scottish botanist who worked for the East India Company in Bengal,

Sumatra, and Malaysia.[16]

43.1.1 Synonym

Artocarpus integer (Thunb.) Merr.[17] is currently accepted name, whereas *Artocarpus integrifolius* L.f. is synonym. However, in *Flora of British India*, Volume 5 (Page 541), J.D. Hooker mentions it as *Artocarpus integrifolia* L.f. Moreover, *Artocarpus heterophyllus* Lam. is a different species.[18]

43.2 Cultivation

Jackfruit flesh

Developing jackfruit in Bangladesh

Opened jackfruit

The jackfruit has played a significant role in Indian agriculture for centuries. Archeological findings in India have revealed that jackfruit was cultivated in India 3000 to 6000 years ago.[19] It has also been widely cultivated in southeast Asia.

43.3 Aroma

Jackfruit have a distinctive, sweet and fruity aroma. In a study of flavour volatiles in five jackfruit cultivars, the main volatile compounds that were detected were: ethyl isovalerate, propyl isovalerate, butyl isovalerate, isobutyl isovalerate, 3-methylbutyl acetate, 1-butanol and 2-methylbutanol.[20]

43.4 Culinary uses of jackfruit

The flesh of the jackfruit is starchy and fibrous and is a source of dietary fiber. The flavor is comparable to a com-

bination of apple, pineapple, mango, and banana.[21] Varieties are distinguished according to characteristics of the fruit's flesh.

- In Bangladesh the fruit is consumed on its own. The unripe fruit is used in curry. The seed is often dried and preserved to be later used in curry.[11] Thailand and Vietnam are major producers of jackfruit, which are often cut, prepared, and canned in a sugary syrup (or frozen in bags/boxes without syrup), and exported overseas, frequently to North America and Europe.

- In Andhra pradesh, Jack fruit is known as Panasa (Pana-sa). It is a popular desert after hearty meals and the skin of the fruit is used to prepare festive curry.

- In Brazil, three varieties are recognized: *jaca-dura*, or the "hard" variety, which has a firm flesh and the largest fruits that can weigh between 15 and 40 kg each, *jaca-mole*, or the "soft" variety, which bears smaller fruits with a softer and sweeter flesh, and *jaca-manteiga*, or the "butter" variety, which bears sweet fruits whose flesh has a consistency intermediate between the "hard" and "soft" varieties.*[22] In Indochina, the two varieties are the "hard" version (more crunchy, drier and less sweet but fleshier), and the "soft" version (more soft, moister, much sweeter with a darker gold-color flesh than the hard variety).

- In Indonesia, jackfruit is called *nangka*. The ripe fruit is usually sold separately and consumed on its own; or sliced and mixed with shaved ice as a sweet concoction dessert, such as *es campur* and *es teler*. The ripe fruit might be dried and fried as *kripik nangka* or jackfruit cracker. The seeds are boiled and consumed with salt as it contains edible starchy content, this is called *beton*. Young (unripe) jackfruit is used in several kinds of curry, such as gulai nangka and gudeg.

- In Tamil Nadu(India) Panruti the sleepy coastal taluk in Cuddalore district is the heaven of jackfruit. It produces the best jackfruits in the country – fat, sweet and tasty. You can buy the fruit round the year. The biggest jackfruit in Panruti weighs over 70 kg which actually makes it a hot contender for the Guinness Book of Records. A Hawaiian jackfruit, at a mere 34 kg holds the title at present. By not staking a claim, Panruti is depriving itself of a world record title every year.*[7]*[23] From here, jack fruits are exported to many states in India as well as exported to many countries across the world. A place known as 'Chakka Gramam'(Tamil: சக்க கிராமம்)(Jackfruit village), it has large jackfruit tree plantations covering a total area of 1,084 hectares and they yield about 40 tonnes per hectare, earning a revenue of around 18 crore rupees(INR) a year for Panruthi farmers. The village also has the highest jackfruit consumption per annum in the country.*[24]

- In Kerala, India two varieties of jackfruit predominate: *varikka* (വരിക്ക) and *koozha* (കൂഴ). *Varikka* has a slightly hard inner flesh when ripe, while the inner flesh of the ripe *koozha* fruit is very soft and almost dissolving. A sweet preparation called *chakka varattiyathu* (jackfruit jam) is made by seasoning pieces of *varikka* fruit flesh in jaggery, which can be preserved and used for many months. Huge jackfruits up to four feet in length with a corresponding girth are sometimes seen in Kerala. The young fruit is *idichakka* or *idianchakka* in Kerala.

- In West Bengal, India the two varieties are called *khaja kathal* and *moja kathal*. The fruits are either eaten alone or as a side to rice, *roti*, *chira*, or *muri*. Sometimes, the juice is extracted and either drunk straight or as a side with *muri*. The extract is sometimes condensed into rubbery delectables and eaten as candies. The seeds are either boiled or roasted and eaten with salt and hot chillies. They are also used to make spicy side-dishes with rice or *roti*.

- In Mangalore, Karnataka, India the varieties are called *bakke* and *imba*. The pulp of the *imba* jackfruit is ground and made into a paste, then spread over a mat and allowed to dry in the sun to create a natural chewy candy.

- In Coorg, Karnataka, India the culinary items made out of jackfruit are aplenty. Jackfruit is known as *Chakke*. Jackfruit seeds are fried and a curry is made.

- In Maharashtra, the hard variety is called *kaapa* and the soft variety is called *barka*. The juice of the *barka* is extracted and spread on greased metal dishes which are then kept for sun-drying. Within 2–3 days, a tasty dried pancake-like dried jackfruit juice called as *phansacha saath* or *phanas poli* results.*[25]

- In Sri Lanka the young fruit is called *polos* - 𝄞𝄞𝄞𝄞𝄞𝄞 ripened fruit is called *waraka* - 𝄞𝄞𝄞 and *wela* - 𝄞𝄞𝄞.

- In Indochina, jackfruit is a frequent ingredient in sweets and desserts.

- In Vietnam, jackfruit is used to make jackfruit *chè* (*chè* is a sweet dessert soup, similar to the Chinese derivative, *bubur chacha*). The Vietnamese also use jackfruit puree as part of pastry fillings, or as a topping on *xôi ngọt* (sweet version of sticky rice portions).

- Jackfruit is known as Rukh-Katahar (= tree katahar) in Nepal, while Bhui-Katahar (= Ground Katahar) denotes pineapple. The ripe fruite is eaten itself (sometimes with a pinch of salt sprinkled) as a delicacy while the unripe fruit is used to prepare savory curry. The ripe fruit is also used to brew alcoholic beverage in some parts of the country.

Jackfruit is commonly used in South and Southeast Asian cuisines.*[26]*[21]

43.4.1   Culinary uses for ripe fruit

- Extracting the jackfruit arils and separating the seeds from the sweet flesh.

- *Kripik nangka*, Indonesian jackfruit chips.

- *Es teler*, Indonesian dessert made from shaved ice, condensed milk, coconut, avocado, and jackfruit.

- *Halo-halo*, an ice dessert from the Philippines with different fruits and toppings.

Ripe jackfruit is naturally sweet with subtle flavoring. It can be used to make a variety of dishes, including custards, cakes, or mixed with shaved ice as *es teler* in Indonesia or *halo-halo* in the Philippines. In India, when the jackfruit is in season, an ice cream chain store called "Naturals" carries jackfruit flavored ice cream.

Ripe jackfruit arils are sometimes seeded, fried, or freeze-dried and sold as jackfruit chips.

The seeds from ripe fruits are edible, are said to have a milky, sweet taste, and may be boiled, baked, or roasted. When roasted, the flavor of the seeds is comparable to chestnuts. Seeds are used as snacks either by boiling or fire roasting, or to make desserts. For making the traditional breakfast dish in southern India: *idlis*, the fruit is used with rice as an ingredient and jackfruit leaves are used as a wrapping for steaming. Jackfruit *dosas* can be prepared by grinding jackfruit flesh along with the batter.

43.4.2   Culinary uses for unripe fruit

- *Gudeg* (left), the unripe jackfruit curry in reddish color acquired from teak leaf, a specialty of Yogyakarta in Java.

- *Ginataang langka*, jackfruit cooked in coconut milk.

- Green jackfruit and potato curry, Kolkata.

- Baby Jackfruit Masala.

Developing jackfruit

The cuisines of India, Nepal, Bangladesh, Sri Lanka, Indonesia, Cambodia, Thailand and Vietnam use cooked young jackfruit.[21] In Indonesia, young jackfruit is cooked with coconut milk as *gudeg*. In many cultures, jackfruit is boiled and used in curries as a staple food. In northern Thailand, the boiled young jackfruit is used in the Thai salad called *tam kanun*. In West Bengal, the unripe green jackfruit called *aechor* or *ichor* is used as a vegetable to make various spicy curries and side dishes, and as fillings for cutlets and chops. It is especially sought after by vegetarians who substitute this for meat, hence is nicknamed as *gacch-patha* (tree-mutton). In the Philippines, it is cooked with coconut milk (*ginataang langka*). In Réunion Island, it is cooked either alone or with meat, such as shrimp or smoked pork. In southern India, unripe jackfruit slices are deep fried to make chips. In Udipi cuisine, jackfruit is used make *appa* and *addae*.

Because unripe jackfruit has a meat-like taste, it is used in curry dishes with spices, in Bihar, Jharkhand, Sri Lankan, Andhran, eastern Indian (Bengali) and (Odisha) and Keralan cuisines. The skin of unripe jackfruit must be peeled first, then the remaining whole jackfruit can be chopped into edible portions and cooked before serving. Young jackfruit has a mild flavor and distinctive meat-like texture and is compared to poultry. Meatless sandwiches have been suggested and are popular with both vegetarian and non-vegetarian populations. Unripe jackfruit is widely known as *panasa katha* in Odisha.

43.5   Nutrition

The edible jackfruit is made of easily digestible flesh (bulbs); a 100-g portion of edible raw jackfruit provides about 95 calories and is a good source of the antioxidant vitamin C, providing about 13.7 mg.[27] Jackfruit seeds are rich in protein. The fruit is also rich in vitamin B_6, potassium, calcium, and iron.[28]

43.6   Seeds

In general, the seeds are gathered from the ripe fruit, sun-dried, then stored for use in rainy season in many parts of South Indian states. They are extracted from fully matured fruits and washed in water to remove the slimy part. Seeds should be stored immediately in closed polythene bags for one or two days to prevent them from drying out. Germination is improved by soaking seeds in clean water for 24 hours. During transplanting, sow seeds in line, 30 cm apart, in a nursery bed filled with 70% soil mixed with 30% organic matter.[29] The seedbed should be shaded partially from direct sunlight to protect emerging seedlings.

Boiled jackfruit seeds are also edible. Often compared to Brazil nuts, they are quite commonly used in curry in the Indian state of Kerala. In Java, the seeds are commonly cooked and seasoned with salt as a snack.

43.7 Wood

Jackfruit tree

The wood of the tree is used for the production of musical instruments. In Indonesia, hardwood from the trunk is carved out to form the barrels of drums used in the *gamelan*, and in the Philippines, its soft wood is made into the body of the *kutiyapi*, a type of boat lute. It is also used to make the body of the Indian string instrument *veena* and the drums *mridangam*, *thimila*, and *kanjira*; the golden, yellow timber with good grain is used for building furniture and house construction in India. The ornate wooden plank called *avani palaka* made of the wood of jackfruit tree is used as the priest's seat during Hindu ceremonies in Kerala. In Vietnam, jackfruit wood is prized for the making of Buddhist statuaries in temples,[30] and fish sauce barrels.[31]

Jackfruit wood is widely used in the manufacture of furniture, doors and windows, and in roof construction. The heartwood is used by Buddhist forest monastics in Southeast Asia as a dye, giving the robes of the monks in those traditions their distinctive light-brown color.[32]

43.8 Commercial availability

Outside of its countries of origin, fresh jackfruit can be found at Asian food markets, especially in the Philippines, Thailand, Vietnam, Malaysia, Cambodia, and Bangladesh. It is also extensively cultivated in the Brazilian coastal region, where it is sold in local markets. It is available canned in sugary syrup, or frozen, already prepared and cut. Dried jackfruit chips are produced by various manufacturers. In northern Australia, particularly in Darwin, jackfruit can be found on the outdoor produce markets during the dry season. Outside of countries where it is grown, jackfruit can be obtained year-round both canned or dried. It has a ripening season in Asia of late spring to late summer.[33]

Jackfruit industries are established in Sri Lanka and Vietnam, where the fruit is processed into products such as flour, noodles, *papad*, and ice cream. It is also canned and sold as a vegetable for export.[28]

43.9 Production and marketing

The marketing of jackfruit involves three groups: producers, traders (middlemen) including wholesalers, and retailers.[34] The marketing channels are rather complex. Large farms sell immature fruits to wholesalers which help cash flow and reduce risk, whereas medium-sized farms sell fruits directly to local markets or retailers.

In Kerala, a large amount of jackfruit production occurs naturally, but around 97% of its production is wasted because of lack of processing units and marketing.

- Leaves of the jackfruit
- Selling jackfruit in Bangkok
- Jackfruit at a fruit stand in Manhattan's Chinatown
- Cut jackfruit

43.10 Cultural significance

The national fruit of Bangladesh is the Jackfruit.[11] The Jackfruit is the state fruit of the Indian states of Kerala and Tamil Nadu, one of the three auspicious fruits of Tamil Nadu, along with the mango and banana.[35]

43.11 Invasive species

In Brazil the jackfruit can become an invasive species as in Brazil's Tijuca Forest National Park in Rio de Janeiro.

The Tijuca is mostly an artificial secondary forest, whose planting began during the mid-19th century, and jackfruit trees have been a part of the park's flora since its founding. Recently, the species has expanded excessively; its fruits, which naturally fall to the ground and open, are eagerly eaten by small mammals such as the common marmoset and coati. The seeds are dispersed by these animals, which allows the jackfruit to compete for space with native tree species. Additionally, as the marmoset and coati also prey opportunistically on bird's eggs and nestlings, the supply of jackfruit as a ready source of food has allowed them to expand their populations, to the detriment of the local bird populations. Between 2002 and 2007, 55,662 jackfruit saplings were destroyed in the Tijuca Forest area in a deliberate culling effort by the park's management.*[36]

43.11.1 Production trends

the top 5 producers of Jackfruits (in 1000 tonnes) were as follows:*[37]

43.12 References

[1] Under its accepted name *Artocarpus heterophyllus* (then as *heterophylla*) this species was described in *Encyclopédie Méthodique, Botanique* 3: 209. (1789) by Jean-Baptiste Lamarck, from a specimen collected by botanist Philibert Commerson. Lamarck said of the fruit that it was coarse and difficult to digest. "Larmarck's original description of *tejas*". Retrieved 2012-11-23. On mange la chair de son fruit, ainsi que les noyaux qu'il contient; mais c'est un aliment grossier et difficile à digérer.

[2] "Name - !*Artocarpus heterophyllus* Lam." . *Tropicos*. Saint Louis, Missouri: Missouri Botanical Garden. Retrieved 2012-11-23.

[3] "TPL, treatment of *Artocarpus heterophyllus*". *The Plant List; Version 1. (published on the internet)*. Royal Botanic Gardens, Kew and Missouri Botanical Garden. 2010. Retrieved 2012-11-23.

[4] "Name – *Artocarpus heterophyllus* Lam. synonyms". *Tropicos*. Saint Louis, Missouri: Missouri Botanical Garden. Retrieved 2012-11-23.

[5] GRIN (2006-11-02). "*Artocarpus heterophyllus* information from NPGS/GRIN" . *Taxonomy for Plants*. National Germplasm Resources Laboratory, Beltsville, Maryland: USDA, ARS, National Genetic Resources Program. Retrieved 2012-11-23.

[6] "*Artocarpus heterophyllus*". Tropical Biology Association. October 2006. Retrieved 2012-11-23.

[7] "Jackfruit Paradise" . *Civil Society*.

[8] Boning, Charles R. (2006). *Florida's Best Fruiting Plants: Native and Exotic Trees, Shrubs, and Vines.* Sarasota, Florida: Pineapple Press, Inc. p. 107.

[9] "Jackfruit, Breadfruit & Relatives". Know & Enjoy Tropical Fruit. 2012. Retrieved 2012-11-23.

[10] "Jackfruit Fruit Facts". California Rare Fruit Growers, Inc. 1996. Retrieved 2012-11-23.

[11] Matin, Abdul. "A poor man's fruit: Now a miracle food!". *The Daily Star*. Retrieved 2015-06-12.

[12] Pradeepkumar, T.; Jyothibhaskar, B. Suma; Satheesan, K. N. (2008). Prof. K. V. Peter, ed. *Management of Horticultural Crops.* Horticulteral Science Series **11**. New Delhi, India: New India Publishing. p. 81. ISBN 978-81-89422-49-3. The English name jackfruit is derived from Portuguese *jaca*, which is derived from Malayalam *chakka*.

[13] Friar Jordanus, 14th century, as translated from the Latin by Henry Yule (1863). *Mirabilia descripta: the wonders of the East*. Hakluyt Society. p. 13. Retrieved 2012-11-23.

[14] *Oxford English Dictionary*, Second Edition, 1989, online edition

[15] Anon. (2000) The American Heritage Dictionary of the English Language: Fourth Edition.

[16] Stewart, Ralph R. (1984). "How Did They Die?". *Taxon* **33** (1): 48–52. doi:10.2307/1222028.

[17] "*Artocarpus integer* (Thunb.) Merr. The Plant List" . Theplantlist.org. Retrieved 2014-06-17.

[18] "*Artocarpus heterophyllus* Lam. The Plant List" . Theplantlist.org. 2012-03-23. Retrieved 2014-06-17.

[19] Preedy, Victor R.; Watson, Ronald Ross; Patel, Vinood B., eds. (2011). *Nuts and Seeds in Health and Disease Prevention* (1st ed.). Burlington, MA: Academic Press. p. 678. ISBN 978-0-12-375689-3.

[20] Ong, B.T.; Nazimah, S.A.H.; Tan, C.P.; Mirhosseini, H.; Osman, A.; Hashim, D. Mat; Rusul, G. (August 2008). "Analysis of volatile compounds in five jackfruit (*Artocarpus heterophyllus* L.) cultivars using solid-phase microextraction (SPME) and gas chromatography-time-of-flight mass spectrometry (GC-TOFMS)". *Journal of Food Composition and Analysis* **21** (5): 416–422. doi:10.1016/j.jfca.2008.03.002. Retrieved 2013-02-02.

[21] *The encyclopedia of fruit & nuts*, By Jules Janick, Robert E. Paull, p. 155

[22] General information, Department of Agriculture, State of Bahia. seagri.ba.gov.br (in Portuguese)

[23] http://www.indiawaterportal.org/news/ jackfruit-paradise-panruti-tn-breaks-all-records-terms-production

[24] http://unparalleledindia.blogspot.in/2013/07/ jackfruit-village-panruti.html

[25] Morton, J. "Jackfruit – Artocarpus heterophyllus".

[26] *The encyclopedia of fruit & nuts*, By Jules Janick, Robert E. Paull, pp.481–485

[27] "Show Foods". Ndb.nal.usda.gov. Retrieved 2014-06-17.

[28] Goldenberg, Suzanne (2014-04-23). "Jackfruit heralded as 'miracle' food crop". *The Guardian*.

[29] *Jackfruit Artocarpus heterophyllus. Field Manual for Extension Workers and Farmers* (PDF). Southampton, UK: Southampton Centre for Underutilised Crops. 2006. ISBN 0-85432-834-3.

[30] "Gỗ mít nài". Nhagoviethung.com. Retrieved 2014-06-17.

[31] "Nam O fish sauce village". *Danang Today*. 2014-02-26. Retrieved 2015-09-22.

[32] Forest Monks and the Nation-state: An Anthropological and Historical Study in Northeast Thailand J.L. Taylor 1993 p. 218

[33] Jackfruit. Hort.purdue.edu. Retrieved on 2011-10-17.

[34] Haq, Nazmul (2006). *Jackfruit: Artocarpus heterophyllus* (PDF). Southampton, UK: Southampton Centre for Underutilised Crops. p. 129. ISBN 0-85432-785-1.

[35] Subrahmanian, N.; Hikosaka, Shu; Samuel, G. John; Thiagarajan, P. (1997). *Tamil social history*. Institute of Asian Studies. p. 88. Retrieved 2010-03-23.

[36] Livia de Almeida, "Guerra contra as jaqueiras" ("War on Jackfruit"), *Revista Veja Rio*, 2007-05-05; see also [http:/,/www.jbrj.gov.br/enbt/posgraduacao/resumos/2008/rodolfo_de_abreu.htm]

[37] "Jackfruit: Improvement in the Asia-Pacific Region" (PDF). *Asia-Pacific Association of Agricultural Research Institutions*.

43.13   External links

- Germplasm Resources Information Network: *Artocarpus heterophyllus*
- Fruits of Warm Climates: Jackfruit and Related Species
- California Rare Fruit Growers: Jackfruit Fruit Facts
- Jackfruit (Artocarpus heterophyllus) on Wayne's Word
- Science in India with Special Reference to Agriculture
- How to Select and Prepare a Jackfruit (Online Video)
- Crops for the Future: Jackfruit (*Artocarpus heterophyllus*)
- Ayurvedic medicinal plants and their uses
- Video Cutting Up A Jack Fruit on YouTube
- Culinary uses of ripe Jackfruit in Southern India
- Jackfruit Seed as ingredient
- How to open and proceed the Jackfruit

- *Artocarpus heterophylla* in West African plants – A Photo Guide.

Chapter 44

Japan wax

Japan wax also known as sumac wax, China green tallow, and Japan tallow. This material is a pale-yellow, waxy, water-insoluble solid with a gummy feel, obtained from the berries of certain sumacs native to Japan and China, such as *Toxicodendron vernicifluum* (lacquer tree) and *Toxicodendron succedanea* (Japanese wax tree).[1]

Japan wax is a byproduct of lacquer manufacture. It is not a true wax but a fat that contains 95% palmitin.[1] Japan wax is sold in flat squares or disks and has a rancid odor. It is extracted by expression and heat, or by the action of solvents.

44.1   Uses

Japan wax is a used candles, furniture polishes, floor waxes, wax matches, soaps, food packaging, pharmaceuticals, cosmetics, pastels, crayons, buffing compounds, metal lubricants, adhesives, thermoplastic resins, and as a substitute for beeswax. Because it undergoes rancidification, it is not often used in foods.

44.2   Other names

Japan tallow; sumac wax; sumach wax; vegetable wax; Japan tallow; China green tallow.

44.3   Properties

Melting point = 124°F (51°C) [2] or 45–53 °C.[1]

Specific gravity ≈ 0.975 [2]

Soluble in benzene, ether, naphtha and alkalis. Insoluble in water and cold ethanol.

Iodine value = 4.5–12.6

Acid value = 6–209

Saponification value = 220 [2]

44.4   References

[1] Claude Leray "Waxes" in Kirk-othmer encyclopedia of chemical technology 2006, Wiley-VCH, Weinheim. doi:10.1002/0471238961.2301240503152020.a01.pub2

[2] Brady, George S.; Clauser, Henry R. ; Vaccari A., John (1997). *Materials Handbook* (14th ed.). New York, NY: McGraw-Hill. ISBN 0-07-007084-9.

Chapter 45

Juniper berry

Juniper berries, here still attached to a branch, are actually modified conifer cones.

A **juniper berry** is the female seed cone produced by the various species of junipers. It is not a true berry but a cone with unusually fleshy and merged scales, which give it a berry-like appearance. The cones from a handful of species, especially *Juniperus communis*, are used as a spice, particularly in European cuisine, and also give gin its distinctive flavour. According to one FAO document, juniper berries are the only spice derived from conifers,[*][1] although tar and inner bark (used as a sweetener in Apache cuisines) from pine trees is sometimes considered a spice as well.

Mature purple and younger green juniper berries can be seen growing alongside one another on the same plant.

45.1 Species

All juniper species grow berries, but some are considered too bitter to eat. In addition to *J. communis*, other edible species include *Juniperus drupacea*,[*][2][*][3] *Juniperus phoenicea*,[*][4] *Juniperus deppeana*, and *Juniperus californica*.[*][5] Some species, for example *Juniperus sabina*, are toxic and consumption is inadvisable.[*][6]

45.2 Characteristics

Juniperus communis berries vary from four to twelve millimeters in diameter; other species are mostly similar in size, though some are larger, notably *J. drupacea* (20–28 mm). Unlike the separated and woody scales of a typical pine cone, those in a juniper berry remain fleshy and merge into a unified covering surrounding the seeds. The berries are green when young, and mature to a purple-black colour over about 18 months in most species, including *J. communis* (shorter, 8–10 months in a few species, and about 24 months in *J. drupacea*).[*][2] The mature, dark berries are usually but not exclusively used in cuisine, while gin is flavoured with fully grown but immature green berries.*[1]

45.3 Uses

Dried juniper berries at a market in Syracuse, Sicily

The flavor profile of young, green berries is dominated by pinene; as they mature this piney, resinous backdrop is joined by what Harold McGee describes as "green-fresh" and citrus notes.*[7] The outer scales of the berries are relatively flavourless, so the berries are almost always at least lightly crushed before being used as a spice. They are used both fresh and dried, but their flavour and odour are at their strongest immediately after harvest and decline during drying and storage.

Juniper berries are used in northern European and particularly Scandinavian cuisine to "impart a sharp, clear flavor" *[1] to meat dishes, especially wild birds (including thrush, blackbird, and woodcock) and game meats (including boar and venison).*[8] They also season pork, cabbage, and sauerkraut dishes. Traditional recipes for choucroute garnie, an Alsatian dish of sauerkraut and meats, universally include juniper berries.*[9] Besides Norwegian and Swedish dishes, juniper berries are also sometimes used in German, Austrian, Czech, Polish and Hungarian cuisine, often with roasts (such as German sauerbraten). Northern Italian cuisine, especially that of the South Tyrol, also incorporates juniper berries.

Juniper, typically *Juniperus communis*, is used to flavor gin, a liquor developed in the 17th century in the Netherlands. The name *gin* itself is derived from either the French *genièvre* or the Dutch *jenever*, which both mean "juniper" .*[1] Other juniper-flavoured beverages include the Finnish rye-and-juniper beer known as sahti, which is flavored with both juniper berries and branches.*[10] The brand Dry Soda produces a juniper-berry soda as part of its lineup. Recently, some American distilleries have begun

using 'New World' varieties of juniper such as *Juniperus occidentalis*.*[11]

Juniper berry was first intended as a medication since juniper berries are a diuretic and were also thought to be an appetite stimulant and a remedy for rheumatism and arthritis. Native Americans are reported to have used the juniper berry as an appetite suppressant in times of hunger. Juniper berry is being researched as a treatment for diet-controlled diabetes, as it releases insulin from the pancreas, hence alleviating hunger. It is also said to have been used by some tribes as a female contraceptive.

A few North American juniper species produce a seed cone with a sweeter, less resinous flavor than those typically used as a spice. For example, one field guide describes the flesh of the berries of *Juniperus californica* as "dry, mealy, and fibrous but sweet and without resin cells" .*[12] Such species have been used not just as a seasoning but as a nutritive food by some Native Americans.*[13] In addition to medical and culinary purposes, Native Americans have also used the seeds inside juniper berries as beads for jewellery and decoration.*[13]

An essential oil extracted from juniper berries is used in aromatherapy and perfumery.*[4] The essential oil can be distilled out of berries which have already been used to flavour gin.*[1]

45.4 History

Juniper berries, including *Juniperus phoenicea* and *Juniperus oxycedrus* have been found in ancient Egyptian tombs at multiple sites. *J. oxycedrus* is not known to grow in Egypt, and neither is *Juniperus excelsa*, which was found along with *J. oxycedrus* in the tomb of Tutankhamun.*[14] The berries imported into Egypt may have come from Greece; the Greeks record using juniper berries as a medicine long before mentioning their use in food.*[15] The Greeks used the berries in many of their Olympics events because of their belief that the berries increased physical stamina in athletes.*[16] The Romans used juniper berries as a cheap domestically produced substitute for the expensive black pepper and long pepper imported from India.*[4] It was also used as an adulterant, as reported in Pliny the Elder's *Natural History*: "Pepper is adulterated with juniper berries, which have the property, to a marvellous degree, of assuming the pungency of pepper." *[17] Pliny also incorrectly asserted that black pepper grew on trees that were "very similar in appearance to our junipers" .

45.5   Notes and references

[1] Ciesla, William M (1998). *Non-wood forest products from conifers.* Food and Agriculture Organization of the United Nations. ISBN 92-5-104212-8. Chapter 8: Seeds, Fruits, and Cones. Retrieved July 27, 2006.

[2] Farjon, A. (2005). *A Monograph of Cupressaceae and Sciadopityaceae.* Royal Botanic Gardens, Kew. pp. 228–400. ISBN 1-84246-068-4.

[3] Adams, R. P. (2004). *Junipers of the World: The genus Juniperus.* Trafford. ISBN 1-4120-4250-X.

[4] Dalby, A. (2002). *Dangerous Tastes: The Story of Spices.* University of California Press. p. 33. ISBN 0-520-23674-2.

[5] Peattie, D., & Landacre, P. H. (1991). *A Natural History of Western Trees.* Houghton Mifflin. p. 226. ISBN 0-395-58175-3.

[6] Grieve, M. (1984). *A Modern Herbal.* Penguin. ISBN 0-14-046440-9.

[7] McGee, Harold (2004). *On Food and Cooking (Revised Edition).* Scribner. p. 410. ISBN 0-684-80001-2.

[8] Montagne, Prosper. *The Concise Larousse Gastronomique.* Octopus. p. 691. ISBN 0-600-60863-8.

[9] Steingarten, Jeffrey (1997). "True Choucroute" . *The Man Who Ate Everything.* Vintage Books. p. 244. ISBN 0-375-70202-4. The chapter is an essay first published in 1989.

[10] Jackson, Michael (1995). Sweating up a suitable thirst. Michael Jackson's Beer Hunter. Retrieved 30 July 2006.

[11] Bend Distillery. Cascade Mountain Gin. Bend Distillery. Retrieved 10 Dec 2010.

[12] Peattie, Donald; Paul (1991). *A Natural History of Western Trees.* Houghton Mifflin Field Guides. p. 226. ISBN 0-395-58175-3.

[13] Moerman, Daniel E (1998). *Native American Ethnobotany.* Timber Press. pp. 282–290. ISBN 0-88192-453-9.

[14] Manniche, Lisa (1999). *Sacred Luxuries: Fragrance, Aromatherapy, and Cosmetics in Ancient Egypt.* Cornell University Press. p. 21. ISBN 0-8014-3720-2.

[15] Dalby, Andrew (1997). *Siren Feasts: A History of Food and Gastronomy in Greece.* Routledge. p. 142. ISBN 0-415-15657-2.

[16] James, Lorman. (1997) *Greek Life.* Gregory House: New York. 76-77.

[17] From Bostock and Riley's 1855 translation. Text online.

45.6   External links

- Medicinal uses of Juniper in Armenia

Chapter 46

Kino (gum)

For other uses, see Red Gum and bloodwood.

Kino is the name of the plant gum produced by various

Copious flow of kino from a wound near the base of the trunk of a Corymbia calophylla *(Marri)*

plants and trees, particularly *Eucalyptus*, in reaction to mechanical damage,[1] and which can be tapped by incisions made in the trunk or stalk. Its red colour, together with the tendency of some species to ooze large amounts of it from wounds, is the source of the common names "red gum" and "bloodwood". The word "kino" is of West African origin.[2]

46.1 Composition

Astringent tannin compounds are a major active component of kinos.[3] The chief constituent of kino is kinotannic acid, of which it contains 70 to 80 per cent. It also contains kino red, a phlobaphene produced from kinotannic acid by oxidation.[4] Kino also yields kinoin, a crystalline neutral principle.[2]

In cold water it is only partially dissolved, leaving a pale flocculent residue which is soluble in boiling water but deposited again upon cooling. It is soluble in alcohol and caustic alkalis, but not in ether.[2]

When exuding from the tree, it resembles red-currant jelly, but hardens in a few hours after exposure to the air and sun.[2] Kinos typically dry to an amber-like material.[5] It consists of dark red angular fragments, rarely larger than a pea.[6] Of the small angular glistening fragments, the smaller are reddish, and the larger are almost black; thin pieces are ruby red. It is brittle and easily powdered. It has no smell, but a very astringent taste.[7]

46.2 Applications and Sources

Kinos are used in medicine, tanning,[3] and as dyes.[7] Kino was introduced to European medicine in 1757 by John Fothergill. When described by him, it was believed to have been brought from the river Gambia in West Africa, and when first imported it was sold in England as *Gummi rubrum astringens gambiense*. It was obtained from *Pterocarpus erinaceus*. In the early 20th century, the drug recognized as the legitimate kind was East Indian, Malabar or Amboyna kino which is the evaporated juice obtained from incisions in the trunk of *Pterocarpus marsupium*.[2] In addition to kinos from these two species, Bengal or Butea kino from *Butea frondosa* and Australian, Botany Bay, or Eucalyptus kino from *Eucalyptus resinifera*, the brown gum tree, were imported into the United States.[6] A West Indian or Jamaica kino is believed to be the product of

Coccoloba uvifera, or seaside grape. It is possible that the same plant is the source of the South American kino.*[8]

Kino is not absorbed at all from the stomach and only very slowly from the intestine. The drug was frequently used in diarrhoea, its value being due to the relative insolubility of kinotannic acid, which enabled it to affect the lower part of the intestine. In this respect it is similar to catechu. It ceased being used as a gargle when antiseptics became recognized as the rational treatment for sore throat.*[2] A medicinal tincture of kino was used as a gargle for the relaxation of the uvula; it contained kino, glycerin, alcohol, and water.*[7]

As they are usually soluble in water, kinos found use in traditional remedies: Eucalyptus kino is used by Australian aborigines in a tea for treating colds.*[5]

Kino was employed to a considerable extent in the East Indies as a cotton dye, giving to the cotton the yellowish-brown color known as nankeen.*[7]

46.3   Notes

[1] A Critical Revision of the Genus Eucalyptus

[2] Chisholm, Hugh, ed. (1911). "Kino". *Encyclopædia Britannica* (11th ed.). Cambridge University Press.

[3] Edited by Pearsall, J., and Trumble, B., *The Oxford English Reference Dictionary*, Oxford University Press, Second Edition, 1996, ISBN 0-19-860046-1

[4] Kino on www.henriettesherbal.com

[5] Aboriginal People and Their Plants, *by Philip A. Clarke, p.104*

[6] "Kino". *Collier's New Encyclopedia*. 1921.

[7] "Kino". *New International Encyclopedia*. 1905.

[8] "Kino". *The American Cyclopædia*. 1879.

46.4   References

- Kino (East Indian, Malabar, Madras, Or Cochin Kino) on chestofbooks.com

46.5   Further reading

- Jean H. Langenheim. *Plant Resins: Chemistry, Evolution, Ecology, and Ethnobotany* (2003).

46.6   External links

- "Kino (gum)". *Encyclopedia Americana*. 1920.

Chapter 47

Urushiol lacquer

For items made with lacquer, see Lacquerware.

Lacquer is a clear or coloured wood finish that dries by

Lacquer box with inlaid mother of pearl peony decor, Ming Dynasty, 16th century

Armorial screen

solvent evaporation or a curing process that produces a hard, durable finish. This finish can be of any sheen level from ultra matte to high gloss, and it can be further polished as required. It is also used for "lacquer paint", which is a paint that typically dries better on a hard and smooth surface.

The term *lacquer* originates from the Sanskrit word *lākshā'* (लाकूषा) meaning "wax", which was used for both the Lac insect (because of their enormous number) and the scarlet resinous secretion it produces that was used as wood finish in ancient India and neighbouring areas.[*][1] In terms of modern products for coating finishes, lac-based finishes are likely to be referred to as shellac, while lacquer often refers to other polymers dissolved in volatile organic compounds (VOCs), such as nitrocellulose, and later acrylic compounds dissolved in *lacquer thinner*, a mixture of several solvents typically containing butyl acetate and xylene or toluene. Lacquer is more durable than shellac.

In the decorative arts, lacquer or lacquerware refers to a variety of techniques used to decorate wood, metal or other surfaces, especially carving into deep coatings of many layers of lacquer.

47.1 Etymology

The archaic French word *lacre* "a kind of sealing wax", from Portuguese *lacre*, unexplained variant of *lacca* "resinous substance", from Arabic *lakk*, from Telugu 🯄🯄🯄 Persian *lak*, the verb *lac* meaning "to cover or coat with laqueur".[*][2] The root of the word is the Sanskrit word *lākshā'* (लाकूषा), which was used for both the Lac insect (because of their enormous number) and the scarlet resinous secretion it produces that was used as wood finish in ancient India and neighbouring areas.[*][1][*][3] Lac resin was once imported in sizeable quantity into Europe from India along with Eastern woods.[*][4][*][5]

47.2 Sheen measurement

Lacquer sheen is a measurement of the shine for a given lacquer.[*][6] Different manufacturers have their own names and standards for their sheen.[*][6] The most common names from least shiny to most shiny are: flat, matte, egg shell, satin, semi-gloss, and gloss (high).

47.3   Urushiol-based lacquers

A Chinese six-pointed tray, red lacquer over wood, from the Song Dynasty (960–1279), 12th-13th century, Metropolitan Museum of Art.

Ming Dynasty Chinese lacquerware container, dated 16th century.

Urushiol-based lacquers differ from most others, being slow-drying, and set by oxidation and polymerization, rather than by evaporation alone. In order for it to set properly it requires a humid and warm environment. The phenols oxidize and polymerize under the action of an enzyme laccase, yielding a substrate that, upon proper evaporation of its water content, is hard. These lacquers produce very hard, durable finishes that are both beautiful and very resistant to damage by water, acid, alkali or abrasion. The active ingredient of the resin is urushiol, a mixture of various phenols suspended in water, plus a few proteins. The

resin is derived from a tree indigenous to China, species *Toxicodendron vernicifluum*, commonly known as the Lacquer Tree.[7] The fresh resin from the *T. vernicifluum* trees causes urushiol-induced contact dermatitis and great care is required in its use. The Chinese treated the allergic reaction with crushed shellfish, which supposedly prevents lacquer from drying properly.[8] Lacquer skills became very highly developed in Asia, and many highly decorated pieces were produced.

During the Shang Dynasty (1600–1046 B.C.), the sophisticated techniques used in the lacquer process were first developed and it became a highly artistic craft,[9] although various prehistoric lacquerwares have been unearthed in China dating back to the Neolithic period and objects with lacquer coating in Japan from the late Jōmon period.[9] The earliest extant lacquer object, a red wooden bowl,[10] was unearthed at a Hemudu culture (5000-4500 B.C.) site in China.[11] By the Han Dynasty (206 B.C. – 220 a.C), many centres of lacquer production became firmly established.[9] The knowledge of the Chinese methods of the lacquer process spread from China during the Han, Tang and Song dynasties, eventually it was introduced to Korea, Japan, Southeast and South Asia.[12]

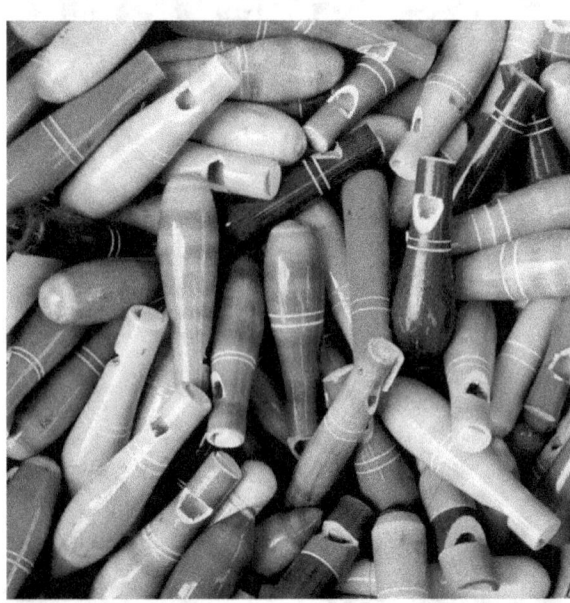

Wooden lacquer-finished whistles made in Channapatna, Karnataka, India

There are two types of lacquer in India: one obtained from the *T. vernicifluum* tree and the other from an insect. In India the insect lac was once used from which a red dye was first extracted, later what was left of the insect was a grease that was used for lacquering objects. Trade of lacquer objects travelled through various routes to the Middle East. Known applications of lacquer in China included coffins, music instruments, furniture, and various household items.*[9] Lacquer mixed with powdered cinnabar is used to produce the traditional red lacquerware from China.

Lacquer mixed with water and turpentine, ready for applying to surface.

The trees must be at least ten years old before cutting to bleed the resin. It sets by a process called "aqua-polymerization", absorbing oxygen to set; placing in a humid environment allows it to absorb more oxygen from the evaporation of the water.

Lacquer-yielding trees in Thailand, Vietnam, Burma and Taiwan, called Thitsi, are slightly different; they do not contain urushiol, but similar substances called "laccol" or "thit-siol". The end result is similar but softer than the Chinese or Japanese lacquer. Unlike Japanese and Chinese *Toxico-dendron verniciflua* resin, Burmese lacquer does not cause allergic reactions; it sets slower, and is painted by crafts-men's hands without using brushes.

Raw lacquer can be "coloured" by the addition of small amounts of iron oxides, giving red or black depending on the oxide. There is some evidence that its use is even older than 8,000 years from archaeological digs in China. Later, pigments were added to make colours. It is used not only as a finish, but mixed with ground fired and unfired clays applied to a mould with layers of hemp cloth, it can produce objects without need for another core like wood. The process is called "kanshitsu" in Japan. Advanced decorative techniques using additional materials such as gold and silver powders and flakes ("makie") were refined to very high standards in Japan also after having been introduced from China. In the lacquering of the Chinese musical instrument, the guqin, the lacquer is mixed with deer horn powder (or ceramic powder) to give it more strength so it can stand up to the fingering.

There are more than four forms of urushiol which is written as thus:

A Chinese lacquer coffin decorated with birds and dragons, from the State of Chu, 4th century B.C.

$R = (CH_2)_{14}CH_3$ or
$R = (CH_2)_7CH=CH(CH_2)_5CH_3$ or
$R = (CH_2)_7CH=CHCH_2CH=CH(CH_2)_2CH_3$ or
$R = (CH_2)_7CH=CHCH_2CH=CHCH=CHCH_3$ or
$R = (CH_2)_7CH=CHCH_2CH=CHCH_2CH=CH_2$ and others.

47.3.1 Types of lacquer

Types of lacquer vary from place to place but they can be divided into unprocessed and processed categories.

The basic unprocessed lacquer is called *raw lacquer* (生漆: *ki-urushi* in Japanese, *shengqi* in Chinese). This is directly from the tree itself with some impurities filtered out. Raw lacquer has a water content of around 25% and appears in a light brown colour. This comes in a standard grade made from Chinese lacquer, which is generally used for ground layers by mixing with a powder, and a high quality grade made from Japanese lacquer called *kijomi-urushi* (生正味漆) which is used for the last finishing layers.

The processed form (in which the lacquer is stirred continuously until much of the water content has evaporated) is called *guangqi* (光漆) in Chinese but comes under many different Japanese names depending on the variation, for example, *kijiro-urushi* (木地呂漆) is standard transparent lacquer sometimes used with pigments and *roiro-urushi* (呂色漆) is the same but pre-mixed with iron hydroxide to produce a black coloured lacquer. *Nashiji-urushi* (梨子地漆) is the transparent lacquer but mixed with gamboge to create an even clearer lacquer and is especially used for the sprinkled-gold technique. These lacquers are generally used for the middle layers. Japanese lacquers of this type are generally used for the top layers and are prefixed by the word *jo-* (上) which means 'top (layer)'.

Processed lacquers can have oil added to them to make them glossy, for example, *shuai-urushi* (朱合漆) is mixed with linseed oil. Other specialist lacquers include *ikkake-urushi* (漆) which is thick and used mainly for applying gold or silver leaf.

47.4   Nitrocellulose lacquers

Slow-drying solvent-based lacquers that contain nitrocellulose, a resin obtained from the nitration of cotton and other cellulostic materials, were developed in the early 1920s, and extensively used in the automobile industry for 30 years. Prior to their introduction, mass-produced automotive finishes were limited in colour, with Japan Black being the fastest drying and thus most popular. General Motors Oakland automobile brand automobile was the first (1923) to introduce one of the new fast drying nitrocelluous lacquers, a bright blue, produced by DuPont under their Duco tradename.

These lacquers are also used on wooden products, furniture primarily, and on musical instruments and other objects. Nitrocellulose lacquers are also used to make firework fuses waterproof. The nitrocellulose and other resins and plasticizers are dissolved in the solvent, and each coat of lacquer dissolves some of the previous coat. These lacquers were a huge improvement over earlier automobile and furniture finishes, both in ease of application and in colour retention. The preferred method of applying quick-drying lacquers is by spraying, and the development of nitrocellulose lacquers led to the first extensive use of spray guns. Nitrocellulose lacquers produce a hard yet flexible, durable finish that can be polished to a high sheen. Drawbacks of these lacquers include the hazardous nature of the solvent, which is flammable and toxic, and the hazards of nitrocellulose in the manufacturing process. Lacquer grade of soluble nitrocellulose is closely related to the more highly nitrated form which is used to make explosives. They become relatively non-toxic after approximately a month since at this point, the lacquer has evaporated most of the solvents used in its production.

47.5   Acrylic lacquers

Lacquers using acrylic resin, a synthetic polymer, were developed in the 1950s. Acrylic resin is colourless, transparent thermoplastic, obtained by the polymerization of derivatives of acrylic acid. Acrylic is also used in enamel paints, which have the advantage of not needing to be buffed to obtain a shine. Enamels, however, are slow drying. The advantage of acrylic lacquer is its exceptionally fast drying time. The use of lacquers in automobile finishes was discontinued when tougher, more durable, weather- and chemical-resistant two-component polyurethane coatings were developed. The system usually consists of a primer, colour coat and clear topcoat, commonly known as clear coat finishes.

47.6   Water-based lacquers

Due to health risks and environmental considerations involved in the use of solvent-based lacquers, much work has gone into the development of water-based lacquers. Such lacquers are considerably less toxic and more environmentally friendly, and in many cases, produce acceptable results. More and more water-based coloured lacquers are replacing solvent-based clear and coloured lacquers in under-hood and interior applications in the automobile and other similar industrial applications. Water based lacquers are used extensively in wood furniture finishing as well.

47.7   Japanning

Main article: Japanning

Just as *china* is a common name for porcelain, *japanning* is an old name to describe the European technique to imitate Asian lacquerware.*[13] As Asian lacquer work became popular in England, France, the Netherlands, and Spain in the 17th century the Europeans developed imitations that were effectively a different technique of lacquering. The European technique, which is used on furniture and other objects, uses finishes that have a resin base similar to shellac. The technique, which became known as japanning, involves applying several coats of varnish which are each heat-dried and polished. In the 18th century, this type of lacquering gained a large popular following. Although traditionally a pottery and wood coating, japanning was the popular (mostly black) coating of the accelerating metalware industry. By the twentieth century, the term was freely applied to coatings based on various varnishes and lacquers besides the traditional shellac.

47.8   Lakshagraha or The House of Lacquer

Main article: Lakshagraha

The Hindu epic, Mahabharatha, describes the building of a house made of lac. Lakshagraha or Lakshagriha (Sanskrit: लाक्षागृहम्) (The House of Lacquer) is a book or parva from the Mahabharata, one of the two major Sanskrit epics of ancient India, the other being the Ramayana. This house was built under the orders of Duryodhana and his evil uncle and mentor Shakuni in a plot to kill the Pandavas along with their mother Kunti. The architect Purochana was employed in the building of Lakshagraha in the forest of Varnavrat. The house was meant to be a death trap, since lacquer is highly flammable. The plot itself was such that nobody would suspect foul play and the eventual death of the Pandavas would pass off as an accident. In the Mahabharata this incident is considered a major turning point, since the Pandavas were considered dead by their cousins, the Kauravas, which gave them ample opportunity to prepare themselves for an upcoming and unavoidable war. However, an escape route was prepared for the Pandavas who had been warned of the plot.

Lakshgraha Varanavat, is located in modern day Handia in Allahabad in Uttar Pradesh, India. The site at Varnavrat has since become a tourist location.*[14]

47.9 See also

- Lacquerware

- Varnish

- Acetate disc

- Lacquer painting

47.10 References

*[15]

[1] Franco Brunello (1973), *The art of dyeing in the history of mankind*, AATCC, 1973, ... *The word lacquer derives, in fact, from the Sanskrit 'Laksha' and has the same meaning as the Hindi word 'Lakh' which signifies one-hundred thousand ... enormous number of those parasitical insects which infest the plants Acacia catecu, Ficus and Butea frondosa ... great quantity of reddish colored resinous substance ... used in ancient times in India and other parts of Asia ...*

[2] http://www.etymonline.com/index.php?term=lacquer

[3] Ulrich Meier-Westhues (November 2007), *Polyurethanes: coatings, adhesives and sealants*, Vincentz Network GmbH & Co KG, 2007, ISBN 978-3-87870-334-1, ... *Shellac, a natural resin secreted by the scaly lac insect, has been used in India for centuries as a decorative coating for surfaces. The word lacquer in English is derived from the Sanskrit word laksha. which means one hundred thousand ...*

[4] Donald Frederick Lach, Edwin J. Van Kley (1994-02-04), *Asia in the making of Europe, Volume 2, Book 1*, University of Chicago Press, 1971, ISBN 978-0-226-46730-6, ... *Along with valuable woods from the East, the ancients imported lac, a resinous incrustation produced on certain trees by the puncture of the lac insect. In India, lac was used as sealing wax, dye and varnish ... Sanskrit, laksha; Hindi, lakh; Persian, lak; Latin, lacca. The Western word "lacquer" is derived from this term ...*

[5] Thomas Brock, Michael Groteklaes, Peter Mischke (2000), *European coatings handbook*, Vincentz Network GmbH & Co KG, 2000, ISBN 978-3-87870-559-8, ... *The word "lacquer" itself stems from the term "Laksha", from the pre-Christian, sacred Indian language Sanskrit, and originally referred to shellac, a resin produced by special insects ("lac insects") from the sap of an Indian fig tree ...*

[6] Wood Finishers Depot: Lacquer Sheen

[7] Britannica Online Encyclopedia: Oriental lacquer

[8] Major, John S., Sarah Queen, Andrew Meyer, Harold D. Roth, (2010), *The Huainanzi: A Guide to the Theory and Practice of Government in Early Han China*, Columbia University Press, p. 219.

[9] Webb, Marianne (2000). *Lacquer: Technology and conservation*. Oxford: Butterworth-Heinemann. p. 3. ISBN 978-0-7506-4412-9.

[10] Stark, Miriam T. (2005). *Archaeology of Asia*. Malden, MA : Blackwell Pub. Page 30. ISBN 1-4051-0213-6.

[11] Wang, Zhongshu. (1982). *Han Civilization*. Translated by K.C. Chang and Collaborators. New Haven and London: Yale University Press. Page 80. ISBN 0-300-02723-0.

[12] Institute of the History of Natural Sciences and Chinese Academy of Sciences, ed. (1983). *Ancient China's technology and science*. Beijing: Foreign Languages Press. p. 211. ISBN 978-0-8351-1001-3.

[13] Niimura, Noriyasu; Miyakoshi, Tetsuo (2003) Characterization of Natural Resin Films and Identification of Ancient Coating . *J. Mass Spectrom. Soc. Jpn.* 51, 440. JOI:JST.JSTAGE/massspec/51.439

[14] http://www.easternuptourism.com/Lakshagrih.jsp

[15] The Black Lacquer Coffin Black Coffin Treasure, *"The artistic design of the black lacquer coffin reflects the painting skills of the Western Han Dynasty. They often paint mystical and grotesque themes about their myth and legends. But through the technique which involve the use of embossing lacquer application..."*

47.11 Further reading

- Kimes, Beverly R., Editor. Clark, Henry A. (1996), *The Standard Catalog of American Cars 1805–1945*, Kraus Publications, ISBN 0-87341-428-4 p. 1050

- Paolo Nanetti (2006), *Coatings from A to Z*, Vincentz Verlag, Hannover, ISBN 3-87870-173-X - A concise compilation of technical terms. Attached is a register of all German terms with their corresponding English terms and vice versa, in order to facilitate its use as a means for technical translation from one language to the other.

- Webb, Marianne (2000), *Lacquer: Technology and Conservation*, Butterworth Heinemann, ISBN 0-7506-4412-5 A Comprehensive Guide to the Technology and Conservation of Asian and European Lacquer

- Michiko, Suganuma. "Japanese lacquer" .

Chapter 48

Lingzhi mushroom

The **lingzhi mushroom** or **reishi mushroom** (traditional Chinese: 靈芝; pinyin: *língzhī*; Japanese: reishi; Vietnamese: *linh chi*; literally: "supernatural mushroom") is a species complex that encompasses several fungal species of the genus *Ganoderma*, most commonly the closely related species *Ganoderma lucidum*, *Ganoderma tsugae*, and *Ganoderma sichuanense*. *G. sichuanense* enjoys special veneration in East Asia, where it has been used as a medicinal mushroom in traditional Chinese medicine for more than 2,000 years,[1] making it one of the oldest mushrooms known to have been used medicinally. Lingzhi is listed in the *American Herbal Pharmacopoeia and Therapeutic Compendium*.

48.1 Taxonomy and naming

Names for the lingzhi fungus have a two thousand year history. The Chinese term *lingzhi* 灵芝 was first recorded during the Eastern Han dynasty (25–220 CE). Petter Adolf Karsten named the genus *Ganoderma* in 1881.[2]

48.1.1 Botanical names

The fungus was given its first binomial name, *Boletus lucidus*, by English botanist William Curtis in 1781. The lingzhi's botanical names have Greek and Latin roots. The generic name *Ganoderma* derives from the Greek *ganos* γανος "brightness; sheen", hence "shining" and *derma* δερμα "skin".[3] The specific epithet *lucidum* is Latin for "shining." *Tsugae* is derived from the Japanese word for "hemlock" (*tsuga*).

There are multiple species of lingzhi, scientifically known to be within the *Ganoderma lucidum* species complex and mycologists are still researching the differences among species within this complex.[4]

48.1.2 Chinese names

In the Chinese language, *lingzhi* is made up of the compounds *ling* "spirit, spiritual; soul; miraculous; sacred; divine; mysterious; efficacious; effective" (cf. Lingyan Temple) and *zhi* 芝 "(traditional) plant of longevity; fungus; seed; branch; mushroom; excrescence". Fabrizio Pregadio notes, "The term *zhi*, which has no equivalent in Western languages, refers to a variety of supermundane substances often described as plants, fungi, or "excrescences"."[5] *Zhi* occurs in other Chinese plant names such as *zhima* 芝麻 "sesame" or "seed", and was anciently used a phonetic loan character for *zhi* 芷 "Angelica iris". Chinese differentiates *Ganoderma* species between *chizhi* 赤芝 "red mushroom" *G. lucidum* and *zizhi* 紫芝 "purple mushroom" *G. japonicum*.

Lingzhi has several synonyms. *ruicao* 瑞草 "auspicious plant" (with *rui* 瑞 "auspicious; felicitous omen" and the suffix *cao* "plant; herb") is the oldest; the (c. 3rd century BCE) *Erya* dictionary defines *qiu* (interpreted as a miscopy of *jun* 菌 "mushroom") as *zhi* 芝 "mushroom" and the commentary of Guo Pu (276–324) says, "The [*zhi*] flowers three times in one year. It is a [*ruicao*] felicitous plant." [6] Other Chinese names for *Ganoderma* include *ruizhi* 瑞芝 "auspicious mushroom", *shenzhi* 神芝 "divine mushroom" (with *shen* "spirit; god' supernatural; divine"), *mulingzhi* 木 芝 (with "tree; wood"), *xiancao* 仙草 "immortality plant" (with *xian* "(Daoism) transcendent; immortal; wizard"), and *lingzhicao* 芝草 or *zhicao* 芝草 "mushroom plant".

Since both Chinese *ling* and *zhi* have multiple meanings, *lingzhi* has diverse English translations. Renditions include "[zhi] possessed of soul power",[7] "Herb of Spiritual Potency" or "Mushroom of Immortality",[8] "Numinous Mushroom",[9] "divine mushroom",[10] "divine fungus",[11] "Magic Fungus",[12] and "Marvelous Fungus".[13]

48.1.3   Japanese names

Japanese language *Reishi* 芝 is a Sino-Japanese loan word from *lingzhi*. This modern Japanese kanji is the shinjitai "new character form" for the kyūjitai "old character form" .

Reishi synonyms divide between Sino-Japanese borrowings and native Japanese coinages. Sinitic loanwords include literary terms such as *zuisō* 瑞草 (from *ruicao*) "auspicious plant" and *sensō* 仙草 (from *xiaocao*) "immortality plant". A common native Japanese name is *mannentake* 万年茸 "10,000 year mushroom". The Japanese writing system uses *shi* or *shiba* 芝 for "grass; lawn; turf" and *take* or *kinoko* 茸 for "mushroom" (e.g., shiitake). Other Japanese terms for *reishi* include *kadodetake* 門出茸 "departure mushroom", *hijiridake* 聖茸 "sage mushroom", and *magoshakushi* 孫杓子 "grandchild ladle".

48.1.4   Korean names

"Ganoderma motif" with Dancheong

Korean language *Yeong Ji* or *Yung Gee* (영지, 芝) is a word from hanja of *lingzhi*. It is also called *seon-cho* (선초, 仙草), *gil-sang-beo-seot* (길상버섯, 吉祥茸), *yeong ji cho* (영지초, 芝草) or *jeok ji* (적지, 赤芝). It can be classified by its color such as *ja-ji* (자지, 紫芝) for purple one, *heuk-ji* (흑지, 黑芝) for black, *cheong-ji* (청지, 靑芝) for blue or green, *baek-ji* (백지, 白芝) for white, *hwang-ji* (황지, 黃芝) for yellow.

48.1.5   Vietnamese names

The Vietnamese language *linh chi* is a Chinese loanword used in tiếng Việt. It is often used with the Vietnamese word for mushroom nấm (nấm Linh Chi) which is the equivalent of *Ganoderma lucidum* or reishi mushroom.

48.1.6   English names

English *lingzhi* or *ling chih* (sometimes spelled "*ling chi*" from French EFEO Chinese transcription) is a Chinese loanword. The *Oxford English Dictionary* gives Chinese "*líng* divine + *zhī* fungus" as the origin of *ling chih* or *lingzhi*, and defines, "The fungus *Ganoderma lucidum*, believed in China to confer longevity and used as a symbol of this on Chinese ceramic ware." [14] The *OED* notes the earliest recorded usage of the Wade–Giles romanization *ling chih* in 1904, [15] and of the Pinyin *lingzhi* in 1980. In addition to the transliterated loanword, English names include "glossy ganoderma" and "shiny polyporus". [16]

48.2   Description

Lingzhi is a polypore mushroom that is soft (when fresh), corky, and flat, with a conspicuous red-varnished, kidney-shaped cap and, depending on specimen age, white to dull brown pores underneath. [8] It lacks gills on its underside and releases its spores through fine pores, leading to its morphological classification as a polypore.

Young sporocarp

48.2.1   Varieties

With the advent of genome sequencing, the Ganoderma genus has been undergoing taxonomic reclassification. Prior to genetic analyses of fungi, they were classified ac-

Depending on the environmental or cultivation conditions the reishi can lose its umbrella shape and resemble antlers instead.

G. lucidum was previously thought to occupy the Asian and European continents. However, with ITS sequencing, it is now known that *G. lucidum* sensu stricto is confined to Europe, and *G. lucidum* comprises two distinct species.[21] Mycologists now consider *G. lucidum* to be a worldwide distributed species-complex in which the development of each strain is shaped by unique physiological state and metabolic demands.[22]

48.3 Biochemistry

cording to their morphological characteristics. The ITS region of the Ganoderma genome is considered to be a standard barcode marker.[17]

It was once thought that *Ganoderma lucidum* generally occurs in two growth forms- a large specimen with a small or nonexistent stalk found in North America and is sessile, and the other a smaller specimen with a long, narrow stalk found mainly in the tropics. However recent molecular evidence has shown the first form to be a distinct species called *G.sessile*, a name given to North American specimens by William Alfonso Murrill in 1902.[18] [19]Environmental conditions play a substantial role in the lingzhi's manifest morphological characteristics. For example, elevated carbon dioxide levels result in stem elongation in lingzhi. Other formations include antlers without a cap; these may be affected by carbon dioxide levels as well. The species were also formerly differentiated by their colors, of which the red reishi was considered to be the most researched. The three main factors that most greatly influence fruitbody development morphology are light, temperature, and humidity. (Water and air quality are impactful but to a lesser degree).[20]

Ganoderic acid A, a compound isolated from lingzhi

Ganoderma lucidum produces a group of triterpenes, called ganoderic acids, which have a molecular structure similar to steroid hormones.[23] It also contains other compounds often found in fungal materials, including polysaccharides (such as beta-glucan), coumarin,[24] mannitol, and alkaloids.[23] Sterols isolated from the mushroom include, ganoderol, ganoderenic acid, ganoderiol, ganodermanontriol, lucidadiol, and ganodermadiol.

48.4   Habitat

Ganoderma lucidum, and its close relative *Ganoderma tsugae*, grow in the northern Eastern Hemlock forests. These two species of bracket fungus have a worldwide distribution in both tropical and temperate geographical regions, growing as a parasite or saprotroph on a wide variety of trees.*[8] Similar species of *Ganoderma* have been found growing in the Amazon.*[25] In nature, lingzhi grows at the base and stumps of deciduous trees, especially maple.*[26] Only two or three out of 10,000 such aged trees will have lingzhi growth, and therefore its wild form is extremely rare. Today, lingzhi is effectively cultivated both indoors under sterile conditions and outdoors on either logs or woodchip beds.

48.5   History

The Chinese classics first used *zhi* during the Warring States period (475–221 BCE) and *lingzhi* during the Han Dynasty (206 BCE-220 CE).

The word *zhi* 芝 occurs approximately 100 times in classical texts.*[27] Occurrences in early Chinese histories, such as the (91 BCE) *Shiji* "Records of the Grand Historian" and (82 CE) *Hanshu* "Book of Han", predominantly refer to the "Mushroom of Immortality; elixir of life". They record that fangshi "masters of esoterica; alchemists; magicians", supposedly followers of Zou Yan (305–240 BCE), claimed to know secret locations like Mount Penglai where the magic *zhi* mushroom grew. Some sinologists propose that the mythical *zhi* 芝 derived from Indian legends about soma that reached China around the 3rd century BCE.*[28] *Fangshi* courtiers convinced Qin and Han emperors, most notably Qin Shi Huang (r. 221–210 BCE) and Emperor Wu of Han (r. 141–87 BCE), to dispatch large expeditions (e.g., Xu Fu in 219 BCE) seeking the *zhi* Plant of Immortality, but none produced tangible results. *Zhi* occurrences in other classical texts often refer to an edible fungus. The *Liji* "Record of Ritual" lists *zhi* "lichens" as a type of condiment.*[29] The *Chuci* "Song of the South" metaphorically mentions, "The holy herb is weeded out".*[30] The *Huainanzi* "Philosophers of Huainan" records a *zizhi* 紫芝 "Purple Mushroom" Aphorism, "The *zhi* fungus grows on mountains, but it cannot grow on barren boulders." *[31]

The word *lingzhi* 靈芝 was first recorded in a *fu* 賦 "rhapsody; prose-poem" by the Han dynasty polymath Zhang Heng (CE 78–139). His *Xijing fu* 西京賦 "Western

Man holding ganoderma *by Chen Hongshou*

Metropolis Rhapsody" description of Emperor Wu of Han's (104 BCE) Jianzhang Palace parallels *lingzhi* with *shijun* 石菌 "rock mushroom": "Raising huge breakers, lifting waves, That drenched the stone mushrooms on the high bank, And soaked the magic fungus on vermeil boughs." *[32] The commentary by Xue Zong (d. 237) notes these fungi were eaten as drugs of immortality.

The (ca. 1st–2nd century CE) *Shennong bencao jing* "Divine Farmer's Classic of Pharmaceutics" classifies *zhi* into six color categories, each of which is believed to benefit the *qi* "Life Force" in a different part of the body: *qingzhi* 芝 "Green Mushroom" for Liver, *chizhi* 赤芝 "Red Mushroom" for heart, *huangzhi* 黃芝 "Yellow Mushroom" for spleen, *baizhi* 白芝 "White Mushroom" for Lung, *heizhi* "Black Mushroom" 黑芝 for kidney, and *zizhi* 紫芝 "Purple Mushroom" for Essence. Commentators identify this red *chizhi* (or *danzhi* 丹芝 "cinnabar mushroom") as the *lingzhi*.

> *Chi Zhi* (*Ganoderma rubra*) is bitter and balanced. It mainly treats binding in the chest, boosts the heart qi, supplements the center,

sharpens the wits, and [causes people] not to forget [i.e., improves the memory]. Protracted taking may make the body light, prevent senility, and prolong life so as to make one an immortal. Its other name is *Dan Zhi* (Cinnabar Ganoderma). It grows in mountains and valleys.[33][34]

While Chinese texts have recorded medicinal uses of *lingzhi* for more than 2,000 years, a few sources erroneously claim more than 4,000 years.[35] Modern scholarship accepts neither the historicity of Shennong "Divine Farmer" (legendary inventor of agriculture, traditionally r. 2737–2697 BCE) nor that he wrote the *Shennong bencao jing*.

The (ca. 320 CE) *Baopuzi*, written by the Jin Dynasty Daoist scholar Ge Hong, has the first classical discussion of *Zhi*.[36] Based upon no-longer extant texts, Ge distinguishes five categories of *zhi*, each with 120 varieties: *Shizhi* 石芝 "stone Zhi", *Muzhi* 木芝 "wood Zhi", *Caozhi* 草芝 "Plant Zhi", *Rouzhi* 肉芝 "flesh *zhi*", and *junzhi* 菌芝 "mushroom *zhi*. For example, the "mushroom *zhi*".

> *Tiny excresences.* These grow deep in the mountains, at the base of large trees or beside springs. They may resemble buildings, palanquins and horses, dragon and tigers, human beings, or flying birds. They may be any of the five colors. They too number 120 for which there exist illustrations. All are to be sought and gathered while using Yu's Pace, and they are to be cut with a bone knife. When dried in the shade, powdered, and taken by the inch-square spoonful, they produce geniehood. Those of the intermediate class confer several thousands of years, and those of the lowest type a thousand years of life.[37]

Yu's Pace is a Daoist ritual walking technique. Pregadio concludes, "While there may be no better term than "mushrooms" or "excresences" to refer to them, and even though Ge Hong states that they "are not different from natural mushrooms (*ziran zhi* 自然芝) (*Baopuzi* 16.287)", the *Zhi* pertain to an intermediate dimension between mundane and transcendent reality." [38]

The (1596) *Bencao Gangmu* ("Compendium of Materia Medica") has a *zhi* 芝 category that includes six types of *Zhi* (calling the green, red, yellow, white, black, and purple ones from the *Shennong bencao jing* the *liuzhi* 六芝 "six mushrooms") and sixteen other fungi, mushrooms, and lichens (e.g., *mu'er* 木耳 "wood ear" "Cloud ear fungus; *Auricularia auricula-judae*"). The author Li Shizhen classified these six differently colored *Zhi* as *Xiancao* 仙草 "immortality herbs", and described the effects of *Chizhi* "red mushroom":

It positively affects the life-energy, or *Qi* of the heart, repairing the chest area and benefiting those with a knotted and tight chest. Taken over a long period of time, agility of the body will not cease, and the years are lengthened to those of the Immortal Fairies.[39][40]

Stuart and Smith's classical study of Chinese herbology describes the *zhi*.

> 芝 (Chih) is defined in the classics as the plant of immortality, and it is therefore always considered to be a felicitous one. It is said to absorb the earthy vapors and to leave a heavenly atmosphere. For this reason it is called 靈芝 (Ling-chih.) It is large and of a branched form, and probably represents Clavaria or Sparassis. Its form is likened to that of coral.[41]

The *Bencao Gangmu* does not list *lingzhi* as a variety of *zhi*, but as an alternate name for the *shi'er* 石耳 "stone ear" "Umbilicaria esculenta" lichen. According to Stuart and Smith,

> [The 石耳 Shih-erh is] edible, and has all of the good qualities of the 芝 (Chih), being also used in the treatment of gravel, and said to benefit virility. It is specially used in hemorrhage from the bowels and prolapse of the rectum. While the name of this would indicate that it was one of the Auriculariales, the fact that the name 靈芝 (Ling-chih) is also given to it might place it among the Clavariaceae.[42]

Chinese pharmaceutical handbooks on *Zhi* mushrooms were the first illustrated publications in the history of mycology. The historian of Chinese science Joseph Needham discussed a no-longer extant Liang Dynasty (502–587) illustrated text called *Zhong Shenzhi* 種神芝 "On the Planting and Cultivation of Magic Mushrooms".

> The pictures of mushrooms in particular must have been an extremely early landmark in the history of mycology, which was a late-developing science in the West. The title of [this book] shows that fungi of some kind were being regularly cultivated – hardly as food, with that special designation, more probably medicinal, conceivably hallucinogenic." [43]

The (1444) Ming Dynasty edition *Daozang* "Daoist canon" contains the *Taishang lingbao zhicao pin* 太上靈寶芝草品 "Classifications of the Most High Divine Treasure

Mushroom Plant" ,*[44] which categorizes 127 varieties of *Zhi*.*[45] A (1598) Ming reprint includes woodblock pictures.*[46]

In Chinese art, the *lingzhi* symbolizes great health and longevity, as depicted in the imperial Forbidden City and Summer Palace.*[47] It was a talisman for luck in the traditional culture of China, and the goddess of healing Guanyin is sometimes depicted holding a *reishi* mushroom.*[48]

48.6 Preparation

Due to its bitter taste, lingzhi is traditionally prepared as a hot water extract product.*[49] Thinly sliced or pulverized lingzhi (either fresh or dried) is added to a pot of boiling water, the water is then reduced to a simmer, and the pot is covered; the lingzhi is then simmered for two hours. The resulting liquid is fairly bitter in taste and dark, with the more active red lingzhi more bitter than the black. The process is sometimes repeated for additional concentration. Alternatively, it can be used as an ingredient in a formula decoction or used to make an extract (in liquid, capsule, or powder form). The more active red forms of lingzhi are far too bitter to be consumed in a soup.

48.7 Medical Uses

Anti-Allergic/Anti-Inflammatory Activity: Studies showed that Reishi extract significantly inhibited all four types of allergic reactions [14], including positive effects against asthma and contact dermatitis and effectively used in treating stiff necks, stiff shoulders, conjunctivitis (inflammation of the fine membrane lining the eye and eyelids), bronchitis, rheumatism and improving "competence" of the immune system without any significant side-effects.*[50]

Anticonvulsant Effects: A water extract from Reishi mycelium significantly increased the threshold for psychomotor seizures in mice.*[51] Mycelial extracts also confer anti-inflammatory activity as evidenced by inhibitory activity of lipopolysaccharide (LPS)-induced nitric oxide (NO) production in murine macrophage-like cell line RAW264.7 cells.*[52]*[53]

Cancer: The use of *G. lucidum* has also been explored as a complementary adjunct treatment in patients undergoing chemotherapy treatment. A recent meta-analysis of five randomized control trials showed that patients responded more positively when given *G. lucidum* alongside their chemotherapy regimen, and the studies also showed that patients had improved immune functions that was measured by their elevated levels of immune response cells.*[54] Several compounds in *G. lucidum* have been studied for apop-

totic activity in colon cancer cells,*[55] antiproliferative effects in ovarian cancer cells, *[56] and induction of apoptosis in human gastric carcinoma cells.*[57]

Cardiovascular Risk Factors: Previous clinical evidence suggested that *G. lucidum* may have antioxidant, cardioprotective, and glycemic regulatory effects.*[58] However, a 2015 Cochrane review did not find evidence to support the use of or treatment of cardiovascular risk factors in people with type 2 diabetes mellitus.*[59]

Diabetes: Several compounds in *G. lucidum* (including polysaccharides, proteoglycans, proteins and triterpenoids) may have hypoglycemic effects. In vitro evidence suggests that protein tyrosine phosphatase 1B is a promising therapeutic target in diabetes, and a *G. lucidum* proteoglycan can inhibit this enzyme. Secondly, *G. lucidum* demonstrates inhibition of aldose reductase and α-glucosidase, which can suppress postprandial hyperglycemia.*[60]A proteoglycan enhanced insulin secretion and decreasing hepatic glucose output (along with increased adipose and skeletal muscle glucose disposal)*[61] and normalized serum lipids in a murine model of diabetes.*[62] A polysaccharide also demonstrated hypoglycemic effects in type 2 diabetic mice.*[63]

Gastrointestinal Health: Recent murine studies suggest that *G. lucidum* may positively impact gut microflora to attenuate metabolic risk factors contributing to obesity.*[64]

Hepatoprotection: *G. lucidum* significantly decreased serum ALT and AST levels in mice livers injured with α-amanitin.*[65] A proteoglycan also demonstrated hepatoprotective effects in carbon tetrachloride-induced liver injury in vitro and in vivo.*[66]

Immunostimulation: *G. lucidum* contains beta glucans and other polysaccharides to stimulate innate immune function and signaling*[67]*[68]*[69]*[70] and activate dendritic cells.*[71]*[72]

Neuroprotection: *G. lucidum* protected dopaminergic neurons through inhibition of microglia.*[73]

48.8 See also

- Medicinal mushrooms

48.9 References

[1] Jones, Kenneth (1990), *Reishi: Ancient Herb for Modern Times*, Sylvan Press, p. 6.

[2] Karsten PA. (1881). "Enumeratio Boletinearum et Polyporearum Fennicarum, systemate novo dispositarum" . *Revue mycologique, Toulouse* (in Latin) **3** (9): 16–19.

[3] Liddell, Henry George and Robert Scott (1980). *A Greek-English Lexicon (Abridged Edition)*. United Kingdom: Oxford University Press. ISBN 0-19-910207-4.

[4] Hseu RS, Wang HH, Wang HF, Moncalvo JM (1996). "Differentiation and grouping of isolates of the *Ganoderma lucidum* complex by random amplified polymorphic DNA-PCR compared with grouping on the basis of internal transcribed spacer sequences". *Appl. Environ. Microbiol.* **62** (4): 1354–63. PMC 167902. PMID 8919797.

[5] Pregadio, Fabrizio (2008). "Zhi 芝 numinous mushrooms; excrescences", in *The Encyclopedia of Taoism*, Fabrizio Pregadio, ed., Routledge, p. 1271.

[6] Tr. by E. Bretschneider (1893), *Botanicon Sinicum; Notes on Chinese Botany from Native and Western Sources*, Kelly & Walsh, p. 40.

[7] Groot, Jan Jakob Maria (1892–1910), *The Religious System of China: Its Ancient Forms, Evolution, History and Present Aspect, Manners, Customs and Social Institutions Connected Therewith*, Brill Publishers, Vol. IV, p. 307.

[8] David Arora (1986). *Mushrooms demystified, 2nd edition*. Ten Speed Press. ISBN 0-89815-169-4.

[9] Pregadio (2008), p. 1271.

[10] .Hu, Shiu-ying (2006), Food plants of China, Chinese University Press.

[11] Bedini, Silvio A. (1994), The Trail of Time, Cambridge University Press, p. 113.

[12] Knechtges, David R. (1996), ' Wen Xuan or Selections of Refined Literature, Volume III, Princeton University Press, p. 211.

[13] Schipper, Kristofer M. (1993), *The Taoist Body*, University of California Press, p. 174.

[14] *Oxford English Dictionary* (2009), CD-ROM edition (v. 4.0), s.v. *ling chih*.

[15] Stephen Wootton Bushell (1904), *Chinese Art*, Victoria and Albert Museum, p. 148. This context describes the *lingzhi* fungus and ruyi scepter as Daoist symbols of longevity on a jade vase.

[16] Names of a Selection of Asian Fungi, multilingual multiscript plant name database.

[17] Pawlik, A (2015). "Genetic and Metabolic Intraspecific Biodiversity of Ganoderma lucidum". *BioMed Research International*: 1-13. doi:10.1155/2015/726149.

[18] http://www.researchgate.net/publication/267456447_Global_diversity_of_the_Ganoderma_lucidum_complex_%28Ganodermataceae_Polyporales%29_inferred_from_morphology_and_multilocus_phylogeny

[19] http://www.mycobank.org/Biolomics.aspx?Table=Mycobank&MycoBankNr_=237038

[20] http://www.dl.begellhouse.com/journals/708ae68d64b17c52,72e9ed69099c0eef,6b7a7dab0ec964e7.html

[21] Wang, D.-M. (2009). "Ganoderma multipileum, the correct name for 'G. lucidum' in tropical Asia". *Botanical Studies* **50** (4): 451–458.

[22] Zakaria, L (2009). "Molecular analysis of Ganoderma species from different hosts in Peninsula Malaysia". *Journal of Biological Sciences* **9** (1): 12–20.

[23] Paterson RR (2006). "Ganoderma – a therapeutic fungal biofactory". *Phytochemistry* **67** (18): 1985–2001. doi:10.1002/chin.200650268. PMID 16905165.

[24] Biosci.Biotechnol.Biochem.,68(4),881–887,2004

[25] *Medicinal Mushrooms: An Exploration of Tradition, Healing, & Culture* (Herbs and Health Series) by Christopher Hobbs (author), Harriet Beinfield

[26] (National Audubon Society; *Field Guide to Mushrooms*, 1993)

[27] Pre-Qin and Han texts, Chinese Text Project.

[28] Unschuld, Paul U. (1985), *Medicine in China: A History of Ideas*, University of California Press, p. 112.

[29] Tr. by Legge, James (1885), *The Li Ki*, 2 vols, Oxford University Press, vol. 1, p. 461.

[30] Tr. by Hawkes, David (1959), *Ch'u Tz'u: The Songs of the South*, Clarendon, p. 258.

[31] Tr. by Major, John S., Sarah Queen, Andrew Meyer, and Harold D. Roth (2010), *The Huainanzi: A Guide to the Theory and Practice of Government in Early Han China*, Columbia University Press, p. 634.

[32] Tr. Knechtges (1996), 201.

[33] 神農本草經, 草上品. 赤芝 °苦, 平, 無毒 °胸中結, 益心氣, 補中, 增智慧, 不忘 °久食, 輕身不老, 延年神仙 °一名丹芝 °延年神仙 °

[34] Tr. by Yang Shouzhong (1998) *The Divine Farmer's Materia Medica: A Translation of the Shen Nong Ben Cao Jing*, Blue Poppy, pp. 17–18

[35] Reishi mushroom, Reishiessence.com.

[36] 抱朴子/卷 11; tr. by Ware, James R. (1966). *Alchemy, Medicine and Religion in the China of A.D. 320: The Nei Pien of Ko Hung*. Dover. pp. 258.

[37] Tr. by Ware (1966), p. 185. This word *jun* 菌 means "mushroom; fungus; bacterium; germ" – not "tiny".

[38] Pregadio (2008), p. 1273.

[39] 本草綱目/菜之三. 胸中結, 益心氣, 補中, 增智慧, 不忘 °久食, 輕身不老, 延年神仙 °

[40] Tr. by Halpern, George M. (2007), *Healing Mushrooms*, Square One, p. 59.

[41] Stuart, G. A. and F. Porter Smith (1911), Chinese Materia Medica, Pt. 1, Vegetable Kingdom, Presbyterian Mission Press. p. 271.

[42] Stuart and Smith (1911), p. 274.

[43] Tr. by Needham, Joseph and Lu Gwei-Djen (1986), Science and Civilisation in China: Biology and biological technology. Botany, Volume 6, Part 1, p. 261.

[44] DZ 1406, 太上靈寶芝草品, Daoist Studies.

[45] Tr. "Catalogue of Mushrooms of Immortality" by , Chapter 1, Introduction, NCCU Institutional Repository, p. 107.

[46] 太上　芝草品, online *Taishang lingbao zhicao pin* illustrated reprint.

[47] Smith JE, Rowan NJ, and Sullivan R (2001) *Medicinal Mushrooms: Their Therapeutic Properties and Current Medical Usage with Special Emphasis on Cancer Treatments* Cancer Research UK, p. 28.

[48] Halpern (2007), p.59.

[49] Smith, Rowan, and Sullivan (2001), p. 31.

[50] Idosi Publication

[51] Socala, Katarzyna (2015). "Evaluation of Anticonvulsant, Antidepressant-, and Anxiolytic-like Effects of an Aqueous Extract from Cultured Mycelia of the Lingzhi or Reishi Medicinal Mushroom Ganoderma lucidum (Higher Basidiomycetes) in Mice.". *Int J Med Mushrooms* **17** (3): 209–18. PMID 25954905.

[52] Geng, Y (2014). "Anti-inflammatory Activity of Mycelial Extracts from Medicinal Mushrooms.". *Int J Med Mushrooms* **16** (4): 319–25. PMID 25271860.

[53] Hasnat, MA (2014). "Anti inflammatory activity on mice of extract of Ganoderma lucidum grown on rice via modulation of MAPK and NF-κB pathways". *Phytochemistry*. doi:10.1016/j.phytochem.2014.10.019.

[54] Jin X, Ruiz Beguerie J, Sze DMY, Chan GCF. *Ganoderma lucidum* (Reishi mushroom) for cancer treatment. Cochrane Database of Systematic Reviews 2012, Issue 6. Art. No.: CD007731. DOI: 10.1002/14651858.CD007731.pub2.

[55] Liang, Z (2014). "Ganoderma lucidum polysaccharides target a Fas/Caspase dependent pathway to induce apoptosis in human colon cancer cells". *Asian Pacific Journal of Cancer Prevention* **15** (9): 3981–3986. doi:10.7314/apjcp.2014.15.9.3981.

[56] Shuyan, Dai (2014). "Ganoderma lucidum inhibits proliferation of human ovarian cancer cells by suppressing VEGF expression and up-regulating the expression of connexin 43.". *BMC Complement Altern Med* **14** (1): 434. doi:10.1186/1472-6882-14-434.

[57] Jang, KJ (2010). "Induction of apoptosis by ethanol extracts of Ganoderma lucidum in human gastric carcinoma cells.". *J Acupunct Meridian Stud* **3** (1). doi:10.1016/S2005-2901(10)60004-0.

[58] Chu, TT (2012). "Study of potential cardioprotective effects of Ganoderma lucidum (Lingzhi): results of a controlled human intervention trial.". *Br J Nutr* **107** (7). doi:10.1017/S0007114511003795.

[59] Klupp, NL (2015). "Ganoderma lucidum mushroom for the treatment of cardiovascular risk factors.". *Cochrane Database Syst Rev.* doi:10.1002/14651858.CD007259.pub2. PMID 25686270.

[60] Ma, Haou-Tzong (2015). "Anti-diabetic effects of Ganoderma lucidum.". *Phytochemistry*. doi:10.1016/j.phytochem.2015.02.017. PMID 25790910.

[61] Pan, D (2013). "Antidiabetic, antihyperlipidemic and antioxidant activities of a novel proteoglycan from ganoderma lucidum fruiting bodies on db/db mice and the possible mechanism.". *PLoS One* **8** (7). doi:10.1371/journal.pone.0068332. PMID 23874589.

[62] Teng, BS (2012). "Hypoglycemic effect and mechanism of a proteoglycan from ganoderma lucidum on streptozotocin-induced type 2 diabetic rats.". *Eur Rev Med Pharmacol Sci* **16** (2): 166–75. PMID 22428467.

[63] Xiao, C (2012). "Hypoglycemic effects of Ganoderma lucidum polysaccharides in type 2 diabetic mice.". *Arch Pharm Res* **35** (10): 1793–801. doi:10.1007/s12272-012-1012-z.

[64] Chang, CJ (2015). "Ganoderma lucidum reduces obesity in mice by modulating the composition of the gut microbiota.". *Nat Commun.* doi:10.1038/ncomms8489. PMID 26102296.

[65] Wu, X (2013). "Hepatoprotective effects of aqueous extract from Lingzhi or Reishi medicinal mushroom Ganoderma lucidum (higher basidiomycetes) on α-amanitin-induced liver injury in mice.". *Int J Med Mushrooms* **15** (4): 383–91. PMID 23796220.

[66] Yang, XJ (2006). "In vitro and in vivo protective effects of proteoglycan isolated from mycelia of Ganoderma lucidum on carbon tetrachloride-induced liver injury.". *World J Gastroenterol* **12** (9): 1379-85. PMID 16552805.

[67] Zhu, L (2013). "Isolation, purification, and immunological activities of a low-molecular-weight polysaccharide from the Lingzhi or Reishi medicinal mushroom Ganoderma lucidum (higher Basidiomycetes).". *Int J Med Mushroom* **15** (4): 407–14. PMID 23796222.

[68] Kuo, MC (2006). "Ganoderma lucidum mycelia enhance innate immunity by activating NF-kappaB.". *J Ethnopharmacol* **103** (2): 217-22. doi:10.1016/j.jep.2005.08.010.

[69] Wang, G (2007). "Enhancement of IL-2 and IFN-gamma expression and NK cells activity involved in the anti-tumor effect of ganoderic acid Me in vivo.". *Int Immunopharmacol* **7** (6): 864-70. doi:10.1016/j.intimp.2007.02.006.

[70] Hsu, HY (2008). "Reishi immuno-modulation protein induces interleukin-2 expression via protein kinase-dependent signaling pathways within human T cells." . *J Cell Physiol* **215** (1): 15–26. doi:10.1002/jcp.21144.

[71] Meng, J (2011). "Analysis of maturation of murine dendritic cells (DCs) induced by purified Ganoderma lucidum polysaccharides (GLPs).". *Int J Biol Macromol* **49** (4): 693–9. doi:10.1016/j.ijbiomac.2011.06.029.

[72] Lin, YL (2005). "Polysaccharide purified from Ganoderma lucidum induced activation and maturation of human monocyte-derived dendritic cells by the NF-kappaB and p38 mitogen-activated protein kinase pathways." . *J Leukoc Biol* **78** (2): 533-43.

[73] Zhang, R (2011). "Ganoderma lucidum Protects Dopaminergic Neuron Degeneration through Inhibition of Microglial Activation." . *Evid Based Complement Alternat Med*. doi:10.1093/ecam/nep075. PMID 19617199.

48.10 External links

- *Lingzhi mushroom* in Index Fungorum.

- Media related to *Ganoderma lucidum* at Wikimedia Commons

Chapter 49

Madhuca longifolia

Madhuca longifolia is an Indian tropical tree found largely in the central and north Indian plains and forests. It is commonly known as **mahua**, **mahwa** or **Iluppai**. It is a fast-growing tree that grows to approximately 20 meters in height, possesses evergreen or semi-evergreen foliage, and belongs to the family Sapotaceae.[1] It is adapted to arid environments, being a prominent tree in tropical mixed deciduous forests in India in the states of West Bengal, Chhattisgarh, Jharkhand, Uttar Pradesh, Bihar, Maharashtra, Telangana, Madhya Pradesh, Kerala, Gujarat and Orissa.[2]

49.1   Uses

It is cultivated in warm and humid regions for its oleaginous seeds (producing between 20 and 200 kg of seeds annually per tree, depending on maturity), flowers and wood. The fat (solid at ambient temperature) is used for the care of the skin, to manufacture soap or detergents, and as a vegetable butter. It can also be used as a fuel oil. The seed cakes obtained after extraction of oil constitute very good fertilizer. The flowers are used to produce an alcoholic drink in tropical India. This drink is also known to affect the animals.[3] Several parts of the tree, including the bark, are used for their medicinal properties. It is considered holy by many tribal communities because of its usefulness.

The tree is considered a boon by the tribals who are forest dwellers and keenly conserve this tree. However, conservation of this tree has been marginalized, as it is not favoured by nontribals.[4]

The leaves of *Madhuca indica* (= *M. longifolia*) are fed on by the moth *Antheraea paphia*, which produces tassar silk (*tussah*), a form of wild silk of commercial importance in India.[5]

The Tamils have several uses for *M. longifolia* (*iluppai* in Tamil). The saying "*aalai illaa oorukku iluppaip poo charkkarai*" indicates when there is no cane sugar available, the flower of *M. longifolia* can be used, as it is very sweet.

M. longifolia *in Hyderabad, India*

However, Tamil tradition cautions that excessive use of this flower will result in imbalance of thinking and may even lead to lunacy.[6]

The alkaloids in the press cake of Madhuca seeds is reportedly used in killing fishes in aquaculture ponds in some parts of India. The cake serves to fertilize the pond, which can be drained, sun dried, refilled with water and restocked with fish fingerlings.[7][8]

49.2   *Mahuwa* flowers

The *mahuwa* flower is edible and is a food item for tribals. They are used to make syrup for medicinal purposes.[2]

They are also fermented to produce the alcoholic drink *mahuwa*, a country liquor. Tribals of Bastar in Chhattisgarh and Orissa, Santhals of Santhal Paraganas (Jharkhand), Koya tribals of North-East Andhra Pradesh (vippa saara: ▯▯▯ ▯▯▯and tribals of North Maharashtra consider the tree and the *mahuwa* drink as part of their cultural heritage. *Mahuwa* is an essential drink for tribal men and women during celebrations.[9] The main ingredients used

244

Mahua *flowers*

for making it are *chhowa gud* (granular molasses) and dried *mahuwa* flowers.

The liquor produced from the flowers is largely colourless, with a whitish tinge and not very strong. The taste is reminiscent of sake with a distinctive smell of mahua flowers. It is inexpensive and the production is largely done in home stills.

Mahua flowers are also used to manufacture jam, which is being made by tribal cooperatives in the Gadchiroli district of Maharashtra.[*][10]

Mahua

49.3 Oil

- Refractive index: 1.452

- Fatty acid composition (acid, %) : palmitic (c16:0) : 24.5, stearic (c18:0) : 22.7, oleic (c18:1) : 37.0, linoleic (c18:2) : 14.3

Mahua for sale in local market which is usually used to make wine at home.

Trifed, a web site of the Ministry of Tribal Affairs, Government of India reports: "Mahuwa oil has emollient properties and is used in skin disease, rheumatism and headache. It is also a laxative and considered useful in habitual constipation, piles and haemorrhoids and as an emetic. Tribals also used it as an illuminant and hair fixer." [*][2]

It has also been used as biodiesel.[*][11]

49.4 Other names

- Other botanical names: *Bassia longifolia* L., *B. latifolia* Roxb., *Madhuca indica* J. F. Gmel., *M. latifolia* (Roxb.) J.F.Macbr., *Illipe latifolia* (Roxb.) F.Muell., *Illipe malabrorum* (Engl.) Note: the authentic genus *Bassia* is in the Chenopodiaceae. The names *B. longifolia* and *B. latifolia* are illegitimate.

- Varieties:

  - *M. longifolia* var. *latifolia* (Roxb.) A.Chev. (=*B. latifolia* (Roxb))

  - *M. longifolia* var. *longifolia*

- Vernacular names:

- Bengali:*mohua*
- Oriya:"Mahula"
- English: honey tree, butter tree
- French: *illipe, arbre à beurre, bassie, madhuca*
- India: *moha, mohua, madhuca, illuppai, kuligam, madurgam, mavagam, nattiluppai, tittinam, mahwa, mahua, mowa, moa, mowrah*
- Sri Lanka: *mee*

- Synonymous names for this tree in some of the Indian states are *mahua* and *mohwa* in Hindi-speaking belt, *mahwa, mahula, Mahula* in Oriya and *maul* in Bengal, *mahwa* and *mohwro* in Maharashtra, *mahuda* in Gujarat, *ippa puvvu* (Telugu: ఇప్ప) in Andhra Pradesh, *ippe* or *hippe* in Karnataka (Kannada), *illupei* or இலுப்பை in Tamil, *poonam* and *ilupa* in Kerala (Malayalam) and *mahula, moha* and *modgi* in Orissa (Oriya).*[2]

49.5 Different views and aspects of *M. longifolia* var. *latifolia*

- Fruit in Narsapur, Medak district, India
- Fruit with leaves in Narsapur, Medak district, India
- Trunk in Narsapur, Medak district, India
- Tree in Narsapur, Medak district, India
-
- Leaves in Umaria district, Madhya Pradesh
- Mahua Tree in Thrissur, Kerala, India

*[12]

49.6 References

[1] Pankaj Oudhia, Robert E. Paull. Butter tree Madhuca latifolia Roxb. Sapotaceae p827-828. Encyclopedia of Fruit and Nuts - 2008, J. Janick and R. E. Paull -editors, CABI, Wallingford, United Kingdom

[2] "Product profile, Mahuwa, Trifed, Ministry of Tribal Affairs, Government of India". Trifed.nic.in. Retrieved 2013-11-21.

[3] Mark Duell (2012-11-07). "Trunk and disorderly! Herd of 50 drunken elephants ransack village after gulping down 500 LITRES of alcohol in shop". London: Dailymail.co.uk. Retrieved 2013-11-21.

[4] "Mahuwa tree and the aborigines of North Maharashtra, D.A.Patil, et al". Niscair.res.in. Retrieved 2013-11-21.

[5] "Non-Wood Forest Products in 15 Countries Of Tropical Asia : An Overview". Fao.org. Retrieved 2013-11-21.

[6] Dr. J.Raamachandran, HERBS OF SIDDHA MEDICINES-The First 3D Book on Herbs, pp38

[7] Keenan, G.I., 1920. The microscopical identification of mohraw meal in insecticides. J. American Pharmaceutical Assoc., Vol. IX, No. 2, pp.144-147

[8] T.V.R.Pillay and M.N.Kutty, 2005. Aquaculture: Principles and Practices. 2nd Edition. Blackwell Publishing Ltd., p.623

[9] "Mahuwah". India9.com. 2005-06-07. Retrieved 2013-11-21.

[10] "Forest department, LIT develop new products from mahua - The Times of India". *The Times Of India*. 2012-12-04.

[11] "Farm Query - Mahua oil". *The Hindu* (Chennai, India). 2014-01-22.

[12] "File:Madhuca longifolia var latifolia (Mahua) fruits Melghat Tiger Reserve Maharashtra 56 249.jpg - Wikimedia Commons". Commons.wikimedia.org. Retrieved 2013-11-21.

49.7 External links

- "*Madhuca longifolia* (J. Konig) J. F. Macbr.". Integrated Taxonomic Information System.
- Alternative edible oil from mahua seeds, The Hindu
- Mowrah Butter, OilsByNature.com
- Famine Foods
- Use of Mahua Oil (*Madhuca indica*) as a Diesel Fuel Extender
- WWF India Mahua

49.8 Bibliography

- Boutelje, J. B. 1980. Encyclopedia of world timbers, names and technical literature.
- Duke, J. A. 1989. Handbook of Nuts. CRC Press.
- Encke, F. et al. 1993. Zander: Handwörterbuch der Pflanzennamen, 14. Auflage.
- Govaerts, R. & D. G. Frodin. 2001. World checklist and bibliography of Sapotaceae.

- Hara, H. et al. 1978–1982. An enumeration of the flowering plants of Nepal.

- Matthew, K. M. 1983. The flora of the Tamil Nadu Carnatic.

- McGuffin, M. et al., eds. 2000. Herbs of commerce, ed. 2.

- Nasir, E. & S. I. Ali, eds. 1970–. Flora of [West] Pakistan.

- Pennington, T. D. 1991. The genera of the Sapotaceae.

- Porcher, M. H. et al. Searchable World Wide Web Multilingual Multiscript Plant Name Database (MMPND) - on-line resource.

- Saldanha, C. J. & D. H. Nicolson. 1976. Flora of Hassan district.

- Saldanha, C. J. 1985–. Flora of Karnataka.

Chapter 50

Maple syrup

Maple syrup is a syrup usually made from the xylem sap of sugar maple, red maple, or black maple trees, although it can also be made from other maple species. In cold climates, these trees store starch in their trunks and roots before the winter; the starch is then converted to sugar that rises in the sap in the spring. Maple trees can be tapped by boring holes into their trunks and collecting the exuded sap. The sap is processed by heating to evaporate much of the water, leaving the concentrated syrup.

Maple syrup was first collected and used by the indigenous peoples of North America. The practice was adopted by European settlers, who gradually refined production methods. Technological improvements in the 1970s further refined syrup processing. The Canadian province of Quebec is by far the largest producer, responsible for about three-quarters of the world's output; Canadian exports of maple syrup exceed C$145 million (approximately US$130.5 million) per year. Vermont is the largest producer in the United States, generating about 5.5 percent of the global supply.

Maple syrup is graded according to the Canada, United States, or Vermont scales based on its density and translucency. Sucrose is the most prevalent sugar in maple syrup. In Canada, syrups must be made exclusively from maple sap to qualify as maple syrup and must also be at least 66 percent sugar.[1] In the United States, a syrup must be made almost entirely from maple sap to be labelled as "maple" , though states such as Vermont and New York[2] have more restrictive definitions (see below).

Maple syrup is often eaten with pancakes, waffles, French toast, or oatmeal and porridge. It is also used as an ingredient in baking, and as a sweetener or flavouring agent. Culinary experts have praised its unique flavour, although the chemistry responsible is not fully understood.[3]

50.1   Sources

Three species of maple trees are predominantly used to produce maple syrup: the sugar maple (*Acer saccha-*

A sugar maple tree

rum), the black maple (*A. nigrum*), and the red maple (*A. rubrum*),[4] because of the high sugar content (roughly two to five percent) in the sap of these species.[5] The black maple is included as a subspecies or variety in a more broadly viewed concept of *A. saccharum*, the sugar maple, by some botanists.[6] Of these, the red maple has a shorter season because it buds earlier than sugar and black maples, which alters the flavour of the sap.[7]

A few other (but not all) species of maple (*Acer*) are also sometimes used as sources of sap for producing maple syrup, including the box elder or Manitoba maple (*Acer negundo*),[8] the silver maple (*A. saccharinum*),[9] and the bigleaf maple (*A. macrophyllum*).[10] Similar syrups may also be produced from birch or palm trees, among other sources.[11][12]

50.2 History

50.2.1 Indigenous peoples

"Sugar-Making Among the Indians in the North" (*19th-century illustration*)

Indigenous peoples living in the northeastern part of North America were the first groups known to have produced maple syrup and maple sugar. According to aboriginal oral traditions, as well as archaeological evidence, maple tree sap was being processed into syrup long before Europeans arrived in the region.[13][14] There are no authenticated accounts of how maple syrup production and consumption began,[15] but various legends exist; one of the most popular involves maple sap being used in place of water to cook venison served to a chief.[14] Other stories credit the development of maple syrup production to Nanabozho, Glooskap, or the squirrel. Aboriginal tribes developed rituals around sugar-making, celebrating the Sugar Moon (the first full moon of spring) with a Maple Dance.[16] Many aboriginal dishes replaced the salt traditional in European cuisine with maple sugar or syrup.[14]

The Algonquians recognized maple sap as a source of energy and nutrition. At the beginning of the spring thaw, they used stone tools to make V-shaped incisions in tree trunks; they then inserted reeds or concave pieces of bark to run the sap into buckets, which were often made from birch bark.[15] The maple sap was concentrated either by dropping hot cooking stones into the buckets[17] or by leaving them exposed to the cold temperatures overnight and disposing of the layer of ice that formed on top. While there was widespread agriculture in Mesoamerica and the Southeast and Southwest regions of the United States, the production of maple syrup is one of only a few agricultural processes in the Northeast that is not a European colonial import.[15]

50.2.2 Europeans

In the early stages of European colonization in northeastern North America, indigenous peoples showed the arriving colonists how to tap the trunks of certain types of maples during the spring thaw to harvest the sap.[18] André Thevet, the "Royal Cosmographer of France", wrote about Jacques Cartier drinking maple sap during his Canadian voyages.[19] By 1680, European settlers and fur traders were involved in harvesting maple products.[20] However, rather than making incisions in the bark as the indigenous inhabitants did, the Europeans used the method of drilling tapholes in the trunks with augers. During the 17th and 18th centuries, processed maple sap was used primarily as a source of concentrated sugar, in both liquid and crystallized-solid form, as cane sugar had to be imported from the West Indies.[15][16]

Maple sugaring parties typically began to operate at the start of the spring thaw in regions of woodland with sufficiently large numbers of maples.[18] Syrup makers first bored holes in the trunks, usually more than one hole per large tree; they then inserted wooden spouts into the holes and hung a wooden bucket from the protruding end of each spout to collect the sap. The buckets were commonly made by cutting cylindrical segments from a large tree trunk and then hollowing out each segment's core from one end of the cylinder, creating a seamless, watertight container.[15] Sap filled the buckets, and was then either transferred to larger holding vessels (barrels, large pots, or hollowed-out wooden logs), often mounted on sledges or wagons pulled by draft animals, or carried in buckets or other convenient containers.[21] The sap-collection buckets were returned to the spouts mounted on the trees, and the process was repeated for as long as the flow of sap remained "sweet". The specific weather conditions of the thaw period were, and still are, critical in determining the length of the sugaring season.[22] As the weather continues to warm, a maple tree's normal early spring biological process eventually alters the taste of the sap, making it unpalatable, perhaps due to an increase in amino acids.[9]

The boiling process was very time-consuming. The harvested sap was transported back to the party's base camp, where it was then poured into large vessels (usually made from metal) and boiled to achieve the desired consistency.[15] The sap was usually transported using large barrels pulled by horses or oxen to a central collection point, where it was processed either over a fire built out in the open or inside a shelter built for that purpose (the "sugar shack").*[15]*[23]

50.2.3   Modern era

A bucket used to collect sap, built circa 1820

Around the time of the American Civil War, syrup makers started using large, flat sheet metal pans as they were more efficient for boiling than heavy, rounded iron kettles, because of a greater surface area for evaporation.*[23] Around this time, cane sugar replaced maple sugar as the dominant sweetener in the US; as a result, producers focused marketing efforts on maple syrup. The first evaporator, used to heat and concentrate sap, was patented in 1858. In 1872, an evaporator was developed that featured two pans and a metal arch or firebox, which greatly decreased boiling time.*[15] Around 1900, producers bent the tin that formed the bottom of a pan into a series of flues, which increased the heated surface area of the pan and again decreased boiling time. Some producers also added a finishing pan, a separate batch evaporator, as a final stage in the evaporation process.*[23]

Buckets began to be replaced with plastic bags, which allowed people to see at a distance how much sap had been collected. Syrup producers also began using tractors to haul vats of sap from the trees being tapped (the sugarbush) to the evaporator. Some producers adopted motor-powered tappers and metal tubing systems to convey sap from the tree to a central collection container, but these techniques were not widely used.*[15] Heating methods also diversified: modern producers use wood, oil, natural gas, propane,

or steam to evaporate sap.*[23] Modern filtration methods were perfected to prevent contamination of the syrup.*[24]

Two taps in a maple tree, using plastic tubing for sap collection

A large number of technological changes took place during the 1970s. Plastic tubing systems that had been experimental since the early part of the century were perfected, and the sap came directly from the tree to the evaporator house.*[25] Vacuum pumps were added to the tubing systems, and preheaters were developed to recycle heat lost in the steam. Producers developed reverse-osmosis machines to take a portion of water out of the sap before it was boiled, increasing processing efficiency.*[15]

Improvements in tubing and vacuum pumps, new filtering techniques, "supercharged" preheaters, and better storage containers have since been developed. Research continues on pest control and improved woodlot management.*[15] In 2009, researchers at the University of Vermont unveiled a new type of tap that prevents backflow of sap into the tree, reducing bacterial contamination and preventing the tree from attempting to heal the bore hole.*[26] Experiments show that it may be possible to use saplings in a plantation instead of mature trees dramatically boosting productivity per acre.*[27]

50.3   Processing

Production methods have been streamlined since colonial days, yet remain basically unchanged. Sap must first be collected and boiled down to obtain pure syrup without chemical agents or preservatives. Maple syrup is made by boiling between 20 and 50 volumes of sap (depending on its

A traditional bucket tap and a plastic-bag tap

concentration) over an open fire until 1 volume of syrup is obtained, usually at a temperature 4.1 °C (7.4 °F) over the boiling point of water. As the boiling point of water varies with changes in air pressure the correct value for pure water is determined at the place where the syrup is being produced, each time evaporation is begun and periodically throughout the day.[23][28] Syrup can be boiled entirely over one heat source or can be drawn off into smaller batches and boiled at a more controlled temperature.[29]

Boiling the syrup is a tightly controlled process, which ensures appropriate sugar content. Syrup boiled too long will eventually crystallize, whereas under-boiled syrup will be watery, and will quickly spoil. The finished syrup has a density of 66° on the Brix scale (a hydrometric scale used to measure sugar solutions).[30] The syrup is then filtered to remove sugar sand, crystals made up largely of sugar and calcium malate.[31] These crystals are not toxic, but create a "gritty" texture in the syrup if not filtered out.[32] The filtered syrup is graded and packaged while still hot, usually at a temperature of 82 °C (180 °F) or greater. The containers are turned over after being sealed to sterilize the cap with the hot syrup. Packages can be made of metal, glass, or coated plastic, depending on volume and target market.[33] The syrup can also be heated longer and further processed to create a variety of other maple products, including maple sugar, maple butter or cream, and maple candy or taffy.[34]

50.3.1 Off-flavours

Off-flavours can sometimes develop during the production of maple syrup; causes include contaminants in the boiling apparatus, such as paint or cleanser; changes in the sap, such as fermentation when it has been left sitting too long; and changes in the tree, such as "buddy sap" late in the season when budding has begun.[35] In some circumstances it is possible to remove off-flavours through processing.[36]

Maple Syrup harvesting

50.4 Production

A "sugar shack" where sap is boiling.

Maple syrup in Quebec is typically sold in cans with this distinctive design

Maple sap being transformed to syrup

Maple syrup production is centred in northeastern North America; however, given the correct weather conditions, it can be made wherever suitable species of maple trees grow.

A maple syrup production farm is called a "sugarbush" or "sugarwood". Sap is often boiled in a "sugar house" (also known as a "sugar shack," "sugar shanty," or *cabane à sucre*), a building louvered at the top to vent the steam from the boiling sap.[37]

Maples are usually tapped beginning at 30 to 40 years of age. Each tree can support between one and three taps, depending on its trunk diameter. The average maple tree will produce 35 to 50 litres (9.2 to 13.2 US gal) of sap per sea-son, up to 12 litres (3.2 US gal) per day.[38] This is roughly equal to 7% of its total sap. Seasons last for four to eight weeks, depending on the weather.[39] During the day, starch stored in the roots for the winter rises through the trunk as sugary sap, allowing it to be tapped.[22] Sap is not tapped at night because the temperature drop inhibits sap flow, although taps are typically left in place overnight.[40] Some producers also tap in autumn, though this practice is less common than spring tapping. Maples can continue to be tapped for sap until they are over 100 years old.[38]

50.5   Commerce

Until the 1930s, the United States produced most of the world's maple syrup.[41] Today, after rapid growth in the 1990s, Canada produces more than 80 percent of the world's maple syrup, producing about 26,500,000 litres (7,000,000 US gal) in 2004. The vast majority of this comes from the province of Quebec, which is the world's largest producer, with about 75 percent of global production totalling 24,660,000 litres (6,510,000 US gal) in 2005.[42] As of 2003, Quebec had more than 7,000 producers, collectively making over 24,000,000 litres (6,300,000 US gal) of syrup.[43] Production in Quebec is controlled through a supply management system, with producers receiving quota allotments from the Federation of Quebec Maple Syrup Producers (Fédération des producteurs acéricoles du Québec), which also maintains reserves of syrup.[44] Canada exports more than 9,400,000 litres (2,500,000 US gal) of maple syrup per year, valued at more than C$145 million.[25][45] The provinces of Ontario, Nova Scotia, New Brunswick, and Prince Edward Island produce smaller amounts of syrup.[42]

The Canadian provinces of Manitoba and Saskatchewan produce maple syrup using the sap of the box elder or Manitoba maple (*Acer negundo*).[8] A Manitoba maple tree's yield is usually less than half that of a similar sugar maple tree.[46] Manitoba maple syrup has a slightly different flavour from sugar-maple syrup, because it contains less sugar and the tree's sap flows more slowly.

Vermont is the biggest US producer, with over 1,320,000 US gallons (5,000,000 L) during the 2013 season, followed by New York with 574,000 US gallons (2,170,000 L) and Maine with 450,000 US gallons (1,700,000 L). Wisconsin, Ohio, New Hampshire, Michigan, Pennsylvania, Massachusetts, and Connecticut all produced marketable quantities of maple syrup of less than 265,000 US gallons (1,000,000 L) each in 2013.[47] As of 2003, Vermont produced about 5.5 percent of the global syrup supply.[43]

Maple syrup has been produced on a small scale in some

other countries, notably Japan and South Korea.*[48] However, in South Korea in particular, it is traditional to consume maple sap, called *gorosoe*, instead of processing it into syrup.*[49]

In 2013, 65% of Canadian maple syrup exports went to the United States (a value of C$178 million), 9% to Japan (C$25 million), 8% to Germany (C$22 million) and 4.3% to the United Kingdom (C$12 million).*[50]

50.6 Grades

See also: Food grading

Following an effort from the International Maple Syrup Institute (IMSI) and many maple syrup producer associations, both Canada and the United States have altered their laws regarding the classification of maple syrup to be uniform. Whereas in the past each state or province had their own laws on the classification of maple syrup, now those laws state the same grades throughout. This had been a work in progress for several years, and most of the finalization of the new grading system was made in 2014. The Canadian Food Inspection Agency announced in the Canada Gazette on 28 June 2014 that rules for the sale of maple syrup would be amended to include new descriptors, at the request of the IMSI.*[51]

As of December 31, 2014, the Canadian Food Inspection Agency (CFIA)*[52] and as of March 2, 2015, the United States Department of Agriculture (USDA) Agricultural Marketing Service (AMS)*[53] issued revised standards on the classification of maple syrup as follows:

- Grade A
    - Golden Colour and Delicate Taste
    - Amber Colour and Rich Taste
    - Dark Colour and Robust Taste
    - Very Dark Colour and Strong Taste
- Processing Grade
- Substandard

As long as maple syrup does not have an off-flavor and is of a uniform color and clean and free from cloudiness, turbidity, sediment, it can be identified as one of the A grades above. If it does exhibit any of the problems mentioned earlier, it does not meet Grade A requirements and must be labeled as Processing Grade maple syrup and may not be sold to the consumer. If maple syrup does not meet the requirements of Processing Grade maple syrup (including a

fairly characteristic maple taste), it is classified as Substandard.*[53]

As of February 2015, this new grading system has been accepted and made law by most maple-producing states and provinces, other than Ontario, Quebec, and Ohio. Vermont, in an effort to "jump-start" the new grading regulations, adopted the new grading system as of January 1, 2014, after the grade changes passed the Senate and House in 2013. Maine passed a bill to take effect as soon as both Canada and the United States adopted the new grades. They are allowing a one-year grace period. In New York, the new grade changes became law on January 1, 2015, with a one-year grace period. New Hampshire did not require legislative approval and so the new grade laws became effective as of December 16, 2014, and must be complied with as of January 1, 2016 at the latest.*[54]

Golden and Amber grades typically have a milder flavour than Dark and Very dark, which are both dark and have an intense maple flavour.*[55] The darker grades of syrup are used primarily for cooking and baking, although some specialty dark syrups are produced for table use.*[56] Syrup harvested earlier in the season tends to yield a lighter color.*[57] With the new grading system, the classification of maple syrup depends ultimately on its translucence. Golden has to be more than 75 percent translucent, Amber has to be 50.0 to 74.9 percent translucent, Dark has to be 25.0 to 49.9 percent translucent, and Very Dark is any product less than 25.0 percent translucent.*[53]

50.6.1 Old grading system

Old US maple syrup grades, left to right: Grade A Light Amber ("Fancy"), Grade A Medium Amber, Grade A Dark Amber, Grade B

In Canada, maple syrup was classified prior to December 31, 2014, by the Canadian Food Inspection Agency (CFIA) as one of three grades, each with several colour classes:

Canada No. 1, including Extra Light, Light, and Medium; No. 2 Amber; and finally No. 3 Dark or any other ungraded category. Producers in Ontario or Québec may have followed either federal or provincial grading guidelines. Québec's and Ontario's guidelines differed slightly from the federal: there were two "number" categories in Québec (Number 1, with four colour classes, and 2, with five colour classes).*[58] As in Québec, Ontario's producers had two "number" grades: 1, with three colour classes; and 2, with one colour class, which was typically referred to as "Ontario Amber" when produced and sold in that province only.*[59] A typical year's yield for a maple syrup producer will be about 25 to 30 percent of each of the #1 colours, 10 percent #2 Amber, and 2 percent #3 Dark.*[30]

The United States used (some states still do, as they await state regulation) different grading standards. Maple syrup was divided into two major grades: Grade A and Grade B. Grade A was further divided into three subgrades: Light Amber (sometimes known as Fancy), Medium Amber, and Dark Amber. The Vermont Agency of Agriculture Food and Markets used a similar grading system of colour, and is roughly equivalent, especially for lighter syrups, but using letters: "AA", "A", etc.*[60]*[61] The Vermont grading system differed from the US system in maintaining a slightly higher standard of product density (measured on the Baumé scale). New Hampshire maintained a similar standard, but not a separate state grading scale. The Vermont-graded product had 0.9 percent more sugar and less water in its composition than US-graded. One grade of syrup not for table use, called commercial or Grade C, was also produced under the Vermont system.*[55]

50.7   Food and nutrition

The basic ingredient in maple syrup is the sap from the xylem of sugar maple or various other species of maple trees. It consists primarily of sucrose and water, with small amounts of the monosaccharides glucose and fructose from the invert sugar created in the boiling process.*[62] Accordingly, sugars comprise 90% of total carbohydrates which contribute nearly all of the 261 calories per 100 g serving (right table).

Maple syrup generally is devoid of micronutrient content (right table), excepting appreciable amounts of zinc and manganese which contribute 44% and 157% of the Daily Value, respectively, per 100 g of syrup consumed (right table).*[63]

Maple syrup also contains trace amounts of amino acids which increase in content as sap flow occurs.*[64] Additionally, maple syrup contains a wide variety of volatile organic compounds, including vanillin, hydroxybutanone,

and propionaldehyde. It is not yet known exactly what compounds are responsible for maple syrup's distinctive flavour,*[31] however its primary flavour contributing compounds are maple furanone, strawberry furanone, and maltol.*[65]

New compounds have been identified in maple syrup, one of which is quebecol, a natural phenolic compound created when the maple sap is boiled to create syrup.*[66]

One author described maple syrup as "a unique ingredient, smooth- and silky-textured, with a sweet, distinctive flavour – hints of caramel with overtones of toffee will not do – and a rare colour, amber set alight. Maple flavour is, well, maple flavour, uniquely different from any other." *[40] Agriculture Canada has developed a "flavour wheel" that details 91 unique flavours that can be present in maple syrup. These flavours are divided into 13 families: vanilla, empyreumatic (burnt), milky, fruity, floral, spicy, foreign deterioration or environment, maple, confectionery, plants forest-humus-cereals, herbaceous, or ligneous.*[67] These flavours are evaluated using a procedure similar to wine tasting.*[68] Other culinary experts praise its unique flavour.*[69]*[70]*[71]*[72]*[73]*[74]*[75]*[76]

Maple syrup and its various artificial imitations are widely used as toppings for pancakes, waffles, and French toast in North America. They can also be used to flavour a variety of foods, including fritters, ice cream, hot cereal, fresh fruit, and sausages. It is also used as sweetener for granola, applesauce, baked beans, candied sweet potatoes, winter squash, cakes, pies, breads, tea, coffee, and hot toddies. Maple syrup can also be used as a replacement for honey in wine (mead).*[77]

50.8   Imitations and substitutions

In the United States, "maple syrup" must be made almost entirely from maple sap, although small amounts of substances such as salt may be added.*[78] "Maple-flavoured" syrups include maple syrup but may contain additional ingredients.*[79] "Pancake syrup", "waffle syrup", "table syrup", and similarly named syrups are substitutes which are less expensive than maple syrup. In these syrups, the primary ingredient is most often high fructose corn syrup flavoured with sotolon; they have no genuine maple content, and are usually thickened far beyond the viscosity of maple syrup.*[80] The fenugreek seed, a spice with high amounts of sotolon, can be prepared to have a maple-like flavour, and is used to make a very strong commercial flavouring that is similar to maple syrup, but much less expensive; one such syrup, Mapleine, was popular during the Great Depression.*[81]*[82] American labelling laws prohibit imitation syrups from having "maple" in their names.*[83]

In Canada, maple syrup must be made entirely from maple sap, and syrup must have a density of 66° on the Brix scale to be marketed as maple syrup.*[30] Québécois sometimes refer to imitation maple syrup as *sirop de poteau* ("pole syrup"), a joke referring to the syrup as having been made by tapping telephone poles.*[84]

Imitation syrups are generally cheaper than maple syrup, but tend to taste artificial. A 2009 *Cook's Illustrated* comparison between top-selling maple and imitation syrups consistently rated the real maple brands (Maple Grove Farms, Highland Sugarworks, Camp Maple, Spring Tree, and Maple Gold) above the imitation brands tested (Eggo, Aunt Jemima, Mrs. Butterworth's, Log Cabin, and Hungry Jack).*[85] In the United States, consumers generally prefer imitation syrups, likely because of the significantly lower cost.*[86]

50.9   Cultural significance

The motif on the flag of Canada is a maple leaf.

Maple syrup and maple sugar were used during the American Civil War and by abolitionists in the years prior to the war because most cane sugar and molasses were produced by Southern slaves.*[87]*[88] Because of food rationing during the Second World War, people in the northeastern United States were encouraged to stretch their sugar rations by sweetening foods with maple syrup and maple sugar,*[15] and recipe books were printed to help housewives employ this alternative source.*[89]

Maple products are considered emblematic of Canada, in particular Quebec, and are frequently sold in tourist shops and airports as souvenirs from Canada. The sugar maple's leaf has come to symbolize Canada, and is depicted on the country's flag.*[90] Several US states, including New York, Vermont and Wisconsin, have the sugar maple as their state tree.*[91] A scene of sap collection is depicted on the Vermont state quarter, issued in 2001.*[92]

50.10   See also

- Birch syrup

- List of foods made from maple

50.11   References

50.11.1   Notes

[1]   "Chapter 13 – Labelling of Maple Products" . Canadian Food Inspection Agency. Retrieved 9 December 2011.

[2]   "New York State Department of Agriculture and Markets" . *NY.gov*. New York State. Retrieved 17 June 2014.

[3]   Amy Christine Brown (June 2010). *Understanding Food: Principles and Preparation*. Cengage Learning. p. 441. ISBN 978-0-538-73498-1. Maple Syrup Colors The flavor and color of maple syrup develop during the boiling of the initially colorless sap. Government standards ... but real maple syrup has a unique flavor and smoothness not duplicated by substitutes. Pure or blended

[4]   Elliot 2006, pp. 8–10.

[5]   Ciesla 2002, pp. 37–38.

[6]   "*Acer saccharum* subsp. *nigrum*". Germplasm Resources Information Network. Retrieved 10 December 2011.

[7]   Heilingmann, Randall B. "Hobby Maple Syrup Production (F-36-02)". Ohio State University. Retrieved 20 September 2010.

[8]   Ehman, Amy Jo (25 April 2011). "Sask. sap too sweet to waste" . *The StarPhoenix*. p. B1.

[9]   Heiligmann, Randall B; Winch, Fred E (1996). "Chapter 3: The Maple Resource" . In Koelling, Melvin R; Heiligmann, Randall B. *North American Maple Syrup Producers Manual*. Bulletin **856**. Ohio State University. Archived from the original on 29 April 2006. Retrieved 20 September 2010.

[10]   Ruth, Robert H; Underwood, J Clyde; Smith, Clark E; Yang, Hoya Y (1972). "Maple sirup production from bigleaf maple" (PDF). *Pnw-181* (US Department of Agriculture, Forest Service, Pacific Northwest Forest and Range Experiment Station): 12.

[11]   Leung, Wency (7 June 2011). "Why settle for maple when you could have birch syrup?". *The Globe and Mail*. Retrieved 12 December 2011.

[12]   Food (1989). *Utilization of tropical foods: trees : compendium on technological and nutritional aspects of processing and utilization of tropical foods, both animal and plant, for purposes of training and field reference*. Food and Agriculture Organization of the United Nations. p. 5. ISBN 978-92-5-102776-9.

[13] Ciesla 2002, pp. 37, 104.

[14] "History". Michigan Maple Syrup Association. Retrieved 20 November 2010.

[15] Koelling, Melvin R; Laing, Fred; Taylor, Fred (1996). "Chapter 2: History of Maple Syrup and Sugar Production". In Koelling, Melvin R; Heiligmann, Randall B. *North American Maple Syrup Producers Manual*. Bulletin **856**. Ohio State University (OSU). Archived from the original on 29 April 2006. Retrieved 20 September 2010.

[16] Eagleson & Hasner 2006, p. 15.

[17] Larkin, David (1998). *Country Wild*. Houghton Mifflin. pp. 146–147. ISBN 978-0-395-77190-7.

[18] Ciesla 2002, p. 37.

[19] Quoted in Lawrence, James M; Martin, Rux (1993). *Sweet maple*. Chapters Publishing Ltd. p. 57. ISBN 978-1-881527-00-8.

[20] Ciesla 2002, pp. 37, 39.

[21] Ciesla 2002, pp. 37–39.

[22] Heiligmann, Randall B; et al. (1996). "Chapter 6: Maple Sap Production". In Koelling, Melvin R; Heiligmann, Randall B. *North American Maple Syrup Producers Manual*. Bulletin **856**. Ohio State University. Archived from the original on 29 April 2006. Retrieved 20 September 2010.

[23] Heiligmann, Randall B; Staats, Lewis (1996). "Chapter 7: Maple Syrup Production". In Koelling, Melvin R; Heiligmann, Randall B. *North American Maple Syrup Producers Manual*. Bulletin **856**. Ohio State University. Archived from the original on 29 April 2006. Retrieved 20 September 2010.

[24] Koelling, Melvin R; et al. (1996). "Chapter 8: Syrup Filtration, Grading, Packing, and Handling". In Koelling, Melvin R; Heiligmann, Randall B. *North American Maple Syrup Producers Manual*. Bulletin **856**. Ohio State University. Archived from the original on 29 April 2006. Retrieved 20 September 2010.

[25] Ciesla 2002, p. 40.

[26] Perkins, Timothy D (October 2009). "Development and testing of the check-valve spout adapter" (PDF). *Maple Digest* **21A**: 21–29. Retrieved 21 September 2010.

[27] Sorkin, Laura (20 January 2014). "Maple Syrup Revolution: A New Discovery Could Change the Business Forever". *Modern Farmer*. Retrieved 20 January 2014.

[28] Eagleson & Hasner 2006, p. 55.

[29] Eagleson & Hasner 2006, p. 53.

[30] Elliot 2006, p. 12.

[31] Ball, David (10 October 2007). "The Chemical Composition of Maple Syrup". *Journal of Chemical Education* **84** (10): 1647–1650. Bibcode:2007JChEd..84.1647B. doi:10.1021/ed084p1647. Retrieved 19 September 2010.

[32] Eagleson & Hasner 2006, p. 56.

[33] Eagleson & Hasner 2006, p. 59.

[34] Eagleson & Hasner 2006, pp. 65–67.

[35] Childs, Stephen. "Maple Flavors and Syrup Grading". Cornell University. Retrieved 19 September 2010.

[36] van den Berg, Abby K; Perkins, Timothy D; Isselhardt, Mark L; Godshall, Mary An; Lloyd, Steven W (October 2009). "Metabolism Off-Flavor in Maple Syrup". *Maple Digest* **21A**: 11–18.

[37] Koelling, Melvin R; Staats, Lewis (1996). "Appendix 1: Maple Production and Processing Facilities". In Koelling, Melvin R; Heiligmann, Randall B. *North American Maple Syrup Producers Manual*. Bulletin **856**. Ohio State University. Archived from the original on 29 April 2006. Retrieved 20 September 2010.

[38] Ciesla 2002, p. 39.

[39] Koelling, Melvin R; Davenport, Russell (1996). "Chapter 1: Introduction". In Koelling, Melvin R; Heiligmann, Randall B. *North American Maple Syrup Producers Manual*. Bulletin **856**. Ohio State University. Archived from the original on 29 April 2006. Retrieved 20 September 2010.

[40] Werner, Leo H. "Maple Sugar Industry". *Canadian Encyclopedia*. Historica-Dominion Institute. Retrieved 20 September 2010.

[41] http://maple.dnr.cornell.edu/pubs/MaplePartOfStimulusPackage.pdf

[42] "Production, Price, & Value, 2002–2004, U.S. & Canadian Provinces" (PDF). *Maple Syrup*. United States Department of Agriculture. September 2005. p. 12. Retrieved 19 September 2010.

[43] Eagleson & Hasner 2006, p. 27.

[44] "Actions de la FPAQ" (in French). Fédération des producteurs acéricoles du Québec. Retrieved 22 September 2010.

[45] Elliot 2006, p. 13.

[46] Kendrick, Jenny. "Tapping the Manitoba Maple" (PDF). Statistics Canada. Retrieved 19 September 2010.

[47] "Maple Syrup Production" (PDF). *Maple Syrup 2013*. United States Department of Agriculture. 13 June 2013. p. 1. Retrieved 11 August 2013.

[48] Watanabe, Toshiyuki; Aso, Kiyoshi (1962). "On the Sugar Composition of Maple Syrup". *Tohoku Journal of Agricultural Research* **13** (2): 175–181.

[49] Sang-Hun, Choe (5 March 2009). "In South Korea, drinks are on the maple tree". *Hadong Journal*. Retrieved 21 September 2010.

[50] "Statistical Overview of the Canadian Maple Industry 2013 - Agriculture and Agri-Food Canada (AAFC)". Agr.gc.ca. Retrieved 2015-10-15.

[51] "Maple syrup labelling changes on tap". CBC News. 27 June 2014. Retrieved 29 June 2014.

[52] "Harper Government Strengthens Competitiveness of Canada's Maple Syrup Industry" (PDF). Canadian Food Inspection Agency. 16 December 2014. Retrieved 10 March 2015.

[53] "United States Standards for Grades of Maple Syrup". US Department of Agriculture. 2 March 2015. Retrieved 10 March 2015.

[54] "United States Standards for Grades of Maple Syrup". International Maple Syrup Institute. 29 January 2015. Retrieved 10 March 2015.

[55] "Frequently Asked Questions". Cornell Sugar Maple Research & Extension Program. Retrieved 22 September 2010.

[56] McGee, Harold (2004). *On food and cooking: the science and lore of the kitchen* (2nd ed.). Simon & Schuster. pp. 668–669. ISBN 978-0-684-80001-1.

[57] Thompson, Jennifer (2003). *Very Maple Syrup*. Celestial Arts. p. 2. ISBN 978-1587611810.

[58] "Maple Syrup Grades". Fédération des producteurs acéricoles du Québec. Retrieved 27 March 2012.

[59] "Maple Syrup Grades". Ontario Maple Syrup Producers Association. Retrieved 19 September 2010.

[60] "Maple Syrup Grades Vermont". Vermont Maple Syrup. Retrieved 27 March 2012.

[61] Ciesla 2002, p. 41.

[62] van den Berg, Abby; Perkins, Timothy; Isselhardt, Mark (December 2006). "Sugar Profiles of Maple Syrup Grades" (PDF). *Maple Syrup Digest*: 12–13.

[63] "Maple Syrup". *National Nutrient Database*. United States Department of Agriculture. Retrieved 19 September 2010.

[64] Morselli, Mariafranca; Whalen, M Lynn (1996). "Appendix 2: Maple Chemistry and Quality". In Koelling, Melvin R; Heiligmann, Randall B. *North American Maple Syrup Producers Manual*. Bulletin **856**. Ohio State University.

[65] Chartier, Francois (30 March 2012). *Taste Buds and Molecules: The Art and Science of Food, Wine, and Flavor*. Houghton Mifflin Harcourt.

[66] Li, Liya; Seeram, Navindra P. (2011). "Quebecol, a novel phenolic compound isolated from Canadian maple syrup". *Journal of Functional Foods* **3** (2): 125. doi:10.1016/j.jff.2011.02.004.

[67] Eagleson & Hasner 2006, pp. 71, 73.

[68] Eagleson & Hasner 2006, p. 74.

[69] Evelyn Roehl (1996). *Whole Food Facts: The Complete Reference Guide*. Inner Traditions * Bear & Company. pp. 135–136. ISBN 978-0-89281-635-4. The unique flavor of maple syrup comes from trace amounts of minerals. sugars, and other substances in the syrup. It is very difficult to synthesize this flavor artificially. To make maple sugar, a crystalline sweetener, maple sap, is boiled until ...

[70] Dinah Bucholz (September 2010). *The Unofficial Harry Potter Cookbook: From Cauldron Cakes to Knickerbocker Glory--More Than 150 Magical Recipes for Muggles and Wizards*. Adams Media. p. 15. ISBN 978-1-4405-0325-2. For golden syrup, use light or dark corn syrup, light molasses, or pure maple syrup. Maple syrup will impart a unique flavor to the finished product, so use it with discretion.

[71] Ada Henne Koene (November 2005). *Food shopper's guide to Holland: a comprehensive review of the finest local and international food products in the Dutch marketplace*. Eburon Uitgeverij BV. p. 25. ISBN 978-90-5972-092-3. Siroop is a term which applies to the light and liquid berry syrups used to flavor drinks. Maple syrup is made from the sap of the sugar maple tree. Its unique flavor makes it the favorite American topping for pancakes and waffles

[72] Alan Davidson (1981). *Oxford Symposium 1981: National & Regional Styles of Cookery*. Oxford Symposium. p. 251. ISBN 978-0-907325-07-9. As the sap flow progresses, sugar content in the sap falls, and the resulting syrup is darker, with a richer maple flavor. ... Now that people buy maple syrup specifically for its "unique" flavor, they might be advised to look for Grade A...

[73] Julie Van Rosendaal (January 2004). *One Smart Cookie: All Your Favorite Cookies, Squares, Brownies and Biscotti ... With Less Fat!*. Rodale. p. 21. ISBN 978-1-57954-944-2. Although it is more expensive than other sweeteners, it adds a deliciously unique flavor to baked goods and frostings. Less expensive grade B syrup is fine for baking, or use corn syrup with a small amount of maple extract. Store maple syrup

[74] Mcwilliams, Margaret. *Nutrition and Dietetics* (2007 ed.). Rex Bookstore, Inc. p. 184. ISBN 978-971-23-4738-2. Light and dark brown sugars contain impurities that alter their color and flavor. Maple syrup and sugar have unique flavors attributable to the impurities in the maple sap from which they are made.

[75] Paula I. Figoni (October 2010). *How Baking Works: Exploring the Fundamentals of Baking Science*. John Wiley & Sons. p. 182. ISBN 978-0-470-39813-5. This makes maple syrup

an extremely expensive sweetener. It is prized for its unique and very sweet aroma, which develops from the Maillard reactions that occur as sap is boiled over high heat. Do not confuse maple-flavored pancake syrup

[76] Active Interest Media, Inc. (February 2001). *Better Nutrition*. Active Interest Media, Inc. p. 47. ISSN 0405-668X. Maple syrup. Nearly twice as sweet as white sugar, maple syrup adds rich flavor and trace minerals to nearly any recipe. Maple sugar is made by ...

[77] Elliot 2006.

[78] Sweeteners and table sirups: maple sirup. 21 CFR §168.140 (USA). Food and Drug Administration.

[79] Sweeteners and table sirups: table sirup. 21 CFR §168.180 (USA). Food and Drug Administration.

[80] Harris, NE; et al. (1975). *Replacement of Sugar Syrup with High-Fructose Syrup in Imitation Maple Syrup*. Defense Technical Information Center. pp. 1–13.

[81] Wilhelm, Honor L (1908). *The Coast* **16**. The Coast Publishing Co. p. 57.

[82] Warren, James R. "Crescent Manufacturing Company". HistoryLink. Retrieved 14 December 2011.

[83] Sweeteners and table sirups: maple sirup: definition, naming. 21 CFR §§168.140(a), 168.180(c) (USA). Food and Drug Administration.

[84] MacInnis, Craig (6 July 2008). "Not just for breakfast anymore". *The Ottawa Citizen*. Retrieved 19 September 2010.

[85] "Maple and Pancake Syrup". *Cook's Illustrated*. January 2009. Retrieved 1 August 2011.

[86] Ingraham, Christopher (27 March 2015). "Why Americans overwhelmingly prefer fake maple syrup". *The Washington Post*. Retrieved 30 March 2015.

[87] "Making the Grade: Why the Cheapest Maple Syrup Tastes Best". *The Atlantic*. 1 November 2011. Retrieved 4 November 2011.

[88] Gellmann, D (2001). "Pirates, Sugar, Debtors, and Slaves: Political Economy and the case for Gradual Abolition in New York". *Slavery & Abolition: A Journal of Slave and Post-Slave Studies* **22** (2): 51–68. doi:10.1080/714005193.

[89] Driver, Elizabeth (2008). *Culinary landmarks: a bibliography of Canadian cookbooks, 1825–1949*. University of Toronto Press. p. 1070. ISBN 978-0-8020-4790-8.

[90] "The maple leaf". Canadian Heritage. 17 November 2008. Retrieved 18 November 2010.

[91] "State Trees & State Flowers". United States National Arboretum. 14 July 2010. Retrieved 18 November 2010.

[92] "The 50 State Quarters Program Summary Report" (PDF). Retrieved 20 October 2013.

50.11.2   Cited works

- Ciesla, William M (2002). *Non-wood Forest Products from Temperate Broad-leaved Trees*. Food and Agriculture Organization of the United Nations. ISBN 978-92-5-104855-9.

- Eagleson, Janet; Hasner, Rosemary (2006). *The Maple Syrup Book*. The Boston Mills Press. ISBN 978-1-55046-411-5.

- Elliot, Elaine (2006). *Maple Syrup: Recipes from Canada's Best Chefs*. Formac Publishing Company. ISBN 978-0-88780-697-1.

50.12   Further reading

- Nearing, Helen; Nearing, Scott (2000). *The Maple Sugar Book* (50th anniversary ed.). Chelsea Green Publishing. ISBN 978-1-890132-63-7.

- Whynott, Douglas (2014). *The Sugar Season: A Year in the Life of Maple Syrup and One Family's Quest for the Sweetest Harvest*. New York: Da Capo Press. ISBN 9780306822056. OCLC 868488316.

50.13   External links

- *Maple Syrup Quality Control Manual*, University of Maine

- "UVM Center for Digital Initiatives: The Maple Research Collection" by the Vermont Agricultural Experiment Station.

- US Food and Drug Administration description of table syrup

Chapter 51

Matsutake

For the fictional character, see Matsutake Kaoru.

Matsutake (Japanese: 松茸, pine mushroom, *Tricholoma matsutake* = syn. *T. nauseosum*) is the common name for a highly sought-after mycorrhizal mushroom that grows in Asia, Europe, and North America. It is prized by the Japanese and Chinese for its distinct spicy-aromatic odor.[*][1][*][2]

51.1 Habitat and distribution

Matsutake grow under trees and are usually concealed under duff on the forest floor free of non-symbiotic trees, e.g. broad-leaved. It forms a symbiotic relationship with the roots of a limited number of tree species. Matsutake are known to grow in China, Japan, Korea, Laos, Canada, Finland, the United States, Sweden, among other countries. In Japan it is most commonly associated with Japanese Red Pine.[*][3]

51.2 Similar species

In the North American Pacific Northwest *Tricholoma magnivelare* is found in coniferous forests made up of one or more of the following species: Douglas-fir, Noble Fir, Shasta Red Fir, Sugar Pine, Ponderosa Pine, or Lodgepole Pine. In California and parts of Oregon, it is also associated with hardwoods, including Tanoak, Madrone, Rhododendron, Salal, and Manzanita. In northeastern North America, the mushroom is generally found in Jack Pine forests. *T. magnivelare* is typically called White Matsutake as it does not feature the brown coloration of the Asian specimen.

In 1999, N. Bergius and E. Danell reported that Swedish (*Tricholoma nauseosum*) and Japanese *matsutake* (*T. matsutake*) are the same species.[*][4] The report led to increased import of matsutake from Northern Europe to Japan be-

cause of the comparable flavor and taste.

51.3 Cost and availability

Songi gui (송이구이), *grilled* matsutake *in Korean cuisine*

Though simple to harvest, matsutake are hard to find because of their specific growth requirements and the rarity of appropriate forest and terrain, combined with competition from local folk and wild animals such as squirrel, rabbits and deer for the once-yearly harvest of mushrooms, causing the price to be very high at times or as low as $2 per pound for pickers when the market will bear it. Domestic production of matsutake in Japan has been sharply reduced over the last 50 years due to a pine nematode *Bursaphelenchus xylophilus*, and it has influenced the price a great deal. The annual harvest of matsutake in Japan is now less than 1,000 tons, and the Japanese mushroom supply is largely made up by imports from China, Korea, the North American Pacific Northwest (Northern California, Oregon, Washington, and British Columbia), and Northern Europe (Sweden and Finland).[*][5] The price for matsutake in the Japanese market is highly dependent on quality, availability, and origin. The Japanese matsutake at the beginning of the season, which is the highest grade, can go up to $2,000 per kilogram. In contrast, the average value for imported matsutake is about $90 per kilogram.*[6]

51.4   See also

- Himematsutake: the "princess matsutake"

- List of *Tricholoma* species

- Medicinal fungi

51.5   References

[1] Ashkenazi, Michael; Jacob, Jeanne (2003). *Food culture in Japan*. Greenwood Publishing Group. p. 49. ISBN 0-313-32438-7.

[2] Play That Fungi Music

[3] Ashburne, John, "In search of the Holy Grail of mushrooms", *Japan Times*, 16 October 2011, p. 7.

[4] Eric Danell, The Swedish matsutake and the Japanese matsutake are the same species!, The Edible Mycorrhizal Mushroom Research Group, Department of Forest Mycology and Pathology, Swedish University of Agricultural Sciences.

[5] (Japanese) 輸入マツタケに異 中 産激減 `フィンランド , J-CAST, 2007/9/26.

[6] Matsutani, Minoru, "Japan's long love affair with 'matsutake'", *Japan Times*, 9 November 2010, p. 3.

51.6   External links

- *Matsutake* in Index Fungorum.

- Winema National Forest matsutake-information

Chapter 52

Metroxylon sagu

Metroxylon sagu (**true sago palm**) is a species of palm in the genus *Metroxylon*, native to tropical southeastern Asia in Indonesia (western New Guinea, and the Moluccas), Papua New Guinea, Malaysia (both Peninsular Malaysia and Sarawak) and possibly also the Philippines (though may have been introduced there).[*][1] It is also naturalised in Thailand, Java, Kalimantan, Sumatra, and the Solomon Islands[*][2]

52.1 Description

True sago palm is a suckering (multiple-stemmed) palm, each stem only flowering once (hapaxanthic) with a large upright terminal inflorescence. A stem grows 7–25 m tall before it ends in an inflorescence. Before flowering, a stem bears about 20 pinnate leaves up to 10 m long. Each leaf has about 150-180 leaflets up to 175 cm long. The inflorescence, 3-7.5 m tall and wide, consists of the continuation of the stem and 15-30 upwardly-curving (first-order) branches spirally arranged on it. Each first-order branch has 15 25 rigid, distichously arranged second-order branches; each second-order branch has 10-12 rigid, distichously arranged third-order branches. Flower pairs are spirally arranged on the third-order branches, each pair consisting of one male and one hermaphrodite flower. The fruit is drupe-like, about 5 cm in diameter, covered in scales which turn from bright green to straw-coloured upon ripening.[*][3]

52.2 Cultivation and uses

Main article: Sago

The tree is of commercial importance as the main source of sago, a starch obtained from the trunk by washing the starch kernels out of the pulverized pith with water. This starch is used in cooking for puddings, noodles, breads, and as a thickener. In the Sepik River region of New Guinea, pancakes made from sago are a staple food, often served with fresh fish. Its leaflets are also used as thatching which can remain intact for up to five years.[*][4] The dried petioles (called *gaba-gaba* in Indonesian) are used to make walls and ceilings; they are very light, and therefore also used in the construction of rafts.

The sago palm reproduces by fruiting. Each stem (trunk) in a sago palm clump flowers and fruits at the end of its life, but the sago palm as an individual organism lives on through its suckers (shoots that are continuously branching off a stem at or below ground level). To harvest the starch in the stem, it is felled shortly before or early during this final flowering stage when starch content is highest. Sago palm is propagated by man by collecting (cutting) and replanting young suckers rather than by seed.[*][3]

Recent research indicates that the sago palm was an important food source for the ancient people of coastal China, in the period prior to the cultivation of rice.[*][5]

52.3 References

[1] Germplasm Resources Information Network: *Metroxylon sagu*

[2] Kew World Checklist of Selected Plant Families, *Metroxylon sagu*

[3] Schuiling, D.L. (2009) *Growth and development of true sago palm* (Metroxylon sagu Rottbøll) *with special reference to accumulation of starch in the trunk: a study on morphology, genetic variation and ecophysiology, and their implications for cultivation.* (PhD thesis Wageningen University).

[4] Palm and Cycad Societies of Australia. Palms: Metroxylon sagu. Retrieved 28 February 2012

[5] http://www.sciencedaily.com/releases/2013/05/130508172138.htm

Chapter 53

Mushroom

This article is about fungi. For use in food, see Edible mushroom. For other uses, see Mushroom (disambiguation).

"Toadstool" redirects here. For other uses, see Toadstool (disambiguation).

A **mushroom** (or **toadstool**) is the fleshy, spore-bearing

Amanita muscaria, *commonly known as "fly agaric"*

fruiting body of a fungus, typically produced above ground on soil or on its food source.

The standard for the name "mushroom" is the cultivated white button mushroom, *Agaricus bisporus*; hence the

word "mushroom" is most often applied to those fungi (Basidiomycota, Agaricomycetes) that have a stem (stipe), a cap (pileus), and gills (lamellae, sing. lamella) on the underside of the cap. These gills produce microscopic spores that help the fungus spread across the ground or its occupant surface.

"Mushroom" describes a variety of gilled fungi, with or without stems, and the term is used even more generally, to describe both the fleshy fruiting bodies of some Ascomycota and the woody or leathery fruiting bodies of some Basidiomycota, depending upon the context of the word.

Forms deviating from the standard morphology usually have more specific names, such as "bolete", "puffball", "stinkhorn", and "morel", and gilled mushrooms themselves are often called "agarics" in reference to their similarity to *Agaricus* or their place Agaricales. By extension, the term "mushroom" can also designate the entire fungus when in culture; the thallus (called a mycelium) of species forming the fruiting bodies called mushrooms; or the species itself.

53.1 Identification

Identifying mushrooms requires a basic understanding of their macroscopic structure. Most are Basidiomycetes and gilled. Their spores, called basidiospores, are produced on the gills and fall in a fine rain of powder from under the caps as a result. At the microscopic level the basidiospores are shot off basidia and then fall between the gills in the dead air space. As a result, for most mushrooms, if the cap is cut off and placed gill-side-down overnight, a powdery impression reflecting the shape of the gills (or pores, or spines, etc.) is formed (when the fruit body is sporulating). The color of the powdery print, called a spore print, is used to help classify mushrooms and can help to identify them. Spore print colors include white (most common), brown, black, purple-brown, pink, yellow, and creamy, but almost never blue, green, or red.[*][1]

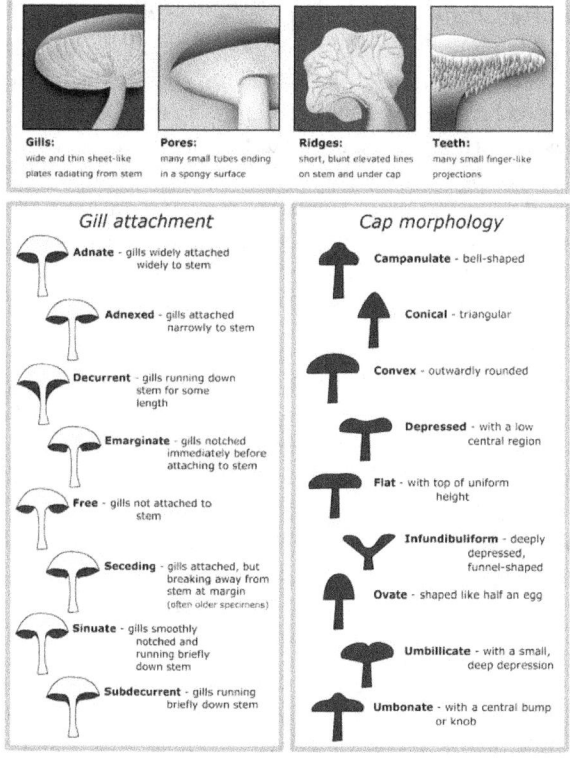

Morphological characteristics of the caps of mushrooms

While modern identification of mushrooms is quickly becoming molecular, the standard methods for identification are still used by most and have developed into a fine art harking back to medieval times and the Victorian era, combined with microscopic examination. The presence of juices upon breaking, bruising reactions, odors, tastes, shades of color, habitat, habit, and season are all considered by both amateur and professional mycologists. Tasting and smelling mushrooms carries its own hazards because of poisons and allergens. Chemical tests are also used for some genera.*[2]

In general, identification to genus can often be accomplished in the field using a local mushroom guide. Identification to species, however, requires more effort; one must remember that a mushroom develops from a button stage into a mature structure, and only the latter can provide certain characteristics needed for the identification of the species. However, over-mature specimens lose features and cease producing spores. Many novices have mistaken humid water marks on paper for white spore prints, or discolored paper from oozing liquids on lamella edges for colored spored prints.

53.2 Classification

Main articles: Sporocarp (fungi), Basidiocarp and Ascocarp

Typical mushrooms are the fruit bodies of members of

Trametes versicolor, *a polypore mushroom*

the order Agaricales, whose type genus is *Agaricus* and type species is the field mushroom, *Agaricus campestris*. However, in modern molecularly defined classifications, not all members of the order Agaricales produce mushroom fruit bodies, and many other gilled fungi, collectively called mushrooms, occur in other orders of the class Agaricomycetes. For example, chanterelles are in the Cantharellales, false chanterelles such as *Gomphus* are in the Gomphales, milk-cap mushrooms (*Lactarius*, *Lactifluus*) and russulas (*Russula*), as well as *Lentinellus*, are in the Russulales, while the tough, leathery genera *Lentinus* and *Panus* are among the Polyporales, but *Neolentinus* is in the Gloeophyllales, and the little pin-mushroom genus, *Rickenella*, along with similar genera, are in the Hymenochaetales.

Within the main body of mushrooms, in the Agaricales, are common fungi like the common fairy-ring mushroom, shiitake, enoki, oyster mushrooms, fly agarics and other amanitas, magic mushrooms like species of *Psilocybe*, paddy straw mushrooms, shaggy manes, etc.

An atypical mushroom is the lobster mushroom, which is a deformed, cooked-lobster-colored parasitized fruitbody of a *Russula* or *Lactarius*, colored and deformed by the mycoparasitic Ascomycete *Hypomyces lactifluorum*.*[3]

Other mushrooms are not gilled, so the term "mushroom" is loosely used, and giving a full account of their classifications is difficult. Some have pores underneath (and are usually called boletes), others have spines, such as the hedgehog mushroom and other tooth fungi, and so on. "Mushroom" has been used for polypores, puffballs, jelly fungi, coral fungi, bracket fungi, stinkhorns, and cup fungi. Thus, the term is more one of common application to macroscopic fungal fruiting bodies than one having precise taxonomic meaning. Approximately 14,000 species of mushrooms are described.[4]

53.3   Etymology

Amanita muscaria, *the most easily recognised "toadstool", is frequently depicted in fairy stories and on greeting cards. It is often associated with gnomes.*[5]

The terms "mushroom" and "toadstool" go back centuries and were never precisely defined, nor was there consensus on application. The term "toadstool" was often, but not exclusively, applied to poisonous mushrooms or to those that have the classic umbrella-like cap-and-stem form. Between 1400 and 1600 AD, the terms *tadstoles, frogstooles, frogge stoles, tadstooles, tode stoles, toodys hatte, paddockstool, puddockstool, paddocstol, toadstoole, and paddockstooles* sometimes were used synonymously with *mushrom, mushrum, muscheron, mousheroms, mussheron, or musserouns.*[6]

The word has apparent analogies in Dutch *padde(n)stoel* (toad-stool/chair, mushroom) and German *Krötenschwamm* (toad-fungus, alt. word for panther cap). In German folklore and old fairy tales, toads are often depicted sitting on toadstool mushrooms and catching, with their tongues, the flies that are said to be drawn to the *Fliegenpilz*, a German name for the toadstool, meaning "flies' mushroom". This is how the mushroom got another of its names, *Krötenstuhl* (a less-used German name for the mushroom), literally translating to "toad-stool".

The term "mushroom" and its variations may have been derived from the French word *mousseron* in reference to moss (*mousse*). The toadstool's connection to toads may be direct, in reference to some species of poisonous toad,[7] or may just be a case of phonosemantic matching from the German word.[8] However, delineation between edible and poisonous fungi is not clear-cut, so a "mushroom" may be edible, poisonous, or unpalatable. The term "toadstool" is nowadays used in storytelling when referring to poisonous or suspect mushrooms. The classic example of a toadstool is *Amanita muscaria*.

Cultural or social phobias of mushrooms and fungi may be related. The term "fungophobia" was coined by William Delisle Hay of England, who noted a national superstition or fear of "toadstools".[9][10] He described the "fungus-hunter" as being contemptible and detailed the larger demographic's attitude toward mushrooms as "abnormal, worthless, or inexplicable".[10] Fungophobia spread to the United States and Australia, where it was inherited from England.[10][11] The underlying cause of a cultural fungophobia may also be related to the exaggerated importance placed on the few deadly and poisonous mushrooms found in the region of that culture.[12] In these regions, mushrooms were also sometimes regarded as magic or satanic, their fruiting bodies appearing quickly overnight from underground. Some believed they were the Devil's fruit, and others that mushroom rings were magical portals.

53.4   Morphology

A mushroom develops from a nodule, or pinhead, less than two millimeters in diameter, called a primordium, which is typically found on or near the surface of the substrate. It is formed within the mycelium, the mass of threadlike hyphae that make up the fungus. The primordium enlarges into a roundish structure of interwoven hyphae roughly resembling an egg, called a "button". The button has a cottony roll of mycelium, the universal veil, that surrounds the developing fruit body. As the egg expands, the universal veil ruptures and may remain as a cup, or volva, at the base of the stalk, or as warts or volval patches on the cap. Many mushrooms lack a universal veil, therefore they do

Amanita jacksonii *buttons emerging from their universal veils*

The blue gills of Lactarius indigo, *a milk-cap mushroom*

not have either a volva or volval patches. Often, a second layer of tissue, the partial veil, covers the bladelike gills that bear spores. As the cap expands, the veil breaks, and remnants of the partial veil may remain as a ring, or annulus, around the middle of the stalk or as fragments hanging from the margin of the cap. The ring may be skirt-like as in some species of *Amanita*, collar-like as in many species of *Lepiota*, or merely the faint remnants of a cortina (a partial veil composed of filaments resembling a spiderweb), which is typical of the genus *Cortinarius*. Mushrooms lacking partial veils do not form an annulus.[13]

The stalk (also called the stipe, or stem) may be central and support the cap in the middle, or it may be off-center and/or lateral, as in species of *Pleurotus* and *Panus*. In other mushrooms, a stalk may be absent, as in the polypores that form shelf-like brackets. Puffballs lack a stalk, but may have a supporting base. Other mushrooms, such as truffles, jellies, earthstars, and bird's nests, usually do not have stalks, and a specialized mycological vocabulary exists to describe their parts.

The way the gills attach to the top of the stalk is an important feature of mushroom morphology. Mushrooms in the genera *Agaricus*, *Amanita*, *Lepiota* and *Pluteus*, among others, have free gills that do not extend to the top of the stalk. Others have decurrent gills that extend down the stalk, as in the genera *Omphalotus* and *Pleurotus*. There are a great number of variations between the extremes of free and decurrent, collectively called attached gills. Finer distinctions are often made to distinguish the types of attached gills: adnate gills, which adjoin squarely to the stalk; notched gills, which are notched where they join the top of the stalk; adnexed gills, which curve upward to meet the stalk, and so on. These distinctions between attached gills are sometimes difficult to interpret, since gill attachment may change as the mushroom matures, or with different environmental conditions.[14]

53.4.1 Microscopic features

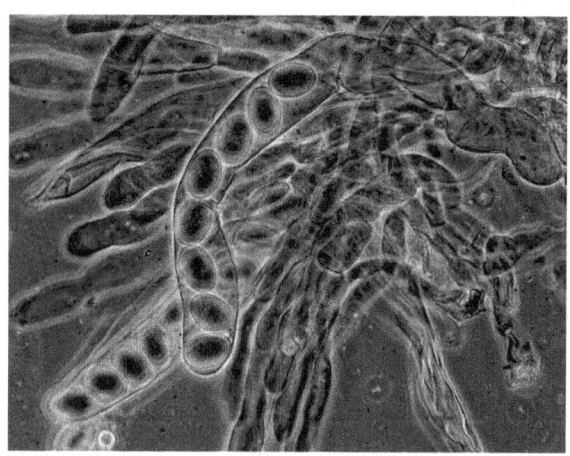

Morchella elata *asci viewed with phase contrast microscopy*

A hymenium is a layer of microscopic spore-bearing cells that covers the surface of gills. In the nongilled mushrooms, the hymenium lines the inner surfaces of the tubes of boletes and polypores, or covers the teeth of spine fungi and the branches of corals. In the Ascomycota, spores develop within microscopic elongated, sac-like cells called asci, which typically contain eight spores in each ascus. The Discomycetes, which contain the cup, sponge, brain, and some club-like fungi, develop an exposed layer of asci, as on the inner surfaces of cup fungi or within the pits of morels. The Pyrenomycetes, tiny dark-colored fungi that live on a wide range of substrates including soil, dung, leaf litter, and decaying wood, as well as other fungi, produce minute, flask-shaped structures called perithecia, within which the asci develop.[15]

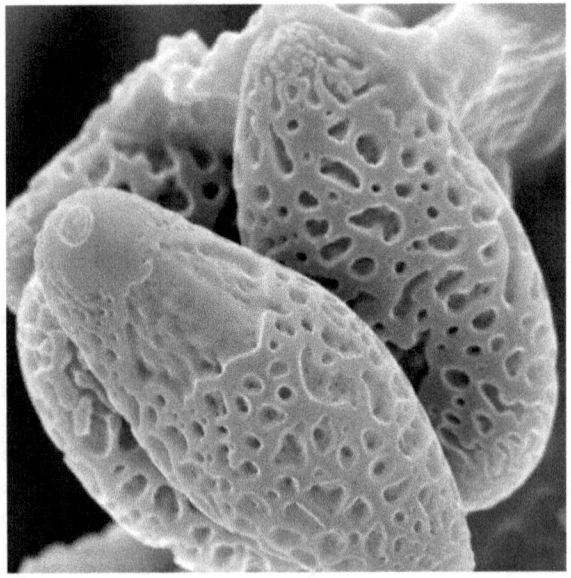

Agaricus bitorquis *mushroom emerging through asphalt concrete in summer*

Austroboletus mutabilis *spores viewed using electron microscopy*

In the Basidiomycetes, usually four spores develop on the tips of thin projections called sterigmata, which extend from club-shaped cells called a basidia. The fertile portion of the Gasteromycetes, called a gleba, may become powdery as in the puffballs or slimy as in the stinkhorns. Interspersed among the asci are threadlike sterile cells called paraphyses. Similar structures called cystidia often occur within the hymenium of the Basidiomycota. Many types of cystidia exist, and assessing their presence, shape, and size is often used to verify the identification of a mushroom.*[15]

The most important microscopic feature for identification of mushrooms is the spores. Their color, shape, size, attachment, ornamentation, and reaction to chemical tests often can be the crux of an identification. A spore often has a protrusion at one end, called an apiculus, which is the point of attachment to the basidium, termed the apical germ pore, from which the hypha emerges when the spore germinates.*[15]

53.5 Growth

Many species of mushrooms seemingly appear overnight, growing or expanding rapidly. This phenomenon is the source of several common expressions in the English language including "to mushroom" or "mushrooming" (expanding rapidly in size or scope) and "to pop up like a mushroom" (to appear unexpectedly and quickly). In reality all species of mushrooms take several days to form primordial mushroom fruit bodies, though they do expand rapidly by the absorption of fluids.

The cultivated mushroom as well as the common field mushroom initially form a minute fruiting body, referred to as the pin stage because of their small size. Slightly expanded they are called buttons, once again because of the relative size and shape. Once such stages are formed, the mushroom can rapidly pull in water from its mycelium and expand, mainly by inflating preformed cells that took several days to form in the primordia.

Similarly, there are even more ephemeral mushrooms, like *Parasola plicatilis* (formerly *Coprinus plicatlis*), that literally appear overnight and may disappear by late afternoon on a hot day after rainfall.*[16] The primordia form at ground level in lawns in humid spaces under the thatch and after heavy rainfall or in dewy conditions balloon to full size in a few hours, release spores, and then collapse. They "mushroom" to full size.

Not all mushrooms expand overnight; some grow very slowly and add tissue to their fruitbodies by growing from the edges of the colony or by inserting hyphae. For example, *Pleurotus nebrodensis* grows slowly, and because of this combined with human collection, it is now critically endangered.*[17]

Though mushroom fruiting bodies are short-lived, the underlying mycelium can itself be long-lived and massive. A colony of *Armillaria solidipes* (formerly known as *Armillaria ostoyae*) in Malheur National Forest in the United States is estimated to be 2,400 years old, possibly older, and spans an estimated 2,200 acres (8.9 km^2). Most of the fungus is underground and in decaying wood or dying tree roots in the form of white mycelia combined with black shoelace-like rhizomorphs that bridge colonized separated woody substrates.*[18]

It has been suggested the electrical stimulus of a lightning bolt striking mycelia in logs accelerates the production of

Yellow flower pot mushrooms (Leucocoprinus birnbaumii*) at various states of development*

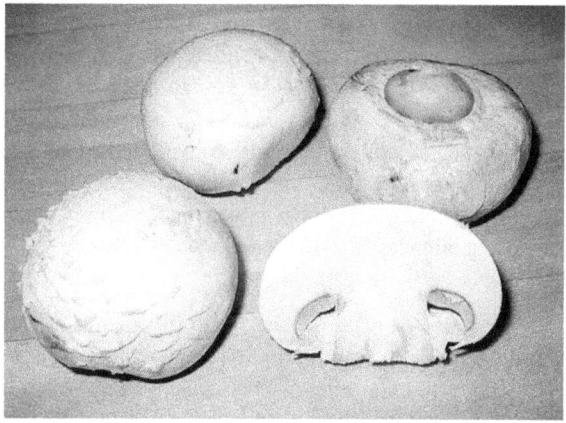

The Agaricus bisporus, *one of the most widely cultivated and popular mushrooms in the world*

mushrooms.*[19]

53.6 Nutrition

Mushrooms are a low-calorie food eaten cooked, raw or as a garnish to a meal. In a 100 g (3.5 ounce) serving, mushrooms are an excellent source (higher than 20% of the daily value, DV) of B vitamins, such as riboflavin, niacin and pantothenic acid, an excellent source of the essential minerals, selenium (37% DV) and copper (25% DV), and a good source (10-19% DV) of phosphorus and potassium. Fat, carbohydrate and calorie content are low, with absence of vitamin C and sodium. There are 27 calories in a typical serving of fresh mushrooms (table).

When exposed to ultraviolet (UV) light even after harvesting,*[20] natural ergosterols in mushrooms produce vitamin D_2,*[21] a process now used to supply fresh vitamin D mushrooms for the functional food grocery market.

In a comprehensive safety assessment of producing vitamin D in fresh mushrooms, researchers showed that artificial UV light technologies were equally effective for vitamin D production as in mushrooms exposed to natural sunlight, and that UV light has a long record of safe use for production of vitamin D in food.*[22]

53.7 Human use

Further information: Ethnomycology

53.7.1 Edible mushrooms

Main articles: Edible mushroom, Mushroom hunting and Fungiculture

Mushrooms are used extensively in cooking, in many cuisines (notably Chinese, Korean, European, and Japanese). Though neither meat nor vegetable, mushrooms are known as the "meat" of the vegetable world.*[23]

Most mushrooms sold in supermarkets have been commercially grown on mushroom farms. The most popular of these, *Agaricus bisporus*, is considered safe for most people to eat because it is grown in controlled, sterilized environments. Several varieties of *A. bisporus* are grown commercially, including whites, crimini, and portobello. Other cultivated species now available at many grocers include shiitake, maitake or hen-of-the-woods, oyster, and enoki. In recent years, increasing affluence in developing countries has led to a considerable growth in interest in mushroom cultivation, which is now seen as a potentially important economic activity for small farmers.*[24]

Mushroom and truffle output in 2005

A number of species of mushrooms are poisonous; although some resemble certain edible species, consuming them could be fatal. Eating mushrooms gathered in the wild is risky and should only be undertaken by individuals knowledgeable in mushroom identification. Common best practice is for wild mushroom pickers to focus on collecting a small number of visually distinctive, edible mushroom species that cannot be easily confused with poisonous varieties. *A. bisporus* contains carcinogens called hydrazines, the most abundant of which is agaritine. However, the carcinogens are destroyed by moderate heat when cooking.[25]

More generally, and particularly with gilled mushrooms, separating edible from poisonous species requires meticulous attention to detail; there is no single trait by which all toxic mushrooms can be identified, nor one by which all edible mushrooms can be identified. Additionally, even edible mushrooms may produce allergic reactions in susceptible individuals, from a mild asthmatic response to severe anaphylactic shock.[26][27]

People who collect mushrooms for consumption are known as mycophagists,[28] and the act of collecting them for such is known as mushroom hunting, or simply "mushrooming".

China is the world's largest edible mushroom producer.[29] The country produces about half of all cultivated mushrooms, and around 2.7 kilograms (6.0 lb) of mushrooms are consumed per person per year by over a billion people.[30]

53.7.2 Toxic mushrooms

Main article: Mushroom poisoning

Many mushroom species produce secondary metabolites

Young Amanita phalloides, *"death cap" mushrooms*

that can be toxic, mind-altering, antibiotic, antiviral, or bioluminescent. Although there are only a small number of deadly species, several others can cause particularly severe and unpleasant symptoms. Toxicity likely plays a role

in protecting the function of the basidiocarp: the mycelium has expended considerable energy and protoplasmic material to develop a structure to efficiently distribute its spores. One defense against consumption and premature destruction is the evolution of chemicals that render the mushroom inedible, either causing the consumer to vomit the meal (see emetics), or to learn to avoid consumption altogether. In addition, due to the propensity of mushrooms to absorb heavy metals, including those that are radioactive, European mushrooms may, to date, include toxicity from the 1986 Chernobyl disaster and continue to be studied.[31][32][33][34][35]

53.7.3 Psychoactive mushrooms

Psilocybe zapotecorum, *a hallucinogenic mushroom*

Mushrooms with psychoactive properties have long played a role in various native medicine traditions in cultures all around the world. They have been used as sacrament in rituals aimed at mental and physical healing, and to facilitate visionary states. One such ritual is the *velada* ceremony. A practitioner of traditional mushroom use is the *shaman* or *curandera* (priest-healer).[36]

Psilocybin mushrooms possess psychedelic properties. Commonly known as "magic mushrooms" or "'shrooms," they are openly available in smart shops in many parts of the world, or on the black market in those countries that have outlawed their sale. Psilocybin mushrooms have been reported as facilitating profound and life-changing insights often described as mystical experiences. Recent scientific work has supported these claims, as well as the long-lasting effects of such induced spiritual experiences.[37]

Psilocybin, a naturally occurring chemical in certain psychedelic mushrooms such as *Psilocybe cubensis*, is being studied for its ability to help people suffering from psychological disorders, such as obsessive-compulsive disorder. Minute amounts have been reported to stop cluster and migraine headaches.[38] A double-blind study, done by the Johns Hopkins Hospital, showed psychedelic mushrooms could provide people an experience with substantial personal meaning and spiritual significance. In the study, one third of the subjects reported ingestion of psychedelic mushrooms was the single most spiritually significant event of their lives. Over two-thirds reported it among their five most meaningful and spiritually significant events. On the other hand, one-third of the subjects reported extreme anxiety. However, the anxiety went away after a short period of time.[39][40] Psilocybin mushrooms have also shown to be successful in treating addiction, specifically with alcohol and cigarettes.[41]

A few species in the *Amanita* genus, most recognizably *A. muscaria*, but also *A. pantherina*, among others, contain the psychoactive compound muscimol. The muscimol-containing chemotaxonomic group of *Amanitas* contains no amatoxins or phallotoxins, and as such are not hepatoxic, though if not properly cured will be non-lethally neurotoxic due to the presence of ibotenic acid. The *Amanita* intoxication is similar to Z-drugs in that it includes CNS depressant and sedative-hypnotic effects, but also dissociation and delirium in high doses.

53.7.4 Medicinal properties

Main article: Medicinal mushrooms

Some mushrooms or extracts are used or studied as

Ganoderma lucidum

possible treatments for diseases, such as cardiovascular disorders.[42] Some mushroom materials, including polysaccharides, glycoproteins and proteoglycans are under basic research for their potential to modulate immune system responses and inhibit tumor growth,[43] whereas other isolates show potential antiviral, antibacterial, antiparasitic, anti-inflammatory, and antidiabetic properties in preliminary studies.[44] Currently, several extracts

have widespread use in Japan, Korea and China, as adjuncts to radiation treatments and chemotherapy,[45][46] even though clinical evidence of efficacy in humans has not been confirmed.

Historically, mushrooms have long been thought to hold medicinal value, especially in traditional Chinese medicine.[47] They have been studied in modern medical research since the 1960s, where most studies use extracts, rather than whole mushrooms. Only a few specific extracts have been tested for efficacy in laboratory research. Polysaccharide-K and lentinan are among extracts best understood from *in vitro* research, animal models such as mice, or early-stage human pilot studies.[46]

Preliminary experiments show glucan-containing mushroom extracts may affect function of the innate and adaptive immune systems, functioning as bioresponse modulators.[46] In some countries, extracts of polysaccharide-K, schizophyllan, polysaccharide peptide, or lentinan are government-registered adjuvant cancer therapies.[45][48]

As of June 2014, whole mushrooms or mushroom ingredients are being studied in 32 human clinical trials registered with the US National Institutes of Health for their potential effects on a variety of diseases and normal physiological conditions, including vitamin D deficiency, cancer, bone metabolism, glaucoma, immune functions and inflammatory bowel disease.[49]

53.7.5 Other uses

A tinder fungus, Fomes fomentarius

Mushrooms can be used for dyeing wool and other natural fibers. The chromophores of mushroom dyes are organic compounds and produce strong and vivid colors, and all colors of the spectrum can be achieved with mushroom dyes. Before the invention of synthetic dyes, mushrooms were the source of many textile dyes.[*][50]

Some fungi, types of polypores loosely called mushrooms, have been used as fire starters (known as tinder fungi).

Mushrooms and other fungi play a role in the development of new biological remediation techniques (e.g., using mycorrhizae to spur plant growth) and filtration technologies (e.g. using fungi to lower bacterial levels in contaminated water).[*][51]

53.8   References

[1] Dickinson C, Lucas J. (1982). *VNR Color Dictionary of Mushrooms*. Van Nostrand Reinhold. pp. 9–11. ISBN 978-0-442-21998-7.

[2] Ammirati *et al.*, 1985, pp. 40–41.

[3] Volk T. (2001). "*Hypomyces lactifluorum*, the lobster mushroom". *Fungus of the Month*. University of Wisconsin-La Crosse, Department of Biology. Retrieved 2008-10-13.

[4] Miles PG, Chang S-T. (2004). *Mushrooms: Cultivation, Nutritional Value, Medicinal Effect, and Environmental Impact*. Boca Raton, Florida: CRC Press. ISBN 0-8493-1043-1.

[5] Harding, Patrick (2008). *Mushroom Miscellany*. HarperCollins. p. 149. ISBN 978-0-00-728464-1.

[6] Ramsbottom J. (1954). *Mushrooms & Toadstools: a study of the activities of fungi*. London: Collins.

[7] "Botany". Ontarioprofessionals.com. 2009-03-26. Retrieved 2010-05-30.

[8] "Yahoo! Babel Fish – Text Translation and Web Page Translation". Babelfish.yahoo.com. Archived from the original on July 18, 2011. Retrieved 2010-05-30.

[9] Hay, William Deslisle (1887). "An Elementary Text-Book of British Fungi". London, S. Sonnenschein, Lowrey. pp. 6–7.

[10] Arora, David (1986). *Mushrooms Demystified, A Comprehensive Guide to the Fleshy Fungi*. Ten Speed Press. pp. 1–3. ISBN 978-0-89815-169-5.

[11] Leschyn, Wade. "Identifying Mushrooms". Peninsula Mycological Circle. Retrieved 2012-01-02.

[12] Hunter, Jessica. "The Mushroom Hunt". Synergy Magazine. Retrieved 2012-01-02.

[13] Stuntz *et al.*, 1978, pp. 12–13.

[14] Stuntz *et al.*, 1978, pp. 28–29.

[15] Ammirati *et al.*, 1985, pp. 25–34.

[16] Nelson N. (2006-08-13). "*Parasola plicatilis*". Retrieved 2008-10-13.

[17] Venturella, G. 2006. *Pleurotus nebrodensis*. In: IUCN. 2009. IUCN Red List of Threatened Species. Version 2009.1. http://www.iucnredlist.org/apps/redlist/details/full/61597/0 Downloaded on 15 October 2009.

[18] Dodge SR. "And the Humongous Fungus Race Continues". US Forest Service: Pacific Northwest Research Station. Retrieved 2011-02-28.

[19] "IEEE Xplore – Development of an Automatic Electrical Stimulator for Mushroom Sawdust Bottle". Ieeexplore.ieee.org. 2005-06-17. doi:10.1109/PPC.2005.300675. Retrieved 2014-01-24.

[20] Kalaras, M. D.; Beelman, R. B.; Elias, R. J. (2012). "Effects of postharvest pulsed UV light treatment of white button mushrooms (Agaricus bisporus) on vitamin D2 content and quality attributes". *Journal of Agricultural and Food Chemistry* **60** (1): 220–5. doi:10.1021/jf203825e. PMID 22132934.

[21] Koyyalamudi SR, Jeong SC, Song CH, Cho KY, Pang G. (2009). "Vitamin D2 formation and bioavailability from *Agaricus bisporus* button mushrooms treated with ultraviolet irradiation" (PDF). *Journal of Agricultural and Food Chemistry* **57** (8): 3351–5. doi:10.1021/jf803908q. PMID 19281276.

[22] Simon, R. R.; Borzelleca, J. F.; Deluca, H. F.; Weaver, C. M. (2013). "Safety assessment of the post-harvest treatment of button mushrooms (Agaricus bisporus) using ultraviolet light". *Food and Chemical Toxicology* **56**: 278–89. doi:10.1016/j.fct.2013.02.009. PMID 23485617.

[23] Haas EM, James P. (2009). *More Vegetables, Please!: Delicious Recipes for Eating Healthy Foods Each & Every Day*. Oakland, California: New Harbinger Publications. p. 22. ISBN 978-1-57224-590-7.

[24] Making Money by growing Mushrooms

[25] Sieger AA (ed.) (1998-01-01). "Spore Prints #338". *Bulletin of the Puget Sound Mycological Society*. Retrieved 2010-07-04.

[26] Hall *et al.*, 2003, pp. 22–24.

[27] Ammirati *et al.*, 1985, pp. 81–83.

[28] Metzler V, Metzler S. (1992). *Texas Mushrooms: a Field Guide*. Austin, Texas: University of Texas Press. p. 37. ISBN 0-292-75125-7. Retrieved 2010-08-04.

[29] "China Becomes World's Biggest Edible Mushroom Producer". Allbusiness.com. August 21, 2003. Archived from the original on September 24, 2009. Retrieved 2010-08-04.

[30] Hall *et al.*, 2003, p. 25.

[31] "Belarus exports radioactive mushrooms, April 2008". Freshplaza.com. Retrieved 2014-01-24.

[32] Radioactivity levels in some wild edible mushroom species in Turkey by Seref Turhan *et al.* in *Isotopes in Environmental and Health Studies*, Volume 43, Issue 3 September 2007, pages 249–56

[33] Archived March 14, 2012 at the Wayback Machine

[34] Hawley C. (July 30, 2010). "A Quarter Century after Chernobyl: Radioactive Boar on the Rise in Germany". *Spiegel Online International.* Retrieved 2010-08-04.

[35] Archived January 1, 1970 at the Wayback Machine

[36] Hudler GW. (2000). *Magical Mushrooms, Mischievous Molds.* Princeton, New Jersey: Princeton University Press. p. 175. ISBN 0-691-07016-4. Retrieved 2010-08-04.

[37] Griffiths R, Richards W, Johnson M, McCann U, Jesse R. (2008). "Mystical-type experiences occasioned by psilocybin mediate the attribution of personal meaning and spiritual significance 14 months later". *Journal of psychopharmacology (Oxford, England)* **22** (6): 621–32. doi:10.1177/0269881108094300. PMC 3050654. PMID 18593735.

[38] Sewell RA, Halpern JH, Pope HG. (2006). "Response of cluster headache to psilocybin and LSD". *Neurology* **66** (12): 1920–22. doi:10.1212/01.wnl.0000219761.05466.43. PMID 16801660.

[39] Griffiths RR, Richards WA, McCann U, Jesse R. (2006). "Psilocybin can occasion mystical-type experiences having substantial and sustained personal meaning and spiritual significance". *Psychopharmacology (Berl).* **187** (3): 268–83. doi:10.1007/s00213-006-0457-5. PMID 16826400.

[40] Weil A. (2006-10-16). "Looking for Mushroom Magic?". Retrieved 2010-08-04.

[41] "Clinical Sunday". *maps.org.*

[42] Guillamón, E; García-Lafuente, A; Lozano, M; d'Arrigo, M; Rostagno, M. A.; Villares, A; Martínez, J. A. (2010). "Edible mushrooms: Role in the prevention of cardiovascular diseases". *Fitoterapia* **81** (7): 715–23. doi:10.1016/j.fitote.2010.06.005. PMID 20550954.

[43] Borchers, A. T.; Krishnamurthy, A; Keen, C. L.; Meyers, F. J.; Gershwin, M. E. (2008). "The immunobiology of mushrooms". *Experimental Biology and Medicine* **233** (3): 259–76. doi:10.3181/0708-MR-227. PMID 18296732.

[44] Lull, C.; Wichers, J.; Savelkoul, F. (Jun 2005). "Anti-inflammatory and Immunomodulating Properties of Fungal Metabolites". *Mediators of Inflammation* (Free full text) **2005** (2): 63–80. doi:10.1155/MI.2005.63. ISSN 0962-9351. PMC 1160565. PMID 16030389.

[45] "Mushrooms in cancer treatment". Cancer Research UK. 2013. Retrieved 25 June 2014.

[46] Borchers AT, Krishnamurthy A, Keen CL, Meyers FJ, Gershwin ME. (2008). "The immunobiology of mushrooms". *Experimental Biology and Medicine* **233** (3): 259–76. doi:10.3181/0708-MR-227. PMID 18296732.

[47] Khan, M. A.; Tania, M; Liu, R; Rahman, M. M. (2013). "Hericium erinaceus: An edible mushroom with medicinal values". *Journal of Complementary and Integrative Medicine* **10**. doi:10.1515/jcim-2013-0001. PMID 23735479.

[48] "Coriolus Versicolor". American Cancer Society. 1 November 2008. Retrieved 2011-03-01.

[49] "Clinical trials, "mushrooms" as search term". *Clinicaltrials.gov.* US National Institutes of Health, Clinical Trial Registry. June 2014. Retrieved 27 June 2014.

[50] Mussak R, Bechtold T. (2009). *Handbook of Natural Colorants.* New York: Wiley. pp. 183–200. ISBN 0-470-51199-0.

[51] Kulshreshtha S, Mathur N, Bhatnagar P. (2014). "Mushroom as a product and their role in mycoremediation". *AMB Express* **4**: 29. doi:10.1186/s13568-014-0029-8. PMC 4052754. PMID 24949264.

53.8.1 Literature cited

- Ammirati JF, Traquair JA, Horgen PA. (1985). *Poisonous Mushrooms of Canada: Including other Inedible Fungi.* Markham, Ontario: Fitzhenry & Whiteside in cooperation with Agriculture Canada and the Canadian Government Publishing Centre, Supply and Services Canada. ISBN 0-88902-977-6.

- Hall IR, Stephenson SL, Buchanan PK, Yun W, Cole ALJ. (2003). *Edible and Poisonous Mushrooms of the World.* Portland, Oregon: Timber Press. ISBN 0-88192-586-1.

- Stuntz DE, Largent DL, Thiers HD, Johnson DJ, Watling R. (1978). *How to Identify Mushrooms to Genus I.* Eureka, California: Mad River Press. ISBN 0-916422-00-3.

53.9 External links

53.9.1 Identification

- Mushroom Observer, a collaborative mushroom recording and identification project

- An Aid to Mushroom Identification, Simon's Rock College

- Online Edible Wild Mushroom Field Guide

53.9.2   Research associations

- North American Mycological Association
- Pacific Northwest Fungi Online Journal

Chapter 54

Mushroom hunting

"Mushrooming" redirects here. For the 2012 Estonian film, see Mushrooming (film).

Mushroom hunting, **mushrooming**, **mushroom pick-**

Mushroom picking - *Franciszek Kostrzewski*

collection of edible mushrooms from Ukraine

ing, **mushroom foraging**, and similar terms describe the activity of gathering mushrooms in the wild, typically for eating. This is popular in most of Europe, including the Nordic, Baltic, and Slavic countries and the Mediterranean Basin, as well as in Australia, Japan, Korea, Canada, the Indian subcontinent, and the northwestern, northeastern, Midwestern and Appalachian United States.

54.1   Identifying mushrooms

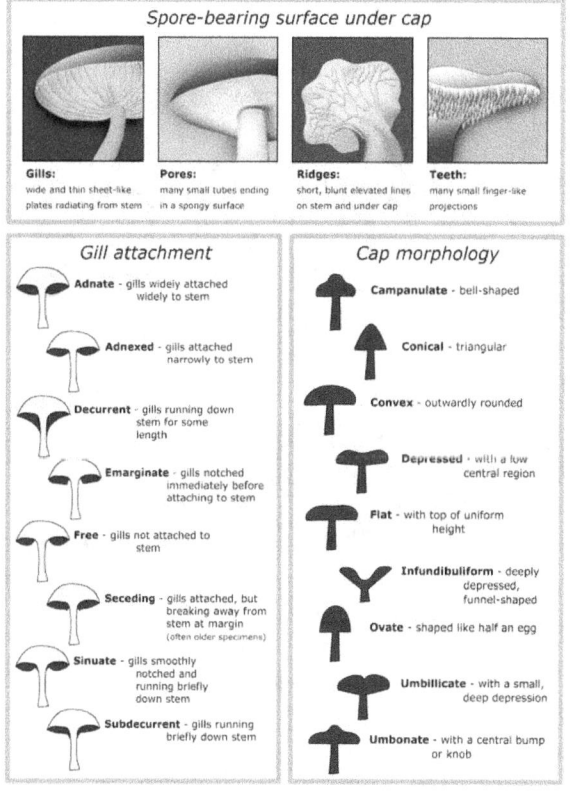

Morphological characteristics of the caps of mushroom, such as those illustrated in the above chart, are essential for correct mushroom identification.

A large number of mushroom species are favored for eating by mushroom hunters. The king bolete is a popular delicacy. Sulphur shelf (also known as chicken mushroom and chicken of the woods) is often gathered because it occurs in bulk, recurs year after year, is easily identified, and has a wide variety of culinary uses. Pine mushrooms, chanterelles, morels, oyster mushrooms, puffballs and polypores are among the most popular types of mushrooms to gather, most of these being fairly simple to properly identify by anyone with practice. Much more care, education, and experience is typically required to make a positive identification of many species, however, and as such, few collect from more dangerous groups, such as *Amanita*, which include some of the most toxic mushrooms in existence.

Many field guides on mushrooms are available, but the ability to identify and prepare edible mushrooms is often passed down through generations, especially in the Slavic countries.

Identification is not the only element of mushroom hunting that takes practice; knowing where and when to search does as well. Most mushroom species require very specific conditions. Some only grow at the base of a certain type of tree, for example. Finding a desired species that is known to grow in a certain region can be a challenge.

54.1.1 Safety issues

Clitocybe rivulosa *is an example of a deadly mushroom species sometimes misidentified as an edible species.*

For more details on this topic, see Mushroom poisoning and List of deadly fungi.

A Czech adage warns that "všechny houby jsou jedlé, ale některé jenom jednou." Translated, that "every mushroom is edible, but some only once." Some mushrooms are deadly or extremely hazardous when consumed. Some that are not deadly can nevertheless cause permanent organ damage. The literature strongly advises:

- That only positively identified mushrooms should be eaten

- That mushrooms be identified a second time during preparation and to cook them unless it can be verified that the species can be eaten raw

- That mushroom types not be combined

- That a sample of any mushroom not well-experienced with be retained for analysis in case of poisoning

- Familiarity with information about deadly mushrooms that are look-alikes of edible ones, as "deadly twins" differ regionally.

- Not gathering mushrooms that are difficult to identify. This applies especially to the mushrooms of the genus *Amanita* and *Cortinarius* and "little brown mushrooms".

- Consuming only a small amount the first time a new species is tried. People react differently to different mushrooms, and all mushroom species can cause an adverse reaction in a few individuals, even the common champignon.*[1]

54.1.2 Little brown mushrooms

Inocybe lacera *is a typical little brown mushroom, and is easily identifiable only by distinctive microscopic features.*

"Little brown mushrooms" (or LBMs) refers to any of a large number of small, dull-coloured agaric species, with few macromorphological uniquely distinguishing characteristics.*[2] As a result, LBMs typically range from difficult to impossible for mushroom hunters to identify. Experienced mushroomers may discern more subtle identifying traits that help narrow the mushroom down to a particular genus or group of species, but exact identification of LBMs often requires close examination of microscopic characteristics plus a certain degree of familiarity or specialization in that particular group.

For mycologists, LBMs are the equivalent of LBJs ("little brown jobs") and DYCs ("damned yellow composites") that are the bane of ornithologists and botanists, respectively.

"Big white mushroom" (or BWM) is also sometimes used to describe groups of difficult to identify larger and paler agarics, many of which are in the genus *Clitocybe*.

54.1.3 Psychotropics

Psilocybe semilanceata *is hunted for its psychotropic properties.*

For more details on this topic, see Psilocybin mushrooms.

The *Amanita muscaria*'s psychotropic properties have been traditionally used by shamans in Siberia in their rituals. However, its use for such purposes today is very rare, despite the mushroom's abundance. Instead, the *Psilocybe semilanceata*, being the only psilocybin-containing mushroom common in Slavic countries, is sought after for its hallucinogenic properties, the latter being more desirable with fewer side effects than those of *A. muscaria*. The use of *P. semilanceata* is however significantly hindered by its small size, requiring larger quantities and being hard to spot. Other Psilocybe species are abundant in the American south and west, as well as Mexico, where they have been used by traditional shamans for centuries. In the west, one can often find mushroom pickers in cow pastures in a stereotypical stoop looking in the grass for psilocybes. This can be quite dangerous, as many species grow in pastures and amateurs often misidentify psilocybes.

- *Amanita muscaria* (Мухомор Красный [Mukhomor Krasniy] - Red Fly-Killer; *Fly Agaric, Toadstool*)

- *Psilocybe semilanceata* (Псилоциба Сосочковидная [Psilotsiba Sosochkovidnaya] - Nipple-Like Psylocybe; *Liberty Cap*)

54.2   Regional importance

Locals are selling mushrooms and berries collected in the Dainava Forest, Lithuania

Forest-picked mushrooms at a Ukrainian market in Kolomyia, Ukraine

- In the United States mushroom picking is popular in the Appalachian area and on the west coast from San Francisco Bay northward, in northern California, Oregon and Washington, and in many other regions.

- British enthusiasts today enjoy an extended average picking season of 75 days compared to just 33 in the 1950s.[*][3]

- In Slavic countries and Baltic countries, mushroom picking is a common family activity. After a heavy rain during the mushroom season whole families often venture into the nearest forest, picking bucketfuls of mushrooms, which are cooked and eaten for dinner upon return (mostly like omelette with eggs or fried on butter) or alternatively dried or marinated for later consumption.

54.2.1 Festivals

The popularity of mushroom picking in some parts of the world has led to mushroom festivals. The festivals are usually between September and October, depending on the mushrooms available in a particular region. Festivals in North America include:

- Aerie Resort on Vancouver Island Great Fall Mushroom Hunt

- Bamfield, Vancouver Island Bamfield Mushroom festival

- Boyne City, Michigan Annual National Morel Mushroom Festival[*][4]

- Buena Vista, Colorado Buena Vista Heritage's Mushroom Festival

- Washington's Long Beach Peninsula Wild Mushroom Celebration

- Lake Quinault Lodge in Washington's Olympic National Forest Quinault Rain Forest Mushroom Festival

- Mendocino County (North of San Francisco)--Mushroom Festival

- Madisonville, Texas Mushroom Festival[*][5]

- Telluride, Colorado Telluride Mushroom Festival[*][6]

- Kennett Square, Pennsylvania Mushroom Festival

- Girdwood, Alaska Fungus Fair

- Muscoda, Wisconsin Morel Mushroom Festival

- Eugene, Oregon Mushroom Festival

- Richmond, Missouri Mushroom Festival

- New Plymouth, Taranaki, New Zealand-Mushroom Ball

54.2.2 Radiation

Nuclear fallout from the Chernobyl disaster is an important issue concerning mushroom picking in Europe. Due to the wide spread of their mycelium, mushrooms tend to accumulate more radioactive caesium-137 than surrounding soil and other organisms. State agencies (e.g. Bellesrad in Belarus) monitor and analyze the degree of radionuclide accumulation in various wild species of plants and animals. In particular, Bellesrad claims that Svinushka (*Paxillus* ssp.), Maslenok (*Suillus* ssp.), Mokhovik (*Xerocomus* ssp.), and

Horkushka (*Lactarius rufus*) are the worst ones in this respect. The safest one is Opyonok Osyenniy (*Armillaria mellea*). *See also: Russian joke.*

This is an issue not only in Poland, Belarus, Ukraine and Russia: the fallout also reached western Europe, and until recently the German government discouraged people gathering certain mushrooms.

54.3 Guidelines for mushroom picking

54.3.1 Poisonous mushrooms commonly confused with edible ones

See also: Mushroom poisoning § Poisonous species and List of deadly fungi

Many mushroom guidebooks call attention to similarities between species, especially significant if an edible species is similar to, or commonly confused with, one that is potentially harmful.

Examples:

- False chanterelles (*Hygrophoropsis aurantiaca*), as the name suggests, can look like real chanterelles (*Cantharellus cibarius*) to the inexperienced eye. The latter do not have sharp gills, but rather blunt veins on the underside. Misidentification in this case is not likely to prove significantly dangerous, as false chanterelles are considered edible, but unpleasant tasting. Mild symptoms have been reported from consuming them.[*][7] Conversely, the Jack O'Lantern mushroom is often mistaken for a chanterelle, and it is potently toxic.

- True morels are distinguished from false morels (*Gyromitra spp.* and *Verpa spp.*). False morels have caps attached at the top of the stalk, while true morels have a honeycombed cap and a single, continuous hollow chamber within.

- Immature *Chlorophyllum molybdites* can be confused with edible *Agaricus* mushrooms.

- Immature puffballs are generally edible, but care must be taken to avoid species such as *Scleroderma citrinum* and immature *Amanita*s. These can be identified by cutting a puffball in half and looking for a dark reticulated gleba or the articulated, nonhomogeneous structures of a gilled mushroom, respectively.

- Highly poisonous *Conocybe filaris* and some *Galerina* species can resemble *Psilocybe*, and the species are observed growing alongside each other. *Psilocybe* species are not deadly but contain the alkaloids psilocybin and psilocin, known to cause hallucinogenic effects; therefore it is often sought for use as a recreational psychedelic drug.

54.3.2 Eating poisonous species

There are treatments to reduce or eliminate the toxicity of certain (but not all) poisonous species to the point where they may be edible.[8] For instance, false morels are deadly poisonous when eaten raw or incorrectly prepared, but their toxins can be reduced by a proper method of parboiling. Prepared in this way, this mushroom is widely used and considered a delicacy in many European countries, although recent research suggests that there may still be long-term health consequences from eating it.[9]

54.4 Commonly gathered mushrooms

A large hen of the woods (Maitake) specimen found in New York state.

Commonly gathered species, grouped by their order taxa, are as follows: **mushroom species mentioned in each group are listed at the end of the paragraph using the following convention:** Latin name (common English names, if any).

54.4.1 Agaricaceae

The *Macrolepiota* genus, usually the *Macrolepiota procera*, and, to a lesser extent, the *M. rhacodes* are highly regarded,

especially in Europe, being very palatable and very large, with specimens of *M. procera* as high as one metre being reported.

- *Agaricus bisporus* also known as the table or button mushroom. Sales of this mushroom in 1996 reached $209 million in Canada.[10] Another well known mushroom known as the *portobello* is a large brown strain of this fungus.

- *Coprinus comatus* (shaggy ink cap) decomposes into ink, and hence are prepared soon after picking and only young specimens are collected. While being a general mushroom hunting guideline, the avoidance of specimens growing in areas with high pollution is especially important with this family, as it is a very effective pollutant absorber.

- *Macrolepiota procera* (parasol mushroom)

54.4.2 Amanitaceae

While the family of Amanitas are approached with extreme caution, as it contains the lethal *Amanita phalloides* and *Amanita virosa*, those confident in their skills often pick the *Amanita rubescens*, which is highly prized in Europe and to a much lesser extent in Russia, accounted by some not to superior taste, but to its relation to the *Amanita caesarea*, which is not found in Russia, but was considered a delicacy worthy of the emperor in Ancient Rome.

- *Amanita rubescens* (European blusher)

- *Amanita caesarea* (Caesar's mushroom)

54.4.3 Boletaceae

A collection of Boletus edulis

This order is often viewed as the order of "noble" mushrooms, containing few poisonous species, identifiable with relative ease, and having superior palatability. The most notable species is the *Boletus edulis*, the "mushroom king", an almost legendary, relatively rare mushroom, edible in almost any (even raw) form, and commonly considered *the* best-tasting mushroom. (It is common to confuse the Russian name, literally "white mushroom", with champignons, often known in English as "white mushrooms".)

- *Boletus edulis* (*Hřib Smrkový, Dubák, Borowik szlachetny, Porcino, King Bolete, Cep, Steinpilz*)

The *Leccinum* genus includes two well-known mushroom species named after the trees they can usually be found next to. The *Leccinum aurantiacum* (as well as the *Leccinum versipelle*), found under aspen trees, and the *Leccinum scabrum* (as well as the *L. holopus*), found under birch trees. The secondary mentioned species, are significantly different in cap colour only. Both types are very sought after, being highly palatable, while more common than the B. edulis.

- *Leccinum aurantiacum* (red-capped scaber stalk)

- *Leccinum scabrum* (birch bolete)

The *Suillus* genus, characterised by its slimy cap, is another prized mushroom, the *Suillus luteus* and *Suillus granulatus* being its most common varieties, and while abundant in some parts of Eurasia, is a rare occurrence in others. It is easy to identify and very palatable.

- *Suillus* (klouzek, slippery Jack, butter mushroom)

The *Xerocomus* genus is generally considered a less desirable (though mostly edible) mushroom group, due to common abundant mould growth on their caps, which can make them poisonous. The *Xerocomus badius*, however is an exception, being moderately sought after, especially in Europe. Some scientific classifications now consider species in the *Xerocomus* genus as members of *Boletus*.

- *Xerocomus* (mossiness mushroom)

- *Xerocomus badius* (*hřib hnědý*)

54.4.4   Cantharellaceae

The *Cantharellus cibarius*, a common and popular mushroom, especially in Europe, is a choice edible and unique mushroom. It is very rarely infested by worms or larvae, has a unique appearance, and when rotting, the decomposed parts are easily distinguishable and separable from those that are edible.

Chanterelles

- *Cantharellus cibarius* (chanterelle, yellow chanterelle, pfifferling)

54.4.5   Helvellaceae

The *Gyromitra esculenta* is considered poisonous, but can be consumed if dried and stored for over a year, according to Slavic literature, and can be used to supplement or replace morel (see *Morchellaceae* below) mushrooms, while Western literature claims that even the fumes of the mushroom are dangerous. It is similar to morels both in appearance and palatability.

- *Gyromitra esculenta* (false morel, beefsteak morel, lorchel)

54.4.6   Morchellaceae

The morel, *Morchella esculenta* is highly prized in Western Europe, India and North America. It is significantly less prized in Slavic countries where, like the *Gyromitra esculenta*, is considered marginally edible with mediocre palatability. Boiling the mushroom and discarding the water is often recommended.

A basket of morels

- *Morchella esculenta* (morel, yellow morel)

54.4.7 Lactarius

Members of the genus *Lactarius*, as the name suggests, lactate a milky liquid when wounded and are often scoffed upon by Western literature. The *Lactarius deliciosus* is however regarded as one of the most palatable mushrooms in Slavic culture, comparable to the *Boletus edulis*. Also considered as similarly palatable are the species *Lactarius necator* and particularly *Lactarius resimus*. Thermal treatment may however be necessary in some cases. Slightly less appealing due to its bitter taste is the *Lactarius pubescens*.

- *Lactarius deliciosus* (saffron milk-cap)

- *Lactarius resimus* (pepper cap)

- *Lactarius necator* (black pepper cap)

- *Lactarius pubescens* (wooly milk-cap)

54.4.8 Russulaceae

The Russula family includes over 750 species and is one of the most common and abundant mushrooms in Eurasia. Their cap colours include red, brown, yellow, blue and green and can be easily spotted. The *Russula vesca* species,

one of the many red-capped varieties, is one of the most common, is reasonably palatable and can be eaten raw. The edible Russulas have a mild taste, compared to many inedible or poisonous species that have a strong hot or bitter taste. The *Russula emetica* (the sickener) is known to cause gastrointestinal upset and has a very hot taste when a small bit is placed on the tongue. Due to their abundance they are however often regarded as an inferior mushroom for hunting.

- *Russula vesca* (Russula)

54.4.9 Tricholomataceae

- *Armillaria* (honey mushroom, shoestring rot). The genus *Armillaria*, with the popular species *A. gallica* and *A. mellea*, being so similar that they are rarely differentiated, are palatable, highly abundant mushrooms. Generally found on decaying tree stumps, they grow in very large quantities and are easy to spot and identify, arguably reducing the fun and challenge in mushroom hunting.

- *Pleurotus ostreatus* (oyster mushroom). It is the most commonly picked tree-dwelling mushroom and is often also artificially cultivated for sale in grocery stores. This sturdy mushroom can be quite palatable when young. Growing these mushrooms at home can be a profitable enterprise and some Russians engage in the activity.

Matsutake, *the highly sought-after pine mushroom, found in coniferous forests in Hiroshima in autumn*

- *Tricholoma matsutake* - = syn. *T. nauseosum*, the rare red pine mushroom that has a very fine aroma. Its fragrance is both sweet and spicy. They grow under trees and are usually concealed under fallen leaves and the duff layer. It forms a symbiotic relationship with the roots of a limited number of tree species. In Japan it is most commonly associated with Japanese red pine. However, in the Pacific Northwest it is found in coniferous forests of Douglas fir, noble fir, sugar pine, and Ponderosa pine. Farther south, it is also associated with hardwoods, namely tanoak and madrone forests. The Pacific Northwest and other similar temperate regions along the Pacific Rim also hold great habitat producing these and other quality wild mushrooms. In 1999, N. Bergius and E. Danell reported that Swedish (*Tricholoma nauseosum*) and Japanese matsutake (*T. matsutake*) are the same species. The report aroused the import from Northern Europe to Japan because of the comparable flavor and taste. Matsutake are difficult to find and are therefore very expensive. Moreover, domestic productions of Matsutake in Japan have been sharply reduced over the last fifty years due to a pine nematode *Bursaphelenchus xylophilus*, and it has influenced the price a great deal. The annual harvest of matsutake in Japan has since further decreased. The price for matsutake in the Japanese market is highly dependent on quality, availability and origin. The Japanese matsutake at the beginning of the season, which is the highest grade, can go up to $2000 per kilogram, while the average value for imported matsutake from China, Europe, and the United States is only about $90 per kilogram.*[11]

- The *Tricholoma magnivelare* is a prized mushroom in North America. British Columbia exports large quantities of this mushroom overseas to Asia where it is in high demand.*[12]

54.5 See also

- Edible mushroom

- Medicinal mushrooms

- Mushroom poisoning

54.6 References

[1] Ho, Marco H. K.; Hill, David J. (2006). "White button mushroom food hypersensitivity in a child". *Journal of Paediatrics and Child Health* **42** (9): 555–556. doi:10.1111/j.1440-1754.2006.00922.x.

[2] IMA Glossary: LBM

[3] Gange, A. C.; Gange, E. G.; Sparks, T. H.; Boddy, L. (2007). "Rapid and recent changes in fungal fruiting patterns". *Science* **317** (5821): 71. doi:10.1126/science.1137489.

[4] http://www.morelfest.com/

[5] http://www.texasmushroomfestival.com/

[6] http://www.telluridemushroomfest.org/

[7] Arora, David. Mushrooms Demystified. Ten Speed Press, 1986

[8]

[9] Michael W. Beug, Marilyn Shaw, and Kenneth W. Cochran. Thirty plus Years of Mushroom Poisoning: Summary of the Approximately 2,000 Reports in the NAMA Case Registry.

[10] Hans E. Gruen

[11] Finding and Preparing The Elusive Matsutake Mushroom

[12] Tom Volk's Fungus of the Month for September 2000

54.7 Further reading

- *Edible and Medicinal Mushrooms of New England and Eastern Canada* (2009) ISBN 978-1-55643-795-3 (1-55643-795-1)

- *Edible Wild Mushrooms of North America: A Field-to-kitchen Guide* (1992) ISBN 978-0-292-72080-0

- *Mushrooms of Northeastern North America* (1997) ISBN 0-8156-0388-6

- *All That the Rain Promises, and More* (1991) ISBN 0-89815-388-3

- *Mushrooms Demystified: A Comprehensive Guide to the Fleshy Fungi* (1986) ISBN 0-89815-169-4

- *100 Edible Mushrooms: With Tested Recipes* (2007) ISBN 0-472-03126-0

- *North American Mushrooms: A Field Guide to Edible and Inedible Fungi* (2006) ISBN 0-7627-3109-5

- *How to Identify Edible Mushrooms* (2007) ISBN 0-00-725961-1

- *Mushrooming Without Fear* (2007) ISBN 1-60239-160-2

- *The Mushroom Rainbow: Only the most delicious or deadly mushrooms sorted by color* (2011) ISBN 978-0-9869409-0-3 (0986940909)

54.8 External links

- The Mushroom Forager

- Mushroom database

- Bill O'Dea Mushroom hunts and festivals in Ireland, Europe and Turkey

- Mushroom-collecting.com A Maine and New England Edible and Medicinal Mushroom Resource

- Polish Mushroom Hunters Club - Photos, Find reports etc.

- David Fischer's AmericanMushrooms.com

- Current Mushroom Find Reports With Photos

- Mushroom descriptive fields

- List of North American Mycological Societies

- A finnish amateur site with photos (partly in English)

- Fungi of Kaluga region (Russian) Site of Russian mushroom hunter with a lot of photos (with titles in English).

- Mushrooms found in continental climate areas of Russia (Russian)

- MushroomExpert.com

- Czech amateur site with photos (Czech)

- Canadian Government Mushroom Picking Site

Chapter 55

Myrobalans

The common name **myrobalan** can refer to several unrelated fruit-bearing plant species:

- Cherry plum, myrobalan plum (*Prunus cerasifera*)

- Amla, Amalaki, emblic myrobalans (*Phyllanthus emblica*)

- Bibhitaki, Belliric myrobalans (*Terminalia bellirica*)

- Haritaki, Chebulic myrobalans (*Terminalia chebula*)

- Arjuna, Arjun myrobalans (*Terminalia arjuna*)

55.1   See also

- Mirabelle

- Triphala

Chapter 56

Natural dye

Naturally dyed skeins made with madder root, Colonial Williamsburg, VA

Natural dyes are dyes or colorants derived from plants, invertebrates, or minerals. The majority of natural dyes are **vegetable dyes** from plant sources roots, berries, bark, leaves, and wood and other organic sources such as fungi and lichens.

Archaeologists have found evidence of textile dyeing dating back to the Neolithic period. In China, dyeing with plants, barks and insects has been traced back more than 5,000 years.[1] The essential process of dyeing changed little over time. Typically, the dye material is put in a pot of water and then the textiles to be dyed are added to the pot, which is heated and stirred until the color is transferred. Textile fibre may be dyed before spinning ("dyed in the wool"), but most textiles are "yarn-dyed" or "piece-dyed" after weaving. Many natural dyes require the use of chemicals called mordants to bind the dye to the textile fibres; tannin from oak galls, salt, natural alum, vinegar, and ammonia from stale urine were used by early dyers. Many

mordants, and some dyes themselves, produce strong odors, and large-scale dyeworks were often isolated in their own districts.

Throughout history, people have dyed their textiles using common, locally available materials, but scarce dyestuffs that produced brilliant and permanent colors such as the natural invertebrate dyes, Tyrian purple and crimson kermes, became highly prized luxury items in the ancient and medieval world. Plant-based dyes such as woad (*Isatis tinctoria*), indigo, saffron, and madder were raised commercially and were important trade goods in the economies of Asia and Europe. Across Asia and Africa, patterned fabrics were produced using resist dyeing techniques to control the absorption of color in piece-dyed cloth. Dyes such as cochineal and logwood (*Haematoxylum campechianum*) were brought to Europe by the Spanish treasure fleets, and the dyestuffs of Europe were carried by colonists to America.

The discovery of man-made synthetic dyes in the mid-19th century triggered a long decline in the large-scale market for natural dyes. Synthetic dyes, which could be produced in large quantities, quickly superseded natural dyes for the commercial textile production enabled by the industrial revolution, and unlike natural dyes, were suitable for the synthetic fibres that followed. Artists of the *Arts and Crafts Movement* preferred the pure shades and subtle variability of natural dyes, which mellow with age but preserve their true colors, unlike early synthetic dyes,[1] and helped ensure that the old European techniques for dyeing and printing with natural dyestuffs were preserved for use by home and craft dyers. Natural dyeing techniques are also preserved by artisans in traditional cultures around the world.

In the early 21st century, the market for natural dyes in the fashion industry is experiencing a resurgence.[2] Western consumers have become more concerned about the health and environmental impact of synthetic dyes in manufacturing and there is a growing demand for products that use natural dyes. The European Union, for example, has encouraged Indonesian batik cloth producers to switch to natural dyes to improve their export market in Europe.[*][3]

56.1 Dyes in use in the fashion industry

Oaxaca artisan Fidel Cruz Lazo dying yarn for rug making

Fibre content determines the type of dye required for a fabric:

- Cellulose fibres: cotton, linen, hemp, ramie, bamboo, rayon

- Protein fibres: wool, angora, mohair, cashmere, silk, soy, leather, suede

Cellulose fibres require fibre-reactive, direct/substantive, and vat dyes, which are colourless, soluble dyes fixed by light and/or oxygen. Protein fibres require vat, acid, or indirect/mordant dyes, that require a bonding agent. Each synthetic fibre requires its own dyeing method, for example, nylon requires acid, disperse and pigment dyes, rayon acetate requires disperse dyes, and so on. The types of natural dyes currently in use by the global fashion industry include:[*][4]

56.1.1 Animal

- Cochineal insect (red)

- Cow urine (Indian yellow)

- Lac insect (red, violet)

- Murex snail (purple)

- Octopus/Cuttlefish (sepia brown)

56.1.2 Plant

- Catechu or Cutch tree (brown)

- Gamboge tree resin (dark mustard yellow)

- Himalayan rubhada root (yellow)

- Indigofera plant (blue)

- Kamala tree (red)

- Larkspur[*][5] plant (yellow)

- Madder root (red, pink, orange)

- Myrabolan fruit (yellow, green, black)

- Pomegranate peel (yellow)

- Weld herb (yellow)

56.2 Origins

Colors in the "ruddy" range of reds, browns, and oranges are the first attested colors in a number of ancient textile sites ranging from the Neolithic to the Bronze Age across the Levant, Egypt, Mesopotamia and Europe, followed by evidence of blues and then yellows, with green appearing somewhat later. The earliest surviving evidence of textile dyeing was found at the large Neolithic settlement at Çatalhöyük in southern Anatolia, where traces of red dyes, possible from ochre (iron oxide pigments from clay), were found.[*][6] Polychrome or multicolored fabrics seem to have been developed in the 3rd or 2nd millennium BCE.[*][6] Textiles with a "red-brown warp and an ochre-yellow weft" were discovered in Egyptian pyramids of the Sixth Dynasty (2345–2180 BCE).[*][7]

The chemical analysis that would definitively identify the dyes used in ancient textiles has rarely been conducted, and even when a dye such as indigo blue is detected it is impossible to determine which of several indigo-bearing plants was used.[*][8] Nevertheless, based on the colors of surviving textile fragments and the evidence of actual dyestuffs found in archaeological sites, reds, blues, and yellows from plant sources were in common use by the late Bronze Age and Iron Age.[*][9]

56.3 Processes

For more details on this topic, see Glossary of dyeing terms. The essential process of dyeing requires soaking the ma-

Dyeing wool cloth, 1482, from British Library Royal MS 15.E.iii, f. 269.

terial containing the dye (the *dyestuff*) in water, adding the textile to be dyed to the resulting solution (the *dyebath*), and bringing the solution to a simmer for an extended period, often measured in days or even weeks, stirring occasionally until the color has evenly transferred to the textiles.[*][10]

Some dyestuffs, such as indigo and lichens, will give good color when used alone; these dyes are called *direct dyes* or *substantive dyes*. The majority of plant dyes, however, also require the use of a mordant, a chemical used to "fix" the color in the textile fibres. These dyes are called *adjective dyes*. By using different mordants, dyers can often obtain a variety of colors and shades from the same dye. Fibres or cloth may be pretreated with mordants, or the mordant may be incorporated in the dyebath. In traditional dyeing, the common mordants are vinegar, tannin from oak bark, sumac or oak galls, ammonia from stale urine, and wood-ash liquor or potash (potassium carbonate) made by leaching wood ashes and evaporating the solution.[*][11][*][12][*][13]

We shall never know by what chances primitive man discovered that salt, vinegar from fermenting fruit, natural alum, and stale urine helped to fix and enhance the colours of his yarns, but for many centuries these four substances were used as mordants.[*][11]

Salt helps to "fix" or increase "fastness" of colors, vinegar improves reds and purples, and the ammonia in stale urine assists in the fermentation of indigo dyes.[*][11] Natural alum (aluminum sulfate) is the most common metallic salt mordant, but tin (stannous chloride), copper (cupric sulfate), iron (ferrous sulfate, called *copperas*) and chrome (potassium dichromate) are also used. Iron mordants "sadden" colors, while tin and chrome mordants brighten colors. The iron mordants contribute to fabric deterioration, referred to as "dye rot". Additional chemicals or *alterants* may be applied after dying to further alter or reinforce the colors.[*][14][*][15][*][16]

A dye-works with baskets of dyestuffs, skeins of dyed yarn, and heated vats for dyeing.

Textiles may be dyed as raw fibre (*dyed in the fleece* or *dyed in the wool*), as spun yarn (*dyed in the hank* or *yarn-dyed*), or after weaving (*piece-dyed*).[*][17] Mordants often leave residue in wool fibre that makes it difficult to spin, so wool was generally dyed after spinning, as yarn or woven cloth. Indigo, however, requires no mordant, and cloth manufacturers in medieval England often dyed wool in the fleece with the indigo-bearing plant woad and then dyed the cloth again after weaving to produce deep blues, browns, reds, purples, blacks, and tawnies.[*][18][*][19]

In China, Japan, India, Pakistan, Nigeria, Gambia, and other parts of West Africa and southeast Asia, patterned silk and cotton fabrics were produced using resist dyeing techniques in which the cloth is printed or stenciled with starch or wax, or tied in various ways to prevent even penetration of the dye when the cloth is piece-dyed. The Chinese *ladao* process is dated to the 10th century; other traditional techniques include tie-dye, batik, Rōketsuzome, katazome, bandhani and leheria.*[20]

The mordants used in dyeing and many dyestuffs themselves give off strong and unpleasant odors, and the actual process of dyeing requires a good supply of fresh water, storage areas for bulky plant materials, vats which can be kept heated (often for days or weeks) along with the necessary fuel, and airy spaces to dry the dyed textiles. Ancient large-scale dyeworks tended to be located on the outskirts of populated areas, on windy promontories.*[21]

56.4 Common dyestuffs

The Hunt of the Unicorn Tapestry, dyed with weld (yellow), madder (red), and woad (blue).

56.4.1 Reds and pinks

A variety of plants produce red dyes, including a number of lichens, henna, alkanet or dyer's bugloss (*Alkanna tinctoria*), asafoetida and dyer's madder *Rubia tinctorum*.*[22] Madder and related plants of the genus *Rubia* are native to many temperate zones around the world, and were already used as sources of good red dye, such as rose madder, in prehistory. Madder has been identified on linen in the tomb of Tutankhamun,*[22] and Pliny the Elder records madder growing near Rome.*[23] Madder was a dye of commercial importance in Europe, being cultivated in the Netherlands and France to dye the red coats of military uniforms until the market collapsed following the development of syn-

thetic alizarin dye in 1869. Madder was also used to dye the "hunting pinks" of Great Britain.*[23]

Turkey red was a strong, very fast red dye for cotton obtained from madder root via a complicated multistep process involving "sumac and oak galls, calf's blood, sheep's dung, oil, soda, alum, and a solution of tin." *[24] Turkey red was developed in India and spread to Turkey. Greek workers familiar with the methods of its production were brought to France in 1747, and Dutch and English spies soon discovered the secret. A sanitized version of Turkey red was being produced in Manchester by 1784, and roller-printed dress cottons with a Turkey red ground were fashionable in England by the 1820s.*[25]*[26]

Munjeet or Indian madder (*Rubia cordifolia*) is native to the Himalayas and other mountains of Asia and Japan. Munjeet was an important dye for the Asian cotton industry and is still used by craft dyers in Nepal.*[27]

Puccoon or bloodroot (*Sanguinaria canadensis*) is a popular red dye among Southeastern Native American basketweavers.*[28] Choctaw basketweavers additionally use sumac for red dye.*[29] Coushattas artists from Texas and Louisiana used the water oak (*Quercus nigra* L.) to produce red.*[30]

A delicate rose color in Navajo rugs comes from fermented prickly pear cactus fruit, *Opuntia polyacantha*.*[31] Navajo weavers also use rainwater and red dirt to create salmon-pink dyes.*[32]

56.4.2 Oranges

Dyes that create reds and yellows can also yield oranges. Navajo dyers create orange dyes from one-seeded juniper, *Juniperus monosperma*, Navajo tea, *Thelesperma gracile*,*[33] or alder bark.*[34]

56.4.3 Yellows

Yellow dyes are "about as numerous as red ones",*[35] and can be extracted from saffron, pomegranate rind, turmeric, safflower, onionskins, and a number of weedy flowering plants.*[35]*[36] Limited evidence suggests the use of weld (*Reseda luteola*), also called mignonette or dyer's rocket*[37] before the Iron Age,*[35] but it was an important dye of the ancient Mediterranean and Europe and is indigenous to England.*[38] Two brilliant yellow dyes of commercial importance in Europe from the 18th century are derived from trees of the Americas: quercitron from the inner bark of Eastern Black Oak (*Quercus velutina*),] native to eastern North America and fustic from the dyer's mulberry tree (*Maclura tinctoria*) of the West Indies and Mexico.*[36]

In rivercane basketweaving among Southeastern Woodlands tribes in the Americas, butternut (*Juglans cinerea*) and yellow root (*Xanthorhiza simplicissima*) provide a rich yellow color.*[28] Chitimacha basket weavers have a complex formula for yellow that employs a dock plant (most likely *Rumex crispus*) for yellow.*[39] Navajo artists create yellow dyes from small snake-weed, brown onion skins, and rubber plant (*Parthenium incanum*). Rabbitbush (*Chrysothamnus*) and rose hips produce pale, yellow-cream colored dyes.*[34]

56.4.4 Greens

If plants that yield yellow dyes are common, plants that yield green dyes are rare. Both woad and indigo have been used since ancient times in combination with yellow dyes to produce shades of green. Medieval and Early Modern England was especially known for its green dyes. The dyers of Lincoln, a great cloth town in the high Middle Ages, produced the Lincoln green cloth associated with Robin Hood by dyeing wool with woad and then overdyeing it yellow with weld or dyer's greenweed (*Genista tinctoria*), also known as dyer's broom.*[40] Woolen cloth mordanted with alum and dyed yellow with dyer's greenweed was overdyed with woad and, later, indigo, to produce the once-famous Kendal green.*[38] This in turn fell out of fashion in the 18th century in favor of the brighter Saxon green, dyed with indigo and fustic.

Soft olive greens are also achieved when textiles dyed yellow are treated with an iron mordant. The dull green cloth common to the Iron Age Halstatt culture shows traces of iron, and was possibly colored by boiling yellow-dyed cloth in an iron pot.*[41] Indigenous peoples of the Northwest Plateau in North America used lichen to dye corn husk bags a sea green.*[42]

Navajo textile artist Nonabah Gorman Bryan developed a two-step process for creating green dye. First the Churro wool yarn is dyed yellow with sagebrush, *Artemisia tridentata*, and then it is soaked in black dye afterbath.*[31] Red onion skins are also used by Navajo dyers to produce green.*[34]

56.4.5 Blues

Blue colorants around the world were derived from indigo dye-bearing plants, primarily those in the genus *Indigofera*, which are native to the tropics. The primary commercial indigo species in Asia was true indigo (*Indigofera tinctoria*). India is believed to be the oldest center of indigo dyeing in the Old World. It was a primary supplier of indigo dye to Europe as early as the Greco-Roman era. The association of India with indigo is reflected in the Greek word for the

dye, which was *indikon* (ινδικόν). The Romans used the term *indicum*, which passed into Italian dialect and eventually into English as the word *indigo*.*[43]

In Central and South America, the important blue dyes were Añil (*Indigofera suffruticosa*) and Natal indigo (*Indigofera arrecta*).*[43]*[44]

In temperate climates including Europe, indigo was obtained primarily from woad (*Isatis tinctoria*), an indigenous plant of Assyria and the Levant which has been grown in Northern Europe over 2,000 years, although from the 18th century it was mostly replaced by superior Indian indigo imported by the British East India Company. Woad was carried to New England in the 17th century and used extensively in America until native stands of indigo were discovered in Florida and the Carolinas. In Sumatra, indigo dye is extracted from some species of *Marsdenia*. Other indigo-bearing dye plants include dyer's knotweed (*Polygonum tinctorum*) from Japan and the coasts of China, and the West African shrub *Lonchocarpus cyanescens*.*[43]*[45]

Natural dyeing with Indigo, Jaipur (Rajasthan, India)

- Badshah Miyan is a traditional dyer from Jaipur who specializes in traditional natural dyeing methods

- A traditional brass container used to dye cloth in quantity

- Indigo stains skin a deep blue for many days

56.4.6 Purples

In medieval Europe, purple, violet, murrey and similar colors were produced by dyeing wool with woad or indigo in the fleece and then piece-dyeing the woven cloth with red dyes, either the common madder or the luxury dyes kermes and cochineal. Madder could also produce purples when used with alum. Brazilwood also gave purple shades with vitriol (sulfuric acid) or potash.*[46]

Choctaw artists traditionally used maple (*Acer* sp.) to create lavender and purple dyes.*[29] Purples can also be derived from lichens, and from the berries of White Bryony from the northern Rocky Mountain states and mulberry (*morus nigra*) (with an acid mordant).*[47]

56.4.7 Browns

Cutch is an ancient brown dye from the wood of acacia trees, particularly *Acacia catechu*, used in India for dyeing cotton. Cutch gives gray-browns with an iron mordant and olive-browns with copper.*[48]

Black walnut (*Juglans nigra*) is used by Cherokee artists to produce a deep brown approaching black.[*][28] Today black walnut is primarily used to dye baskets but has been used in the past for fabrics and deerhide. Juniper, *Juniperus monosperma*, ashes provide brown and yellow dyes for Navajo people,[*][31] as do the hulls of wild walnuts (*Juglans major*).[*][49] Khaki, which translates a Hindustani word signifying "soil-coloured", was introduced into British uniforms in India, which were dyed locally with a dye prepared from the native mazari palm *Nannorrhops*.

56.4.8 Greys and blacks

Choctaw dyers use maple (*Acer* sp.) for a grey dye.[*][29] Navajo weavers create black from mineral yellow ochre mixed with pitch from the piñon tree(*Pinus edulis*) and the three-leaved sumac (*Rhus trilobata*).[*][31] They also produce a cool grey dye with blue flower lupine and a warm grey from Juniper mistletoe (*Phoradendron juniperinum*).[*][34]

56.4.9 Lichen

Dye-bearing lichen produce a wide range of greens,[*][42] oranges, yellows, reds, browns, and bright pinks and purples. The lichen *Rocella tinctoria* was found along the Mediterranean Sea and was used by the ancient Phoenicians. In recent times, lichen dyes have been an important part of the dye traditions of Wales, Ireland, Scotland, and among native peoples of the southwest and Intermontane Plateaus of the United States.[*][42] Scottish lichen dyes include cudbear (also called archil in England and litmus in the Netherlands), and crottle.[*][50]

56.4.10 Fungi

Miriam C. Rice, (1918 2010) of Mendocino, California, pioneered research into using various mushrooms for natural dyes. She discovered mushroom dyes for a complete rainbow palette. Swedish and American mycologists, building upon Rice's research, have discovered sources for true blues (*Sarcodon squamosus*) and mossy greens (*Hydnellum geogenium*).[*][51] *Hypholoma fasciculare* provides a yellow dye, and fungi such as *Phaeolus schweinitzii* and *Pisolithus tinctorius* are used in dyeing textiles and paper.[*][52]

56.5 Luxury dyestuffs

From the second millennium BC to the 19th century, a succession of rare and expensive natural dyestuffs came in and out of fashion in the ancient world and then in Europe. In many cases the cost of these dyes far exceeded the cost of the wools and silks they colored, and often only the finest grades of fabrics were considered worthy of the best dyes.

56.5.1 Royal purple

The premier luxury dye of the ancient world was Tyrian purple or royal purple, a purple-red dye which is extracted from several genera of sea snails, primarily the spiny dye-murex *Murex brandaris* (currently known as *Bolinus brandaris*). Murex dye was greatly prized in antiquity because it did not fade, but instead became brighter and more intense with weathering and sunlight. Murex dyeing may have been developed first by the Minoans of East Crete or the West Semites along the Levantine coast, and heaps of crushed murex shells have been discovered at a number of locations along the eastern Mediterranean dated to the mid-2nd millennium BC. The classical dye known as Phoenician Red was also derived from murex snails.[*][53]

Murex dyes were fabulously expensive – one snail yields but a single drop of dye – and the Roman Empire imposed a strict monopoly on their use from the reign of Alexander Severus (AD 225–235) that was maintained by the succeeding Byzantine Empire until the Early Middle Ages.[*][54] The dye was used for imperial manuscripts on purple parchment, often with text in silver or gold, and *porphyrogenitos* or "born in the purple" was a term for Byzantine offspring of a reigning Emperor. The color matched the increasing rare purple rock porphyry, also associated with the imperial family.

56.5.2 Crimson and scarlet

Tyrian purple retained its place as the premium dye of Europe until it was replaced "in status and desirability"[*][55] by the rich crimson reds and scarlets of the new silk-weaving centers of Italy, colored with kermes. Kermes is extracted from the dried unlayed eggs of the insect *Kermes vermilio* or *Kermococcus vermilio* found on species of oak (especially the Kermes oak of the Mediterranean region). The dye is of ancient origin; jars of kermes have been found in a Neolithic cave-burial at Adaoutse, Bouches-du-Rhône.[*][56] Similar dyes are extracted from the related insects *Porphyrophora hamelii* (Armenian cochineal) of the Caucasus region, *Porphyrophora polonica* (Polish cochineal or Saint John's blood) of Eastern Europe, and the lac-producing insects of India, Southeast Asia, China, and Tibet.[*][57][*][58][*][59]

When kermes-dyed textiles achieved prominence around the mid-11th century, the dyestuff was called "grain" in all Western European languages because the desiccated eggs resemble fine grains of wheat or sand.[*][54] Textiles dyed with kermes were described as *dyed in the grain.*[58] Woollens were frequently dyed in the fleece with woad and then piece-dyed in kermes, producing a wide range colors from blacks and grays through browns, murreys, purples, and sanguines.*[58] By the 14th and early 15th century, brilliant *full grain* kermes scarlet was "by far the most esteemed, most regal" color for luxury woollen textiles in the Low Countries, England, France, Spain and Italy.*[54]

Cochineal (*Dactylopius coccus*) is a scale insect of Central and North America from which the crimson-coloured dye carmine is derived. It was used by the Aztec and Maya peoples. Moctezuma in the 15th century collected tribute in the form of bags of cochineal dye.*[60] Soon after the Spanish conquest of the Aztec Empire cochineal began to be exported to Spain, and by the seventeenth century it was a commodity traded as far away as India. During the colonial period the production of cochineal (in Spanish, *grana fina*) grew rapidly. Produced almost exclusively in Oaxaca by indigenous producers, cochineal became Mexico's second most valued export after silver.*[61] Cochineal produces purplish colors alone and brilliant scarlets when mordanted with tin, and cochineal, which produced a stronger dye and could thus be used in smaller quantities, replaced kermes dyes in general use in Europe from the 17th century.*[62]*[63]

56.5.3 The rise of formal black

During the course of the 15th century, the civic records show brilliant reds falling out of fashion for civic and high-status garments in the Duchy of Burgundy in favor of dark blues, greens, and most importantly of all, black.*[64]*[65] The origins of the trend for somber colors are elusive, but are generally attributed to the growing influence of Spain and possibly the importation of Spanish merino wools. The trend spread in the next century: the Low Countries, German states, Scandinavia, England, France, and Italy all absorbed the sobering and formal influence of Spanish dress after the mid-1520s.*[65]*[66]

Producing fast black in the Middle Ages was a complicated process involving multiple dyeings with woad or indigo followed by mordanting, but at the dawn of Early Modern period, a new and superior method of dyeing black dye reached Europe via Spanish conquests in the New World. The new method used logwood (*Haematoxylum campechianum*), a dyewood native to Mexico and Central America. Although logwood was poorly received at first, producing a blue inferior to that of woad and indigo, it was discovered to produce a fast black in combination with a ferrous sulfate(copperas) mordant.*[55]*[65] Despite changing fashions in color, logwood was the most widely used dye by the 19th century, providing the sober blacks of formal and

mourning clothes.*[55]

56.6 Decline and rediscovery

The first synthetic dyes were discovered in the mid-19th century, starting with William Henry Perkin's mauveine in 1856, an aniline dye derived from coal tar.*[67] Alizarin, the red dye present in madder, was the first natural pigment to be duplicated synthetically, in 1869,*[68] leading to the collapse of the market for naturally grown madder.*[69] The development of new, strongly colored aniline dyes followed quickly: a range of reddish-purples, blues, violets, greens and reds became available by 1880. These dyes had great affinity for animal fibres such as wool and silk. The new colors tended to fade and wash out, but they were inexpensive and could be produced in the vast quantities required by textile production in the industrial revolution. By the 1870s commercial dyeing with natural dyestuffs was fast disappearing.*[67]

At the same time the Pre-Raphaelite artist and founding figure of the Arts and Crafts movement William Morris took up the art of dyeing as an adjunct to his manufacturing business, the design firm of Morris & Co. Always a medievalist at heart, Morris loathed the colors produced by the fashionable aniline dyes. He spent much of his time at his Staffordshire dye works mastering the processes of dyeing with plant materials and making experiments in the revival of old or discovery of new methods. One result of these experiments was to reinstate indigo dyeing as a practical industry and generally to renew the use of natural dyes like madder which had been driven almost out of use by the commercial success of the anilines. Morris saw dyeing of wools, silks, and cottons as the necessary preliminary to the production of woven and printed fabrics of the highest excellence; and his period of incessant work at the dye-vat (1875–76) was followed by a period during which he was absorbed in the production of textiles (1877–78), and more especially in the revival of carpet- and tapestry-weaving as fine arts. Morris & Co. also provided naturally dyed silks for the embroidery style called art needlework.*[70]*[71]

Scientists continued to search for new synthetic dyes that would be effective on cellulose fibres like cotton and linen, and that would be more colorfast on wool and silk than the early anilines. Chrome or mordant dyes produced a muted but very fast color range for woollens. These were followed by acid dyes for animal fibres (from 1875) and the synthesis of indigo in Germany in 1880. The work on indigo led to the development of a new class of dyes called vat dyes in 1901 that produced a wide range of fast colors for vegetable fibres.*[72] Disperse dyes were introduced in 1923 to color the new textiles of cellulose acetate, which could not be colored with any existing dyes. Today disperse dyes are the only effective means of coloring many synthetics. Reactive dyes for both wool and cotton were introduced in the mid-1950s, and are used both in commercial textile production and in craft dyeing.*[72]

In America, synthetic dyes became popular among a wide range of Native American textile artists; however, natural dyes remained in use, as many textile collectors prefer natural dyes over synthetics. Today, dyeing with natural materials is often practiced as an adjunct to handspinning, knitting and weaving.*[73] It remains a living craft in many traditional cultures of North America, Africa, Asia, and the Scottish Highlands.*[74]

56.7   Notes

[1] Goodwin (1982), p. 11.

[2] Calderin, Jay (2009). *Form, Fit, Fashion*. Rockport. p. 125. ISBN 978-1-59253-541-5.

[3] Faizal, Elly Burhaini (October 29, 2011). "Indonesia told to produce more 'green' products" . *The Jakarta Post*. Retrieved November 9, 2011.

[4] Calderin, Jay (2009). *Form, Fit, Fashion*. Rockport. pp. 125–26. ISBN 978-1-59253-541-5.

[5] "Larkspur" . *Conservation and Art Materials Encyclopedia Online (CAMEO)*. Museum of Fine Arts Boston. Retrieved August 31, 2015.

[6] Barber (1991), pp. 223–25.

[7] Rogers, Penelope Walton, "Dyes and Dyeing" . In Jenkins (2003), pp. 25–29.

[8] Barber (1991), pp. 227, 237.

[9] Barber (1991), pp. 228–29.

[10] Goodwin (1982), pp. 29–31.

[11] Goodwin (1982), p. 12

[12] Goodwin (1982), p. 32

[13] Kerridge (1988), pp. 165–66

[14] Barber (1991), pp. 235–36, 239.

[15] Goodwin (1982), pp. 32–34.

[16] "informational site about antique quilts and vintage textiles including article and an interactive chat group" . Quilt History. Retrieved 2013-04-22.

[17] Kerridge (1988), pp. 15, 16, 135

[18] Munro (2003), p. 210

[19] Kerridge (1988), pp. 15, 17

[20] Gillow & Sentence (1999), pp. 122–36

[21] Barber (1991), p. 239.

[22] Barber (1991), p. 232.

[23] Goodwin (1982), pp. 64–65.

[24] Goodwin (1982), p. 65.

[25] Tozer & Levitt (1983), pp. 29–30.

[26] Cannon & Cannon (2002), p. 76

[27] Cannon & Cannon (2002), p. 80.

[28] Chancey, 37

[29] Chancey 51

[30] Chancey 66

[31] Bryan and Young 5

[32] Bryan and Young 62

[33] Bryan and Young 6

[34] "12 Plant Navajo Dye Chart, Craftperson: Maggie Begay." *Bair's Indian Trading Company.* (retrieved 9 Jan 2011)

[35] Barber (1991), p. 233

[36] Goodwin (1982), pp. 60–63

[37] *Reseda luteola*

[38] Goodwin (1982), p. 63

[39] Chancey 47

[40] Cannon & Cannon (2002), p. 110.

[41] Barber (1991), p. 228.

[42] Chancey 173

[43] See Indigo dye.

[44] Goodwin (1982), p. 70

[45] Goodwin (1982), pp. 11, 70–76

[46] Kerridge (1988), pp. 166–67

[47] Goodwin (1982), pp. 107, 112

[48] Goodwin (1982), p. 60.

[49] Bryan and Young 61

[50] Goodwin (1982), pp. 87–92.

[51] Beebee, Dorothy M. "Mushrooms for Color." 30 Nov 2010 (retrieved 9 Jan 2011)

[52] Beebee, Dorothy M. "Miriam C. Rice and Mushrooms for Color." *Turkey Red Journal.* Fall 2008 (retrieved 9 Jan 2011)

[53] Barber (1991), pp. 228–29.

[54] Munro, John H. "The Anti-Red Shift – To the Dark Side: Colour Changes in Flemish Luxury Woollens, 1300–1500". In Netherton and Owens-Crocker (2007), pp. 56–57.

[55] Schoeser (2007), p. 118

[56] Barber (1991), pp. 230–31

[57] Barber (1991), p. 231

[58] Munro, John H. "Medieval Woollens: Textiles, Technology, and Organisation". In Jenkins (2003), pp. 214–15.

[59] Goodwin (1982), p. 56

[60] Threads In Tyme, LTD. "Time line of fabrics". Archived from the original on 2005-10-28. Retrieved 5 January 2011.

[61] Behan, Jeff. "The bug that changed history". Retrieved 5 January 2011.

[62] Schoeser (2007), pp. 121, 248

[63] Barber (1982), p. 55.

[64] Munro (2007), pp. 76–77.

[65] , Munro (2007), pp. 87–93.

[66] Boucher, François: *20,000 Years of Fashion*, pp. 219, 244

[67] Thompson & Thompson (1987), p. 10

[68] Hans-Samuel Bien, Josef Stawitz, Klaus Wunderlich "Anthraquinone Dyes and Intermediates" in Ullmann's Encyclopedia of Industrial Chemistry 2005 Wiley-VCH, Weinheim: 2005. doi:10.1002/14356007.a02355.

[69] Goodwin (1982), p. 65

[70] *Dictionary of National Biography* (1901), "William Morris"

[71] Parry (1983), pp. 36–46.

[72] Thompson & Thompson (1987), pp. 11–12

[73] Goodwin (1982), pp. 7–8.

[74] Gillow & Sentance (1999), pp. 118–19.

56.8 References

The section on William Morris incorporates text from the Dictionary of National Biography, *supplemental volume 3 (1901), a publication now in the public domain.*

- Barber, E. J. W. (1991). *Prehistoric Textiles*. Princeton University Press. ISBN 0-691-00224-X.

- Boucher, François (1966). *20,000 Years of Fashion*. Harry Abrams.

- Bryan, Nonabah Gorman & Young, Stella (2002). *Navajo Natives Dyes: Their Preparation and Use*. Dover Publications. ISBN 978-0-486-42105-6.

- Cannon, John; Cannon, Margaret (2002). *Dye Plants and Dyeing* (2nd ed.). A&C Black. ISBN 978-0-7136-6374-7.

- Cardon, Dominique (2007). *Natural Dyes: Sources, Tradition, Technology and Science*. Archetype Publications. ISBN 1-904982-00-X.

- Chancey, Jill R., ed. (2005). *By Native Hands: Woven Treasures from the Lauren Rogers Museum of Art*. Lauren Rogers Museum of Art. ISBN 0-935903-07-0.

- Gillow, John; Sentance, Bryan (1999). *World Textiles*. Bulfinch. ISBN 0-8212-2621-5.

- Goodwin, Jill (1982). *A Dyer's Manual*. Pelham. ISBN 0-7207-1327-7.

- Hofenk de Graaf, Judith (2004). *The Colourful Past: Origins, Chemistry and Identification of Natural Dyestuffs*. Abegg-Stiftung and Archetype Publications. ISBN 1-873132-13-1.

- Jenkins, David, ed. (2003). *The Cambridge History of Western Textiles* (2 vols.). Cambridge University Press. ISBN 0-521-34107-8.

- Kerridge, Eric (1988). *Textile Manufactures in Early Modern England*. Manchester University Press. ISBN 978-0-7190-2632-4.

- Netherton, Robin, & Owen-Crocker, Gale R., eds. (2007). *Medieval Clothing and Textiles* **3**. Boydell Press. ISBN 978-1-84383-291-1.

- Parry, Linda (1983). *William Morris Textiles*. Viking Press. ISBN 0-670-77074-4.

- Schoeser, Mary (2007). *Silk*. Yale University Press. ISBN 0-300-11741-8.

- Thompson, Frances; Thompson, Tony (1987). *Synthetic Dyeing: for Spinners, Weavers, Knitters and Embroiderers*. David & Charles. ISBN 0-7153-8874-6.

- Tozer, Jane; Levitt, Sarah (1983). *Fabric of Society: a Century of People and their Clothes 1770–1870*. Laura Ashley Press. ISBN 0-9508913-0-4.

56.9 External links

- International Mushroom Dye Institute
- Cochineal Master's Thesis-History and Uses

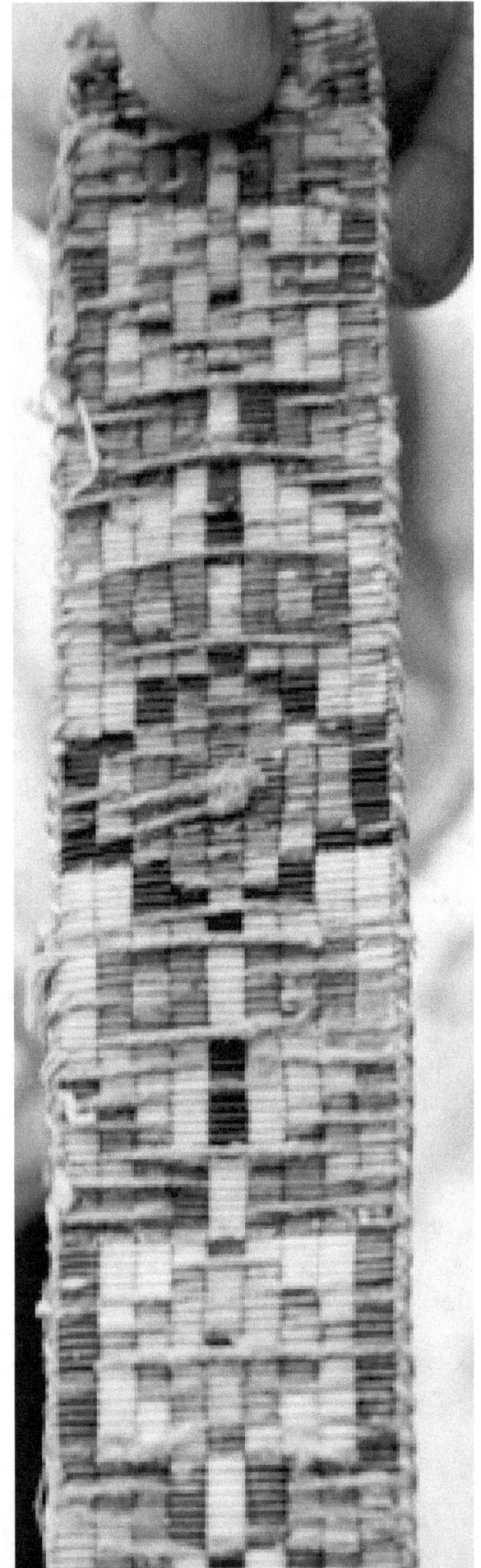

Byzantine Emperor Justinian I clad in Tyrian purple, 6th-century mosaic at Basilica of San Vitale, Ravenna, Italy

Indigo-dyed and discharge-printed textile, William Morris, 1873

Chapter 57

Natural rubber

"Rubber" and "India rubber" redirect here. For other uses, see Rubber (disambiguation).
This article is about the polymeric material "natural rubber". For man-made rubber materials, see Synthetic rubber.
Natural rubber, also called **India rubber** or *caoutchouc*,

Latex being collected from a tapped rubber tree, Cameroon.

as initially produced, consists of polymers of the organic compound isoprene, with minor impurities of other organic compounds plus water. Malaysia is a leading producer of rubber. Forms of polyisoprene that are used as natural rubbers are classified as elastomers. Natural rubber is used by many manufacturing companies for the production of rubber products. Currently, rubber is harvested mainly in the form of the latex from certain trees. The latex is a sticky,

Rubber tree plantation of Thailand.

milky colloid drawn off by making incisions into the bark and collecting the fluid in vessels in a process called "tapping". The latex then is refined into rubber ready for commercial processing. Natural rubber is used extensively in many applications and products, either alone or in combination with other materials. In most of its useful forms, it has a large stretch ratio and high resilience, and is extremely waterproof.[1]

57.1 Uses

Compared to vulcanized rubber, uncured rubber has relatively few uses. It is used for cements; for adhesive, insulating, and friction tapes; and for crepe rubber used in insulating blankets and footwear. Vulcanized rubber, on the other hand, has numerous applications. Resistance to abrasion makes softer kinds of rubber valuable for the treads of vehicle tires and conveyor belts, and makes hard rubber valuable for pump housings and piping used in the handling of abrasive sludge.

The flexibility of rubber is often used in hose, tires, and rollers for a wide variety of devices ranging from domestic clothes wringers to printing presses; its elasticity makes it suitable for various kinds of shock absorbers and for specialized machinery mountings designed to reduce vibration. Being relatively impermeable to gases, rubber is useful in the manufacture of articles such as air hoses, balloons, balls, and cushions. The resistance of rubber to water and to the action of most fluid chemicals has led to its use in rainwear, diving gear, and chemical and medicinal tubing, and as a lining for storage tanks, processing equipment, and railroad tank cars. Because of their electrical resistance, soft rubber goods are used as insulation and for protective gloves, shoes, and blankets; hard rubber is used for articles such as telephone housings, parts for radio sets, meters, and other electrical instruments. The coefficient of friction of rubber, which is high on dry surfaces and low on wet surfaces, leads to the use of rubber both for power-transmission belting and for water-lubricated bearings in deep-well pumps.

57.2   Varieties

The major commercial source of natural rubber latex is the Pará rubber tree (*Hevea brasiliensis*), a member of the spurge family, *Euphorbiaceae*. This species is widely used because it grows well under cultivation and a properly managed tree responds to wounding by producing more latex for several years.

Congo rubber, formerly a major source of rubber, came from vines in the genus *Landolphia* (*L. kirkii*, *L. heudelotis*, and *L. owariensis*).[2] These cannot be cultivated, and the intense drive to collect latex from wild plants was responsible for many of the atrocities committed under the Congo Free State.

Many other plants produce forms of latex rich in isoprene polymers, though not all produce usable forms of polymer as easily as the Pará rubber tree does; some of them require more elaborate processing to produce anything like usable rubber, and most are more difficult to tap. Some produce other desirable materials, for example gutta-percha (*Palaquium gutta*)[3] and chicle from *Manilkara* species. Others that have been commercially exploited, or at least have shown promise as sources of rubber, include the rubber fig (*Ficus elastica*), Panama rubber tree (*Castilla elastica*), various spurges (*Euphorbia* spp.), lettuce (*Lactuca* species), the related *Scorzonera tau-saghyz*, various *Taraxacum* species, including common dandelion (*Taraxacum officinale*) and Russian dandelion (*Taraxacum kok-saghyz*), and perhaps most importantly for its hypoallergenic properties, guayule (*Parthenium argentatum*). The term **gum rubber** is sometimes applied to the tree-obtained version of natural rubber in order to distinguish it from the synthetic version.[1]

57.3   Discovery of commercial potential

The Para rubber tree is indigenous to South America. Charles Marie de La Condamine is credited with introducing samples of rubber to the *Académie Royale des Sciences* of France in 1736.[4] In 1751, he presented a paper by François Fresneau to the Académie (eventually published in 1755) which described many of the properties of rubber. This has been referred to as the first scientific paper on rubber.[4] In England, Joseph Priestley, in 1770, observed that a piece of the material was extremely good for rubbing off pencil marks on paper, hence the name "rubber". Later, it slowly made its way around England.

South America remained the main source of the limited amounts of latex rubber used during much of the 19th century. The trade was well protected and exporting seeds from Brazil was said to be a capital offense, although there was no law against it. Nevertheless, in 1876, Henry Wickham smuggled 70,000 Pará rubber tree seeds from Brazil and delivered them to Kew Gardens, England. Only 2,400 of these germinated after which the seedlings were then sent to India, Ceylon (Sri Lanka), Indonesia, Singapore, and British Malaya. Malaya (now Peninsular Malaysia) was later to become the biggest producer of rubber. In the early 1900s, the Congo Free State in Africa was also a significant source of natural rubber latex, mostly gathered by forced labor. Liberia and Nigeria also started production of rubber.

In India, commercial cultivation of natural rubber was introduced by the British planters, although the experimental efforts to grow rubber on a commercial scale in India were initiated as early as 1873 at the Botanical Gardens, Calcutta. The first commercial *Hevea* plantations in India were established at Thattekadu in Kerala in 1902.

In Singapore and Malaya, commercial production of rubber was heavily promoted by Sir Henry Nicholas Ridley, who served as the first Scientific Director of the Singapore Botanic Gardens from 1888 to 1911. He distributed rubber seeds to many planters and developed the first technique for tapping trees for latex without causing serious harm to the tree.[5] Because of his very fervent promotion of this crop, he is popularly remembered by the nickname "Mad Ridley".[6]

57.4   Properties

Rubber exhibits unique physical and chemical properties. Rubber's stress-strain behavior exhibits the Mullins effect and the Payne effect, and is often modeled as hyperelastic. Rubber strain crystallizes.

Rubber latex

Due to the presence of a double bond in each repeat unit, natural rubber is susceptible to vulcanisation and sensitive to ozone cracking.

The two main solvents for rubber are turpentine and naphtha (petroleum). The former has been in use since 1764 when François Fresnau made the discovery. Giovanni Fabbroni is credited with the discovery of naphtha as a rubber solvent in 1779. Because rubber does not dissolve easily, the material is finely divided by shredding prior to its immersion.

An ammonia solution can be used to prevent the coagulation of raw latex while it is being transported from its collection site.

57.4.1 Elasticity

Main article: Rubber elasticity

On a microscopic scale, relaxed rubber is a disorganized cluster of erratically changing wrinkled chains. In stretched rubber, the chains are almost linear. The restoring force is due to the preponderance of wrinkled conformations over more linear ones. For the quantitative treatment see ideal chain, for more examples see entropic force.

Cooling below the glass transition temperature still permits local conformational changes but a reordering is practically impossible because of the larger energy barrier for the concerted movement of longer chains. "Frozen" rubber's elasticity is low and strain results from small changes of bond lengths and angles: this caused the *Challenger* disaster, when the American Space Shuttle's flattened o-rings failed to relax to fill a widening gap.[7] The glass transition is fast and reversible: the force resumes on heating.

The parallel chains of stretched rubber are susceptible to

crystallization. This takes some time because turns of twisted chains have to move out of the way of the growing crystallites. Crystallization has occurred, for example, when, after days, an inflated toy balloon is found withered at a relatively large remaining volume. Where it is touched, it shrinks because the temperature of the hand is enough to melt the crystals.

Vulcanization of rubber creates disulfide bonds between chains, which limits the degrees of freedom and results in chains that tighten more quickly for a given strain, thereby increasing the elastic force constant and making the rubber harder and less extensible.

57.5 Chemical makeup

Chemical structure of cis-polyisoprene, the main constituent of natural rubber. Synthetic cis-polyisoprene and natural cis-polyisoprene are derived from different precursors, isopentenyl pyrophosphate and isoprene.

Latex is the polymer cis-1,4-polyisoprene – with a molecular weight of 100,000 to 1,000,000 daltons. Typically, a small percentage (up to 5% of dry mass) of other materials, such as proteins, fatty acids, resins, and inorganic materials (salts) are found in natural rubber. Polyisoprene can also be created synthetically, producing what is sometimes referred to as "synthetic natural rubber", but the synthetic and natural routes are completely different.*[1] Some natural rubber sources, such as gutta-percha, are composed of trans-1,4-polyisoprene, a structural isomer that has similar, but not identical, properties.

Natural rubber is an elastomer and a thermoplastic. Once the rubber is vulcanized, it will turn into a thermoset. Most rubber in everyday use is vulcanized to a point where it shares properties of both; i.e., if it is heated and cooled, it is degraded but not destroyed.

The final properties of a rubber item depend not just on the polymer, but also on modifiers and fillers, such as carbon black, factice, whiting, and a host of others.

57.5.1 Biosynthesis

Rubber particles are formed in the cytoplasm of specialized latex-producing cells called laticifers within rubber plants.[8] Rubber particles are surrounded by a single phospholipid membrane with hydrophobic tails pointed inward. The membrane allows biosynthetic proteins to be sequestered at the surface of the growing rubber particle, which allows new monomeric units to be added from outside the biomembrane, but within the lacticifer. The rubber particle is an enzymatically active entity that contains three layers of material, the rubber particle, a biomembrane, and free monomeric units. The biomembrane is held tightly to the rubber core due to the high negative charge along the double bonds of the rubber polymer backbone.[9] Free monomeric units and conjugated proteins make up the outer layer. The rubber precursor is isopentenyl pyrophosphate (an allylic compound), which elongates by Mg^{2+}-dependent condensation by the action of rubber transferase. The monomer adds to the pyrophosphate end of the growing polymer.[10] The process displaces the terminal high-energy pyrophosphate. The reaction produces a cis polymer. The initiation step is catalyzed by prenyltransferase, which converts three monomers of isopentenyl pyrophosphate into farnesyl pyrophosphate.[11] The farnesyl pyrophosphate can bind to rubber transferase to elongate a new rubber polymer.

The required isopentenyl pyrophosphate is obtained from the mevalonate pathway, which derives from acetyl-CoA in the cytosol. In plants, isoprene pyrophosphate can also be obtained from 1-deox-D-xyulose-5-phosphate/2-C-methyl-D-erythritol-4-phosphate pathway within plasmids.[12] The relative ratio of the farnesyl pyrophosphate initiator unit and isoprenyl pyrophosphate elongation monomer determines the rate of new particle synthesis versus elongation of existing particles. Though rubber is known to be produced by only one enzyme, extracts of latex have shown numerous small molecular weight proteins with unknown function. The proteins possibly serve as cofactors, as the synthetic rate decreases with complete removal.[13]

57.6   Current sources

Close to 28 million tons of rubber were produced in 2013, of which approximately 44% was natural. Since the bulk of the rubber produced is of the synthetic variety, which is derived from petroleum, the price of natural rubber is determined, to a large extent, by the prevailing global price of crude oil.[14][15][16] Today, Asia is the main source of natural rubber, accounting for about 94% of output in 2005. The three largest producing countries, Thailand, Indonesia (2.4 million tons)[17] and Malaysia, together account for around 72% of all natural rubber production. Natural rubber is not cultivated widely in its native continent of South America due to the existence of South American leaf blight, and other natural predators of the rubber tree.

Rubber is generally cultivated in large plantations. The image shows a coconut shell used in collecting latex, in plantations in Kerala, India

57.6.1   Cultivation

Rubber latex is extracted from rubber trees. The economic life period of rubber trees in plantations is around 32 years up to 7 years of immature phase and about 25 years of productive phase.

The soil requirement of the plant is generally well-drained, weathered soil consisting of laterite, lateritic types, sedimentary types, nonlateritic red, or alluvial soils.

The climatic conditions for optimum growth of rubber trees are:

- Rainfall of around 250 cm evenly distributed without any marked dry season and with at least 100 rainy days per year

- Temperature range of about 20 to 34 °C, with a monthly mean of 25 to 28 °C

- High atmospheric humidity of around 80%

- Bright sunshine, amounting to about 2000 hours per year at the rate of six hours per day throughout the year

- Absence of strong winds

Many high-yielding clones have been developed for commercial planting. These clones yield more than 2,000 kg of dry rubber per hectare per year, when grown under ideal conditions.

57.6.2   Collection

In places such as Kerala, where coconuts are in abundance, the half shell of coconut is used as the collection container in a spiral to the right. For this reason, tapping cuts usually ascend to the left to cut more tubes.

The trees will drip latex for about four hours, stopping as latex coagulates naturally on the tapping cut, thus blocking the latex tubes in the bark. Tappers usually rest and have a meal after finishing their tapping work, then start collecting the liquid "field latex" at about midday. Some trees will continue to drip after the collection and this leads to a small amount of "cup lump"which is collected at the next tapping. The latex that coagulates on the cut is also collected as "tree lace". Tree lace and cup lump together account for 10–20% of the dry rubber produced. Latex that drips onto the ground, "earth scrap", is also collected periodically for processing of low-grade product.

Field coagula

A woman in Sri Lanka in the process of harvesting rubber.

Mixed field coagula.

for the latex, but glazed pottery or aluminium or plastic cups are more common elsewhere. The cups are supported by a wire that encircles the tree. This wire incorporates a spring so it can stretch as the tree grows. The latex is led into the cup by a galvanised "spout"knocked into the bark. Tapping normally takes place early in the morning, when the internal pressure of the tree is highest. A good tapper can tap a tree every 20 seconds on a standard half-spiral system, and a common daily "task" size is between 450 and 650 trees. Trees are usually tapped on alternate or third days, although many variations in timing, length, and number of cuts are used. The latex, which contains 25–40% dry rubber, is in the bark, so the tapper must avoid cutting right through to the wood, else the growing cambial layer will be damaged and the renewing bark will be badly deformed, making later tapping difficult. It is usual to tap a pannel at least twice, sometimes three times, during the tree's life. The economic life of the tree depends on how well the tapping is carried out, as the critical factor is bark consumption. A standard in Malaysia for alternate daily tapping is 25 cm (vertical) bark consumption per year. The latex tubes in the bark ascend

Smallholder's lump at a remilling factory

There are four types of field coagula, "cuplump", "treelace", "smallholders' lump" and "earth scrap". Each has significantly different properties.[18]

Cup lump is the coagulated material found in the collection cup when the tapper next visits the tree to tap it again. It arises from latex clinging to the walls of the cup after the latex was last poured into the bucket, and from late-dripping latex exuded before the latex-carrying vessels of the tree become blocked. It is of higher purity and of greater value than the other three types.

Tree lace is the coagulum strip that the tapper peels off the previous cut before making a new cut. It usually has higher copper and manganese contents than cup lump. Both copper and manganese are pro-oxidants and can lower the physical properties of the dry rubber.

Smallholders' lump is produced by smallholders who collect rubber from trees far away from the nearest factory. Many Indonesian smallholders, who farm paddies in remote areas, tap dispersed trees on their way to work in the paddy fields and collect the latex (or the coagulated latex) on their way home. As it is often impossible to preserve the latex sufficiently to get it to a factory that processes latex in time for it to be used to make high quality products, and as the latex would anyway have coagulated by the time it reached the factory, the smallholder will coagulate it by any means available, in any container available. Some smallholders use small containers, buckets etc., but often the latex is coagulated in holes in the ground, which are usually lined with plastic sheeting. Acidic materials and fermented fruit juices are used to coagulate the latex a form of assisted biological coagulation. Little care is taken to exclude twigs, leaves, and even bark from the lumps that are formed, which may also include tree lace collected by the smallholder.

Earth scrap is the material that gathers around the base of the tree. It arises from latex overflowing from the cut and running down the bark of the tree, from rain flooding a collection cup containing latex, and from spillage from tappers' buckets during collection. It contains soil and other contaminants, and has variable rubber content, depending on the amount of contaminants mixed with it. Earth scrap is collected by the field workers two or three times a year and may be cleaned in a scrap-washer to recover the rubber, or sold off to a contractor who will clean it and recover the rubber. It is of very low quality and under no circumstances should it be included in block rubber or brown crepe.

57.6.3 Processing

The latex will coagulate in the cups if kept for long. The latex has to be collected before coagulation. The collected latex, "field latex", is transferred into coagulation tanks for the preparation of dry rubber or transferred into air-tight containers with sieving for ammoniation. Ammoniation is necessary to preserve the latex in a colloidal state for longer periods of time.

Removing coagulum from coagulating troughs.

Latex is generally processed into either latex concentrate for manufacture of dipped goods or it can be coagulated under controlled, clean conditions using formic acid. The coagulated latex can then be processed into the higher-grade, technically specified block rubbers such as SVR 3L or SVR CV or used to produce Ribbed Smoke Sheet grades.

Naturally coagulated rubber (cup lump) is used in the manufacture of TSR10 and TSR20 grade rubbers. The processing of the rubber for these grades is a size reduction and cleaning process to remove contamination and prepare the material for the final stage of drying.[19]

The dried material is then baled and palletized for storage and shipment in various methods of transportation.

57.6.4 Transportation

Natural rubber latex is shipped from factories in southwest Asia, South America, and North Africa to destinations around the world. As the cost of natural rubber has risen significantly, the shipping methods which offer the lowest cost per unit of weight are preferred. Depending on the destination, warehouse availability, and transportation conditions, some methods are more suitable to certain buyers than others. In international trade, latex rubber is mostly shipped in 20-foot ocean containers. Inside the ocean container, various types of smaller containers are used by factories to store latex rubber.[20]

57.7 Uses

57.7.1 Contemporary manufacturing

Around 25 million tonnes of rubber is produced each year, of which 42 percent is natural rubber. The remainder is synthetic rubber derived from petrochemical sources. Around

Compression molded (cured) rubber boots before the flashes are removed

70 percent of the world's natural rubber is used in tires. The top end of latex production results in latex products such as surgeons' gloves, condoms, balloons and other relatively high-value products. The mid-range which comes from the technically specified natural rubber materials ends up largely in tires but also in conveyor belts, marine products, windshield wipers and miscellaneous rubber goods. Natural rubber offers good elasticity, while synthetic materials tend to offer better resistance to environmental factors such as oils, temperature, chemicals or ultraviolet light and suchlike. "Cured rubber" is rubber which has been compounded and subjected to the vulcanisation process which creates cross-links within the rubber matrix.

57.7.2 Prehistoric uses

The first use of rubber is thought to have been by the Olmecs, who centuries later passed on the knowledge of natural latex from the *Hevea* tree in 1600 BC to the ancient Mayans. They boiled the harvested latex to make a ball for a Mesoamerican ballgame.*[21]

57.7.3 Pre-World War II manufacturing

Other significant uses of rubber are door and window profiles, hoses, belts, gaskets, matting, flooring, and dampeners (antivibration mounts) for the automotive industry. Gloves (medical, household and industrial) and toy balloons are also large consumers of rubber, although the type of rubber used is concentrated latex. Significant tonnage of rubber is used as adhesives in many manufacturing industries and products, although the two most noticeable are the paper and the carpet industries. Rubber is also commonly used to make rubber bands and pencil erasers.

57.7.4 Pre-World War II textile applications

Rubber produced as a fiber, sometimes called 'elastic', has significant value for use in the textile industry because of its excellent elongation and recovery properties. For these purposes, manufactured rubber fiber is made as either an extruded round fiber or rectangular fibers that are cut into strips from extruded film. Because of its low dye acceptance, feel and appearance, the rubber fiber is either covered by yarn of another fiber or directly woven with other yarns into the fabric. In the early 1900s, for example, rubber yarns were used in foundation garments. While rubber is still used in textile manufacturing, its low tenacity limits its use in lightweight garments because latex lacks resistance to oxidizing agents and is damaged by aging, sunlight, oil, and perspiration. Seeking a way to address these shortcomings, the textile industry has turned to neoprene (polymer of chloroprene), a type of synthetic rubber, as well as another more commonly used elastomer fiber, spandex (also known as elastane), because of their superiority to rubber in both strength and durability.

57.8 Vulcanization

Main article: Vulcanization

Natural rubber is often vulcanized, a process by which the rubber is heated and sulfur, peroxide or bisphenol are added to improve resistance and elasticity, and to prevent it from perishing. The development of vulcanization is most closely associated with Charles Goodyear in 1839.*[22] Before World War II era manufacturing, carbon black was often used as an additive to rubber to improve its strength, especially in vehicle tires.

57.9 Allergic reactions

Main article: Latex allergy

Some people have a serious latex allergy, and exposure to natural latex rubber products such as latex gloves can cause anaphylactic shock. The antigenic proteins found in *Hevea* latex may be deliberately reduced (though not eliminated)*[23] through processing.

Latex from non-*Hevea* sources, such as Guayule, can be used without allergic reaction by persons with an allergy to *Hevea* latex.*[24]

Some allergic reactions are not to the latex itself, but from residues of chemicals used to accelerate the cross-linking process. Although this may be confused with an allergy to latex, it is distinct from it, typically taking the form of Type IV hypersensitivity in the presence of traces of specific processing chemicals.*[23]*[25]

57.10 Alternative sources

Dandelion milk has long been known to contain latex. The latex exhibits the same quality as the natural rubber from rubber trees. Yet in the wild types of dandelion, the latex content is low and varies greatly.

In Nazi Germany, research projects tried to use dandelions as a base for rubber production, but failed.*[26]

In 2013, by inhibiting one key enzyme and using modern cultivation methods and optimization techniques, scientists in the Fraunhofer Institute for Molecular Biology and Applied Ecology (IME) in Germany developed a cultivar that is suitable for commercial production of natural rubber.*[27] In collaboration with Continental Tires, IME is building a pilot facility. The first prototype test tires made with blends from dandelion-rubber are scheduled to be tested on public roads over the next few years.

57.11 Microbial degradation

Natural rubber is susceptible to degradation by a wide range of bacteria.*[28]*[29]*[30]*[31]*[32]*[33]*[34]*[35]

57.12 See also

- Akron, Ohio, center of the rubber industry in the US
- Condoms, also called "rubbers"
- Crepe rubber
- Ebonite
- Emulsion dispersion
- Fordlândia, failed attempt to establish a rubber plantation in Brazil
- Reinforced rubber
- Resilin, a rubber substitute
- Rubber seed oil
- Rubber technology
- Stevenson Plan, historical British plan to stabilize rubber prices
- Charles Greville Williams, researched natural rubber being a polymer of the monomer isoprene

57.13 References

57.13.1 Notes

[1] Heinz-Hermann Greve "Rubber, 2. Natural" in *Ullmann's Encyclopedia of Industrial Chemistry*, 2000, Wiley-VCH, Weinheim. doi:10.1002/14356007.a23_225

[2] "Rubber and Other Latex Products" . Retrieved 31 August 2014.

[3] Burns, Bill. "The Gutta Percha Company" . *History of the Atlantic Cable & Undersea Communications*. Retrieved 14 February 2009.

[4] "Charles Marie de la Condamine" . *http://www.bouncing-balls.com/*.

[5] Cornelius-Takahama, Vernon (2001). "Sir Henry Nicholas Ridley" . *Singapore Infopedia*. Retrieved 9 February 2013.

[6] Leng, Dr Loh Wei; Keong, Khor Jin (19 September 2011). "Mad Ridley and the rubber boom" . *Malaysia History*. Retrieved 9 February 2013.

[7] "Casing Joint Design" (PDF). *Report - Investigation of the Challenger Accident*. US Government Printing Office. Retrieved August 29, 2015.

[8] Koyama, Tanetoshi; Steinbüchel, Alexander, eds. (June 2011). "Biosynthesis of Natural Rubber and Other Natural Polyisoprenoids" . *Polyisoprenoids*. Biopolymers **2**. Wiley-Blackwell. pp. 73–81. ISBN 978-3-527-30221-5.

[9] Paterson-Jones, J.C.; Gilliland, M.G.; Van Staden, J. (June 1990). "The Biosynthesis of Natural Rubber" . *Journal of Plant Physiology* **136** (3): 257–263. doi:10.1016/S0176-1617(11)80047-7. ISSN 0176-1617.

[10] Schulze Gronover, Christian; Wahler, Daniela; Prufer, Dirk (5 Jul 2011). "4. Natural Rubber Biosynthesis and Physic-Chemical Studies on Plant Derived Latex". In Magdy, Elnashar. *Biotechnology of Biopolymers*. ISBN 978-953-307-179-4.

[11] Xie, W.; McMahan, C. M.; Distefano, A.J. DeGraw, M. D.; et al. (2008). "Initiation of rubber synthesis: In vitro comparisons of benzophenone-modified diphosphate analogues in three rubber preducing species". *Phytochemistry* **69**: 2539–2545. doi:10.1016/j.phytochem.2008.07.011.

[12] Casey, P. J.; Seabra, M. C. (1996). "Protein Prenyltransferases". *Journal of Biological Chemistry* **271** (10): 5289–5292. doi:10.1074/jbc.271.10.5289.

[13] Kang, H.; Kang, M. Y.; Han, K. H. (2000). "Identification of Natural Rubber and Characterization of Biosynthetic Activity". *Plant Physiol* **123** (3): 1133–1142. doi:10.1104/pp.123.3.1133.

[14] "Overview of the Causes of Natural Rubber Price Volatility". En.wlxrubber.com. 2010-02-01. Retrieved 2013-03-21.

[15] "STATISTICAL SUMMARY OF WORLD RUBBER SITUATION" (PDF). International Rubber Study Group. Nov 2014. Retrieved 2015-02-10.

[16] Short run and long run effects of the world crude oil prices on the Malaysian natural rubber and palm oil export prices

[17] Listiyorini, Eko (2010-12-16). "Rubber Exports From Indonesia May Grow 6%−8% Next Year". bloomberg.com. Archived from the original on 4 November 2012. Retrieved 2013-03-21.

[18] This section has been copied almost verbatim from the public domain UN Food and Agriculture Organization (FAO), ecoport.com article: Cecil, John; Mitchell, Peter; Diemer, Per; Griffee, Peter (2013). "Processing of Natural Rubber, Manufacture of Latex-Grade Crepe Rubber". *ecoport.org*. FAO, Agricultural and Food Engineering Technologies Service. Retrieved March 19, 2013.

[19] Technical Grades and Basis for Grading by ASTM D2227 - Basic Rubber Testing

[20] Transportation of Natural Rubber - Industry Source

[21] "The Mayan-Olmec Connection". Maya12-21-2012.com. 2012-12-21. Retrieved 2013-03-21.

[22] Slack, Charles. "Noble Obsession: Charles Goodyear, Thomas Hancock, and the Race to Unlock the Greatest Industrial Secret of the Nineteenth Century". Hyperion 2002. [ISBN 9780786867899]

[23] "Premarket Notification [510(k)] Submissions for Testing for Skin Sensitization To Chemicals In Natural Rubber Products" (PDF). FDA. Retrieved 22 September 2013.

[24] http://www.fda.gov/forconsumers/consumerupdates/ucm048052.htm

[25] American Latex Allergy Association. "Allergy Fact Sheet".

[26] Autarkie und Ostexpansion: Pflanzenzucht und Agrarforschung im Nationalsozialismus, (agrarian research during the NS regime) Susanne Heim, Wallstein, 2002, ISBN 389244496X

[27] "Making Rubber from Dandelion Juice". *sciencedaily.com*. sciencedaily.com. Retrieved 22 November 2013.

[28] Rook, J.J. (1955). "Microbiological deterioration of vulcanized rubber". *Appl. Microbiol* **3**: 302–309.

[29] Leeang, K.W.H. (1963). "Microbiologic degradation of rubber". *J. Am. Water Works Assoc* **53**: 1523–1535.

[30] Tsuchii, A.; Suzuki, T.; Takeda, K. (1985). "Microbial degradation of natural rubber vulcanizates". *Appl. Environ. Microbiol* **50**: 965–970.

[31] Heisey, R.M.; Papadatos, S. (1995). "Isolation of microorganisms able to metabolize puri¢ed natural rubber". *Appl. Environ. Microbiol* **61**: 3092–3097.

[32] Jendrossek, D.; Tomasi, G.; Kroppenstedt, R.M. (1997). "Bacterial degradation of natural rubber: a privilege of actinomycetes?". *FEMS Microbiology Letters* **150**: 179–188. doi:10.1016/s0378-1097(97)00072-4.

[33] Linos, A. and Steinbuchel, A. (1998) Microbial degradation of natural and synthetic rubbers by novel bacteria belonging to the genus Gordona. Kautsch. Gummi Kunstst. 51, 496-499.

[34] Linos, Alexandros; Steinbuchel, Alexander; Spröer, Cathrin; Kroppenstedt, Reiner M. (1999). "Gordonia polyisoprenivorans sp. nov., a rubber degrading actinomycete isolated from automobile tire". *Int. J. Syst. Bacteriol* **49**: 1785–1791. doi:10.1099/00207713-49-4-1785.

[35] Linos, Alexandros; Reichelt, Rudolf; Keller, Ulrike; Steinbuchel, Alexander (October 1999). "A Gram-negative bacterium, identified as Pseudomonas aeruginosa AL98, is a potent degrader of natural rubber and synthetic cis-1,4-polyisoprene". *FEMS Microbiology Letters* **182**: 155. doi:10.1111/j.1574-6968.2000.tb08890.x.

57.13.2 Bibliography

- Ascherson, Neal. (1963). *The King Incorporated*. Allen & Unwin. ISBN 1-86207-290-6 (*1999 Granta edition*).

- Brydson, J.A. *Rubbery Materials and their Compounds*

- Hobhouse, Henry (2005) [2003]. *Seeds of Wealth: Five Plants That Made Men Rich*. Shoemaker & Hoard. pp. 125–185. ISBN 1-59376-089-2.

- Hochschild, Adam. (1998). *King Leopold's Ghost: A Story of Greed, Terror, and Heroism in Colonial Africa.* Mariner Books. ISBN 0-330-49233-0.

- Morton, Maurice. *Rubber Technology*

- Petringa, Maria. (2006). *Brazza, A Life for Africa.* Bloomington, IN: AuthorHouse. ISBN 978-1-4259-1198-0

Chapter 58

Naval stores

Naval stores are all products derived from pine sap, which are used to manufacture soap, paint, varnish, shoe polish, lubricants, linoleum, and roofing materials.

The term *naval stores* originally applied to the resin-based components used in building and maintaining wooden sailing ships, a category which includes cordage, mask, turpentine, rosin, pitch and tar.

58.1 History

Ships made of wood required a flexible material, insoluble in water, to seal the spaces between planks. Pine pitch was often mixed with fibers like hemp to caulk spaces which might otherwise leak. Crude gum or oleoresin can be collected from the wounds of living pine trees.

58.1.1 Colonial North America

The Royal Navy relied heavily upon naval stores from American colonies, and naval stores were an essential part of the colonial economy. Masts came from the large white pines of New England, while pitch came from the longleaf pine forests of Carolina, which also produced sawn lumber, shake shingles, and staves.*[1]

Naval stores played a role during the American Revolutionary War. As Britain attempted to cripple French and Spanish capacities through blockade, they declared naval stores to be contraband. At the time Russia was Europe's chief producer of naval stores, leading to the seizure of 'neutral' Russian vessels. In 1780 Catherine the Great announced that her navy would be used against anyone interfering with neutral trade, and she gathered together European neutrals in the League of Armed Neutrality. These actions were beneficial for the struggling colonists as the British were forced to act with greater caution.*[2]

The major producers of naval stores in the 19th and 20th century were the United States of America, and France, where Napoleon encouraged planting of pines in areas of sand dunes. In the 1920s the United States exported eleven million gallons of spirits. By 1927, France exported about 20 percent of the world's resin.*[3]

58.2 Separation techniques

Naval stores are recovered from the tall oil byproduct stream of Kraft process pulping of pines in the United States. Tapping of living pines remains common in other parts of the world. Turpentine and pine oil may be recovered by steam distillation of oleoresin or by destructive distillation of pine wood; solvent extraction of shredded stumps and roots has become more common with the availability of inexpensive naphtha. Rosin remains in the still after turpentine and water have boiled off.*[4]

58.3 See also

- Shipbuilding

- Naval stores industry

- Bark hack

58.4 Footnotes

[1] Greene, Jack P, Pursuits of Happiness, Chapel Hill: University of North Carolina Press, 1988, pp. 144-145

[2] Crosby, Alfred W., Jr., America, Russia, Hemp, and Napoleon, Ohio State University Press, 1965, pp. 8

[3] Outland III, Robert B. (2004). *Tapping the pines : the naval stores industry in the American South.* Baton Rouge: Louisiana State University Press. ISBN 9780807129814.

[4] Kent, James A. *Riegel's Handbook of Industrial Chemistry* (Eighth Edition) Van Nostrand Reinhold Company (1983) ISBN 0-442-20164-8 pp.569-573

58.5   External links

- http://www.srs.fs.usda.gov/organization/history/naval_stores.htm

- http://www.maritime.org/conf/conf-kaye-tar.htm

- http://www.fao.org/documents/show_cdr.asp?url_file=/docrep/V6460E/v6460e04.htm

- http://www.hchsonline.org/places/turpentine.html

- http://www.unctv.org/exploringNC/episode308.html

Chapter 59

Nutmeg

For other uses, see Nutmeg (disambiguation).

Nutmeg (also known as *pala* in Indonesia) is one of the

Nutmeg seeds showing "veins"

Mace (red) within nutmeg fruit

two spices – the other being mace – derived from several species of tree in the genus *Myristica*.*[1] The most important commercial species is *Myristica fragrans*, an evergreen tree indigenous to the Banda Islands in the Moluccas (or Spice Islands) of Indonesia.

Nutmeg is the seed of the tree, roughly egg-shaped and about 20 to 30 mm (0.8 to 1.2 in) long and 15 to 18 mm (0.6 to 0.7 in) wide, and weighing between 5 and 10 g (0.2 and 0.4 oz) dried, while mace is the dried "lacy" reddish covering or aril of the seed. The first harvest of nutmeg trees takes place 7–9 years after planting, and the trees reach full production after twenty years. Nutmeg is usually used in powdered form. This is the only tropical fruit that is the source of two different spices, obtained from different parts of the plant. Several other commercial products are also produced from the trees, including essential oils, extracted oleoresins, and nutmeg butter.

59.1   Botany and cultivation

Myristica fragrans *tree in Goa, India*

The common or fragrant nutmeg, *Myristica fragrans*, is native to the Banda Islands in the Moluccas, Indonesia. It is also cultivated on Penang Island in Malaysia, in the Caribbean, especially in Grenada, and in Kerala, a state in southern India. Other species used to adulterate the spice include Papuan nutmeg *M. argentea* from New Guinea, and

Nutmegs on a tree in Kerala, India

M. malabarica from India. In the 17th-century work *Hortus Botanicus Malabaricus*, Hendrik van Rheede records that Indians learned the usage of nutmeg from the Indonesians through ancient trade routes.

Nutmeg trees are dioecious plants which are propagated sexually and asexually, the latter being the standard. Sexual propagation by seedling yields 50% male seedlings, which are unproductive. As there is no reliable method of determining plant sex before flowering in the sixth to eighth year, and sexual propagation bears inconsistent yields, grafting is the preferred method of propagation. Epicotyl grafting, approach grafting, and patch budding have proved successful, with epicotyl grafting being the most widely adopted standard. Air-layering, or marcotting, is an alternative though not preferred method because of its low (35-40%) success rate.

59.2 Culinary uses

Nutmeg and mace have similar sensory qualities, with nutmeg having a slightly sweeter and mace a more delicate flavour. Mace is often preferred in light dishes for the bright orange, saffron-like hue it imparts. Nutmeg is used for flavouring many dishes, usually in ground or grated form, and is best grated fresh in a nutmeg grater.

In Penang cuisine, dried, shredded nutmeg rind with sugar coating is used as toppings on the uniquely Penang *ais kacang*. Nutmeg rind is also blended (creating a fresh, green, tangy taste and white colour juice) or boiled (resulting in a much sweeter and brown juice) to make iced nutmeg juice.

In Indian cuisine, nutmeg is used in many sweet, as well as savoury, dishes (predominantly in Mughlai cuisine). It is also added in small quantities as a medicine for infants. It may also be used in small quantities in *garam masala*. Ground nutmeg is also smoked in India.[*][2]

Commercial jar of mace

Mature mace of nutmeg, size about 38 mm (1.5 in)

In Indonesian cuisine, nutmeg is used in various dishes,[*][3] mainly in many soups, such as *soto* soup, *baso* soup or *sup kambing*.

In Middle Eastern cuisine, ground nutmeg is often used as a spice for savoury dishes.

In original European cuisine, nutmeg and mace are used especially in potato dishes and in processed meat products; they are also used in soups, sauces, and baked goods. It is also commonly used in rice pudding. In Dutch cuisine, nutmeg is added to vegetables such as Brussels sprouts, cauliflower, and string beans. Nutmeg is a traditional in-

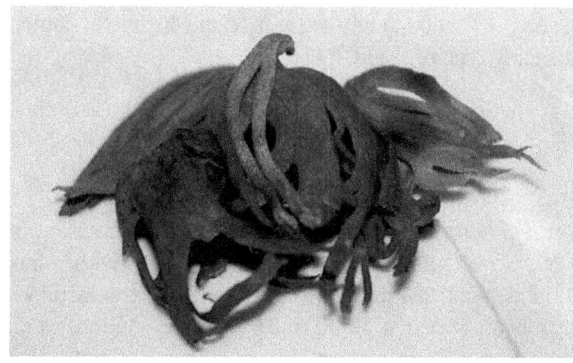

Mace (জয়িত্রি jayitri),Kolkata, West Bengal, India

gredient in mulled cider, mulled wine, and eggnog. In Scotland, mace and nutmeg are usually both essential ingredients in haggis.

In Italian cuisine, nutmeg is almost uniquely used as part of the stuffing for many regional meat-filled dumplings like tortellini, as well as for the traditional meatloaf.

Japanese varieties of curry powder include nutmeg as an ingredient.

In the Caribbean, nutmeg is often used in drinks such as the Bushwacker, Painkiller, and Barbados rum punch. Typically, it is just a sprinkle on the top of the drink.

The pericarp (fruit/pod) is used in Grenada and also in Indonesia to make jam, or is finely sliced, cooked with sugar, and crystallised to make a fragrant candy.

In the US, nutmeg is known as the main pumpkin pie spice and often shows up in simple recipes for other winter squashes such as baked acorn squash.

59.3 Essential oils

The essential oil obtained by steam distillation of ground nutmeg is used widely in the perfumery and pharmaceutical industries. This volatile fraction typically contains 60-80% d-camphene by weight, as well as quantities of d-pinene, limonene, d-borneol, l-terpineol, geraniol, safrol, and myristicin.[4] In its pure form, myristicin is a toxin, and consumption of excessive amounts of nutmeg can result in myristicin poisoning.[5] The oil is colourless or light yellow, and smells and tastes of nutmeg. It contains numerous components of interest to the oleochemical industry, and is used as a natural food flavouring in baked goods, syrups, beverages, and sweets. It is used to replace ground nutmeg, as it leaves no particles in the food. The essential oil is also used in the cosmetic and pharmaceutical industries, for instance, in toothpaste, and as a major ingredient in some cough syrups. In traditional medicine, nutmeg and nutmeg

oil were used for disorders related to the nervous and digestive systems.

After extraction of the essential oil, the remaining seed, containing much less flavour, is called "spent". Spent is often mixed in industrial mills with pure nutmeg to facilitate the milling process, as nutmeg is not easy to mill due to the high percentage of oil in the pure seed. Ground nutmeg with a variable percentage of spent (around 10% w/w) is also less likely to clot. To obtain a better running powder, a small percentage of rice flour also can be added.

59.4 Nutmeg butter

Nutmeg butter is obtained from the nut by expression. It is semisolid, reddish-brown in colour, and tastes and smells of nutmeg. About 75% (by weight) of nutmeg butter is trimyristin, which can be turned into myristic acid, a 14-carbon fatty acid, which can be used as a replacement for cocoa butter, can be mixed with other fats like cottonseed oil or palm oil, and has applications as an industrial lubricant.

59.5 History

Nutmeg is known to have been a prized and costly spice in European medieval cuisine as a flavouring, medicinal, and preservative agent. Saint Theodore the Studite (c. 758 – 826) allowed his monks to sprinkle nutmeg on their pease pudding when required to eat it. In Elizabethan times, because nutmeg was believed to ward off the plague, demand increased and its price skyrocketed.[6]

Until the mid-19th century, the small island of Banda (known to early English adventurers as "Run" although it is a different island both are now part of the Banda Islands), was the world's only source of nutmeg and mace. Nutmeg was known as a valuable commodity by Muslim sailors from the port of Basra (including the fictional character Sinbad the Sailor in the *One Thousand and One Nights*). Nutmeg was traded by Arabs during the Middle Ages and sold to the Venetians for high prices, but the traders did not divulge the exact location of their source in the profitable Indian Ocean trade, and no European was able to deduce its location.

In August 1511, Afonso de Albuquerque conquered Malacca, which at the time was the hub of Asian trade, on behalf of the king of Portugal. In November of the same year, after having secured Malacca and learning of Banda's location, Albuquerque sent an expedition of three ships led by his friend António de Abreu to find it. Malay pilots, either recruited or forcibly conscripted, guided them via Java, the Lesser Sundas, and Ambon to Run, arriving in early

1512.[7] The first Europeans to reach the Bandas, the expedition remained in Banda for about a month, buying and filling their ships with Banda's nutmeg and mace, and with cloves in which Banda had a thriving *entrepôt* trade.[8] An early account of Banda is in *Suma Oriental*, a book written by the Portuguese apothecary Tomé Pires, based in Malacca from 1512 to 1515. Full control of this trade by the Portuguese was not possible, and they remained participants without a foothold in the islands.

Nutmeg sorting on the Banda islands, 1899–1900

The trade in nutmeg later became dominated by the Dutch in the 17th century. The English and Dutch, through their competing East India and Dutch East India Companies, engaged in prolonged struggles to gain control of Run Island. At the end of the Second Anglo-Dutch War, the Dutch gained control of Run, while England controlled New Amsterdam (New York) in North America.

The Dutch waged a bloody war, including massacring and enslaving the inhabitants of the island of Banda, to control nutmeg production in the East Indies in 1621. Thereafter, the Banda Islands were run as a series of plantation estates, with the Dutch mounting annual expeditions in local war-vessels to extirpate nutmeg trees planted elsewhere.

As a result of the Dutch interregnum during the Napoleonic Wars, the British took temporary control of the Banda Islands from the Dutch and transplanted nutmeg trees, complete with soil, to Sri Lanka, Penang, Bencoolen, and Singapore.[9] (There is evidence that the tree existed in Sri Lanka even before this.)[10] From these locations they were transplanted to their other colonial holdings elsewhere, notably Zanzibar and Grenada. The national flag of Grenada, adopted in 1974, shows a stylised split-open nutmeg fruit. The Dutch retained control of the spice islands until World War II.

Connecticut received its nickname ("the Nutmeg State" , "Nutmegger") from the legend that some unscrupulous Connecticut traders would whittle "nutmeg" out of wood,

creating a "wooden nutmeg" , a term which later came to mean any type of fraud.[11]

59.6 World production

World production of nutmeg is estimated to average between 10,000 and 12,000 tonnes per year, with annual world demand estimated at 9,000 tonnes; production of mace is estimated at 1,500 to 2,000 tonnes. Indonesia and Grenada dominate production and exports of both products, with world market shares of 75% and 20%, respectively. Other producers include India, Malaysia (especially Penang, where the trees grow wild within untamed areas), Papua New Guinea, Sri Lanka, and Caribbean islands, such as St. Vincent. The principal import markets are the European Community, the United States, Japan, and India. Singapore and the Netherlands are major re-exporters.

59.7 Medical research

Nutmeg has been used in medicine since at least the seventh century. In the 19th century, it was used as an abortifacient, which led to numerous recorded cases of nutmeg poisoning. Although used as a folk treatment for other ailments, unprocessed nutmeg has no proven medicinal value today.[12]

One study has shown that the compound macelignan isolated from *M. fragrans* (Myristicaceae) may exert antimicrobial activity against *Streptococcus mutans*,[13] and another that a methanolic extract from the same plant inhibited Jurkat cell activity in human leukemia,[14] but these are not currently used treatments.

59.8 Psychoactivity and toxicity

59.8.1 Effects

In low doses, nutmeg produces no noticeable physiological or neurological response, but in large doses, raw nutmeg has psychoactive effects. In its freshly ground form (from whole nutmegs), nutmeg contains myristicin, a monoamine oxidase inhibitor and psychoactive substance. Myristicin poisoning can induce convulsions, palpitations, nausea, eventual dehydration, and generalized body pain.[15] It is also reputed to be a strong deliriant.[16] For these reasons, nutmeg has been banned in Saudi Arabia.[17]

Fatal myristicin poisonings in humans are very rare, but three have been reported: one in an 8-year-old child[18] and another in a 55-year-old adult, the latter case attributed to a combination with flunitrazepam.[19]

In case reports, raw nutmeg produced anticholinergic-like symptoms, attributed to myristicin and elemicin.[18][20][21]

Intoxications with nutmeg had effects that varied from person to person, but were often reported to be an excited and confused state with headaches, nausea, dizziness, dry mouth, bloodshot eyes, and memory disturbances. Nutmeg was also reported to induce hallucinogenic effects, such as visual distortions and paranoid ideation. Intoxication took several hours before the maximum effect was reached. Effects and aftereffects lasted up to several days.[15][22][23][24][25][26][27][28][29][30]

Myristicin poisoning is potentially deadly to some pets and livestock, and may be caused by culinary quantities of nutmeg harmless to humans. For this reason it is recommended not to feed eggnog to dogs.[31]

59.8.2 History of use

Peter Stafford's *Psychedelics Encyclopedia* quotes an 1883 report from Mumbai noting that "the Hindus of West India take nutmeg as an intoxicant", and records that the spice has been used for centuries as a form of snuff in rural eastern Indonesia and India, later seeing the ground seed mixed with betel and other kinds of snuff. In 1829, the Czech physiologist Jan Evangelista Purkinje ingested three ground nutmegs with a glass of wine and recorded headaches, nausea, hallucinations, and a sense of euphoria that lasted for several days.[12]

Swiss chemist Albert Hofmann, who discovered LSD, and Harvard ethnobotanist Richard Evans Schultes documented reports of nutmeg's use as an intoxicant by students, prisoners, sailors, alcoholics, and marijuana smokers. In his autobiography, Malcolm X writes about taking nutmeg and other "semi-drugs" while serving time in prison.[12]

The *Angewandte Chemie International Edition* records the use of nutmeg as an intoxicant in the United States in the post-World War II period, notably among young people, bohemians, and prisoners. A 1966 *New York Times* piece named it along with morning glory seeds, diet aids, cleaning fluids, cough medicine, and other substances as "alternative highs" on college campuses.[12]

59.8.3 Toxicity during pregnancy

Nutmeg was once considered an abortifacient, but may be safe for culinary use during pregnancy. However, it inhibits prostaglandin production and contains hallucinogens that may affect the fetus if consumed in large quantities.[32]

59.9 References

[1] "Nutmeg". *Encyclopedia Britannica Online*.

[2] Pat Chapman (2007). *India Food and Cooking: The Ultimate Book on Indian Cuisine*. New Holland Publishers. p. 16. ISBN 978-1-84537-619-2.

[3] Arthur L. Meyer; Jon M. Vann (2008). *The Appetizer Atlas: A World of Small Bites*. Houghton Mifflin Harcourt. p. 196. ISBN 0-544-17738-X.

[4] The Merck Index (1996). 12th edition

[5] *Utilization of Tropical Foods: Sugars, Spices and Stimulants: Compendium on Technological and Nutritional Aspects of Processing and Utilization of Tropical Foods, Both Animal and Plant, for Purposes of Training and Field Reference*. Food & Agriculture Org. 1989. p. 35. ISBN 978-92-5-102837-7.

[6] Milton, Giles. *Nathaniel's Nutmeg* p.3

[7] Hannard (1991), page 7; Milton, Giles (1999). *Nathaniel's Nutmeg*. London: Sceptre. pp. 5 and 7. ISBN 978-0-340-69676-7.

[8] Hannard (1991), page 7

[9] Giles Milton, *Nathaniel's Nutmeg*, 1999, London: Hodder and Stoughton, ISBN 0-340-69675-3

[10] 'Nutmeg', Department of Export Agriculture website

[11] "Connecticut State Library: Nicknames for Connecticut". Cslib.org. Retrieved 2012-09-07.

[12] Shafer, Jack (2010-12-14) Stupid drug story of the week: The nutmeg scare, *Slate.com*

[13] Devi, P. B.; Ramasubramaniaraja, R. (2009). "Dental Caries and Medicinal Plants – An Overview". *Journal of Pharmacy Research* **2** (11): 1669–1675. ISSN 0974-6943.

[14] Chirathaworn, C.; Kongcharoensuntorn, W.; Dechdoungchan, T.; Lowanitchapat, A.; Sa-Nguanmoo, P.; Poovorawan, Y. (2007). "Myristica fragrans Houtt. Methanolic extract induces apoptosis in a human leukemia cell line through SIRT1 mRNA downregulation". *Journal of the Medical Association of Thailand = Chotmaihet thangphaet* **90** (11): 2422–2428. PMID 18181330.

[15] Demetriades, A. K.; Wallman, P. D.; McGuiness, A.; Gavalas, M. C. (2005). "Low Cost, High Risk: Accidental Nutmeg Intoxication" (pdf). *Emergency Medicine Journal* **22** (3): 223–225. doi:10.1136/emj.2002.004168. PMC 1726685. PMID 15735280.

[16] "Nutmeg". *Plants*. Erowid. Retrieved 2012-04-22.

[17] "The Flavors of Arabia". Retrieved 2015-02-23.

[18] Weil, Andrew (1966). "The Use of Nutmeg as a Psychotropic Agent". *Bulletin on Narcotics* (UNODC) **1966** (4): 15–23.

[19] Stein, U.; Greyer, H.; Hentschel, H. (2001). "Nutmeg (myristicin) poisoning--report on a fatal case and a series of cases recorded by a poison information centre". *Forensic Science International* **118** (1): 87–90. doi:10.1016/S0379-0738(00)00369-8. PMID 11343860.

[20] Shulgin, A. T.; Sargent, T.; Naranjo, C. (1967). "The Chemistry and Psychopharmacology of Nutmeg and of Several Related Phenylisopropylamines" (pdf). *Psychopharmacology Bulletin* **4** (3): 13. PMID 5615546.

[21] McKenna, A.; Nordt, S. P.; Ryan, J. (2004). "Acute Nutmeg Poisoning". *European Journal of Emergency Medicine* **11** (4): 240–241. doi:10.1097/01.mej.0000127649.69328.a5. PMID 15249817.

[22] Burroughs, William S. (1957). "Letter from a Master Addict to Dangerous Drugs". *British Journal of Addiction to Alcohol & Other Drugs* **53** (2): 119–132. doi:10.1111/j.1360-0443.1957.tb05093.x.

[23] Quin, G. I.; Fanning, N. F.; Plunkett, P. K. (1998). "Letter: Nutmeg Intoxication" (pdf). *Journal of Accident & Emergency Medicine* **15** (4): 287–288. doi:10.1136/emj.15.4.287-d. PMC 1343156. PMID 9681323.

[24] Brenner, N.; Frank, O. S.; Knight, E. (1993). "Chronic Nutmeg Psychosis" (pdf). *Journal of the Royal Society of Medicine* **86** (3): 179–180. PMC 1293919. PMID 8459391.

[25] Scholefield, J. H. (1986). "Letter: Nutmeg--an Unusual Overdose" (pdf). *Archives of Emergency Medicine* **3** (2): 154–155. doi:10.1136/emj.3.2.154. PMC 1285340. PMID 3730084.

[26] Venables, G. S.; Evered, D.; Hall, R. (1976). "Letter: Nutmeg Poisoning" (pdf). *British Medical Journal* **1** (6001): 96. doi:10.1136/bmj.1.6001.96-c. PMC 1638356. PMID 942686.

[27] Panayotopoulos, D. J.; Chisholm, D. D. (1970). "Correspondence: Hallucinogenic Effect of Nutmeg" (pdf). *British Medical Journal* **1** (5698): 754. doi:10.1136/bmj.1.5698.754-b. PMC 1699804. PMID 5440555.

[28] Williams, E. Y.; West, F. (1968). "The Use of Nutmeg as a Psychotropic Drug. Report of two Cases" (pdf). *Journal of the National Medical Association* **60** (4): 289–290. PMC 2611568. PMID 5661198.

[29] Dale, H. H. (1909). "Note on Nutmeg-Poisoning" (pdf). *Proceedings of the Royal Society of Medicine* **2** (Therapeutical and Pharmacological Section): 69–74. PMC 2046458. PMID 19974070.

[30] Cushny, A. R. (1908). "Nutmeg Poisoning" (pdf). *Proceedings of the Royal Society of Medicine* **1** (Therapeutical and Pharmacological Section): 39–44. PMC 2045778. PMID 19973353.

[31] "Don't Feed Your Dog Toxic Foods".

[32] Herb and drug safety chart Herb and drug safety chart from BabyCentre UK

59.10   Further reading

- Milton, Giles (1999), *Nathaniel's Nutmeg: How One Man's Courage Changed the Course of History*

- Burroughs, William S. (1959). *Naked Lunch*. Paris: Olympia Press. p. 228.

- Gable, R. S. (2006). The toxicity of recreational drugs. *American Scientist* 94: 206–208

- Devereux, P. (1996). *Re-Visioning the Earth: A Guide to Opening the Healing Channels Between Mind and Nature*. New York: Fireside. pp. 261–262.

59.11   External links

- Georgetown Official Website

Chapter 60

Oak

"Quercus" redirects here. For other uses, see Quercus (disambiguation).
"Oak tree" redirects here. For other uses, see Oak Tree (disambiguation).
This article is about oaks (Quercus). For other uses of "Oak", see Oak (disambiguation).

An **oak** is a tree or shrub in the genus *Quercus* (/ˈkwɜrkəs/;*[1] Latin "oak tree") of the beech family, Fagaceae. There are approximately 600 extant species of oaks. The common name "oak" may also appear in the names of species in related genera, notably *Lithocarpus*. The genus is native to the Northern Hemisphere, and includes deciduous and evergreen species extending from cool temperate to tropical latitudes in the Americas, Asia, Europe, and North Africa. North America contains the largest number of oak species, with approximately 90 occurring in the United States. Mexico has 160 species, of which 109 are endemic. The second greatest center of oak diversity is China, which contains approximately 100 species.*[2]

Oaks have spirally arranged leaves, with lobate margins in many species; some have serrated leaves or entire leaves with smooth margins. Many deciduous species are marcescent, not dropping dead leaves until spring. In spring, a single oak tree produces both male flowers (in the form of catkins) and small female flowers.*[3] The fruit is a nut called an acorn, borne in a cup-like structure known as a cupule; each acorn contains one seed (rarely two or three) and takes 6–18 months to mature, depending on species. The live oaks are distinguished for being evergreen, but are not actually a distinct group and instead are dispersed across the genus.

60.1 Classification

Oak trees are a flowering plant. Oaks may be divided into two genera (sometimes referred to as subgenera) and a number of sections:

60.1.1 Genus *Quercus*

Oak at Schönderling

See also: List of Quercus species

The genus *Quercus* is divided into the following sections:

- Sect. *Quercus* (synonyms *Lepidobalanus* and *Leucobalanus*), the white oaks of Europe, Asia and North America. Styles are short; acorns mature in 6 months and taste sweet or slightly bitter; the inside of an acorn shell is hairless. The leaves mostly lack a bristle on their lobe tips, which are usually rounded. The type species is Quercus robur.

- Sect. *Mesobalanus*, Hungarian oak and its relatives of Europe and Asia. Styles long; acorns mature in about 6 months and taste bitter; the inside of this acorn's shell is hairless. The section *Mesobalanus* is closely related to section *Quercus* and sometimes included in it.

- Sect. *Cerris*, the Turkey oak and its relatives of Europe and Asia. Styles long; acorn mature in 18 months and taste very bitter. The inside of the acorn's shell is hairless. Its leaves typically have sharp lobe tips, with bristles at the lobe tip.

- Sect. *Protobalanus*, the canyon live oak and its relatives, in southwest United States and northwest Mexico. Styles short, acorns mature in 18 months and taste very bitter. The inside of the acorn shell appears woolly. Leaves typically have sharp lobe tips, with bristles at the lobe tip.

- Sect. *Lobatae* (synonym *Erythrobalanus*), the red oaks of North America, Central America and northern South America. Styles long; acorns mature in 18 months and taste very bitter. The inside of the acorn shell appears woolly. The actual nut is encased in a thin, clinging, papery skin. Leaves typically have sharp lobe tips, with spiny bristles at the lobe.

60.1.2 Genus *Cyclobalanopsis*

- The ring-cupped oaks of eastern and southeastern Asia. Evergreen trees growing 10–40 m (33–131 ft) tall. They are distinct from subgenus *Quercus* in that they have acorns with distinctive cups bearing concrescent rings of scales; they commonly also have densely clustered acorns, though this does not apply to all of the species. IUCN, ITIS, Encyclopedia of Life and *Flora of China* treats *Cyclobalanopsis* as a distinct genus, but some taxonomists consider it a subgenus of *Quercus*. It contains about 150 species. Species of *Cyclobalanopsis* are common in the evergreen subtropical laurel forests which extend from southern Japan, southern Korea, and Taiwan across southern China and northern Indochina to the eastern Himalayas, in association with trees of genus *Castanopsis* and the laurel family (Lauraceae).

60.2   Hybridization

Interspecific hybridization is quite common among oaks but usually between species within the same section only and most common in the white oak group (subgenus *Quercus*, section *Quercus*; see **List of *Quercus* species**). Intersection hybrids, except between species of sections *Quercus*

A hybrid white oak, possibly Quercus stellata × Q. muhlenbergii

and *Mesobalanus*, are unknown. Recent systematic studies appear to confirm a high tendency of *Quercus* species to hybridize because of a combination of factors. White oaks are unable to discriminate against pollination by other species in the same section. Because they are wind pollinated and they have weak internal barriers to hybridization, hybridization produces functional seeds and fertile hybrid offspring.[*][4] Ecological stresses, especially near habitat margins, can also cause a breakdown of mate recognition as well as a reduction of male function (pollen quantity and quality) in one parent species.[*][4][*][5]

Frequent hybridization among oaks has consequences for oak populations around the world; most notably, hybridization has produced large populations of hybrids with copious amounts of introgression, and the evolution of new species.[*][6] Frequent hybridization and high levels of introgression have caused different species in the same populations to share up to 50% of their genetic information.[*][7] Having high rates of hybridization and introgression produces genetic data that often does not differentiate between two clearly morphologically distinct species, but instead differentiates populations.[*][8] Numerous hypotheses have been proposed to explain how oak species are able to remain morphologically and ecologically distinct with such high levels of gene flow, but the phenomenon is still largely a mystery to botanists.

The Fagaceae, or beech family, to which the oaks belong, is a very slow evolving clade compared to other angiosperms,[*][9][*][10] and the patterns of hybridization and introgression in *Quercus* pose a great challenge to the concept of a species since a species is often defined as a group of "actually or potentially interbreeding populations which are reproductively isolated from other such groups."

*[11] By this definition, many species of *Quercus* would be lumped together according to their geographic and ecological habitat, despite clear distinctions in morphology and, to a large extent, genetic data.

60.3 Uses

Heart of oak beams of the frame of Saint-Girons church in Monein, France

Oak wood has a density of about 0.75 g/cm³ (0.43 oz/cu in) creating great strength and hardness. The wood is very resistant to insect and fungal attack because of its high tannin content. It also has very appealing grain markings, particularly when quartersawn. Oak planking was common on high status Viking longships in the 9th and 10th centuries. The wood was hewn from green logs, by axe and wedge, to produce radial planks, similar to quarter-sawn timber. Wide, quarter-sawn boards of oak have been prized since the Middle Ages for use in interior panelling of prestigious buildings such as the debating chamber of the House of Commons in London and in the construction of fine furniture. Oak wood, from *Quercus robur* and *Quercus petraea*, was used in Europe for the construction of ships, especially naval men of war, until the 19th century, and was the prin-

cipal timber used in the construction of European timber-framed buildings. Today oak wood is still commonly used for furniture making and flooring, timber frame buildings, and for veneer production. Barrels in which wines, sherry, and spirits such as brandy, Irish whiskey, Scotch whisky and Bourbon whiskey are aged are made from European and American oak. The use of oak in wine can add many different dimensions to wine based on the type and style of the oak. Oak barrels, which may be charred before use, contribute to the colour, taste, and aroma of the contents, imparting a desirable oaky vanillin flavour to these drinks. The great dilemma for wine producers is to choose between French and American oakwoods. French oaks (*Quercus robur*, *Q. petraea*) give the wine greater refinement and are chosen for the best wines since they increase the price compared to those aged in American oak wood. American oak contributes greater texture and resistance to ageing, but produces more powerful wine bouquets. Oak wood chips are used for smoking fish, meat, cheeses,*[12] and other foods.

Sherry maturing in oak barrels

Japanese oak is used in the making of professional drums from the manufacturer Yamaha Drums. The higher density of oak gives the drum a brighter and louder tone compared to traditional drum materials such as maple and birch. In hill states of India, besides fuelwood and timber, the local people use oak wood for making agricultural implements. The leaves are used as fodder during lean period and bedding for livestock.*[13]*[14]

The bark of the cork oak is used to produce wine stoppers (corks). This species grows in the Mediterranean Sea region, with Portugal, Spain, Algeria, and Morocco producing most of the world's supply.

Of the North American oaks, the northern red oak is the one of most prized of the red oak group for lumber, much

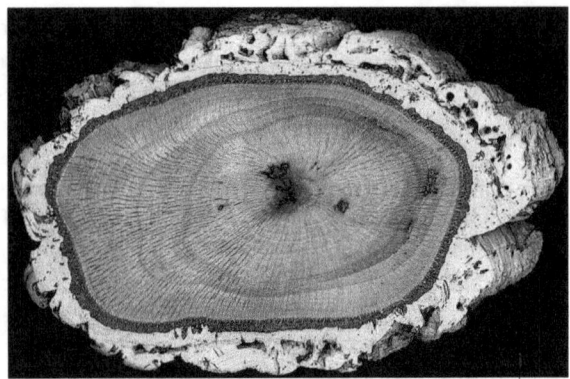

A cross section of the trunk of a cork oak, Quercus suber

60.4 Biodiversity and ecology

Oak forest in Estonia.

of which is marketed as red oak regardless of the species of origin. It is not good for outdoor use due to its open capillaries unless the wood is treated. If the wood is properly treated with preservatives, it will not rot as quickly as cured white oak heartwood. The closed cell structure of white oaks prevent them from absorbing preservatives. With northern red oak, one can blow air through an end grain piece 10 inches long to make bubbles come out in a glass of water. These openings give fungus easy access when the finish deteriorates. Shumard oak, a member of the red oak subgenus, provides timber which is described as "mechanically superior" to Northern Red oak. Cherrybark oak is another type of red oak which provides excellent timber.

The standard for the lumber of the white oak group – all of which is marketed as white oak – is the white oak. White oak is often used to make wine barrels. The wood of the deciduous pedunculate oak and sessile oak accounts for most of the European oak production, but evergreen species, such as Holm oak and cork oak also produce valuable timber.

The bark of the White Oak is dried and used in medical preparations. Oak bark is also rich in tannin, and is used by tanners for tanning leather. Acorns are used for making flour or roasted for acorn coffee.

Oak galls were used for centuries as a main ingredient in iron gall ink, a kind of manuscript ink, harvested at a specific time of year. In Korea, oak bark is used to make shingles for traditional roof construction.

Oak has been listed as one of the 38 substances used to prepare Bach flower remedies,[15] a kind of alternative medicine promoted for its effect on health. However, according to Cancer Research UK, "there is no scientific evidence to prove that flower remedies can control, cure or prevent any type of disease, including cancer" .[16]

Oak on sandy earth.

Oaks are keystone species in a wide range of habitats from Mediterranean semi-desert to subtropical rainforest. For example, oak trees are important components of hardwood forests, and certain species are particularly known to grow in associations with members of the Ericaceae in oak-heath forests.[17][18] A number of kinds of truffles, including the two well known varieties, the black Périgord truffle[19] and the white Piedmont truffle,[20] have symbiotic relationships with oak trees. The European pied flycatcher is an example of an animal species that often depends upon oak trees.

Many species of oaks are under threat of extinction in the wild, largely due to land use changes, livestock grazing and unsustainable harvesting. For example, over the past 200 years, large areas of oak forest in the highlands of Mexico, Central America and the northern Andes have been cleared for coffee plantations and cattle ranching. There is a continuing threat to these forests from exploitation for timber, fuelwood and charcoal.[21] In the USA, entire oak ecosystems have declined due to a combination of factors still imperfectly known, but thought to include fire suppression, increased consumption of acorns by growing mammal populations, herbivory of seedlings, and introduced pests.[22] In a recent survey, 78 wild oak species have been identified as being in danger of extinction, from a global total of over 500 species.[23] The proportion under threat may be much higher in reality, as there is insufficient information about over 300 species, making it near impossible to form any judgement of their status.

In the Himalayan region of India, oak forests are being invaded by pine forests due to the increase in temperature. The associated species of pine forest may cross frontiers and become new elements of the oak forests.[24]

In eastern North America, rare species of oak trees include scarlet oak (*Quercus coccinea*), chinkapin oak (*Quercus muehlenbergii*), and post oak (*Quercus stellata*).[25]

60.5   Diseases and pests

See also: List of Lepidoptera that feed on oaks
  Sudden oak death (*Phytophthora ramorum*) is a water

Oak powdery mildew on pedunculate oak

mould that can kill oaks within just a few weeks. Oak wilt, caused by the fungus *Ceratocystis fagacearum* (a fungus

closely related to Dutch elm disease), is also a lethal disease of some oaks, particularly the red oaks (the white oaks can be infected but generally live longer). Other dangers include wood-boring beetles, as well as root rot in older trees which may not be apparent on the outside, often being discovered only when the trees come down in a strong gale. Oak apples are galls on oaks made by the gall wasp. The female kermes scale causes galls to grow on kermes oak. Oaks are used as food plants by the larvae of Lepidoptera (butterfly and moth) species such as the gypsy moth, *Lymantria dispar*, which can defoliate oak and other broadleaved tree species in North America.[26]

A considerable number of galls are found on oak leaves, buds, flowers, roots, etc. Examples are oak artichoke gall, oak marble gall, oak apple gall, knopper gall, and spangle gall.

A number of species of fungus cause powdery mildew on oak species. In Europe the species *Erysiphe alphitoides* is the most common cause.[27]

A new and yet little understood disease of mature oaks, acute oak decline, has been reported in parts of the UK since 2009.[28]

Oak processionary moth (*Thaumetopoea processionea*) has become a serious threat in the UK since 2006. The caterpillars of this species defoliate the trees, and are hazardous to human health; their bodies are covered with poisonous hairs which can cause rashes and respiratory problems.[29]

60.6   Toxicity

The leaves and acorns of the oak tree are poisonous to cattle, horses, sheep, and goats in large amounts due to the toxin tannic acid, and cause kidney damage and gastroenteritis. Symptoms of poisoning include lack of appetite, depression, constipation, diarrhea (which may contain blood), blood in urine, and colic. The exception to livestock and oak toxicity is the domestic pig, which may be fed entirely on acorns in the right conditions, and has traditionally been pastured in oak woodlands (such as the Spanish *dehesa* and the English system of pannage) for hundreds of years.

Acorns are also edible to humans, after leaching of the tannins.[30]

60.7   Cultural significance

60.7.1   National symbol

The oak is a common symbol of strength and endurance and has been chosen as the national tree of many coun-

Oak branches on the coat of arms of Estonia

tries. Already an ancient Germanic symbol (in the form of the Donar Oak, for instance), certainly since the early nineteenth century, it stands for the nation of Germany and oak branches are thus displayed on some German coins, both of the former Deutsche Mark and the current Euro currency.[31] In 2004 the Arbor Day Foundation[32] held a vote for the official National Tree of the United States of America. In November 2004, the United States Congress passed legislation designating the oak as America's National Tree.[33]

Other countries have also designated the oak as their national tree including Serbia, Cyprus (Golden Oak), England, Estonia, France, Germany, Moldova, Romania, Jordan, Latvia, Lithuania, Poland, the United States, Wales, Galicia and Bulgaria.[34]

Oaks as regional and state symbols

The oak is the emblem of County Londonderry in Northern Ireland, as a vast amount of the county was covered in forests of the tree until relatively recently. The name of the county comes from the city of Derry, which originally in Irish was known as *Doire* meaning *oak*.

The Irish County Kildare derives its name from the town of Kildare which originally in Irish was *Cill Dara* meaning the Church of the Oak or Oak Church.

Iowa designated the oak as its official state tree in 1961; and the White Oak is the state tree of Connecticut, Illinois and Maryland. The Northern Red Oak is the provincial tree of Prince Edward Island, as well as the state tree of New Jersey. The Live Oak is the state tree of Georgia, USA.

The oak is a national symbol from the Basque Country, specially in the province of Biscay.

The coat-of-arms of Vest-Agder, Norway, features an oak tree.

Oak leaves are traditionally an important part of German Army regalia. The Nazi party used the traditional German eagle, standing atop of a swastika inside a wreath of oak leaves. It is also known as the Iron Eagle. During the Third Reich of Nazi Germany, oak leaves were used for military valor decoration on the Knights Cross of the Iron Cross. They also symbolize rank in the United States Armed Forces. A gold oak leaf indicates an O-4 (Major or Lt. Commander), whereas a silver oak leaf indicates an O-5 (Lt. Colonel or Commander).[35] Arrangements of oak leaves, acorns and sprigs indicate different branches of the United States Navy Staff corps officers.[35] Oak leaves are embroidered onto the covers (hats) worn by field grade officers and flag officers in the United States armed services.

If a member of the United States Army or Air Force earns multiple awards of the same medal, then instead of wearing a ribbon or medal for each award, he or she wears one metal representation of an "oak leaf cluster" attached to the appropriate ribbon for each subsequent award.

Political use

The oak tree is used as a symbol by a number of political parties. It is the symbol of Toryism (on account of the Royal Oak) and the Conservative Party in the United Kingdom,[36] and formerly of the Progressive Democrats in Ireland[37] and the Democrats of the Left in Italy. In the cultural arena, the oakleaf is the symbol of the National Trust (UK), The Woodland Trust, and The Royal Oak Foundation.[35]

60.7.2   Religious

In Greek mythology, the oak is the tree sacred to Zeus, king of the gods. In Zeus's oracle in Dodona, Epirus, the sacred oak was the centerpiece of the precinct, and the priests would divine the pronouncements of the god by interpreting the rustling of the oak's leaves.[38]

In Baltic mythology, the oak is the sacred tree of Latvian Pērkons, Lithuanian Perkūnas and Prussian Perkūns. Pērkons is the god of thunder and one of the most important deities in the Baltic pantheon.

In Celtic polytheism, the name of the oak tree was part of the Proto-Celtic word for 'druid': **derwo-weyd- > *druwid-*; however, Proto-Celtic **derwo-* (and **dru-*) can also be adjectives for 'strong' and 'firm', so Ranko Matasovic interprets that **druwid-* may mean 'strong knowledge'. As in other Indo-European faiths, Taranus, being a Thunder God, was associated with the oak tree.[39] The Indo-Europeans worshiped the oak and connected it with a thunder or lightning god; "tree" and *drus* may also be cognate with "Druid," the Celtic priest to whom the oak was sacred. There has even been a study that shows that oaks are more likely to be struck by lightning than any other tree of the same height.[40]

In Norse mythology, the oak was sacred to the thunder god, Thor. Thor's Oak was a sacred tree of the Germanic Chatti tribe. According to legend, the Christianisation of the heathen tribes by Saint Boniface was marked by the oak's being replaced by the fir (whose triangular shape symbolizes the Trinity) as a "sacred" tree.[41]

> Thrice on my bossy shield I struck my spear;
>
> And thrice a ghost's shrill voice was heard in air;
>
> The sacred oaks that skirt this sloping wood
>
> Are dead--revive their withered roots with blood;
>
> The blood of foes shall fertilze the plain,
>
> and Odin's spirt feast on heaps of slain.
>
> Hark! now I hear his mighty voice from far--
>
> Rise, sons of Odin, and prepare for war[42]

In the Bible, the oak tree at Shechem is the site where Jacob buries the foreign gods of his people (Gen. 35:4). In addition, Joshua erects a stone under an oak tree as the first covenant of the Lord (Josh. 24.25–7). In Isaiah 61, the prophet refers to the Israelites as "Oaks of Righteousness".

The badnjak is central tradition in Serbian Orthodox Church Christmas celebration where young and straight oak, is ceremonially felled early on the morning of Christmas Eve.

In Slavic mythology, the oak was the most important tree of the god Perun.

60.7.3 Historical

Several singular oak trees, such as the Royal Oak in Britain and the Charter Oak in the United States, are of great historical or cultural importance; for a list of important oaks, see Individual oak trees.

"The Proscribed Royalist, 1651", a famous painting by John Everett Millais, depicted a Royalist fleeing from Cromwell's forces and hidden in an oak. Millais painted the picture in Hayes, Kent, from a local oak tree that became known as the Millais Oak.[43][44]

Approximately 50 km west of Toronto, Canada is the town of Oakville, ON, famous for its history as a shipbuilding port on Lake Ontario.

The city of Raleigh, N.C., is known as "The City of Oaks."

The Jurupa Oak tree – a clonal colony of *Quercus palmeria* or Palmer's oak found in Riverside County, California – is believed to be the world's oldest organism at 13,000 years.[45]

Large groups of very old oak trees are rare. One of the oldest groups of oak trees, found in Poland, is about 480 years old, which was assessed by dendrochronological methods.[46]

In Republican Rome a crown of oak leaves was given to those who had saved a life of a citizen in battle; it was called the "civic oak".[40]

60.7.4 Famous oak trees

Main article: List of notable trees

Tamme-Lauri oak is the thickest and oldest tree in Estonia.

- The Emancipation Oak is designated one of the 10 Great Trees of the World by the National Geographic Society and is part of the National Historic Landmark district of Hampton University.

- The Ivenack Oak which is one of the largest trees in Europe is located in Mecklenburg-Vorpommern, Germany, and is approximately 800 years old.[47]

- The Bowthorpe Oak, located in Bourne, Lincolnshire, is thought to be 1,000 years old. It was featured in the Guinness Book of World Records and was filmed for a TV documentary for its astonishing longevity.[47]

- The Minchenden (or Chandos) Oak, in Southgate, London, is said to be the largest oak tree in England (already 27 feet or 8.2 meters in girth in the nineteenth century), and is perhaps 800 years old.*[48]

- The Seven Sisters Oak is the largest certified southern live oak tree. Located in Mandeville, Louisiana, it is estimated to be up to 1,500 years old with a trunk that measures 38 ft (11.6 meters).*[49]*[50]

- The Major Oak is an 800–1000 year old tree located in Sherwood Forest, Nottinghamshire. According to folklore, it was used by Robin Hood for shelter.

- Friendship Oak is a 500-year-old southern live oak located in Long Beach, Mississippi.

- The Crouch Oak is believed to have originated in the 11th Century and is located in Addlestone, Surrey. It is an important symbol of the town with many local businesses adopting its name. It used to mark the boundary of Windsor Great Park. Legend says that Queen Elizabeth I stopped by it and had a picnic.

- The Angel Oak is a southern live oak located in Angel Oak Park on John's Island near Charleston, South Carolina. The Angel Oak is estimated to be in excess of 1400 years old, stands 66.5 ft (20.3 m) tall, and measures 28 ft (8.5 m) in circumference.

- The Kaiser's Oak, located at the village of Gommecourt in Artois, France, named in honour of Kaiser Wilhelm II, symbolically marked from late 1914 to April 1917 the furthest point in the West of the German Imperial Army during World War One.

60.8 Historical note on Linnaean species

Linnaeus described only five species of oak from eastern North America, based on general leaf form. These were white oak, *Quercus alba*; chestnut oak, *Q. montana*; red oak, *Q. rubra*; willow oak *Q. phellos*; and water oak, *Q. nigra*. Because he was dealing with confusing leaf forms, the *Q. montana* and *Q. rubra* specimens actually included mixed foliage of more than one species.

60.9 See also

- Donar's Oak
- Fab Tree Hab
- Foloi oak forest

- Goethe Oak
- List of plants poisonous to equines
- List of Quercus species
- Oak Apple Day
- Thousand Oaks, California

60.10 References

[1] *Sunset Western Garden Book,* 1995, Leisure Arts, pp. 606–607, ISBN 0376038519.

[2] Hogan, C. Michael (2012) *Oak*. ed. Arthur Dawson. Encyclopedia of Earth. National Council for Science and the Environment. Washington DC

[3] Conrad, Jim. "Oak Flowers" . backyardnature.com. 2011-12-12. Retrieved 2013-11-03.

[4] Williams, Joseph H.; Boecklen, William J. & Howard, Daniel J. (2001). "Reproductive processes in two oak (*Quercus*) contact zones with different levels of hybridisation" . *Heredity* **87** (6): 680–690. doi:10.1046/j.1365-2540.2001.00968.x.

[5] Arnold, M. L. (1997). *Natural Hybridization and Evolution*. New York: Oxford University Press. ISBN 0-19-509974-5.

[6] Conte, L.; Cotti, C. & Cristofolini, G. (2007). "Molecular evidence for hybrid origin of *Quercus crenata* Lam. (Fagaceae) from *Q-cerris* L. and *Q-suber* L." . *Plant Biosystems* **141** (2): 181–193. doi:10.1080/11263500701401463.

[7] Gomory, D. & Schmidtova, J. (2007). "Extent of nuclear genome sharing among white oak species (*Quercus* L. subgen. *Lepidobalanus* (Endl.) Oerst.) in Slovakia estimated by allozymes" . *Plant Systematics and Evolution* **266** (3–4): 253–264. doi:10.1007/s00606-007-0535-0.

[8] Kelleher, C. T.; Hodkinson, T. R.; Douglas, G. C. & Kelly, D. L. (2005). "Species distinction in Irish populations of *Quercus petraea* and *Q. robur*: Morphological versus molecular analyses" . *Annals of Botany* **96** (7): 1237–1246. doi:10.1093/aob/mci275. PMID 16199484.

[9] Frascaria, N.; Maggia, L.; Michaud, M. & Bousquet, J. (1993). "The RBCL Gene Sequence from Chestnut Indicates a Slow Rate of Evolution in the Fagaceae" . *Genome* **36** (4): 668–671. doi:10.1139/g93-089. PMID 8405983.

[10] Manos, P. S.; Stanford, A. M. (2001). "The historical biogeography of Fagaceae: Tracking the tertiary history of temperate and subtropical forests of the Northern Hemisphere" . *International Journal of Plant Sciences* **162** (Suppl. 6): S77–S93. doi:10.1086/323280.

[11] Raven, Peter H.; Johnson, George B.; Losos, Jonathan B.; Singer, Susan R. (2005). *Biology* (Seventh ed.). New York: McGraw Hill. ISBN 0-07-111182-4.

[12] Cheese. swaledalecheese.co.uk

[13] Kala, C.P. (2004). Studies on the indigenous knowledge, practices and traditional uses of forest products by human societies in Uttarakhand state of India. GBPIHED, Almora, India

[14] Kala, C.P. (2010). *Medicinal Plants of Uttarakhand: Diversity Livelihood and Conservation.* BioTech Books, Delhi, ISBN 8176222097.

[15] D. S. Vohra (1 June 2004). *Bach Flower Remedies: A Comprehensive Study.* B. Jain Publishers. p. 3. ISBN 978-81-7021-271-3. Retrieved 2 September 2013.

[16] "Flower remedies". Cancer Research UK. Retrieved September 2013.

[17] *The Natural Communities of Virginia Classification of Ecological Community Groups* (Version 2.3), Virginia Department of Conservation and Recreation, 2010. Dcr.virginia.gov. Retrieved on 2011-12-10.

[18] Schafale, M. P. and A. S. Weakley. 1990. *Classification of the natural communities of North Carolina: third approximation.* North Carolina Natural Heritage Program, North Carolina Division of Parks and Recreation.

[19] "Truffle Glossary: Black Truffles". thenibble.com. 2010-07-01. Retrieved 1 July 2010.

[20] "Truffle Glossary: White Truffles". thenibble.com. 2010-07-01. Retrieved 1 July 2010.

[21] Kappelle, M. (2006). "Neotropical montane oak forests: overview and outlook", pp 449–467 in: Kappelle, M. (ed.). *Ecology and conservation of neotropical montane oak forests.* Ecological Studies No. 185. Springer-Verlag, Berlin, doi:10.1007/3-540-28909-7_34 ISBN 978-3-540-28908-1.

[22] Lorimer, C.G. (2003) Editorial: The decline of oak forests. American Institute of Biological Sciences.

[23] Oldfield, S. & Eastwood, A. (2007) The Red List of Oaks Flora & Fauna International (FFI) and Botanic Gardens Conservation International (BGCI) ISBN 978-1-903703-25-0

[24] Kala, C.P. (2012). *Biodiversity, communities and climate change.* Teri Publications, New Delhi, ISBN 817993442X.

[25] Carpenter, Paul (1990). *Plants in the Landscape.* New York: W.H. Freeman and Company. p. 73. ISBN 0716718081.

[26] "Trees: Oak Insects and Diseases: Gypsy Moth". TreeHelp.com. Retrieved 27 April 2010.

[27] Mougou, A.; Dutech, C.; Desprez-Loustau, M. -L. (2008). "New insights into the identity and origin of the causal agent of oak powdery mildew in Europe". *Forest Pathology* **38** (4): 275. doi:10.1111/j.1439-0329.2008.00544.x.

[28] Kinver, Mark (28 April 2010). "Oak disease 'threatens landscape'". *BBC News.* Retrieved 29 April 2010.

[29] "Invasion of toxic moths". *The Northern Echo.* July 10, 2012.

[30] Bainbridge, D. A. (12–14 November 1986), *Use of acorns for food in California: past, present and future,* San Luis Obispo, CA.: Symposium on Multiple-use Management of California's Hardwoods

[31] Schierz, Kai Uwe (2004). "Von Bonifatius bis Beuys, oder: Vom Umgang mit heiligen Eichen". In Hardy Eidam, Marina Moritz, Gerd-Rainer Riedel, Kai-Uwe Schierz. *Bonifatius: Heidenopfer, Christuskreuz, Eichenkult* (in German). Stadtverwaltung Erfurt. pp. 139–45.

[32] "Trees – Arbor Day Foundation". Arborday.org. Retrieved 27 April 2010.

[33] "Oak Trees". arborday.org. Retrieved 27 April 2010.

[34] "Oak as a Symbol". *Venables Oak.* Retrieved 26 September 2012.

[35] "Political or Symbolic". *Extended Definition: oak.* Retrieved 26 September 2012.

[36] Pickles, Eric. "The Conservative Party". Conservatives.com. Retrieved 27 April 2010.

[37] Coalition Government 1989 To 1992. progressivedemocrats.ie

[38] Frazer, James George (1922). *The Golden Bough.* Chapter XV: The Worship of the Oak.

[39] Taylor, John W. (September 1979). "Tree Worship", *Mankind Quarterly,* pp. 79–142.

[40] *Oak. A Dictionary of Literary Symbols* (Cambridge).

[41] von Staufer, Maria. "The Chronological History of the Christmas Tree". *The Christmas Archives.* Retrieved 7 February 2010.

[42] War Song, *The Poetry of various Glees, Songs. and Company, London, 1798*

[43] Millais, J.G. (1899) *Life and Letters of Sir John Everett Millais,* vol. 1, p. 166, London : Methuen.

[44] Arborecology, containing a photograph of the Millais oak. arborecology.co.uk

[45] Thaindian News: Jurupa Oak tree is world's oldest organism at 13,000 years. December 24, 2009. Thaindian.com. Retrieved on 2011-12-10.

[46] Ufnalski K. The oldest groups of oak trees in Poland. Proceedings of EuroDendro 2008 "The long history of wood utilization" News of Forest History Nr. V (39)/2008:83–84

[47] Bermosa, Nobert. "Famous Oak Trees in the World". Retrieved 30 September 2012.

[48] Geograph: TQ2993: Minchenden Oak, Garden of Remembrance, Waterfall Road, N14

[49] Seven Sisters Oak. americanforests.org

[50] "Seven Sisters Oak". *100 Oaks Project*. 4 December 2009.

- Janka Hardness Scale – The Janka Hardness Scale for many Exotic and Domestic species

- Eichhorn, Markus (May 2010). "Oak – A Very English Tree". *Test Tube*. Brady Haran for the University of Nottingham.

60.11 Bibliography

- Byfield, Liz (1990) *An oak tree*, Collins book bus, London : Collins Educational, ISBN 0-00-313526-8

- Philips, Roger. Trees of North America and Europe, Random House, Inc., New York ISBN 0-394-50259-0, 1979.

- Logan, William B. (2005) *Oak : the frame of civilization*, New York ; London : W.W. Norton, ISBN 0-393-04773-3

- Paterson, R.T. (1993) *Use of trees by livestock*, **5**: *Quercus*, Chatham : Natural Resources Institute, ISBN 0-85954-365-X

- Royston, Angela (2000) *Life cycle of an oak tree*, Heinemann first library, Oxford : Heinemann Library, ISBN 0-431-08391-6

- Savage, Stephen (1994) *Oak tree*, Observing nature series, Hove : Wayland, ISBN 0-7502-1196-2

- Tansley, Arthur G., Sir (1952) *Oaks and oak woods*, Field study books, London : Methuen.

- Żukow-Karczewski, Marek. *Dąb – król polskich drzew* (Oak – the king of the Polish trees), "AURA" (A Monthly for the protection and shaping of human environment), 9/88.

60.12 External links

- *Flora of China – Cyclobalanopsis*

- Flora Europaea: *Quercus*

- Oaks from Bialowieza Forest

- Common Oaks of Florida

- Oaks of the world

- The Global Trees Campaign The Red List of Oaks and Global Survey of Threatened Quercus

- Latvia - the land of oaks

Chapter 61

Palm wine

For other uses, see Palm wine (disambiguation).

Palm wine is an alcoholic beverage created from the sap of various species of palm tree such as the palmyra, date palms, and coconut palms.*[1]*[2] It is known by various names in different regions and is common in various parts of Asia, Africa the Caribbean and South America.

Bottles and a glass of palm wine

Palm wine production by small holders and individual farmers may promote conservation as palm trees become a source of regular household income that may economically be worth more than the value of timber sold.*[3]

Palm wine is known as "palm Wine" in [Liberia] *emu*, *nkwu*, *oguro* in Nigeria; *nsamba* in the Democratic Republic of the Congo; *nsafufuo* in Ghana;*[4] *kallu* in South India; *Htan Yay* (□□□□□□in Myanmar; *matango* in Cameroon; *tuak* in North Sumatra, Indonesia; *mnazi* in the Mijikenda language of Kenya; *goribon* (Rungus) in Sabah, Borneo; and *tubâ* in the Philippines, Borneo and Mexico. In the Philippines, *tubâ* refers both to the freshly harvested, sweetish cloudy-white sap and the one with the red lauan-tree tan bark colorant. In Leyte, the red *tubâ* is aged with the tan bark for up to six months to two years, until it gets dark red and tapping its glass container gives a sound that does not suddenly stop. This type of *tubâ* is called *bahal* (for *tubâ* aged this way for up to six months) and *bahalina* (for *tubâ* aged thus for up to a year or more). *Toddy* is also consumed in Sri Lanka and Myanmar.

61.1 Tapping

The sap is extracted and collected by a tapper. Typically the sap is collected from the cut flower of the palm tree. A container is fastened to the flower stump to collect the sap. The white liquid that initially collects tends to be very sweet and non-alcoholic before it is fermented. An alternate method is the felling of the entire tree. Where this is practiced, a fire is sometimes lit at the cut end to facilitate the collection of sap.

Palm sap begins fermenting immediately after collection, due to natural yeasts in the pores of pot and air (often spurred by residual yeast left in the collecting container). Within two hours, fermentation yields an aromatic wine of up to 4% alcohol content, mildly intoxicating and sweet. The wine may be allowed to ferment longer, up to a day, to yield a stronger, more sour and acidic taste, which some people prefer. Longer fermentation produces vinegar instead of stronger wine.*[5]

Toddy collectors at work on Cocos nucifera *palms*

Tapping palm sap in East Timor

Toddy drawer in India, 1870

61.2 Distilled

Palm wine may be distilled to create a stronger drink, which goes by different names depending on the region (e.g., *arrack, village gin, charayam,* and *country whiskey*). Throughout Nigeria, this is commonly called *ogogoro*. In parts of southern Ghana distilled palm wine is called *akpeteshi* or *burukutu*. In Togo and Benin it is called *sodabe*, in the Philippines it is called *lambanog*, while in Tunisia it is called *Lagmi* . In parts of Kenya (coast), it is known as "chang'aa" . Chang'aa can be applied to wounds to stop heavy bleeding (mechanism of action not known). In Ivory Coast, it is called "koutoukou."

61.3 Africa

In Africa, the sap used to create palm wine is most often taken from wild datepalms such as the silver date palm (*Phoenix sylvestris*), the palmyra, and the jaggery palm (*Caryota urens*), or from oil palm such as the African Oil Palm (*Elaeis guineense*) or from *Raffia palms, kithul* palms, or *nipa* palms. In part of central and western Democratic Republic of the Congo, palm wine is called *malafu*. Palm wine tapping is mentioned in the novel *Things Fall Apart* by the Nigerian writer Chinua Achebe and is central to the plot of the novel *The Palm Wine Drinkard* by Nigerian author Amos Tutuola.

Palm wine plays an important role in many ceremonies in parts of Nigeria such as among the Igbo (or Ibo) peoples, and elsewhere in central and western Africa. Guests at weddings, birth celebrations, and funeral wakes are served generous quantities. Palm wine is often infused with medicinal herbs to remedy a wide variety of physical complaints. As a token of respect to deceased ancestors, many drinking sessions begin with a small amount of palm wine spilled on the ground (*Kulosa malafu* in Kikongo ya Leta). Palm wine is enjoyed by men and women, although women usually drink it in less public venues.

In some parts of the Eastern Nigeria, the Igbo Land, palm wine is called "Nkwu Elu" or "Mmanya Ocha" (white drink). For instance, in "Urualla" and other "ideator" towns, it is used for traditional wedding. A young man who is going for the first introduction at his inlaws is required to come with palm wine. There are specific gallons of palm wine required depending on the custom of the various towns in some parts of the Igbo Land.

61.4 India

In India and South Asia, coconut palms and Palmyra palms such as the *Arecaceae* and *Borassus* are preferred. In southern Africa, palm wine (*ubusulu*) is produced in Maputaland, an area in the south of Mozambique between the Lobombo mountains and the Indian Ocean. It is mainly produced from the lala palm (Hyphaene coriacea) by cutting the stem and collecting the sap. In some areas of India, palm wine is evaporated to produce the unrefined sugar called jaggery.

In parts of India, the unfermented sap is called *neera* (*padaneer* in Tamil Nadu) and is refrigerated, stored and distributed by semi-government agencies. A little lime is added to the sap to prevent it from fermenting. *Neera* is said to contain many nutrients including potash.

In India, palm wine or toddy is served as either *neera* or *padaneer* (a sweet, non-alcoholic beverage derived from fresh sap) or *kallu* (a sour beverage made from fermented sap, but not as strong as wine).[*][6] Kallu is usually drunk soon after fermentation by the end of day, as it becomes more sour and acidic day by day. The drink, like vinegar in taste, is considered to have a short shelf life. However, it may be refrigerated to extend its life. Spices are added in order to brew the drink and give it its distinct taste.

In Karnataka, India, palm wine is usually available at toddy shops (known as *Kallu Kadai* in [Tamil], *Kalitha Gadang* in Tulu, *Kallu Dukanam* in Telugu, *Kallu Angadi* in Kannada or "Liquor Shop" in English). In Tamil Nadu, this beverage is currently banned, though the legality fluctuates with politics. In the absence of legal toddy, moonshine distillers of arrack often sell methanol-contaminated alcohol, which can have lethal consequences. To discourage this practice, authorities have pushed for inexpensive "Indian Made Foreign Liquor" (IMFL), much to the dismay of toddy tappers.

A toddy tapper belonging to Goundla caste in the state of Telangana selling toddy under the palm trees

In the state of Andhra Pradesh (India), toddy is a popular drink in rural parts. The *kallu* is collected, distributed and sold by the people of a particular caste called Settibalija or Goud or Gamalla (Goundla). It is a big business in the cities of those districts. In villages, people drink it every day after work.

There are two main types of *kallu* in Andhra Pradesh, namely *Thadi Kallu* (from Toddy Palmyra trees) and *Eetha Kallu* (from silver date palms). *Eetha Kallu* is very sweet and less intoxicating, whereas *Thati Kallu* is stronger (sweet in the morning, becoming sour to bitter-sour in the evening) and is highly intoxicating. People enjoy *kallu* right at the trees where it is brought down. They drink out of leaves by holding them to their mouths while the Goud pours the *kallu* from the *binki* (*kallu* pot). There are different types of toddy (*kallu*) according to the season: 1. *poddathadu*, 2.

parpudthadu, 3. *pandudthadu*, .

In the Indian state of Kerala, toddy is used in leavening (as a substitute for yeast) a local form of hopper called the "Vellayappam" . Toddy is mixed with rice dough and left over night to aid in fermentation and expansion of the dough causing the dough to rise overnight, making the bread soft when prepared. In Kerala, toddy is sold under a licence issued by the excise department and it is an industry having more than 50,000 employees with a welfare board under the labour department. It is also used in the preparation of a soft variety of Sanna, which is famous in the parts of Karnataka and Goa in India.

61.5   Indonesia and Malaysia

Lithograph of a palm wine vendor and a native KNIL soldier consuming tuak (1854)

Tuak is imbibed in Sumatra, Sulawesi, Borneo, Bali and parts of Malaysia such as Penang Island and East Malaysia). The beverage is a popular drink among the Ibans and other Dayaks of Sarawak during the Gawai festivals, weddings, hosting of guests and other special occasions. The Batak people of North Sumatra also consume palm wine. In Northern Sumatra the palm sap is mixed with raru bark to make Tuak. The brew is served at stalls along with snacks.*[1] The same word is used for other drinks in Indonesia, for example those made using fermented rice.

61.6   Democratic Republic of Congo

There are four types of palm wine in the central and southern DRC. From the oil palm comes *ngasi*, *dibondo* comes from the raffia palm, *cocoti* from the coconut palm, and *mahusu* from a short palm which grows in the savannah areas of western Bandundu and Kasai provinces.

61.7   South America

Production of palm wine may have contributed to the endangered status of the Chilean wine palm (*Jubaea chilensis*).*[7]

61.8   Other areas

In Tuvalu, the process of making toddy can clearly be seen with tapped palm trees that line Funafuti International Airport.

Palm wine is collected, fermented and stored in calabashes in Bandundu Province, Democratic Republic of the Congo

61.9   Consumption by animals

Some small pollinating mammals consume large amounts of fermented palm nectar as part of their diet, especially the southeast Asian pen-tailed treeshrew. The inflorescences of the bertam palm contain populations of yeast which ferment the nectar in the flowers to up to 3.8% alcohol (average: 0.6%). The treeshrews metabolize the alcohol very efficiently and do not appear to become drunk from the fermented nectar.*[8]

61.10 Names

There are a variety of regional names for Palm wine:

*a Telugu, Tamil and Malayalam.
*b Marathi.

61.11 Gallery

- Bowl for tuak drinking made from a gourd (late 19th century)

- Tapping the "arènpalm" (Arenga pinnata), one of the palms used to make palm wine, in Ambon, Moluccas (1919). The palm tree also supplies fiber to cover roofs and sugar. In the Moluccas the tree was especially appreciated because of the palm wine that can be made from the sap of the immature flower flasks. This was called toewak (Dutch), tuak or sagoweer (saguer). The fresh sap, "sugar water", was also so drunk. It is fermented to make the alcoholic beverage and can also be made into vinegar.

- Palm wine seller in Bali (1929)

- Taken at Southern Leyte where a tuba gatherer climb the coconut tree to harvest some tuba.

- Sitting on the coconut palm while gathering tuba.

- Locally called "manananggot" for tuba gatherer.

- Gathering tuba from the coconut tree.

61.12 See also

- Desi daru

- Arrack, an alcoholic beverage distilled from coconut palm wine in southeast Asia.

- Palm-wine music, a West African musical genre.

- Tuak, an alcoholic beverage made of fermented rice, yeast and sugar.

- Madurai Veeran, a deity who consumes toddy.

- Sree Muthappan, another deity who consumes toddy.

- List of Indonesian beverages

61.13 References

Notes

[1] Enjoying 'tuak' in Batak country by Wan Ulfa Nur Zuhra, NORTH SUMATRA, Feature, January 21 2013 Jakarta Post

[2] Rundel, Philip W. *The Chilean Wine Palm* in the *Mildred E. Mathias Botanical Garden Newsletter*, Fall 2002, Volume 5(4). Retrieved 2008-08-31

[3] Confirel:Sugar Palm Tree - Conservation of natural heritage retrieved on 15 April 2012

[4] Toddy and Palm Wine – Practical Answers on the Practical Action website. Retrieved 2008-08-31

[5] Fermented and vegetables. A global perspective. Chapter 4

[6] Toddy/Kallu and Neera/Padhaneer

[7] C. Michael Hogan. 2008. *Chilean Wine Palm: Jubaea chilensis*, GlobalTwitcher.com, ed. N. Stromberg

[8] Frank Wiens, Annette Zitzmann, Marc-André Lachance, Michel Yegles, Fritz Pragst, Friedrich M. Wurst, Dietrich von Holst, Saw Leng Guan, and Rainer Spanagel. Chronic intake of fermented floral nectar by wild treeshrews Proceedings of the National Academy of Sciences. Published online before print 2008-07-28. Retriev 2008-08-25

[9] Law, S.V.; et al. (2011). "MiniReview- Popular fermented foods and beverages in Southeast Asia" (PDF). *International Food Research Journal* (18). Retrieved 20 January 2012.

[10] Gnarfgnarf:Palm wine, rice wine, grape wine, beers and other drinks and beverages of Cambodia, 9 April 2012, retrieved on 15 April 2012

[11] Anchimbe - Creating New Names for Common Things in Cameroon English (I-TESL-J)

[12] "English-Chinese Translation of "palm wine"". Websaru Dictionary. Retrieved 20 January 2012.

61.14 External links

- Article on Philippine palm wine

Chapter 62

Peat

For other uses, see Peat (disambiguation).

Peat (**turf**) is an accumulation of partially decayed

Peat gatherers at Westhay, Somerset Levels in 1905

Peat stacks and cutting at Westhay, Somerset Levels

Harvesting the peat at Westhay, Somerset Levels

Peat in Lewis, Scotland

vegetation or organic matter that is unique to natural areas called peatlands or mires.[1][2] The peatland ecosystem is the most efficient carbon sink on the planet[2] because peatland plants capture the CO_2 which is naturally

released from the peat, thus maintaining an equilibrium. In natural peatlands, the "annual rate of biomass production is greater than the rate of decomposition", but it takes "thousands of years for peatlands to develop the deposits of 1.5 to 2.3 m, which is the average depth of the boreal peatlands".[2] One of the most common components is *Sphagnum* moss, although many other plants can contribute.

Soils that contain mostly peat are known as histosols. Peat forms in wetland conditions, where flooding obstructs flows of oxygen from the atmosphere, slowing rates of decomposition.*[3]

Peatlands, also known as mires,*[Notes 1] particularly bogs, are the most important source of peat,*[4] but other less common wetland types also deposit peat, including fens, pocosins, and peat swamp forests. Other words for lands dominated by peat include moors or muskegs. Landscapes covered in peat also have specific kinds of plants, particularly *Sphagnum* moss, ericaceous shrubs, and sedges (see bog for more information on this aspect of peat). Since organic matter accumulates over thousands of years, peat deposits also provide records of past vegetation and climates stored in plant remains, particularly pollen. Hence, they allow humans to reconstruct past environments and changes in human land use.*[5]

Peat is harvested as an important source of fuel in certain parts of the world. By volume, about 4 trillion m^3 of peat are in the world, covering a total of around 2% of global land area (about 3 million km^2), containing about 8 billion terajoules of energy.*[6] Over time, the formation of peat is often the first step in the geological formation of other fossil fuels such as coal, particularly low-grade coal such as lignite.*[7]

Depending on the agency, peat is not generally regarded as a renewable source of energy, as its extraction rate in industrialized countries far exceeds its slow regrowth rate of 1 mm per year,*[8] and as peat regrowth is also reported to take place in only 30-40% of peatlands.*[9] Because of this, the United Nations Framework Convention on Climate Change (UNFCCC),*[10] and another organization affiliated with the United Nations classifies peat as a fossil fuel.*[11] However, the Intergovernmental Panel on Climate Change has begun to classify peat as a "slow-renewable" fuel.*[12] This is also the classification used by many in the peat industry.*[10]

At 106 g CO_2/MJ,*[13] the carbon dioxide emission intensity of peat is higher than that of coal (at 94.6 g CO_2/MJ) and natural gas (at 56.1).

Peat fires have been responsible for some large public health disasters, including the 1997 Southeast Asian haze.

62.1 Peatlands distribution

In a widely cited article, Joosten and Clarke (2002) defined peatlands, or mire (which they claim are the same)*[Notes 2]*[1] as,

> ...the most widespread of all wetland types in the world, representing 50 to 70% of global

wetlands. They cover over four million km2 or 3% of the land and freshwater surface of the planet. In these ecosystems are found one third of the world's soil carbon and 10% of global freshwater resources. These ecosystems are characterized by the unique ability to accumulate and store dead organic matter from Sphagnum and many other non-moss species, as peat, under conditions of almost permanent water saturation. Peatlands are adapted to the extreme conditions of high water and low oxygen content, of toxic elements and low availability of plant nutrients. Their water chemistry varies from alkaline to acidic. Peatlands occur on all continents, from the tropical to boreal and Arctic zones from sea level to high alpine conditions. Joosten and Clarke 2002

Peat extraction in East Frisia, Germany

Peatlands are areas of land with a naturally accumulated layer of peat. Peatlands are found in at least 175 countries and cover around 4 million km^2 or 3% of the world's land area. In Europe, peatlands extend to about 515,000 km^2.*[14]

Peat deposits are found in many places around the world, including northern Europe and North America, principally in Canada and the northern United States. Here, too, occur some of the world's largest peatlands, including the West Siberian Lowland, the Hudson Bay Lowland, and the Mackenzie River Valley.*[15] The amount of peat is smaller in the Southern Hemisphere, partly because there is less land, yet South America (Southern Patagonia/Tierra del Fuego) has one of the world's largest wetlands, the vast Magellanic Moorland, with extensive peat-dominated landscapes.*[15] Peat can be found in New Zealand, Kerguelen, and the Falkland Islands, Indonesia (Kalimantan (Sungai Putri, Danau Siawan, Sungai Tolak), Rasau Jaya (West Kalimantan), and Sumatra). Indonesia has more tropical peat land and mangrove forests than any other nation on earth, but Indonesia is losing wetlands by 100,000 hectares (250,000 acres) per year.*[16]

About 60% of the world's wetlands are peat. About 7% of total peatlands have been exploited for agriculture and forestry. Under proper conditions, peat will turn into lignite coal over geologic periods of time.

A peat stack in Ness on the Isle of Lewis (Scotland)

62.2 Formation

Peat forms when plant material, usually in wet areas, is inhibited from decaying fully by acidic and anaerobic conditions. It is composed mainly of wetland vegetation: principally bog plants including mosses, sedges, and shrubs. As it accumulates, the peat can hold water, thereby slowly creating wetter conditions, and allowing the area of wetland to expand. Peatland features can include ponds, ridges, and raised bogs.*[4]

Most modern peat bogs formed in high latitudes after the retreat of the glaciers at the end of the last ice age some 12,000 years ago.*[17] Peat usually accumulates slowly, at the rate of about a millimeter per year.*[8]

Peat in the world's peatlands is currently believed to have been forming for 360 million years and contains 550 Gt of carbon.*[18]

Worked bank in blanket bog, near Ulsta, Yell, Shetland Islands

62.3 Types of peat material

Peat material is either fibric, hemic, or sapric. Fibric peats are the least decomposed, and comprise intact fiber. Hemic peats are somewhat decomposed, and sapric are the most decomposed. *Phragmites* peat is one composed of reed grass, *Phragmites australis*, and other grasses. It is denser than many other types of peat. Engineers may describe a soil as peat which has a relatively high percentage of organic material. This soil is problematic because it exhibits poor consolidation properties.

62.4 Characteristics and uses

Peat is soft and easily compressed. Under pressure, water in the peat is forced out. Upon drying, peat can be used as fuel. It has industrial importance as a fuel in some countries, such as Ireland and Finland, where it is harvested on an industrial scale. In many countries, including Ireland and Scotland, where trees are often scarce, peat is traditionally used for cooking and domestic heating. Stacks of drying peat dug from the bogs can still be seen in some rural areas. Peat's insulating properties make it of use to industry.

Falkland Islanders shovelling peat in the 1950s

Although peat has many uses for humans, it also presents severe problems at times. Wet or dry, it can be a major

Peat fire

fire hazard, as peat fires can burn almost indefinitely (or at least until the fuel is exhausted). Peat fires can even burn underground, reigniting after the winter, provided a source of oxygen is present. Peat deposits also pose major difficulties to builders of structures, roads, and railways, as they are highly compressible under even small loads. When the West Highland Line was built across Rannoch Moor, in western Scotland, its builders had to float the tracks on a mattress of tree roots, brushwood, and thousands of tons of earth and ashes.

Peat bogs had considerable ritual significance to Bronze Age and Iron Age peoples, who considered them to be home to (or at least associated with), nature gods or spirits. The bodies of the victims of ritual sacrifices have been found in a number of locations in Scotland, England, Ireland, and especially northern Germany and Denmark, almost perfectly preserved by the tanning properties of the acidic water. (See Tollund Man for one of the most famous examples of a bog body). Peat wetlands formerly had a degree of metallurgical importance, as well. During the Dark Ages, peat bogs were the primary source of bog iron, used to create the swords and armour of the Vikings. Many peat swamps along the coast of Malaysia serve as a natural means of flood mitigation. The peat swamps serve like a natural form of water catchment whereby any overflow will be absorbed by the peat. However, this is effective only if the forests are still present, since they prevent peat fires.

62.4.1 In Scotland

Some Scotch whisky distilleries, such as those on Islay, use peat fires to dry malted barley. The drying process takes about 30 hours. This gives the whiskies a distinctive smoky flavour, often called "peatiness".[19] The peatiness, or degree of peat flavor, of a whisky is calculated in ppm of phenol. The normal Highland whiskies have a peat level of up to 30 ppm, and the whiskies on Islay usually have up to

50 ppm. In rare types, like the Octomore,[20] the whisky can have more than 100 ppm of peat.

62.4.2 In Ireland

Industrial milled peat production in a section of the Bog of Allen in the Irish Midlands: The 'turf' in the foreground is machine-produced for domestic use.

In Ireland, large-scale domestic and industrial peat usage is widespread. In the Republic of Ireland, a state-owned company called Bord na Móna is responsible for managing peat production. It produces milled peat which is used in power stations. It sells processed peat fuel in the form of peat briquettes which are used for domestic heating. These are oblong bars of densely compressed, dried, and shredded peat. Peat moss is a manufactured product for use in garden cultivation. Turf (dried out peat sods) is very commonly used in rural areas.

In Northern Ireland there is small-scale domestic turf cutting in rural areas, but areas of bog lands have been diminished because of changes in agriculture. Afforestation has seen the establishment of tentative steps towards conservation, such as at Peatlands Park, County Armagh, which is an Area of Special Scientific Interest.[21]

62.4.3 In England

The extraction of peat from the Somerset Levels is known to have taken place during Roman times, and has been carried out since the Levels were first drained.[22] On Dartmoor there were several commercial distillation plants formed and run by the British Patent Naphtha Company in 1844. These produced naphtha on a commercial scale from the high-quality local peat.[23]

Fenn's, Whixall and Bettisfield Mosses are elements of a post-Ice Age peat bog that straddles the England-Wales border. Only lightly hand-dug, it is now a national nature reserve which is being restored to natural condition and contains many rare plant and animal species due to the acidic environment created by the peat.[*][24]

62.4.4   In Finland

See also: Peat energy in Finland
The climate, geography and environment of Finland favour

The Toppila Power Station, a peat-fired facility in Oulu, Finland

bog and peat bog formation. Peat is available in considerable quantities: some estimates put the amount of peat in Finland alone to be twice the size of the North Sea oil reserves.[*][25] This abundant resource (often mixed with wood at an average of 2.6%) is burned to produce heat and electricity. Peat provides around 6.2% of Finland's annual energy production, second only to Ireland.[*][26] The contribution of peat to greenhouse gas emissions of Finland can exceed a yearly amount of 10 million tonnes carbon dioxide, equal to the total emissions of all passenger car traffic in Finland.

Finland classifies peat as a slowly renewing biomass fuel,[*][27] and that position has also been taken by the European Union. The Intergovernmental Panel on Climate Change has taken the position that peat is not a fossil fuel. Peat producers in Finland often claim that peat is a special form of biofuel because of the relatively fast retake rate of released CO_2 if the bog is not forested for the following 100 years. Also, agricultural and forestry-drained peat bogs actively release more CO_2 annually than is released in peat energy production in Finland. The average regrowth rate of a single peat bog, however, is indeed slow, from 1,000 up to 5,000 years. Furthermore, it is a common practice to forest used peat bogs instead of giving them a chance to renew, leading to lower levels of CO_2 storage than the original

peat bog.

At 106 g CO_2/MJ,[*][13] the carbon dioxide emissions of peat are higher than those of coal (at 94.6 g CO_2/MJ) and natural gas (at 56.1). According to one study, increasing the average amount of wood in the fuel mixture from the current 2.6% to 12.5% would take the emissions down to 93 g CO_2/MJ, though little effort is being made to achieve this.[*][28]

Peat extraction is also seen by some conservationists as the main threat to mire biodiversity in Finland. The International Mire Conservation Group (IMCG) in 2006 urged the local and national governments of Finland to protect and conserve the remaining pristine peatland ecosystems. This includes the cessation of drainage and peat extraction in intact mire sites and the abandoning of current and planned groundwater extraction that may affect these sites. A proposal for a Finnish peatland management strategy was presented to the government in 2011, after a lengthy consultation phase.[*][29]

62.4.5   In Russia

Shatura Power Station. Russia has the largest peat power capacity in the world

Use of peat for energy production was prominent during the Soviet Union, with the peak occurring in 1965 and declining from that point. In 1929, over 40% of the Soviet Union's electric energy came from peat, which dropped to 1% by 1980.

In the 1960s, larger sections of swamps and bogs in Western Russia were drained for agricultural use and to generate peat fields for mining.[*][30] Plans are underway to increase peat output and increase peat's contribution to Russian energy generation.[*][31] However, there is concern about the environmental impact as peat fields are flammable, drainage degrades eco-systems, and burning of peat releases carbon

The Bor Peat Briquette Factory, Russia

dioxide.*[31] Due to 2010 forest and peat fires the Russian government is under heavy pressure to finance re-flooding of the previously drained bogs around Moscow. The initial costs for the programme are estimated to be about 20 to 25 billion rubles, which is close to 500 million euros.

Currently, Russia is responsible for 17% of the world's peat production, and 20% of the peat that it produces, 1.5 million tons, is used for energy purposes.*[32]*[33] Shatura Power Station in Moscow Oblast and Kirov Power Station in Kirov Oblast are the two largest peat power stations in the world.

62.4.6   Use in agriculture

Peat is important for farmers and gardeners, who mix it into soil to improve its structure and to increase acidity. In Sweden, farmers use dried peat to absorb excrement from cattle that are wintered indoors. The most important property of peat is retaining moisture in soil when it is dry and yet preventing the excess of water from killing roots when it is wet. Peat can store nutrients although it is not fertile itself – it is a polyelectrolytic with a high ion exchange capacity due to its oxidized lignin. Peat is discouraged as a soil amendment by the Royal Botanic Gardens, Kew, England, and has been since 2003.*[34] Peat is an important raw material in horticulture. However, it is recommended to treat peat thermally, e.g., through soil steaming, in order to kill inherent pest and reactivate nutrients.

62.4.7   Freshwater aquaria

Peat is sometimes used in freshwater aquaria, most commonly in soft water or blackwater river systems, such as those mimicking the Amazon River basin. In addition to being soft in texture and therefore suitable for demersal (bottom-dwelling) species such as *Corydoras* catfish, peat

is reported to have a number of other beneficial functions in freshwater aquaria. It softens water by acting as an ion exchanger; it also contains substances that are beneficial for plants, and for the reproductive health of fishes. It can even prevent algae growth and kill microorganisms. Peat often stains the water yellow or brown due to the leaching of tannins.*[35]

62.4.8   Water filtration

Peat is used in water filtration, such as for the treatment of septic tank effluent, as well as for urban runoff. Due to its purifying properties, peat also serves as a filter for septic tanks and may be used as a water purifier.

62.4.9   Balneotherapy

Peat is widely used in balneotherapy (the use of bathing to treat disease). Many traditional spa treatments include peat as part of peloids. Such health treatments have a very long tradition in Europe, especially in Poland, the Czech Republic, Germany and Austria. Some of these old spas go back to the 18th century, and they are still active today. The most common types of peat application in balneotherapy are peat muds, poultices, and suspension baths.*[36]

62.4.10   Peat archives

Authors Rydin and Jeglum in *Biology of Habitats* described the concept of peat archives, a phrase coined by influential peatland scientist Harry Godwin in 1981.*[37]*[38] *[39]

> "In a peat profile there is a fossilized record of changes over time in the vegetation, pollen, spores, animals (from microscopic to the giant elk), and archaeological remains that have been deposited in place, as well as pollen, spores and particles brought in by wind and weather. These remains are collectively termed the peat archives.
> "Rydin 2013"

In *Quaternary Palaeoecology*, first published in 1980, Birks and Birks described how paleoecological studies "of peat can be used to reveal what plant communities were present (locally and regionally), what time period each community occupied, how environmental conditions changed, and how environment affected the ecosystem in that time and place."*[38]*[40]

Scientists continue to compare modern mercury (Hg) accumulation rates in bogs with historical natural archives records in peat bogs and lake sediments to estimate the potential human impacts on the biogeochemical cycle of mercury, for example.[*][41] Over the years different dating models and technologies for measuring date sediments and peat profiles accumulated over the last 100–150 years, have been used, including the widely used vertical distribution of 210Pb, the ICP-SMS[*][42] and more recently the Initial Penetration (IP).[*][43]

62.5 Environmental and ecological issues

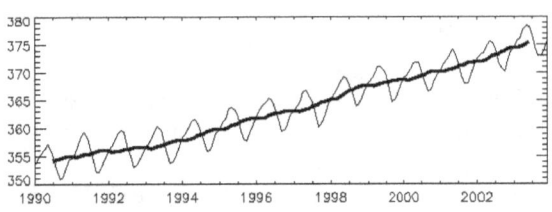

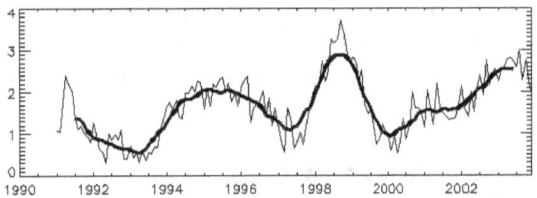

Increase, and change relative to previous year, of the atmospheric concentration of carbon dioxide.

Because of the distinctive ecological conditions of peat wetlands, they provide habitat for a distinctive fauna and flora. For example, whooping cranes nest in North American peatlands, while Siberian cranes nest in the West Siberian peatland. Such habitats also have many species of wild orchids and carnivorous plants. It takes centuries for a peat bog to recover from disturbance. For more on biological communities, see wetland, bog or fen.

Recent studies indicate that the world's largest peat bog, located in Western Siberia and the size of France and Germany combined, is thawing for the first time in 11,000 years. As the permafrost melts, it could release billions of tons of methane gas into the atmosphere. The world's peatlands are thought to contain 180 to 455 billion metric tons of sequestered carbon, and they release into the atmosphere 20 to 45 million metric tons of methane annually. The peatlands' contribution to long-term fluctuations in these atmospheric gases has been a matter of considerable debate.[*][44]

One of the characteristics for peat is that bioaccumulations of metals are often concentrated in the peat, of significant environmental concern is accumulated mercury.[*][45]

62.5.1 Peat drainage

Large areas of organic wetland (peat) soils are currently drained for agriculture, forestry, and peat extraction. This process is taking place all over the world. This not only destroys the habitat of many species, but also heavily fuels climate change. As a result of peat drainage, the organic carbon which was built up over thousands of years and is normally under water is suddenly exposed to the air. It decomposes and turns into carbon dioxide (CO_2), which is released into the atmosphere.[*][46] The global CO_2 emissions from drained peatlands have increased from 1,058 Mton in 1990 to 1,298 Mton in 2008 (>20%). This increase has particularly taken place in developing countries, of which Indonesia, China, Malaysia, and Papua New Guinea, are the fastest growing top emitters. This estimate excludes emissions from peat fires (conservative estimates amount to at least 4,000 Mton/CO_2-eq./yr for south-east Asia). With 174 Mton/CO_2-eq./yr the EU is after Indonesia (500 Mton) and before Russia (161 Mton) the World's 2nd largest emitter of drainage related peatland CO_2 (excl. extracted peat and fires). Total CO_2 emissions from the worldwide 500,000 km^2 of degraded peatland may exceed 2.0 Gtons (including emissions from peat fires) which is almost 6% of all global carbon emissions.[*][47]

62.5.2 Peat fires

See also: Slash and burn and Arctic methane release

Peat has a high carbon content and can burn under low

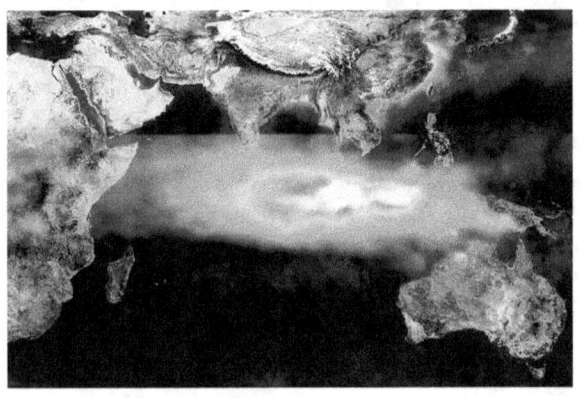

Smoke and ozone pollution from Indonesian fires, 1997.

moisture conditions. Once ignited by the presence of a heat source (e.g., a wildfire penetrating the subsurface), it smolders. These smoldering fires can burn undetected for very long periods of time (months, years, and even centuries) propagating in a creeping fashion through the underground peat layer. Peat fires are emerging as a global threat with significant economic, social, and ecological impacts.

Despite the damage that the burning of raw peat can cause, bogs are naturally subject to wildfires and depend on the wildfires to keep woody competition from lowering the water table and shading out many bog plants. Several families of plants including the carnivourous Sarracenia, Dionaea, Utricularia and even non-carnivorous plants such as the Sandhills Lily, Toothache Grass and many species of orchid are now threatened and in some cases endangered from the combined forces of human drainage, negligence and absence of fire.*[48]*[49]*[50]

Recent burning of peat bogs in Indonesia, with their large and deep growths containing more than 50 billion tons of carbon, has contributed to increases in world carbon dioxide levels. Peat deposits in Southeast Asia could be destroyed by 2040.*[51]*[52]

It is estimated that in 1997, peat and forest fires in Indonesia released between 0.81 and 2.57 Gt of carbon; equivalent to 13–40 percent of the amount released by global fossil fuel burning, and greater than the carbon uptake of the world's biosphere. These fires may be responsible for the acceleration in the increase in carbon dioxide levels since 1998.*[53]*[54] More than 100 peat fires in Kalimantan and East Sumatra have continued to burn since 1997. Each year, the peat fires in Kalimantan and East Sumatra ignite new forest fires above the ground.

In North America, peat fires can occur during severe droughts throughout their occurrence, from boreal forests in Canada to swamps and fens in the subtropical southern Florida Everglades.*[55] Once a fire has burnt through the area, hollows in the peat are burnt out, and hummocks are desiccated but can contribute to *Sphagnum* recolonization.*[56]

In the summer of 2010, an unusually high heat wave of up to 40 °C (104 °F) ignited large deposits of peat in Central Russia, burning thousands of houses and covering the capital of Moscow with a toxic smoke blanket. The situation remained critical until the end of August 2010.*[57]*[58]

62.6   Wise use and protection

In June 2002, the United Nations Development Programme launched the Wetlands Ecosystem and Tropical Peat Swamp Forest Rehabilitation Project. This project was targeted to last for 5 years until 2007 and brings together the efforts of various non-government organisations.

In November 2002, the **International Peat Society** and the International Mire Conservation Group (IMCG) published guidelines on the "Wise Use of Mires and Peatlands Backgrounds and Principles including a framework for decision-making". The aim of this publication is to develop mechanisms that can balance the conflicting demands on the global peatland heritage, to ensure its wise use to meet the needs of humankind.

In June 2008, the International Peat Society published the book *Peatlands and Climate Change*, summarizing the currently available knowledge on the topic. In 2010, IPS presented a "Strategy for Responsible Peatland Management" which can be applied worldwide for decision-making.

62.7   See also

- Acid sulfate soil
- Acrotelm
- Gytta
- Histosols
- Irish Peatland Conservation Council
- List of bogs
- Peat-fired power stations
- Tropical peat
- Turbary
- Unified Soil Classification System

62.8   Notes

[1] "The term 'peatland' includes mires (Joosten and Clarke 2002)."

[2] Supported by the "Dutch Ministry of Foreign Affairs (DGIS) under the [www.wetlands.org/projects/GPI/default.htm Global Peatland Initiative], managed by Wetlands International in co-operation with the IUCN- Netherlands Committee, Alterra, the International Mire Conservation Group and the International Peatland Society."

62.9   References

[1] Joosten, Hans; Clarke, Donal (2002). Wise Use of Mires and Peatlands: Background and Principles including a Framework for Decision-Making (PDF) (Report). Totnes, Devon. ISBN 951-97744-8-3.

[2] Hugron, Sandrine; Bussières, Julie; Rochefort, Line (2013). Tree plantations within the context of ecological restoration of peatlands: practical guide (PDF) (Report). Laval, Québec, Canada: Peatland Ecology Research Group (PERG). Retrieved 22 February 2014.

[3] Keddy, P.A. 2010. Wetland Ecology: Principles and Conservation (2nd edition). Cambridge University Press, Cambridge, UK. 497 p. Chapter 1.

[4] Gorham, E. (1957). The development of peatlands. Quarterly Review of Biology, 32, 145–66.

[5] Keddy, P.A. 2010. Wetland Ecology: Principles and Conservation (2nd edition). Cambridge University Press, Cambridge, UK. 497 p. 323-325

[6] World Energy Council (2007). "Survey of Energy Resources 2007" (PDF). Retrieved 2008-08-11.

[7] http://www.abc.net.au/science/articles/2013/02/18/3691317.htm#.UbZSEUDVDTc

[8] Keddy, P.A. 2010. Wetland Ecology: Principles and Conservation (2nd edition). Cambridge University Press, Cambridge, UK. 497 p. Chapter 7.

[9] http://www.eurosaiwgea.org/Activitiesandmeetings/OtherEUROSAIWGEAmeetings/Documents/Estonia_energy.pdf

[10] http://www.seai.ie/Archive1/Files_Misc/IEABioenergyAgreementTask38CaseStudy.pdf

[11] http://www.un-documents.net/ocf-07.htm Today's primary sources of energy are mainly non-renewable: natural gas, oil, coal, peat, and conventional nuclear power. There are also renewable sources, including wood, plants, dung, falling water, geothermal sources, solar, tidal, wind, and wave energy, as well as human and animal muscle-power. Nuclear reactors that produce their own fuel ("breeders"), and eventually fusion reactors, are also in this category.

[12] http://www.worldenergy.org/publications/survey_of_energy_resources_2007/peat/704.asp

[13] The CO_2 emission factor of peat fuel. Imcg.net. Retrieved on 2011-05-09.

[14] IUCN UK Commission of Inquiry on Peatlands Full Report, IUCN UK Peatland Programme October 2011

[15] Fraser, L.H. Fraser and P.A. Keddy (eds.). 2005. The World's Largest Wetlands: Ecology and Conservation. Cambridge University Press, Cambridge, UK. 488 p. and P.A. Keddy (eds.). 2005. The World's Largest Wetlands: Ecology and Conservation. Cambridge University Press, Cambridge, UK. 488 p.

[16] Waspada.co.ik

[17] Vitt, D.H., L.A. Halsey and B.J. Nicholson. 2005. The Mackenzie River basin. Pp. 166-202 in L.H. Fraser and P.A. Keddy (eds.). The World's Largest Wetlands: Ecology and Conservation. Cambridge University Press, Cambridge, UK. 488 p.

[18] International Mire Conservation Group (2007-01-03). "Peat should not be treated as a renewable energy source" (PDF). Retrieved 2007-02-12.

[19] Peat and its significance in whisky

[20] Octomore peated whisky

[21] "Peatlands Park ASSI". *NI Environment Agency*. Retrieved 14 August 2010.

[22] "Somerset Peat Paper – Issues consultation for the Minerals Core Strategy" (PDF). Somerset County Council. September 2009. p. 7. Retrieved 30 November 2011.

[23] , Dartmoor history

[24] http://www.ccgc.gov.uk/landscape--wildlife/protecting-our-landscape/special-landscapes--sites/protected-landscape/national-nature-reserves/fenns-whixall-and-bettisfiel.aspx

[25] VAPO Company webpage

[26] Renewable energy sources and peat, Ministry of Trade and Industry of Finland, last updated: 04.07.2005

[27] Archived January 21, 2013 at the Wayback Machine

[28] VTT 2004: Wood in peat fuel – impact on the reporting of greenhouse gas emissions according to IPCC guidelines

[29]

[30] Serghey Stelmakovich. "Russia institutes peat fire prevention program". Retrieved August 9, 2010.

[31] MacDermott M (September 9, 2009). "Russia plans mining peat environmental disaster". Retrieved August 9, 2010.

[32] "2007 Survey of Energy Resources" (PDF). World Energy Council 2007. 2007. Retrieved 2011-01-23.

[33] "Peat: Useful Resource or Hazard?". Russian Geographical Society. August 10, 2010. Retrieved 2011-01-29.

[34] "Peat-free compost at Kew". RBG Kew. 2011. Retrieved 2011-06-24.

[35] Scheurmann, Ines (1985). *Natural Aquarium Handbook, The*. (trans. for Barron's Educational Series, Hauppauge, New York: 2000). Munich, Germany: Gräfe & Unzer GmbH.

[36] International Peat Society Peat Balneology, Medicine and Therapeutics

[37] Godwin, Sir Harry (1981). *The archives of the peat bogs*. Cambridge: Cambridge University Press.

[38] Rydin, Håkan; Jeglum, John K. (18 July 2013) [8 Jun 2006]. *The Biology of Peatlands*. Biology of Habitats (2 ed.). University of Oxford Press. p. 400. ISBN 0198528728.

[39] Keddy, P.A. (2010), *Wetland Ecology: Principles and Conservation* (2 ed.), Cambridge, UK.: Cambridge University Press, pp. 323–325

[40] Birks, Harry John Betteley; Birks, Hilary H. (2004) [1980]. *Quaternary Palaeoecology*. Blackburn Press. pp. 289 pages.

[41] Biester, Harald; Bindler, Richard (2009), *Modelling Past Mercury Deposition from Peat Bogs – The Influence of Peat Structure and 210Pb Mobility* (PDF), Working Papers of the Finnish Forest Research Institute (128), retrieved 21 October 2014

[42]

[43]

[44] MacDonald, Glen M.; Beilman, David W.; Kremenetski, Konstantine V.; Sheng, Yongwei; Smith, Laurence C. & Velichko, Andrei A. (2006). "Rapid early development of circumarctic peatlands and atmospheric CH_4 and CO_2 variations". *Science* **314** (5797): 285–288. doi:10.1126/science.1131722.

[45] Mitchell, Carla P. J.; Branfireun, Brian A. and Kolka, Randall K. (2008). "Spatial Characteristics of Net Methylmercury Production Hot Spots in Peatlands" (PDF). *Environmental Science and Technology* (American Chemical Society) **42** (4): 1010–1016. doi:10.1021/es0704986. Archived (PDF) from the original on 31 October 2008.

[46] Wetlands.org, Wetlands International | Peatlands and CO_2 Emissions

[47] Wetlands.org, The Global Peat CO2 Picture, Wetlands International and Greifswald University, 2010

[48] http://www.pitcherplant.org/

[49] http://www.unc.edu/news/archives/may03/lilly050903.html

[50] http://www.dmr.state.ms.us/Coastal-Ecology/preserves/plants/grasses-sedges-rushes/toothache-grass/toothache-grass.htm

[51] "Asian peat fires add to warming". *BBC News*. 2005-09-03. Retrieved 2010-05-22.

[52] Joel S. Levine (31 December 1999). *Wildland fires and the environment: a global synthesis*. UNEP/Earthprint. ISBN 978-92-807-1742-6. Retrieved 9 May 2011. web link

[53] Cat Lazaroff, Indonesian Wildfires Accelerated Global Warming, Environment News Service

[54] Fred Pearce Massive peat burn is speeding climate change, New Scientist, 6 November 2004

[55] "Florida Everglades". U.S. Geological Survey,. 15 January 2013. Retrieved 11 June 2013.

[56] Fenton, Nicole; Lecomte, Nicolas; Légaré, Sonia & Bergeron, Yves (2005). "Paludification in black spruce (*Picea mariana*) forests of eastern Canada: Potential factors and management implications". *Forest Ecology and Management* **213** (1–3): 151–159. doi:10.1016/j.foreco.2005.03.017.

[57] "Fog from peat fires blankets Moscow amid heat wave". *BBC*. 26 July 2010.

[58] "Russia begins to localize fires, others rage". *Associated Press*. 30 July 2010.

62.10 External links

- International Peat Society

- International Mire Conservation Group

- Cutover and Cutaway bogs from IPCC

- Gardening without peat information supplied by Kew gardens in London

- Peat-free gardens from the RSPB

- Massive peat burn is speeding climate change From The New Scientist

- King Class Torf in Turkey

- Meadowview Biological Research Station

- Industry - Peat

- Equipment for peat extraction

Chapter 63

Permaforestry

Permaforestry is an approach to the wildcrafting and harvesting of the forest biomass that uses cultivation to improve the natural harmonious systems. It is a relationship of interdependence between humans and the natural systems in which the amount of biomass available from the forest increases with the health of its natural systems.

Examples of bioproducts derived from biomass created through permaforestry include: honey, maple syrup and other tree saps, gourmet foods, functional foods, berries, wild mushrooms, ginseng, wild rice, herbs, fiddleheads, fish, frogs and crustaceans, pharmaceuticals, natural health products, essential oils, educational products, arts and crafts, decorative products, floral and greenery, garden horticultural products, woodworking, lumber, biochemicals, biofuels and bioenergy.

63.1 History

Permaforestry was extensively practiced by many aboriginal cultures throughout the world prior to colonization. It was replaced by industrial agriculture in most regions where the land could permit the use of machinery, monoculture, or intensive farming and harvesting practices. In the beginning of the 21st century there was a new surge of interest in permaforestry practices to address social issues such as food shortages, rural impoverishment, and changes in the logging industry. Furthermore, climate change and the "green" shift have inspired many individuals to revisit the old resource production methods that worked with nature rather than against it. The high price of agricultural land and machinery has also contributed to the development of permaforestry on land that had been previously classified as unsuitable for agriculture.

63.2 See also

- Biomass

- Biomass (ecology)

- Bioproduct

- Native

- Natural landscape

- Permaculture

- Terra preta

- Traditional ecological knowledge

- Wildness

- Wilderness

- Wildlife

- World Forestry Congress

Chapter 64

Pine tar

Pine tar is a sticky material produced by the high temperature carbonization of pine wood in anoxic conditions (dry distillation or destructive distillation). The wood is rapidly decomposed by applying heat and pressure in a closed container; the primary resulting products are charcoal and pine tar.

Pine tar consists primarily of aromatic hydrocarbons, tar acids and tar bases. Components of tar vary according to the pyrolytic process (e.g. method, duration, temperature) and origin of the wood (e.g. age of pine trees, type of soil and moisture conditions during tree growth). The choice of wood, design of kiln, burning and collection of the tar can vary. Only pine stumps and roots are used in the traditional production of pine tar.

Pine tar has a long history as a wood preservative, as a wood sealant for maritime use, in roofing construction and maintenance, in soaps and in the treatment of carbuncles and skin diseases, such as psoriasis, eczema, and rosacea.

64.1   History and general use

Pine tar has long been used in Scandinavian nations as a preservative for wood which may be exposed to harsh conditions, including outdoor furniture and ship decking and rigging. The high-grade pine tar used in this application is often called **Stockholm Tar**[2] since, for many years, a single company held a royal monopoly on its export out of Stockholm, Sweden.[3] It is also known as "Archangel Tar".[4] Tar and pitch for maritime use was in such demand that it became an important export for the American colonies (later United States), such as North Carolina, which had extensive pine forests. North Carolinians later became known as "Tar Heels."

It was used as a preservative on the bottoms of wooden, Nordic style skis until modern synthetic materials replaced wood in the construction of such skis. The pine tar also helped the adhesion of waxes which aided the grip and glide of such skis.

Pine tar is widely used as a veterinary care product.[5] It is a traditional antiseptic and hoof care product for horses and cattle.[5] Pine tar has been used when chickens start pecking the low hen. Applying a smear of pine tar on the wound gives the attacking hens something else to do. They are distracted by the effort of trying to get the sticky pine tar off their beaks.

Pine tar is now mainly used as a softening solvent in the rubber industry, and for construction material and special paints.

64.2   Use as a wood preservative

Pine tar can be used for preserving wooden boats (and other wood which will be exposed to the elements) by using a mixture of pine tar, gum turpentine and boiled linseed oil. First, a thin coat is applied using a mixture with greater turpentine. This allows it to permeate deeper into the oakum and fibre of the wood and lets the tar seep into any pinholes and larger gaps that might be in the planks. The tar weeps out to the exterior and indicates where the boat needs the most attention. Having the solution in place and the repairs complete, the vessel is ready for the thicker standard mix. Pine tar is also efficacious for properly saturating lead or standard oakum so that the endurance of the sealing capacity is optimal.

Such treatments, while effective, must be continually reapplied.

64.3   Use in weatherproofing rope

Traditionally, hemp and other natural fibers were the norm for rope production. Such rope would quickly rot when exposed to rain, and was typically tarred to preserve it. The tar would stain the hands of ship's crews, and British Navy seamen became known as "tars."

64.4 Use of pine tar in baseball

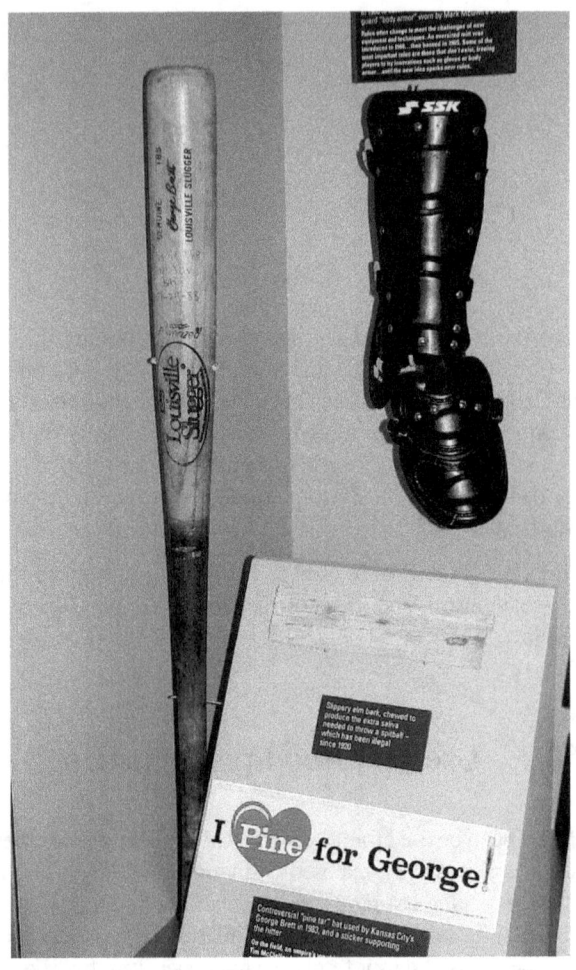

George Brett's pine tar bat at left, from a 2006 exhibit at The Henry Ford in Dearborn, Michigan

In baseball, pine tar is applied to the handles of baseball bats. Because of its texture, pine tar improves a batter's grip on the bat and prevents the bat from slipping out of the batter's hands during hard swings. It also helps hitters, because they do not have to grip the bat as hard and thus the hitter gets more "pop."

Rule 1.10(c) of the 2002 Official rules of Major League Baseball states that batters may apply pine tar only from the handle of the bat extending up 18 inches. The most famous example of the rule being applied is the Pine Tar Game, the July 24, 1983 game between the Kansas City Royals and New York Yankees in which George Brett hit a home run to put the Royals ahead 5–4. Yankees manager Billy Martin immediately protested that Brett's bat had more than 18 inches of pine tar. The umpires called Brett out and nullified the home run. However, league president Lee MacPhail overruled the umpires. MacPhail said that the pine tar restriction wasn't about competitive advantage, but economics. If too much pine tar was on the bat, pine tar would end up on the ball and render it unusable for play. MacPhail said that the umpires shouldn't have taken the home run off the board, but simply discarded the bat. The game was resumed from the point of the home run, and the Royals won.

Pine tar is also sometimes used by pitchers in baseball to improve grip on the ball in cold weather, although it is questionable whether it gives a pitcher any competitive advantage. However, the application of any foreign substance to a ball (except Baseball Rubbing Mud, which is applied by the umpires) is expressly prohibited by 8.02 of the Official rules of Major League Baseball. If a player is caught violating this rule, it results in an automatic ten-game suspension in the minor leagues. There is no mandatory suspension for this infraction at the major league level, although suspensions are often used to discipline offending players.*[6]

64.5 Medical

Pine tar has also been used for treating skin conditions, often as soap, though this use as a drug was banned by the FDA along with many other ingredients, due to a lack of proof of effectiveness.*[7]

Some pine tar products contain creosote, a probable carcinogen. This depends on whether they were produced in an open or closed kiln.*[8] Some soaps are accordingly advertised as "creosote-free."

It is used in veterinary medicine as an expectorant and an antiseptic in chronic skin conditions.*[1]

64.6 See also

- Tarring and feathering (punishment)

64.7 References

[1] *Merck Index*, 11th Edition, **7417**. p. 1182

[2] "Stockholm Tar". MedicAnimal.com. Retrieved 23 Sep 2012.

[3] Theodore P. Kaye. "Pine Tar; History And Uses". San Francisco Maritime Park Association. Retrieved 2010-08-01.

[4] Hugh Chisholm (1911). "Tar". *The Encyclopaedia Britannica* **26** (11 ed.). p. 414. Retrieved 2010-08-01.

[5] Wickstrom, Mark. "Phenols and Related Compounds". *The Merck Veterinary Manual.* Merck Manuals. Retrieved 16 April 2015.

[6] "Pineda Ejected for Pine Tar on Neck". espn.go.com. Retrieved 22 Apr 2014.

[7] Bonnie Aikman (11/07/1990). "Clean-Up of Ineffective Ingredients in OTC Drug Products" (Press release). Food and Drug Administration. Retrieved 2014-04-19. Check date values in: |date= (help)

[8] "Grandpa's Wonder Pine Tar Soap as Shampoo Bar". Badgerandblade.com. Retrieved 2014-04-19.

64.8 External links

- History of Pine Tar

Chapter 65

Pinyon pine

*Single-leaf Pinyon (*Pinus monophylla *subsp.* monophylla*)*

The **pinyon** or **piñon** pine group grows in the southwestern United States and in Mexico. The trees yield edible **pinyon nuts**, which were a staple of the Native Americans (American Indians), and are still widely eaten. Harvesting techniques of the prehistoric Indians are still being used to today to collect the pinyon seeds for personal use or for commercialization. The Pinyon nut or seed is high in fats and calories.

Pinyon wood, especially when burned, has a distinctive fragrance, making it a common wood to burn in chimineas.[*][1] The pinyon pine trees are also known to influence the soil in which they grow by increasing concentrations of both macronutrients and micronutrients.[*][2]

Some of the species are known to hybridize, the most notable ones being *P. quadrifolia* with *P. monophylla*, and *P. edulis* with *P. monophylla*.

The Pinyon Jay (*Gymnorhinus cyanocephalus*) takes its name from the tree, and pinyon nuts form an important part of its diet. It is very important for regeneration of pinyon woods, as it stores large numbers of the seeds in the ground for later use, and excess seeds not used are in an ideal position to grow into new trees. The Mexican Jay is also important for the dispersal of some pinyon species, as, less often,

is the Clark's Nutcracker. Many other species of animal also eat pinyon nuts, without dispersing them.

65.1 Species

A single-leaf pinyon from Mono County, California. The short stature and rounded crown are typical of the Pinyon.

A forest of two-needle pinyons in Grand Canyon National Park, Arizona.

340

Genetic differentiation in the pinyon pine has been observed associated to insect herbivory and environmental stress.[3][4]

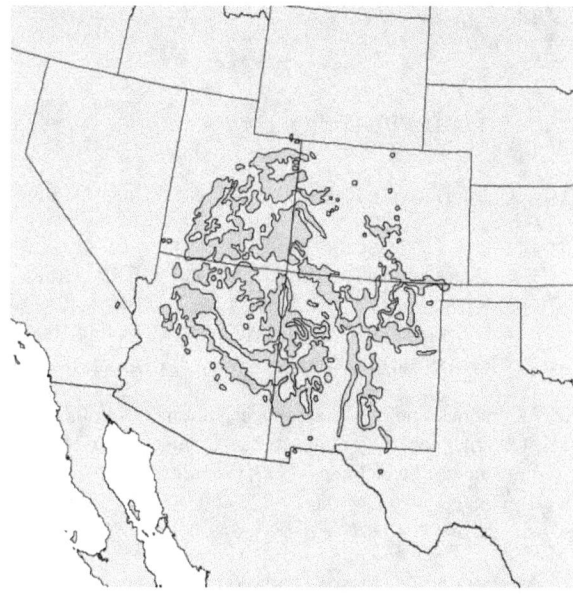

Range of the Colorado Pinyon, on the two most important species in the United States.

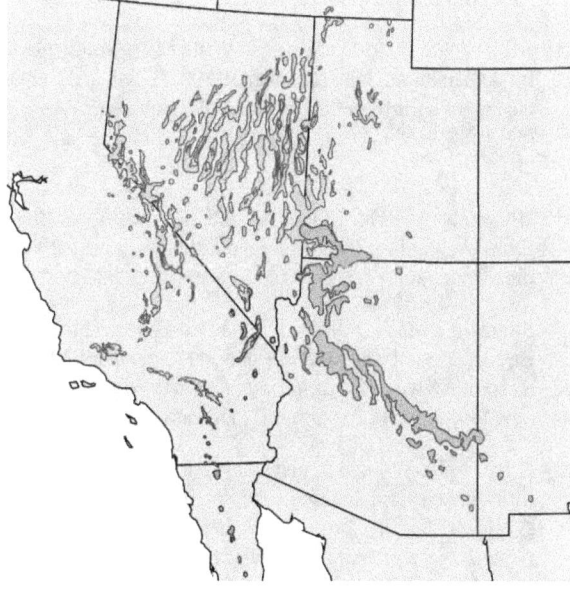

Range of the three sub-species of the Single-leaf Pinyon.

There are eight species of true pinyon (*Pinus* subsection *Cembroides*):[5]

- *Pinus cembroides* – Mexican pinyon

- *Pinus orizabensis* – Orizaba pinyon

- *Pinus johannis* – Johann's pinyon (includes *P. discolor* - Border pinyon)

- *Pinus culminicola* – Potosi pinyon

- *Pinus remota* – Texas pinyon or Papershell pinyon

- *Pinus edulis* – Two-needle piñon or Colorado pinyon (when grown in Colorado)

- *Pinus monophylla* – Single-leaf pinyon

- *Pinus quadrifolia* – Parry pinyon (includes *P. juarezensis*).

These additional Mexican species are also related and mostly called pinyons:

- *Pinus rzedowskii* – Rzedowski's pine

- *Pinus pinceana* – Weeping pinyon

- *Pinus maximartinezii* – Big-cone pinyon

- *Pinus nelsonii* – Nelson's pinyon

The three bristlecone pine species of the high mountains of the southwestern United States, and the lacebark pines of Asia are closely related to the pinyon pines.

65.2 Pinyon seeds as food for prehistoric Native Americans

The seeds of the Pinyon pine, usually called "pine nuts," were an important food for pre-historic Indians living in the mountains of southwestern United States and northern Mexico. The nuts continue to be gathered, eaten, and marketed. All species of pine produce edible seeds, but in North America only pinyon produces seeds large enough to be a major source of food.[6]

The Pinyon has probably been a source of food since shortly after the earliest arrival of *Homo sapiens* in the American southwest, 12,000 or more years ago. In the Great Basin of the United States, archaeological evidence indicates that the range of the pinyon pine expanded northward after the Ice Age, reaching its northernmost (and present) limit in southern Idaho about 4,000 BCE.[7] Hunter/gatherer Indians undoubtedly collected the edible seeds, but, at least in some areas, the pinyon nuts were not harvested and eaten in quantity until about 600 CE. Increased use of pinyon nuts was possibly related to a population increase of humans and a decline in the number of game animals, thereby forcing the Great Basin inhabitants to seek additional sources of food.[8]

The suitability of pinyon seeds as a staple food is reduced because of the unreliability of the harvest. Abundant crops of cones and seeds occur only every two to seven years, averaging a good crop every four years. Years of high production of seed tend to be the same over wide areas of the pinyon range.[*][9]

65.3 Traditional method of harvesting

In 1878, Naturalist John Muir described the Indian method of harvesting pinyon seeds in Nevada. In September and October, the harvesters knocked the cones off the pinyon trees with poles, stacked the cones into a pile, put brushwood on top, lit it, and lightly scorched the pinyon cones with fire. The scorching burned off the sticky resin coating the cones and loosened the seeds. The cones were then dried in the sun until the seeds could be easily extracted. Muir said the Indians closely watched the pinyon trees year-round and could predict the scarcity or abundance of the crop months before harvest time.[*][10] In 1891, B. H. Dutcher observed the harvesting of pinyon seeds by the Panamint Indians (Timbisha) people) in the Panamint Range overlooking Death Valley, California. The harvesting method was similar to the foregoing except that the pinyon seeds were extracted immediately after the cones had been scorched in the brushwood fire.[*][11]

Both the above accounts described a method of extracting the seeds from the green cones. Another method is to leave the cones on the trees until they are dry and brown, then beat the cones with a stick, knocking the cones loose or the seeds loose from the cones which then fall to the ground where they can be collected.[*][12] The nomadic hunter-gathering people of the Great Basin usually consumed their pinyon seeds during the winter following harvest; the agricultural Pueblo people of the Rio Grande valley of New Mexico could store them for two or three years in underground pits.[*][13]

Each pinyon cone produces 10 to 30 seeds and a productive stand of pinyon trees in a good year can produce 250 pounds (110 kg) on 1 acre (0.40 ha) of land. An average worker can collect about 22 pounds (10.0 kg) of unshelled pinyon seed in a day's work. Production per worker of 22 pounds of unshelled pinyon seeds more than one-half that in shelled seeds amounts to nearly 30,000 calories of nutrition. That is a high yield for the effort expanded by hunter-gatherers. Moreover, the pinyon seeds are high in fat, often in short supply for hunter-gatherers.[*][14]

65.4 See also

- Pinus *classification*
- Pine nut

65.5 References

[1] "Chiminea Woods: Pinon, Apple, and Hickory Oh My!".

[2] Barth, R.C. (January 1980). "Influence of Pinyon Pine Trees on Soil Chemical and Physical Properties". *Soil Science Society of America Journal* **44** (1): 112–114. doi:10.2136/sssaj1980.03615995004400010023x. Retrieved 6 August 2012.

[3] Whitham, Thomas G.; Mopper, Susan (1985-05-31). "Chronic Herbivory: Impacts on Architecture and Sex Expression of Pinyon Pine" (PDF). *Science* **228** (4703): 1089–1091. doi:10.1126/science.228.4703.1089. Retrieved 6 August 2012.

[4] Mopper, Susan; Mitton, Jeffry B.; Whitham, Thomas G.; Cobb, Neil S.; Christensen, Kerry M. (June 1991). "Genetic Differentiation and Heterozygosity in Pinyon Pine Associated with Resistance to Herbivory and Environmental Stress". *Evolution* **45** (4): 989–999. doi:10.2307/2409704. Retrieved 6 August 2012.

[5] Bentancourt, Julio L.; Schuster, William S.; Mitton, Jeffry B.; Anderson, R. Scott (October 1991). "Fossil and Genetic History of a Pinyon Pine (Pinus Edulis) Isolate". *Ecology* **72** (5): 1685–1697. doi:10.2307/1940968. Retrieved 6 August 2012.

[6] "Piñon Nuts: The Manna of the Mountains", *Mother Earth News*, http://www.motherearthnews.com/real-food/pinon-nuts-zmaz77jazgoe.aspx, accessed 29 Jul 2015

[7] Simms, Steven R. (1985), "Pine Nut Use in Three Great Basin Cases: Data, Theory, and a Fragmentary Material Record," *Journal of California and Great Basin Anthropology*, Vol. 7, No. 2, pp 166-167. Downloaded from JSTOR.

[8] Hildebrandt, William R. and Ruby, Allika (2006), "Prehistoric Pinyon Exploitation in the Southwestern Great Basin: A View from the Coso Range," *Journal of California and Great Basin Anthropology*, Vol. 26, No. 1, p. 11. Downloaded from JSTOR.

[9] Jeffers, Richard M., "Piñon PIne Seed Production, Collection, and Storage," United States Forest Service, http://www.fs.fed.us/rm/pubs_rm/rm_gtr258/rm_gtr258_191_197.pdf, accessed 30 July 2015

[10] Rhode, David (1988), "Two Nineteenth-Century Reports of Great Basin Subsistence Practices," *Journal of California and Great Basin Anthropology*, Vol. 10, No. 2, pp. 156-157. Downloaded from JSTOR.

[11] Dutcher, B. H. (Oct. 1893), "Piñon Gathering among the Panamint Indians", *American Anthropologist*, Vol. 6, No. 4, pp. 377-380. Downloaded from JSTOR.

[12] "Singleleaf Pinyon", USDA/NRCS, http://plants.usda.gov/plantguide/pdf/cs_pimo.pdf, accessed 30 Jul 2015

[13] "Indian Use of Pinyon-Juniper Woodlands" http://mojavedesert.net/plant-us/pinyon-juniper.html, accessed 30 Jul 2015

[14] Jeffers, pp. 195–196; "Piñon nuts, roasted (Navajo)", http://nutritiondata.self.com/facts/ethnic-foods/10473/2, accessed 30 July 2015

65.6 External links

- Pine classification

- Arboretum de Villardebelle Images of the cones of all the pinyons and allied pines

- *Pinus monophylla* U.S. Forest Service

- *Pinus cembroides* U.S. Forest Service

- *Pinus quadrifolia* U.S. Forest Service

- *Pinus edulis* U.S. Forest Service

Chapter 66

Pitch (resin)

For other uses, see Pitch.

 Pitch is a name for any of a number of viscoelastic, solid

The pitch shown in this pitch drop experiment has a viscosity approximately 230 billion times that of water.

polymers. Pitch can be natural or manufactured, derived from petroleum, coal tar[*][1] or plants. Various forms of pitch may also be called tar, bitumen or asphalt. Pitch produced from plants is also known as resin. Some products made from plant resin are also known as rosin.

Pitch was traditionally used to help caulk the seams of wooden sailing vessels (see shipbuilding). Pitch was also used to waterproof wooden containers, and in the making of torches. Petroleum-derived pitch is black in colour, hence the adjectival phrase, "pitch-black".

66.1 Viscoelastic properties

Naturally occurring asphalt/bitumen, a type of pitch, is a viscoelastic polymer. This means that even though it seems to be solid at room temperature and can be shattered with a hard impact, it is actually fluid and will flow over time, but extremely slowly. The pitch drop experiment taking place at University of Queensland is a long-term experiment which demonstrates the flow of a piece of pitch over many years. For the experiment, pitch was put in a glass funnel and allowed to slowly drip out. Since the pitch was allowed to start dripping in 1930, only nine drops have fallen. It was calculated in the 1980s that the pitch in the experiment has a viscosity approximately 230 billion (2.3×10^{11}) times that of water.[*][2] The eighth drop fell on 28 November 2000, and the ninth drop fell on 17 April 2014.[*][3] Another experiment was begun by a colleague of Nobel Prize winner Ernest Walton in the physics department of Trinity College in Ireland in 1944. Over the years, the pitch had produced several drops, but none had been recorded. On Thursday, July 11, 2013 scientists at Trinity College caught pitch dripping from a funnel on camera for the first time.[*][4]

The viscoelastic properties of pitch make it the vehicle of choice for polishing high-quality optical lenses and mirrors. In use the pitch is formed into a lap or polishing surface, which is charged with iron oxide or cerium oxide. The surface to be polished is pressed into the pitch, then rubbed against the surface so formed. The ability of pitch to flow, albeit slowly, keeps it in constant uniform contact with the optical surface.

66.2 Production

The heating (dry distilling) of wood causes tar and pitch to drip away from the wood and leave behind charcoal. Birch-bark is used to make birch-tar, a particularly fine tar. The terms tar and pitch are often used interchangeably. However, pitch is considered more solid, while tar is more liquid. Traditionally, pitch that was used for waterproofing buckets, barrels and small boats was drawn from pine. It is used to make Cutler's resin.

66.3 See also

- Asphaltene

- Tar

66.4 References

[1] COAL-TAR PITCH, HIGH TEMPERATURE

[2] The Pitch Drop Experiment

[3] Biever, Celeste; Lisa Grossman (17 April 2014). "Longest experiment sees pitch drop after 84-year wait" . *New Scientist*. Retrieved 23 June 2014.

[4] "Trinity College experiment succeeds after 69 years" . RTE News. 24 July 2013. Retrieved 23 June 2014.

66.5 External links

- The Pitch Drop Experiment

- Pine Tar Production

- Primitive tar and charcoal production

Chapter 67

Putu (mushroom)

Putu is the fleshy, fruiting body of a fungus, typically produced above ground on soil or on its food source in the forest area.

When the Indian mushroom is available, Chhattisgarh and Jashpurian people are very likely to eat it after it rains. When the monsoon starts to wet the forest area and if the sun warms the soil after raining then the mushroom will grow partially above and below the ground.

67.1 References

Chapter 68

Rattan

For the 1944 film, see Rattan (film).
"Ratan" redirects here. For locality situated in Jämtland County, Sweden, see Rätan.

Rattan (from the Malay *rotan*) is the name for the roughly 600 species of palms in the tribe **Calameae** (Greek 'kálamos' = reed), native to tropical regions of Africa, Asia and Australasia. Rattan is also known as *manila*, or *malacca*, named after the ports of shipment Manila and Malacca City, and as *manau* (from the Malay *rotan manau*, the trade name for *Calamus manan* canes in Southeast Asia.).[*][1]

68.1 Structure

Most rattans differ from other palms in having slender stems, 2–5 cm diameter, with long internodes between the leaves; also, they are not trees but are vine-like, scrambling through and over other vegetation. Rattans are also superficially similar to bamboo. Unlike bamboo, rattan stems ("malacca") are solid, and most species need structural support and cannot stand on their own. However, some genera (e.g. *Metroxylon, Pigafetta, Raphia*) are more like typical palms, with stouter, erect trunks. Many rattans have spines which act as hooks to aid climbing over other plants, and to deter herbivores. Rattans have been known to grow up to hundreds of metres long. Most (70%) of the world's rattan population exist in Indonesia, distributed among Borneo, Sulawesi, Sumbawa islands. The rest of the world's supply comes from the Philippines, Sri Lanka, Malaysia and Bangladesh

68.2 Economic and environmental issues

In forests where rattan grows, its economic value can help protect forest land, by providing an alternative to loggers who forgo timber logging and harvest rattan canes instead. Rattan is much easier to harvest, requires simpler tools and is much easier to transport. It also grows much faster than most tropical wood. This makes it a potential tool in forest maintenance, since it provides a profitable crop that depends on rather than replaces trees. It remains to be seen whether rattan can be as profitable or useful as the alternatives.

Rattans are threatened with overexploitation, as harvesters are cutting stems too young and reducing their ability to resprout.[*][2] Unsustainable harvesting of rattan can lead to forest degradation, affecting overall forest ecosystem services. Processing can also be polluting. The use of toxic chemicals and petrol in the processing of rattan affects soil, air and water resources, and also ultimately people's health. Meanwhile, the conventional method of rattan production is threatening the plant's long-term supply, and the income of workers.[*][3]

68.3 Uses

Generally, raw rattan is processed into several products to be used as materials in furniture making.[*][4] The various species of rattan range from several millimetres up to 5–7 cm in diameter. From a strand of rattan, the skin is usually peeled off, to be used as rattan weaving material. The remaining "core" of the rattan can be used for various purposes in furniture making. Rattan is a very good material mainly because it is lightweight, durable, suitable for outdoor use, and to a certain extent flexible.

68.3.1 Furniture making

Rattans are extensively used for making furniture and baskets. When cut into sections, rattan can be used as wood to make furniture. Rattan accepts paints and stains like many other kinds of wood, so it is available in many colours; and it can be worked into many styles. Moreover, the inner core can be separated and worked into wicker.

Indonesians making rattan furniture, circa 1948

Chair, Josephinism style, typical Viennese, around 1780

A rattan chair

68.3.2   Handicraft and arts

Many of the properties of rattan that make it suitable for furniture also make it a popular choice for handicraft and art pieces. Uses include rattan baskets, plant containers and other decorative works.

Due to its durability and resistance to splintering, sections of rattan can be used as staves or canes for martial arts 70 cm-long rattan sticks, called *baston*, are used in Filipino martial arts, especially Arnis/Eskrima/Kali and for the striking weapons in the Society for Creative Anachronism's full-contact "heavy combat".[5][6]

Along with birch and bamboo, rattan is a common material used for the handles in percussion mallets, especially mallets for keyboard percussion (vibraphone, xylophone, marimba, etc.).

It is also used to make walking sticks and crooks for high-end umbrellas.

68.3.3   Rattan as a shelter material

Most natives or locals from the rattan rich countries employ the aid of this sturdy plant in their home building projects. It is heavily used as a housing material in the rural areas. The skin of the plant or wood is primarily used for weaving.[7]

68.3.4   Food source and medicinal potential

The fruit of some rattans exudes a red resin called dragon's blood.Some rattan fruits are edible,with sour taste akin to citrus. This resin was thought to have medicinal properties in antiquity and was also used as a dye for violins, among other things.[8] The resin normally results in a wood with a light peach hue. In the Indian state of Assam, the shoot is also used as vegetable.

68.3.5 Corporal punishment

Thin rattan canes were the standard implement for school corporal punishment in England and Wales, and are still used for this purpose in schools in Singapore, Malaysia and several African countries - and similar canes are used for military punishments in the Singapore Armed Forces,*[9]

Heavier canes, also of rattan, are used for judicial corporal punishments in Malaysia, Aceh, Singapore, and Brunei*[10]

68.3.6 Other uses

Traditionally the women of the Wemale ethnic group of Seram Island, Indonesia wore rattan girdles around their waist.*[11]

In early 2010, scientists in Italy announced that rattan wood would be used in a new "wood to bone" process for the production of artificial bone. The process takes small pieces of rattan and places it in a furnace. Calcium and carbon are added. The wood is then further heated under intense pressure in another oven-like machine and a phosphate solution is introduced. This process produces almost an exact replica of bone material. The process takes about 10 days. At the time of the announcement the bone was being tested in sheep and there had been no signs of rejection. Particles from the sheep's bodies have migrated to the "wood bone" and formed long continuous bones. The new bone-from-wood programme is being funded by the European Union. Implants into humans are anticipated to start in 2015.*[12]

68.4 References

[1] Johnson, Dennis V (2004): *Rattan Glossary: And Compendium Glossary with Emphasis on Africa*. Rome: Food and Agriculture Organization of the United Nations, p. 22

[2] MacKinnon, K. (1998) Sustainable use as a conservation tool in the forests of South-East Asia. Conservation of Biological Resources (E.J. Milner Gulland & R Mace, eds), pp 174–192. Blackwell Science, Oxford.

[3] "WWF Rattan Switch project". WWF. July 2010. Archived from the original on 2010-08-03. Retrieved 16 July 2010.

[4] Rattan, Furniture. "Rattan Furniture". Retrieved 24 December 2011.

[5] "What is the SCA?". Society for Creative Anachronism, Inc. Retrieved 14 July 2012. Since we prefer that no one gets hurt, SCA combatants wear real armor and use rattan swords.

[6] "Marshals' Handbook" (PDF). Society for Creative Anachronism. March 2007 revision. Retrieved 16 March 2010. Check date values in: |date= (help)

[7] All About Rattan at Rattancraft.com.

[8] "Rattan" at Encyclopedia.com.

[9] Singapore: Caning in the military forces at World Corporal Punishment Research (includes a photograph of a military caning in progress).

[10] Judicial caning in Singapore, Malaysia and Brunei at World Corporal Punishment Research.

[11] Jaqueline M. Piper, *Bamboo and rattan, traditional uses and beliefs*, Oxford Univ Press, 1995, ISBN 978-0195889987

[12] "Turning wood into bones" . *BBC News*. 8 January 2010. Retrieved 22 May 2010.

68.5 Further reading

• Siebert, Stephen F. 2012. *The Nature and Culture of Rattan: Reflections on Vanishing Life in the Forests of Southeast Asia*. University of Hawai'i Press. ISBN 978-0-8248-3536-1

68.6 External links

• "Rattan". *Encyclopedia Americana*. 1920.

Chapter 69

Resin

This article is about the plant secretion. For other uses, see Resin (disambiguation).

In polymer chemistry and materials science, **resin** is a

Cedar of Lebanon cone showing flecks of resin as used in the mummification of Egyptian Pharaohs.

"solid or highly viscous substance, usually containing pre-polymers with reactive groups." [1] Such viscous substances can be plant derived or synthetic.

69.1 Examples of natural resins

Notable examples of plant resins include amber, Balm of Gilead, balsam, Canada balsam, Boswellia, copal from

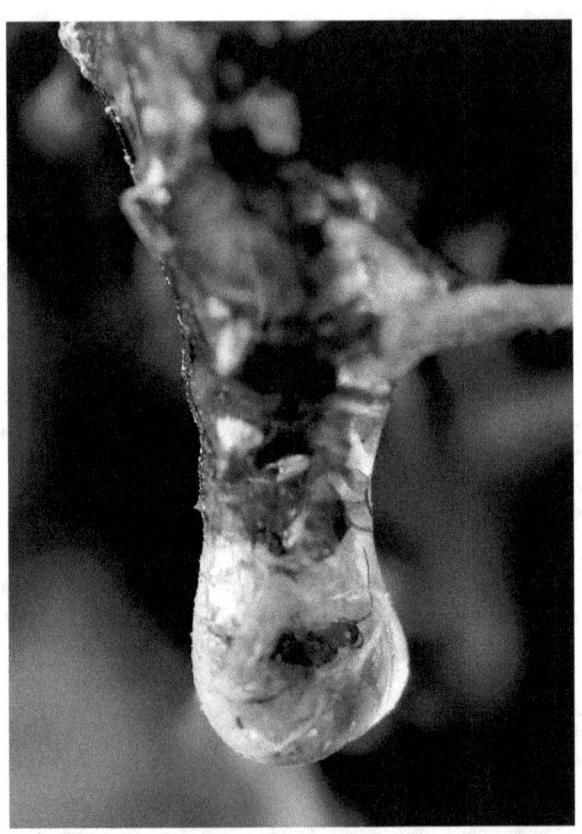

Insect trapped in resin

trees of *Protium copal* and *Hymenaea courbaril*, dammar gum from trees of the family Dipterocarpaceae, Dragon's blood from the dragon trees (*Dracaena* species), elemi, frankincense from *Boswellia sacra*, galbanum from *Ferula gummosa*, gum guaiacum from the lignum vitae trees of the genus *Guaiacum*, kauri gum from trees of *Agathis australis*, labdanum from mediterranean species of *Cistus*, mastic (plant resin) from the mastic tree *Pistacia lentiscus*, myrrh from shrubs of *Commiphora*, sandarac resin from *Tetraclinis articulata*, the national tree of Malta, styrax (a Benzoin resin from various *Styrax* species), Spinifex resin from Australian *Spinifex* grasses, and turpentine, distilled from pine resin. Amber is fossil resin (also called resinite) from coniferous and other tree species. Copal, kauri gum, dammar and other resins may also be found as subfossil deposits. Subfossil copal can be distinguished from genuine fossil amber because it becomes tacky when a drop of a solvent such as acetone or chloroform is placed on it.[2] Rosin is a solid resin obtained from pines and some other plants, mostly conifers, produced by heating fresh liquid resin to vaporize the volatile liquid (terpene) components.[3] Plant resins are generally produced as stem secretions, but in some South American species such as *Euphorbia dalechampia* and *Clusia* species they are produced as pollination rewards, and used by some stingless bee species to construct their nests.[4][5] Propolis, consisting largely of resins collected from plants such as poplars and conifers, is used by honey bees to seal gaps in their hives.

Shellac and lacquer are insect-derived resins.

Asphaltite and Utah resin are petroleum bitumens, not a product secreted by plants, although it was ultimately derived from plants.

69.2   History and etymology

Plant resins have a very long history that was documented in ancient Greece by Theophrastus, in ancient Rome by Pliny the Elder, and especially in the resins known as frankincense and myrrh, prized in ancient Egypt.[6] These were highly prized substances, and required as incense in some religious rites. Amber is a hard fossilized resin from ancient trees.

The word *resin* comes from French *resine*, from Latin *resina* "resin", which either derives from or is a cognate of the Greek ῥητίνη *rhētinē* "resin of the pine", of unknown earlier origin, though probably non-Indo-European.[7][8]

The word "resin" has been applied in the modern world to nearly any component of a liquid that will set into a hard lacquer or enamel-like finish. An example is nail polish, a modern product which contains "resins" that are organic compounds, but not classical plant resins. Certain "casting resins" and synthetic resins (such as epoxy resin) have also been given the name "resin" because they solidify in the same way as some plant resins, but synthetic resins are liquid monomers of thermosetting plastics, and do not derive from plants.

69.3   Plant resins

Plants secrete resins and rosins for their protective benefits. They confound a wide range of herbivores, insects, and pathogens; while the volatile phenolic compounds may attract benefactors such as parasitoids or predators of the herbivores that attack the plant.[9]

The resin produced by most plants is a viscous liquid, composed mainly of terpenes, with lesser components of dissolved non-volatile solids, which make resin thick and sticky. The most common terpenes in resin are the bicyclic terpenes alpha-pinene, beta-pinene, delta-3 carene, and sabinene, the monocyclic terpenes limonene and terpinolene, and smaller amounts of the tricyclic sesquiterpenes, longifolene, caryophyllene and delta-cadinene. Some resins also contain a high proportion of resin acids. The individual components of resin can be separated by fractional distillation. Rosins on the other hand are less volatile and consist inter alia of diterpenes.

The composition of resins varies with the species. A notable case Jeffrey Pine and Gray Pine, the volatile components of their wood resins are largely pure *n*-heptane with little or no terpenes.

Some resins when soft are known as 'oleoresins', and when containing benzoic acid or cinnamic acid they are called balsams. Oleoresins are naturally occurring mixtures of an oil and a resin; they can be extracted from various plants. Other resinous products in their natural condition are a mix with gum or mucilaginous substances and known as gum resins. Several natural resins are used as ingredients in perfumes, e.g., balsams of Peru and tolu, elemi, styrax, and certain turpentines.[3]

Certain resins are obtained in a fossilized condition, amber being the most notable instance of this class; African copal and the kauri gum of New Zealand are also procured in a semi-fossil condition.

69.3.1   Non-resinous exudates

Other liquid compounds found inside plants or exuded by plants, such as sap, latex, or mucilage, are sometimes confused with resin, but are not the same. Saps, in particular, serve a nutritive function that resins do not.

69.3.2   Derivatives

See also: Rosin

Solidified resin from which the volatile terpene components have been removed by distillation is known as rosin. Typical rosin is a transparent or translucent mass, with a vitreous fracture and a faintly yellow or brown colour, non-odorous or having only a slight turpentine odor and taste.

Rosin is insoluble in water, mostly soluble in alcohol, essential oils, ether and hot fatty oils, and softens and melts under the influence of heat, is not capable of sublimation,

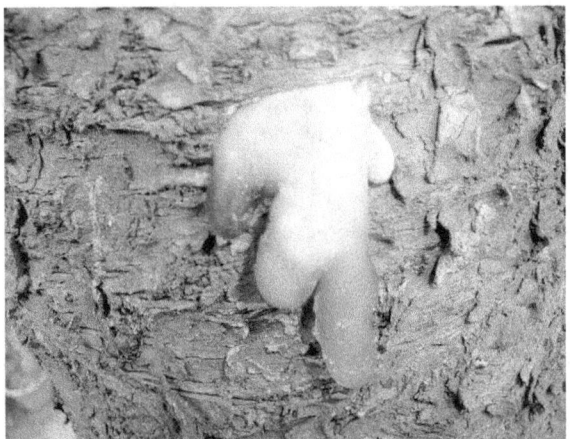

Extremely viscous resin extruding from the trunk of a mature Araucaria columnaris.

Resin of a pine

and burns with a bright but smoky flame. Rosin consists of a complex mixture of different substances including organic acids named the resin acids. These are closely related to the terpenes, and derive from them through partial oxidation. Resin acids can be dissolved in alkalis to form resin soaps, from which the purified resin acids are regenerated by treatment with acids. Examples of resin acids are abietic acid (sylvic acid), $C_{20}H_{30}O_2$, plicatic acid contained in cedar, and pimaric acid, $C_{20}H_{30}O_2$, a constituent of galipot resin. Abietic acid can also be extracted from rosin by means of hot alcohol; it crystallizes in leaflets, and on oxidation yields trimellitic acid, isophthalic acid and terebic acid. Pimaric acid closely resembles abietic acid into which it passes when distilled in a vacuum; it has been supposed to consist of three isomers.

69.4   Uses

69.4.1   Plant resins

Natural plant resins were once valued for the production of varnishes, adhesives, and food glazing agents. They are also prized as raw materials for organic synthesis, and provide constituents of incense and perfume.

The hard transparent resins, such as the copals, dammars, mastic and sandarac, are principally used for varnishes and adhesives, while the softer odoriferous oleoresins (frankincense, elemi, turpentine, copaiba) and gum resins containing essential oils (ammoniacum, asafoetida, gamboge, myrrh, and scammony) are more largely used for therapeutic purposes and incense. The resin of the Aleppo Pine is used to flavour retsina, a Greek resinated wine.[*][10]

Lumps of dried frankincense resin

69.4.2   Synthetic resins

Many materials are produced via the conversion of synthetic resins to solids. Important examples are bisphenol A diglycidyl ether, which is a resin converted to epoxy glue upon the addition of a hardnener. Silicone are often prepared from silicone resins via room temperature vulcanization.

69.5 See also

- Resin extraction – method of harvesting resin from trees

- Balsam of Peru – a balsam used in food and drink for flavoring, in perfumes and toiletries for fragrance, and in medicine and pharmaceutical items.

- Mastic (plant resin) – resin from the Pistacia lentiscus tree

- Pitch (resin)

- Kino (gum) – a plant gum similar to resin

- Biodegradable – plant resins are naturally biodegradable in many circumstances.

- Resin casting – casting with a resin, usually using a synthetic not a natural resin.

- Polyresin – a hard, synthetic resin for casting in molds

69.6 References

[1] http://goldbook.iupac.org/RT07166.html

[2] David Grimaldi, *Amber: Window to the Past*, 1996, p 16-20, American Museum of Natural History

[3] Fiebach, Klemens; Grimm, Dieter (2000). "Resins, Natural". *Ullmann's Encyclopedia of Industrial Chemistry*. doi:10.1002/14356007.a23_073. ISBN 978-3-527-30673-2.

[4] Bittrich, V.; Amaral, Maria C. E. (1996). "Flower morphology and pollination biology of some Clusia species from the Gran Sabana (Venezuela)". *Kew Bulletin* **51** (4): 681–694.

[5] Gonçalves-Alvim, Silmary de Jesus (2001). "Resin-collecting bees (Apidae) on Clusia palmicida (Clusiaceae) in a riparian forest in Brazil". *Journal of Tropical Ecology* **17** (1): 149–153.

[6] "Queen Hatshepsut's expedition to the Land of Punt: The first oceanographic cruise?". Dept. of Oceanography, Texas A&M University. Retrieved 2010-05-08.

[7] "resin, n. and adj." . *OED Online*. Oxford University Press. September 2014. Retrieved 2 December 2014.

[8] "resin (n.)". *Online Etymology Dictionary*. Retrieved 2 December 2014.

[9] "Plant Resins: Chemistry, evolution, ecology, and ethnobotany" , by Jean Langenheim, Timber Press, Portland, OR. 2003

[10] http://www.fao.org/docrep/x0453e/x0453e10.htm

69.7 External links

- The dictionary definition of Resin at Wiktionary

- Media related to Resin at Wikimedia Commons

Chapter 70

Root beer

A glass of root beer with foam.

Root beer is a dark brown sweet beverage traditionally made using the root or bark of the tree *Sassafras albidum* (sassafras) or the vine *Smilax ornata* (sarsaparilla) as the primary flavor. Root beer may be alcoholic or non-alcoholic, and may be carbonated or non-carbonated. Most root beer has a thick foamy head when poured. Modern, commercially produced root beer is generally sweet, foamy, carbonated, and non-alcoholic, and is flavoured using artificial sassafras. It may or may not contain caffeine.

70.1 History

Sassafras root beverages were made by Native Americans for culinary and medicinal reasons before the arrival of Europeans in North America, but European culinary techniques have been applied to making traditional sassafras-based beverages similar to root beer since the 16th and 17th centuries. Root beer was sold in confectionery stores since the 1840s, and written recipes for root beer have been documented since the 1860s. It is possible that it was combined with soda as early as the 1850s, and root beer sold in stores was most often sold as a syrup rather than a ready-made beverage.*[1] The tradition of brewing root beer is thought to have evolved out of other small beer traditions that produced fermented drinks with very low alcohol content that were thought to be healthier to drink than possibly tainted local sources of drinking water, and enhanced by the medicinal and nutritional qualities of the ingredients used. Beyond its aromatic qualities, the medicinal benefits of sassafras were well known to both Native Americans and Europeans, and druggists began marketing root beer for its medicinal qualities.*[2]

Pharmacist Charles Elmer Hires was the first to successfully market a commercial brand of root beer. Hires discovered his root tea made from sassafras in 1875, debuted a commercial version of root beer at the Philadelphia Centennial Exposition in 1876, and began selling his extract. Hires was a teetotaler who wanted to call the beverage "root tea". However, his desire to market the product to Pennsylvania coal miners caused him to call his product "root beer" instead.*[3]*[4] In 1886, Hires began to bottle a beverage made from his famous extract. By 1893, root beer was distributed widely across the United States. Non-alcoholic versions of root beer became commercially successful, especially during Prohibition.*[5]*[6]

Not all traditional or commercial root beers were sassafras-based. One of Hires's early competitors was Barq's, which began selling its sarsaparilla-based root beer in 1898 and was labeled simply as "Barq's". *[7] In 1919, Roy Allen opened his root beer stand in Lodi, California, which led to the development of A&W root beer. One of Allen's innovations was that he served his homemade root beer in cold, frosty mugs. IBC is another brand of commercially produced root beer that emerged during this period and is still well-known today.[5]

Safrole, the aromatic oil found in sassafras roots and bark that gave traditional root beer its distinctive flavour, was banned for commercially mass-produced foods and drugs by the FDA in 1960.[8] Laboratory animals that were given oral doses of sassafras tea or sassafras oil that contained large doses of safrole developed permanent liver damage or various types of cancer.[8] While sassafras is no longer used in commercially produced root beer and is sometimes substituted with artificial flavors, natural extracts with the safrole distilled and removed are available.[9][10]

70.2   Traditional method

One traditional recipe for making root beer involves cooking a syrup from molasses and water, letting the syrup cool for three hours, combining the root ingredients (including sassafras root, sassafras bark, and wintergreen). Yeast was added, and the beverage was left to ferment for 12 hours, after which it was strained and rebottled for secondary fermentation. This recipe would usually result in a beverage of 2% alcohol or less, although the recipe could be modified to produce a more alcoholic beverage.[11]

70.3   Ingredients

Commercial root beer is now produced in every U.S. state.[12] It is a beverage almost exclusive to North America, yet there are a few brands produced in other countries, such as the Philippines and Thailand. The flavor of these beverages often varies from typical North American versions.[13] While there is no standard recipe, the primary ingredients in modern rootbeer are sugar and artificial sassafras flavoring, which complement other flavors. Common flavorings are vanilla, wintergreen, cherry tree bark, licorice root, sarsaparilla root, nutmeg, acacia, anise, molasses, cinnamon, and honey. Soybean protein is sometimes used to create a foamy quality, and caramel-coloring is used to make the beverage brown.[11]

Ingredients in early and traditional root beers include allspice, birch bark, coriander, juniper, ginger, wintergreen, hops, burdock root, dandelion root, spikenard, pipsissewa, guaiacum chips, sarsaparilla, spicewood, wild cherry bark, yellow dock, prickly ash bark, sassafras root, vanilla beans, dog grass, molasses and licorice.[14] Many of these ingredients are still used in traditional and commercially produced root beer today, which is often thickened, foamed, or carbonated. Although most mainstream brands are caffeine-free, Barq's does contain caffeine.[15]

Root beer may be made at home with processed extract obtained from a factory,[16] or it can also be made from herbs and roots that have not yet been processed. Alcoholic and non-alcoholic traditional root beers make a thick and foamy head when poured, often enhanced by the addition of yucca extract or other thickeners.

70.3.1   List of main ingredients

Roots and herbs

- *Sassafras albidum* – Sassafras roots and bark containing the aromatic oil safrole (or an artificial substitute)

- *Smilax regelii* – Sarsaparilla

- *Smilax glyciphylla* – Sweet Sarsaparilla

- *Piper auritum* – Root Beer Plant or Hoja Santa

- *Glycyrrhiza glabra* – Liquorice (root)

- *Aralia nudicaulis* – Wild Sarsaparilla or "Rabbit Root"

- *Gaultheria procumbens* – Wintergreen (leaves and berries)

- *Betula lenta* – Sweet Birch (sap/syrup/resin)

- *Betula nigra* – Black Birch (sap/syrup/resin)

- *Prunus serotina* – Black Cherry

- *Picea rubens* – Red Spruce

- *Picea mariana* – Black Spruce

- *Picea sitchensis* – Sitka Spruce

- *Arctium lappa* – Burdock (root)

- *Taraxacum officinale* – Dandelion (root)

Foam

- *Quillaja saponaria* – Soapbark

- *Manihot esculenta* – Cassava, Manioc or Yucca (root)

Spices

- *Pimenta dioica* – Allspice
- *Theobroma cacao* – Chocolate.
- *Trigonella foenum-graecum* – Fenugreek
- *Myroxylon balsamum* – Tolu balsam
- *Abies balsamea* – Balsam Fir
- *Myristica fragrans* – Nutmeg
- *Cinnamomum verum* – Cinnamon (bark)
- *Cinnamomum aromaticum* – Cassia (bark)
- *Syzygium aromaticum* – Clove
- *Foeniculum vulgare* – Fennel (seed)
- *Zingiber officinale* – Ginger (stem/rhizome)
- *Illicium verum* – Star Anise
- *Pimpinella anisum* – Anise
- *Humulus lupulus* – Hops
- *Mentha* species – Mint

Other ingredients

- *Hordeum vulgare* – Barley (malted)
- *Hypericum perforatum* – St. John's Wort
- Sugar
- Molasses
- Yeast

70.4   See also

- Apple Beer
- Beer
- Beverage
- Birch beer
- Category:Root beer stands
- Dandelion and burdock
- Ginger beer
- Horehound beer

- Julmust
- List of brand name soft drinks products
- List of soft drink flavors
- Root beer float
- Root
- Sarsaparilla (soft drink) – a similar, although distinct, beverage
- Spruce beer
- Malta (soft drink)
- Malzbier
- Sassafras albidum

- Lewis and Clark

70.5   References

[1] Smith, Andrew (August 30, 2006). *Encyclopedia of Junk Food and Fast Food*. Greenwood. p. 231-232. ISBN 978-0313335273.

[2] Cresswell, Stephen (January 6, 1998). *Homemade Root Beer, Soda & Pop*. Storey Publishing. p. 4. ISBN 978-1580170529.

[3] Funderburg, Anne Cooper (2002). *Sundae Best: A History of Soda Fountains*. Popular Press. pp. 93–95. ISBN 978-0879728540.

[4] "Eric's Gourmet Root Beer Site - History" . Retrieved 8 February 2015.

[5] Smith, Andrew (November 30, 2012). *The Oxford Encyclopedia of Food and Drink in America*. p. 1, 188. ISBN 978-0199734962.

[6] Bennett, Eileen (June 28, 1998). "Local Historians Argue Over the Root of Hires" . *The Press of Atlantic City*. Retrieved April 5, 2015.

[7] Boudreaux, Edmond (February 5, 2013). *Legends and Lore of the Mississippi Golden Gulf Coast*. The History Press. p. 145. ASIN B00BBXFJOC.

[8] Dietz, B; Bolton, Jl (Apr 2007). "Botanical dietary supplements gone bad." . *Chemical research in toxicology* **20** (4): 586–90. doi:10.1021/tx7000527. ISSN 0893-228X. PMC 2504026. PMID 17362034.

[9] http://www.accessdata.fda.gov/scripts/cdrh/cfdocs/cfCFR/CFRSearch.cfm?fr=172.580

[10] Higgins, Nadia (August 1, 2013). *Fun Food Inventions (Awesome Inventions You Use Every Day)*. 21st Century. p. 30. ISBN 978-1467710916.

[11] Sokolov, Raymond (April 5, 1993). *Why We Eat What We Eat: How Columbus Changed the Way the World Eats*. Touchstone. p. 174. ISBN 978-0671797911.

[12] "Brands - A World of Root Beer Resources - Root Beer World". Retrieved 8 February 2015.

[13] "anthony's root beer barrel". *anthony's root beer barrel*. Retrieved 8 February 2015.

[14] Bellis, Mary. "The History of Root Beer." About Money. Web. 5 Mar. 2015.

[15] "F.A.Qs". *anthony's root beer barrel*. Retrieved 8 February 2015.

[16] MAKING ROOT BEER AT HOME by David B. Fankhauser

Chapter 71

Rubia

"Madder" redirects here. For other uses, see Madder (disambiguation).

Rubia is a genus of flowering plants in the Rubiaceae family. It contains around 80 species of perennial scrambling or climbing herbs and subshrubs native to the Old World.*[1] The genus and its best-known species are commonly known as **madder**, e.g. *Rubia tinctorum* (common madder), *Rubia peregrina* (wild madder), and *Rubia cordifolia* (Indian madder).*[2]

71.1 Uses

Rubia was an economically important source of a red pigment in many regions of Asia, Europe and Africa. Several species, such as *Rubia tinctorum* in Europe, *Rubia cordifolia* in India, and, *Rubia argyi* in east Asia, were extensively cultivated from antiquity until the mid nineteenth century. The genus name *Rubia* derives from the Latin *ruber* meaning "red".

The plant's roots contain an organic compound called Alizarin, that gives its red colour to a textile dye known as Rose madder. It was also used as a colourant, especially for paint, that is referred to as Madder lake. The invention of a synthesized duplicate, an anthracene compound called alizarin, greatly reduced demand for the natural derivative.*[3]

71.2 Species

- *Rubia agostinhoi* Dans. & P.Silva
- *Rubia aitchisonii* Deb & Malick
- *Rubia alaica* Pachom.
- *Rubia alata* Wall.
- *Rubia albicaulis* Boiss.

- *Rubia angustisissima* Wall. ex G.Don
- *Rubia argyi* (H.Lév. & Vaniot) Hara ex Lauener
- *Rubia atropurpurea* Decne.
- *Rubia balearica* (Willk.) Porta
- *Rubia caramanica* Bornm.
- *Rubia charifolia* Wall. ex G.Don
- *Rubia chinensis* Regel & Maack
- *Rubia chitralensis* Ehrend.
- *Rubia clematidifolia* Blume ex Decne.
- *Rubia cordifolia* L.
- *Rubia crassipes* Collett & Hemsl.
- *Rubia cretacea* Pojark.
- *Rubia danaensis* Danin
- *Rubia davisiana* Ehrend.
- *Rubia deserticola* Pojark.
- *Rubia discolor* Turcz.
- *Rubia dolichophylla* Schrenk
- *Rubia edgeworthii* Hook.f.
- *Rubia falciformis* H.S.Lo
- *Rubia filiformis* F.C.How ex H.S.Lo
- *Rubia florida* Boiss.
- *Rubia fruticosa* Aiton
- *Rubia garrettii* Craib
- *Rubia gedrosiaca* Bornm.
- *Rubia haematantha* Ary Shaw

- *Rubia hexaphylla* (Makino) Makino

- *Rubia himalayensis* Klotzsch

- *Rubia hispidicaulis* D.G.Long

- *Rubia horrida* (Thunb.) Puff

- *Rubia infundibularis* Hemsl. & Lace

- *Rubia jesoensis* (Miq.) Miyabe & Kudo

- *Rubia komarovii* Pojark.

- *Rubia krascheninnikovii* Pojark.

- *Rubia laevissima* Tschern.

- *Rubia latipetala* H.S.Lo

- *Rubia laurae* (Holmboe) Airy Shaw

- *Rubia laxiflora* Gontsch.

- *Rubia linii* J.M.Chao

- *Rubia magna* P.G.Xiao

- *Rubia mandersii* Collett & Hemsl.

- *Rubia manjith* Roxb. ex Fleming

- *Rubia maymanensis* Ehrend. & Schönb.-Tem.

- *Rubia membranacea* Diels

- *Rubia oncotricha* Hand.-Mazz.

- *Rubia oppositifolia* Griff.

- *Rubia ovatifolia* Z.Ying Zhang ex Q.Lin

- *Rubia pallida* Diels

- *Rubia pauciflora* Boiss.

- *Rubia pavlovii* Bajtenov & Myrz.

- *Rubia peregrina* L.

- *Rubia petiolaris* DC.

- *Rubia philippinensis* Elmer

- *Rubia podantha* Diels

- *Rubia polyphlebia* H.S.Lo

- *Rubia pterygocaulis* H.S.Lo

- *Rubia rechingeri* Ehrend.

- *Rubia regelii* Pojark.

- *Rubia rezniczenkoana* Litv.

- *Rubia rigidifolia* Pojark.

- *Rubia rotundifolia* Banks & Sol.

- *Rubia salicifolia* H.S.Lo

- *Rubia schugnanica* B.Fedtsch. ex Pojark.

- *Rubia schumanniana* E.Pritz.

- *Rubia siamensis* Craib

- *Rubia sikkimensis* Kurz

- *Rubia sylvatica* (Maxim.) Nakai

- *Rubia tatarica* (Trevir.) F.Schmidt

- *Rubia tenuifolia* d'Urv.

- *Rubia tenuissima* ined.

- *Rubia thunbergii* DC.

- *Rubia tibetica* Hook.f.

- *Rubia tinctorum* L.

- *Rubia transcaucasica* Grossh.

- *Rubia trichocarpa* H.S.Lo

- *Rubia truppeliana* Loes.

- *Rubia wallichiana* Decne.

- *Rubia yunnanensis* Diels

71.3 References

[1] "*Rubia* in the World Checklist of Rubiaceae". Retrieved April 2014.

[2] Cannon J, Cannon M (2002). *Dye Plants and Dyeing* (2 ed.). A & C Black. pp. 76–80. ISBN 978-0-7136-6374-7.

[3] "Material Name: madder". *material record*. Museum of Fine Arts, Boston. November 2007. Retrieved 2009-01-01.

71.4 External links

- *Rubia* in the World Checklist of Rubiaceae

Chapter 72

Sassafras

This article is about the various species of the sassafras tree of the Northern Hemisphere; for the North American sassafras, see Sassafras albidum. For other uses, see Sassafras (disambiguation).

Sassafras is a genus of the extant and one extinct species of deciduous trees in the family Lauraceae, native to eastern North America and eastern Asia.[*][1][*][2][*][3] The genus is distinguished by its aromatic properties, which have made the tree useful to humans.

72.1 Overview

Male and female Sassafras albidum *flowers. The male flower is on the left; the female is on the right. The male flower has nine stamens (one partially obscured), while the female has a central pistil.*

72.1.1 Name

The name "sassafras", applied by the botanist Nicolas Monardes in 1569, comes from the French *sassafras*. Some sources claim it originates from the Latin *saxifraga* or *saxifragus*: "stone-breaking;" *saxum* "rock" + *frangere* "to break").[*][4][*][5] Early European colonists reported that the plant was called *winauk* by native Americans in Delaware

and Virginia and *pauane* by the Timucua. Native Americans distinguished between white sassafras and red sassafras, which terms referred to the same plant but to different parts of the plant with distinct colors and uses.[*][6] Sassafras was known as fennel wood (German *fenchelholz*) due to its distinctive aroma.[*][7]

72.1.2 Description

Sassafras trees grow from 9–35 m (30–115 ft) tall with many slender sympodial branches, and smooth, orange-brown bark or yellow bark.[*][8] All parts of the plants are fragrant. The species are unusual in having three distinct leaf patterns on the same plant: unlobed oval, bilobed (mitten-shaped), and trilobed (three-pronged); the leaves are hardly ever five-lobed.[*][9] Three-lobed leaves are more common in *sassafras tzumu* and *sassafras randaiense* than in their North American counterparts, although three-lobed leaves do sometimes occurs *sassafras albidum*. The young leaves and twigs are quite mucilaginous, and produce a citrus-like scent when crushed. The tiny, yellow flowers are five-petaled; *sassafras albidum* and *sassafras hesperia* are dioecious, with male and female flowers on separate trees, while *sassafras tzumu* and *sassafras randaiense* have male and female flowers occurring on the same trees. The fruit is a drupe, blue-black when ripe.[*][10]

The largest known sassafras tree in the world is located in Owensboro, Kentucky, and measures over 100 feet high and 21 feet in circumference.[*][11][*][12]

72.1.3 Habitat and range

Many Lauraceae are aromatic, evergreen trees or shrubs adapted to high rainfall and humidity, but the *Sassafras* genus is deciduous. Deciduous sassafras trees lose all of their leaves for part of the year, depending on variations in rainfall. In deciduous tropical Lauraceae, leaf loss coincides with the dry season in tropical, subtropical and arid regions. In temperate climates, the dry season is due to the inability of the plant to absorb water available to it only in the form of ice.

Sassafras is commonly found in open woods, along fences, or in fields. It grows well in moist, well-drained, or sandy loam soils and tolerates a variety of soil types, attaining a maximum in southern and wetter areas of distribution.[*][13]

Sassafras albidum ranges from southern Maine and southern Ontario west to Iowa, and south to central Florida and eastern Texas, in North America. *Sassafras tzumu* may be found in Anhui, Fujian, Guangdong, Guangxi, Guizhou, Hubei, Hunan, Jiangsu, Sichuan, Yunnan, and Zhejiang, China.[*][14] *Sassafras randaiense* is native to Taiwan.[*][15]

72.1.4 Importance to wildlife

The leaves, bark, twigs, stems, and fruits are eaten by birds and mammals in small quantities. For most animals, Sassafras is not consumed in large enough quantities to be important, although it is an important deer food in some areas. Carey and Gill rate its value to wildlife as fair, their lowest rating. Sassafras leaves and twigs are consumed by white-tailed deer. Other sassafras leaf browsers include groundhogs, marsh rabbits, and American black bears. Rabbits eat sassafras bark in winter. American beavers will cut sassafras stems. Sassafras fruits are eaten by many species of birds, including bobwhite quail, eastern kingbirds, great crested flycatchers, phoebes, wild turkeys, gray catbirds, northern flickers, pileated woodpeckers, downy woodpeckers, thrushes, vireos, and northern mockingbirds. Some small mammals also consume sassafras fruits.[*][16]

72.2 Species

The genus sassafras includes four species, three extant and one extinct. Sassafras plants are endemic to North America and East Asia, with two species in each region that are distinguished by some important characteristics, including the frequency of three-lobed leaves (more frequent in East Asian species) and aspects of their sexual reproduction (North American species are dioecious).

Taiwanese sassafras', Taiwan, is treated by some botanists in a distinct genus as *Yushunia randaiensis* (Hayata) Kamikoti, though this is not supported by recent genetic evidence, which shows *Sassafras* to be monophyletic.[*][3][*][17]

72.2.1 North America

- *Sassafras albidum* (Nuttall) Nees – **sassafras, white sassafras, red sassafras** or **silky sassafras**, eastern

Fossil Sassafras hesperia *leaf from Early Ypresian, Klondike Mountain Formation, Washington state, USA.*

North America, from southernmost Ontario, Canada through the eastern United States, south to central Florida, and west to southern Iowa and East Texas. Formerly, Wisconsin.[*][18]

- †*Sassafras hesperia* (Berry) – Western North American, from the Eocene Klondike Mountain Formation of Washington and British Columbia.[*][2]

72.2.2 East Asia

- *Sassafras tzumu* (Hemsl.) Hemsl. – **Chinese sassafras** or ***Tzumu***, central and southwestern China.

- *Sassafras randaiense* (Hayata) Rehd. – Taiwan.

72.3 Human uses of sassafras

All parts of sassafras plants, including roots, stems, twig leaves, bark, flowers, and fruit, have been used for culinary, medicinal, and aromatic purposes, both in areas where they are endemic and in areas where they were imported, such as

Europe. The wood of sassafras trees is also used as a material for building ships and furniture in China, Europe, and the United States, and sassafras played an important role in the history of the European colonization of the American continent in the 16th and 17th centuries. Sassafras twigs have even been used as toothbrushes or fire starters.

72.3.1 Culinary uses of sassafras

See culinary uses of Sassafras albidum *for more information on culinary use specific to the extant North American species, and legislation in the United States restricting the use of products derived from sassafras.*

Sassafras albidum is an important ingredient in some distinct foods of the United States. It is the main ingredient in traditional root beer and sassafras root tea, and ground leaves of sassafras are a distinctive additive in Louisiana Creole cuisine (see the article on Filé powder, and a common thickening and flavoring agent in gumbo). Methods of cooking with sassafras combine this ingredient native to American with traditional North American as well as European culinary techniques, to create a unique blend of Creole cuisine, and are thought by some to be heavily influenced by a blend of cultures.[19] Sassafras is no longer used in commercially produced root beer since safrole oil was banned for use in commercially mass-produced foods and drugs by the FDA in 1960 due to health concerns.[20]

Sassafras leaves and flowers have also been used in salads, and to flavor fats or cure meats.[21][22]

72.3.2 Medicinal uses of sassafras

Numerous Native American tribes used the leaves of sassafras to treat wounds by rubbing the leaves directly into a wound, and used different parts of the plant for many medicinal purposes such as treating acne, urinary disorders, and sicknesses that increased body temperature, such as high fevers.[23] East Asian types of sassafras such as *Sassafras tzumu* (chu mu) and *Sassafras randaiense* (chu shu) are used in Chinese medicine to treat rheumatism and trauma.[24] Some modern researchers conclude that the oil, roots and bark of sassafras have analgesic and antiseptic properties. Different parts of the sassafras plant (including the leaves and stems, the bark, and the roots) have been used to treat

> "scurvy, skin sores, kidney problems, toothaches, rheumatism, swelling, menstrual disorders and sexually transmitted diseases, bronchitis, hypertension, and dysentery. It is also used as a fungicide, dentifrice, rubefacient, di-

aphoretic, perfume, carminative and sudorific."[25]

Before the twentieth century, Sassafras enjoyed a great reputation in the medical literature, but became valued for its power to improve the flavor of other medicines.[26]

Sassafras wood and oil were both used in dentistry. Early toothbrushes were crafted from sassafras twigs or wood because of its aromatic properties.[13] Sassafras was also used as an early dental anesthetic and disinfectant.[27][28]

72.3.3 Uses of sassafras wood

Sassafras albidum is often grown as an ornamental tree for its unusual leaves and aromatic scent. Outside of its native area, it is occasionally cultivated in Europe and elsewhere.[29] The durable and beautiful wood of sassafras plants has been used in shipbuilding and furniture-making in North America, in Asia, and in Europe (once Europeans were introduced to the plant).[30] Sassafras wood was also used by Native Americans in the southeastern United States as a fire-starter because of the flammability of its natural oils found within the wood and the leaves.[31]

72.3.4 Safrole oil and aromatic uses

Steam distillation of dried root bark produces an essential oil consisting mostly of safrole, which once was extensively used as a fragrance in perfumes and soaps, food and for aromatherapy. Safrole is a precursor for the clandestine manufacture of the drug MDMA, as well as the drug MDA (3-4 methylenedioxyamphetamine) and as such, its transport is monitored internationally.

Safrole oil has also been used as a natural insect or pest deterrent, and in cordials (such as the opium-based Godfrey's) and homemade liquor to mask strong or unpleasant smells.[13][21] Safrole oil has also been added to soap and other toiletries.[27] Safrole oil is banned in the United States for use in commercially mass-produced foods and drugs by the FDA as a potential carcinogen.[20]

72.3.5 Exploitation and commodification of the sassafras plant

For a more detailed description of uses by indigenous peoples of North America, and a history of the commodification and exploitation of sassafras albidum by Europeans in the United States in the 16th and 17th centuries, see the article on the extant North American species of sassafras, sassafras albidum.

S. albidum *is a host plant for the spicebush swallowtail.*

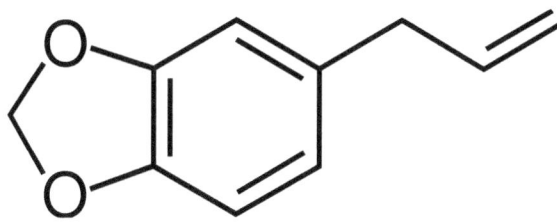

Chemical structure of safrole, a constituent of sassafras essential oil

In modern times, the sassafras plant has been exploited for the extraction of safrole oil, which is used in a variety of commercial products as well as in the manufacture of illegal drugs, and these plants are primarily harvested for commercial purposes in Asia and Brazil.*[32]

72.4 References

[1] Flora of North America: *Sassafras*

[2] Wolfe, Jack A. & Wehr, Wesley C. 1987. The sassafras is an ornamental tree. "Middle Eocene Dicotyledonous Plants from Republic, Northeastern Washington". *United States Geological Survey Bulletin* **1597**:13

[3] Nie, Z.-L., Wen, J. & Sun, H. (2007). "Phylogeny and biogeography of Sassafras (Lauraceae) disjunction between eastern Asia and eastern North America". *Plant Systematics and Evolution* **267**: 191–203. doi:10.1007/s00606-007-0550-1.

[4] *Bibliotheca Americana*. John Carter Brown Library. 1570. pp. 246, 267, 346. Retrieved 2014-12-09.

[5] *Webster's Revised Unabridged Dictionary* (1913 ed.). 1913. pp. 1277–1280. Retrieved 2014-12-09.

[6] Austin, Daviel (November 29, 2004). *Florida Ethnobotany.* CRC Press. p. 606. ISBN 978-0849323324.

[7] Weaver, William (December 19, 2000). *Sauer's Herbal Cures: America's First Book of Botanic Healing, 1762-1778.* Routledge. p. 274. ISBN 978-0415923606.

[8] Dirr, Manual of woody landscape plants. Page 938.

[9] Noble Plant Image Gallery Sassafras (includes photo of five-lobed leaf)

[10] Henk van der Werff, "*Sassafras* J. Presl in F. Berchtold & J. S. Presl., Prir. Rostlin. 2: 30, 67. 1825" , *Flora of North America* **3**

[11] [web head http://www.uky.edu/Ag/Horticulture/kytreewebsite/pdffiles/SASSAFRAprint.pdf "Sassafras albidum"] Check |url= scheme (help) (PDF). Kentucky Cooperative Extension Service.

[12] Whit Bronaugh (May–June 1994). "The biggest sassafras" . *American Forests.*

[13] Small, Ernest (September 23, 2013). *North American Cornucopia: Top 100 Indigenous Food Plants.* CRC Press. pp. 603–606. ISBN 978-1466585928.

[14] Wiersema, John; León, Blanca (February 26, 1999). *World Economic Plants: A Standard Reference.* CRC Press. p. 616. ISBN 978-0849321191.

[15] Nie, Z.-L.; Wen, J.; Sun, H. (2007). "Phylogeny and biogeography of *Sassafras* (Lauraceae) disjunct between eastern Asia and eastern North America". *Plant Systematics and Evolution* **267** (1-4): 0378–2697. doi:10.1007/s00606-007-0550-1.

[16] This section incorporates text from a public domain work of the US government: Sullivan, Janet (1993). "Sassafras albidum" . *Fire Effects Information System, [Online].* U.S. Department of Agriculture, Forest Service, Rocky Mountain Research Station, Fire Sciences Laboratory (Producer).

[17] Kamikoti, S. (1933). *Ann. Rep. Taihoku Bot. Gard.* **3**: 78

[18] U.S. Forest Service Silvics Manual: *Sassafras albidum*

[19] Nobles, Cynthia Lejeune (2009), "Gumbo" , in Tucker, Susan; Starr, S. Frederick, *New Orleans Cuisine: Fourteen Signature Dishes and Their Histories*, University Press of Mississippi, p. 110, ISBN 978-1-60473-127-9

[20] Dietz, B; Bolton, Jl (Apr 2007).   "Botanical dietary supplements gone bad." . *Chemical research in toxicology* **20** (4): 586–90. doi:10.1021/tx7000527. ISSN 0893-228X. PMC 2504026. PMID 17362034.

[21] Duke, James (September 27, 2002).   *CRC Handbook of Medicinal Spices*.   CRC Press.   p.   274.   ISBN 978-0849312793.

[22] Weatherford, Jack (September 15, 1992).   *Native Roots: How the Indians Enriched America*. Ballantine Books. p. 52. ISBN 978-0449907139.

[23] Duke, James (December 15, 2000). *The Green Pharmacy Herbal Handbook: Your Comprehensive Reference to the Best Herbs for Healing*. Rodale Books. p. 195. ISBN 978-1579541842.

[24] Wikibooks:Traditional Chinese Medicine/From Sabal Peregrina To Syzygium Samarangense

[25] Tiffany Leptuck, "Medical Attributes of 'Sassafras albidum' - Sassafras"], Kenneth M. Klemow, Ph.D., Wilkes-Barre University, 2003

[26] Keeler, H. L. (1900). *Our Native Trees and How to Identify Them*. Charles Scriber's Sons, New York.

[27] Barceloux, Donald (March 7, 2012). *Medical Toxicology of Natural Substances: Foods, Fungi, Medicinal Herbs, Plants, and Venomous Animals*. Wiley. ASIN B007KGA15Q.

[28] Dental Protective Association of the United States (June 7, 2010). *Dental Digest* **6**. Nabu Press. p. 546. ISBN 978-1149862315.

[29] U.S. Forest Service: *Sassafras albidum* (pdf file)

[30] De-Yuan, Hong (June 30, 2015). *Plants of China: A Companion to the Flora of China*. Cambridge University Press. p. 313. ISBN 978-1107070172.

[31] Bartram, William (December 1, 2002). *William Bartram on the Southeastern Indians (Indians of the Southeast)*. University of Nebraska Press. p. 270. ISBN 978-0803262058.

[32] Blickman, Tom (February 3, 2009).   "Harvesting Trees" . *Transational Institute*. Transnational Institute. Retrieved April 4, 2015.

72.5   External links

- *Drug Digest*,   "Sassafras"

- U of Arkansas: Division of Agriculture Plant of the Week: Sassafras

- GardenGuides.com Sassafras – Shrub Plant Guide

- Plants for a Future: Plant Portrait – Sassafras albidum

- TVA: Native Plant – Sassafras

- Missouri Plants – Sassafras albidum

- The Jefferson Monticello: The Lucy Meriwether Lewis Marks exhibit – article by Wendy Cortesi

- FossilMuseum.net: Rare Sassafras Plant Fossils

Chapter 73

Serenoa

Serenoa repens, commonly known as **saw palmetto**, is the sole species currently classified in the genus *Serenoa*. It has been known by a number of synonyms, including *Sabal serrulatum*, under which name it still often appears in alternative medicine. It is a small palm, growing to a maximum height of around 7–10 ft (2–3 m).*[3] Its trunk is sprawling, and it grows in clumps or dense thickets in sandy coastal lands or as undergrowth in pine woods or hardwood hammocks. Erect stems or trunks are rarely produced but are found in some populations. It is endemic to the southeastern United States, most commonly along the Atlantic and Gulf Coastal plains, but also as far inland as southern Arkansas. It is a hardy plant; extremely slow growing, and long lived, with some plants, especially in Florida where it is known as simply the palmetto, possibly being as old as 500–700 years.*[4]

Saw palmetto is a fan palm, with the leaves that have a bare petiole terminating in a rounded fan of about 20 leaflets. The petiole is armed with fine, sharp teeth or spines that give the species its common name. The teeth or spines are easily capable of breaking the skin, and protection should be worn when working around a Saw Palmetto. The leaves are light green inland, and silvery-white in coastal regions. The leaves are 1–2 m in length, the leaflets 50–100 cm long. They are similar to the leaves of the palmettos of genus *Sabal*. The flowers are yellowish-white, about 5 mm across, produced in dense compound panicles up to 60 cm long. The fruit is a large reddish-black drupe and is an important food source for wildlife and historically for humans. The plant is used as a food plant by the larvae of some Lepidoptera species such as *Batrachedra decoctor*, which feeds exclusively on the plant. This plant is also edible to human beings, but the greener it is the more bitter tasting it would be.

The generic name honors American botanist Sereno Watson.

73.1 Medical Use

Main article: Saw palmetto extract

The fruits of the saw palmetto are highly enriched with

Saw palmettos beneath the larger evergreen canopy in the Apalachicola National Forest in Florida

fatty acids and phytosterols, and extracts of the fruits have been the subject of intensive research for the symptomatic treatment of benign prostatic hyperplasia (BPH).

Numerous meta-analyses of clinical trials *S. repens* extract in the treatment of BPH have found it safe and effective for mild-to-moderate BPH compared to placebo, finasteride, and tamsulosin*[5]*[6] Two larger trials found the extract no different from placebo.*[7]*[8] An updated meta-analysis including these trials found that saw palmetto extract "was not more effective than placebo for treatment of hyperplasia." *[9]

S. repens extract has been promoted as useful for people with prostate cancer. However, according to the American Cancer Society, "available scientific studies do not support claims that saw palmetto can prevent or treat prostate cancer in humans" .*[10]

73.2 Ethnobotany

Indigenous names reported by Austin[*][11] include: tala (Choctaw); cani (Timucua); ta´:la (Koasati); taalachoba ("big palm" , Alabama); ta:łał a´ kko ("big palm," Creek); talco´:bˆı ("big palm," Mikasuki); talimushi ("palmetto's uncle," Choctaw), and guana (Taino, possibly). Saw palmetto fibers have been found among materials from indigenous people as far north as Wisconsin and New York, strongly suggesting this material was widely traded prior to European contact.[*][12] The leaves are used for thatching by several indigenous groups; so commonly so that there is a location in Alachua County, Florida named Kanapaha ("palm house").[*][13] The fruits may have been used to treat an unclear form of fish poisoning by the Seminoles and Bahamians.[*][14]

73.3 See also

- List of ineffective cancer treatments

73.4 References

[1] "*Serenoa repens* (W. Bartram) Small" . *Germplasm Resources Information Network*. United States Department of Agriculture. 1997-05-22. Retrieved 2010-04-12.

[2] Kew World Checklist of Selected Plant Families

[3] "Conservation Plant Characteristics for Serenoa repens" .

[4] Tanner, George W.; J. Jeffrey Mullahey; David Maehr (July 1996). "Saw-palmetto: An Ecologically and Economically Important Native Palm" (PDF). Circular WEC-109. University of Florida Cooperative Extension Service.

[5] Wilt T, Ishani A, Mac Donald R (2002). Tacklind, James, ed. "Serenoa repens for benign prostatic hyperplasia" . *Cochrane Database Syst Rev* (3): CD001423. doi:10.1002/14651858.CD001423. PMID 12137626.

[6] Boyle, P; Robertson C; Lowe F; Roehrborn C (Apr 2004). "Updated meta-analysis of clinical trials of Serenoa repens extract in the treatment of symptomatic benign prostatic hyperplasia" . *BJU Int* **93** (6): 751–756. doi:10.1111/j.1464-410X.2003.04735.x. PMID 15049985.

[7] Bent S, Kane C, Shinohara K, et al. (February 2006). "Saw palmetto for benign prostatic hyperplasia" . *N. Engl. J. Med.* **354** (6): 557–566. doi:10.1056/NEJMoa053085. PMID 16467543.

[8] Dedhia RC, McVary KT (June 2008). "Phytotherapy for lower urinary tract symptoms secondary to benign prostatic hyperplasia" . *J. Urol.* **179** (6): 2119–2125. doi:10.1016/j.juro.2008.01.094. PMID 18423748.

[9] Tacklind, J; MacDonald R; Rutks I; Wilt TJ (April 2009). Tacklind, James, ed. "Serenoa repens for benign prostatic hyperplasia" . *Cochrane Database Syst Rev* (2): CD001423. doi:10.1002/14651858.CD001423.pub2. PMC 3090655. PMID 19370565.

[10] "Saw Palmetto" . American Cancer Society. 28 November 2008. Retrieved 13 September 2013.

[11] Austin, DF (2004). *Florida Ethnobotany*. Boca Raton, FL: CRC Press. ISBN 978-0-8493-2332-4.

[12] Whitford AC (1941). "Textile fibers used in eastern aboriginal North America" . *Anthropological Papers of the American Museum of Natural History* **38**: 5–22.

[13] Simpson, JC (1956). *A Provisional Gazetteer of Florida Place-Names of Indian Derivation*. Tallahassee: Florida Geological Survey.

[14] Sturtevant, WC (1955). *The Mikasuki Seminole: Medical Beliefs and Practices*. Ann Arbor, MI: University Microfilms.

73.5 Further Reading

Bernichtein, Sophie; Pigat, Natascha; Camparo, Philippe; Latil, Alain; Viltard, Melanie; Friedlander, Gerard; Goffin, Vincent (14 February 2015). "Anti-Inflammatory Properties of Lipidosterolic Extract of Serenoa Repens (Permixon (R)) in a Mouse Model of Prostate Hyperplasia" . *The Prostate (2015)* **75** (7): 706–722. doi:10.1002/pros.22953.

73.6 External links

- *Serenoa* in Flora of North America
- *Serenoa repens*
- *Serenoa repens* from Floridata
- *Scanpalm - Serenoa repens*
- Interactive Distribution Map for *Serenoa repens*
- *Serenoa repens for hair loss*

Chapter 74

Shellac

For the minimalist rock trio, see Shellac (band).

Shellac is a resin secreted by the female lac bug, on trees

Some of the many different colors of shellac

in the forests of India and Thailand. It is processed and sold as dry flakes (pictured at right) and dissolved in ethanol to make liquid shellac, which is used as a brush-on colorant, food glaze and wood finish. Shellac functions as a tough natural primer, sanding sealant, tannin-blocker, odour-blocker, stain, and high-gloss varnish. Shellac was once used in electrical applications as it possesses good insulation qualities and it seals out moisture. Phonograph (gramophone) records were also made of it during the 78-rpm recording era which ended in Western countries during the 1950s.

From the time it replaced oil and wax finishes in the 19th century, shellac was one of the dominant wood finishes in the western world until it was largely replaced by nitrocellulose lacquer in the 1920s and 1930s.

74.1 Etymology

Shellac comes from *shell* and *lac*, a calque of French *laque en écailles*, "lac in thin pieces", later *gomme-laque*, "gum lac".[1] Most European languages (except Romance ones) have borrowed the word for the substance from English or from the German equivalent *Schellack*.

74.2 Production

Lac tubes created by Kerria Lacca

Shellac is scraped from the bark of the trees where the female lac bug, *Kerria lacca* (Order *Hemiptera*, Family *Kerriidae*), also known as *Laccifer lacca*, secretes it to form a tunnel-like tube as it traverses the branches of the tree. Though these tunnels are sometimes referred to as "cocoons", they are not literally cocoons in the entomological sense.[2] This insect is in the same Superfamily as the insect from which cochineal is obtained. The insects suck the sap of the tree and excrete "sticklac" almost constantly. The least coloured shellac is produced when the insects feed on the kusum tree (*Schleichera*).

The number of lac bugs required to produce 1 kilogram (2.2 lb) of shellac has variously been estimated as 50,000,[3] 200,000,[4] or 300,000.[5][6] The root word lakh is a South Asian unit for 100,000 and presumably refers to the huge numbers of insects that swarm on host trees, up to 150 per square inch.[7]

The raw shellac, which contains bark shavings and lac bugs removed during scraping, is placed in canvas tubes (much like long socks) and heated over a fire. This causes the shellac to liquify, and it seeps out of the canvas, leaving the bark and bugs behind. The thick, sticky shellac is then dried into a flat sheet and broken into flakes, or dried into "but-

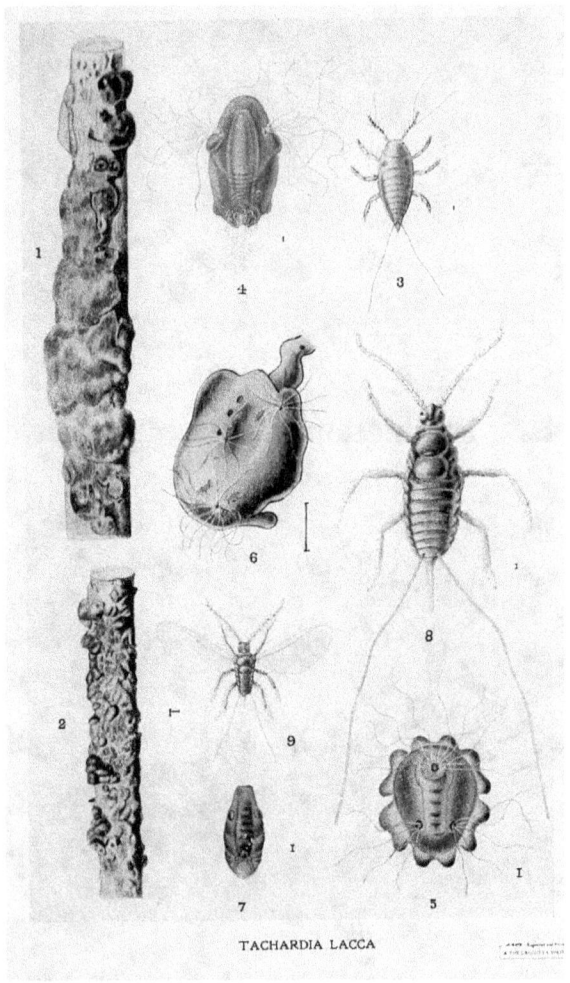

TACHARDIA LACCA

Drawing of the insect Kerria lacca *and its shellac tubes, by Harold Maxwell-Lefroy, 1909*

tons" (pucks/cakes), then bagged and sold. The end-user then crushes it into a fine powder and mixes it with ethyl alcohol prior to use, to dissolve the flakes and make liquid shellac.

Liquid shellac has a limited shelf life (about 1 year), hence it is sold in dry form for dissolution prior to use. Liquid shellac sold in hardware stores is clearly marked with the production (mixing) date, so the consumer can know whether the shellac inside is still good. Alternatively, old shellac may be tested to see if it is still usable: a few drops on glass should quickly dry to a hard surface. Shellac that remains tacky for a long time is no longer usable. Storage life depends on peak temperature, so refrigeration extends shelf life.

The thickness (concentration) of shellac is measured by the unit "pound cut", referring to the amount (in pounds) of shellac flakes dissolved in a gallon of denatured alcohol. For example: a 1-lb. cut of shellac is the strength obtained by dissolving one pound of shellac flakes in a gallon of alcohol. Most pre-mixed commercial preparations come at a 3-lb. cut. Multiple thin layers of shellac produce a significantly better end result than a few thick layers. Thick layers of shellac do not adhere to the substrate or to each other well, and thus can peel off with relative ease; in addition, thick shellac will obscure fine details in carved designs in wood and other substrates.

Shellac naturally dries to a high-gloss sheen. For applications where a flatter (less shiny) sheen is desired, products containing amorphous silica,[8] such as "Shellac Flat," may be added to the dissolved shellac.

Shellac naturally contains a small amount of wax (3%–5% by volume), which comes from the lac bug. In some preparations, this wax is removed (the resulting product being called "dewaxed shellac"). This is done for applications where the shellac will be coated with something else (such as paint or varnish), so the topcoat will adhere. Waxy (non-dewaxed) shellac appears milky in liquid form, but dries clear.

74.3 Colors and availability

Shellac comes in many warm colors, ranging from a very light blond ("platina") to a very dark brown ("garnet"), with many varieties of brown, yellow, orange and red in between. The colour is influenced by the sap of the tree the lac bug is living on and by the time of harvest. Historically, the most commonly sold shellac is called "orange shellac", and was used extensively as a combination stain and protectant for wood paneling and cabinetry in the 20th century.

Shellac was once very common anywhere paints or varnishes were sold (such as hardware stores). However, cheaper and more abrasion- and chemical-resistant finishes, such as polyurethane, have almost completely replaced it in decorative residential wood finishing such as hardwood floors, wooden wainscoting plank paneling, and kitchen cabinets. These alternative products, however, must be applied over a stain if the user wants the wood coloured; clear or blond shellac may be applied over a stain without affecting the color of the finished piece, as a protective topcoat. "Wax over shellac" (an application of buffed-on paste wax over several coats of shellac) is often regarded as a beautiful, if fragile, finish for hardwood floors. Luthiers still use shellac to *French polish* fine acoustic stringed instruments, but it has been replaced by synthetic plastic lacquers and varnishes in many workshops.[9]

74.4 Properties

A decorative medal made in France in early 20th century moulded from shellac compound, the same used for phonograph records of the period.

Shellac is a natural bioadhesive polymer and is chemically similar to synthetic polymers, and thus can be considered a natural form of plastic. It can be turned into a moulding compound when mixed with wood flour and moulded under heat and pressure methods, so it can also be classified as thermoplastic.

Shellac scratches more easily than most lacquers and varnishes, and application is more labor-intensive, which is why it has been replaced by plastic in most areas. But damaged shellac can easily be touched-up with another coat of shellac (unlike polyurethane) because the new coat merges with and bonds to the existing coat(s). Shellac is much softer than Urushi lacquer for instance, which is far superior in regards to both chemical and mechanical resistance.

Shellac is soluble in alkaline solutions such as ammonia, sodium borate, sodium carbonate, and sodium hydroxide, and also in various organic solvents. When dissolved in denatured alcohol or ethanol, shellac yields a coating of good durability and hardness.

Upon mild hydrolysis shellac gives a complex mix of aliphatic and alicyclic hydroxy acids and their polymers that varies in exact composition depending upon the source of the shellac and the season of collection. The major component of the aliphatic component is aleuritic acid, whereas the main alicyclic component is shellolic acid.*[10]

Shellac is UV-resistant, and does not darken as it ages

(though the wood under it may do so, as in the case of pine).*[4]

74.5 History

The earliest written evidence of shellac goes back 3,000 years, but shellac is known to have been used earlier.*[4] According to the Mahabharata, an entire palace was built out of dried shellac.*[4]

Shellac was in rare use as a dyestuff for as long as there was a trade with the East Indies. Merrifield*[11] cites 1220 for the introduction of shellac as an artist's pigment in Spain. Lapis lazuli as ultramarine pigment from Afghanistan was already being imported long before this.

The use of overall paint or varnish decoration on large pieces of furniture was first popularised in Venice (then later throughout Italy). There are a number of 13th century references to painted or varnished cassone, often dowry cassone that were made deliberately impressive as part of dynastic marriages. The definition of varnish is not always clear, but it seems to have been a spirit varnish based on gum benjamin or mastic, both traded around the Mediterranean. At some time, shellac began to be used as well. An article from the *Journal of the American Institute of Conservation* describes the use of infrared spectroscopy to identify a shellac coating on a 16th-century cassone.*[12] This is also the period in history where "varnisher" was identified as a distinct trade, separate from both carpenter and artist.

Another use for shellac is sealing wax. Woods's *The Nature and Treatment of Wax and Shellac Seals*[13] discusses the various formulations, and the period when shellac started to be added to the previous beeswax recipes.

The "period of widespread introduction" would seem to be around 1550 to 1650, when the substance moves from being a rarity on highly decorated pieces to being described in the standard texts of the day.

74.6 Uses

74.6.1 Historical

In the early- and mid-20th century, orange shellac was used as a one-product finish (combination stain and varnish-like topcoat) on decorative wood paneling used on walls and ceilings in homes, particularly in the US. In the American South, use of knotty pine plank paneling covered with orange shellac was once as common in new construction as drywall is today. It was also often used on kitchen cabinets and hardwood floors, prior to the advent of polyurethane.

Until the advent of vinyl in 1949, most gramophone records were pressed from shellac compounds. From 1921 to 1928, 18,000 tons of shellac were used to create 260 million records for Europe.*[7] In the 1930s, it was estimated that half of all shellac was used for gramophone records.*[14] Use of shellac for records was common until the 1950s and continued into the 1970s in some non-Western countries.

Shellac was historically used as a protective coating on paintings.

Sheets of Braille were coated with shellac to help protect them from wear due to being read by hand.

Shellac was used from the mid-19th century to produce small moulded goods such as picture frames, boxes, toilet articles, jewelry, inkwells and even dentures. Advances in plastics have rendered shellac obsolete as a moulding compound.

Shellac (both orange and white varieties) were used both in the field and laboratory to glue and stabilize dinosaur bones until about the mid 1960s. While effective at the time, the long-term negative effects of shellac (being organic in nature) on dinosaur bones and other fossils is debated and shellac is very rarely used by professional conservators and fossil preparators today.

Shellac was once used for fixing inductor, motor, generator and transformer windings, where it was applied directly to single layer windings in an alcohol solution. For multilayer windings, the whole coil was submerged in shellac solution, then drained and placed in a warm place to allow the alcohol to evaporate. The shellac then locks the wire turns in place, provides extra insulation and prevents movement and vibration, reducing buzz and hum. In motors and generators it also helps transfer force generated by magnetic attraction and repulsion from the windings to the rotor or armature. In more recent times, synthetic resins, such as glyptol, (Glyptal), have been substituted for the shellac. Some applications use shellac mixed with other natural or synthetic resins, such as pine resin or phenol-formaldehyde resin, of which Bakelite is the best known, for electrical use. Mixed with other resins, barium sulfate, calcium carbonate, zinc sulfide, aluminium oxide and/or cuprous carbonate (malachite), shellac forms a component of heat-cured capping cement used to fasten the caps or bases to the bulbs of electric lamps.

74.6.2 Current

It is the central element of the traditional "French polish" method of finishing furniture and fine violas, guitars and pianos.

Shellac, edible, is used as a glazing agent on pills (see excipients) and candies, in the form of *pharmaceutical glaze*

(or, *confectioner's glaze*). Because of its acidic properties (resisting stomach acids), shellac-coated pills may be used for a timed enteric or colonic release.*[15] Shellac is used as a 'wax' coating on citrus fruit to prolong its shelf/storage life. It is also used to replace the natural wax of the apple, which is removed during the cleaning process.*[16] When used for this purpose, it has the food additive E number E904.

Shellac coating applied with either a standard or modified Huon-Stuehrer nozzle, can be economically micro-sprayed onto various smooth candies, such as chocolate coated peanuts. Irregularities on the surface of the product being sprayed typically result in the formation of unsightly aggregates ("lac-aggs") which precludes the use of this technique on foods such as walnuts or raisins (however, chocolate-coated raisins being smooth surfaced, are able to be sprayed successfully using a modified Huon-Stuehrer nozzle).

Because it is compatible with most other finishes, shellac is also used as a barrier or primer coat on wood to prevent the bleeding of resin or pigments into the final finish, or to prevent wood stain from blotching.*[2]

Shellac is an odour and stain blocker and so is often used as the base of "solves all problems" primers. Although its durability against abrasives and many common solvents is not very good, shellac provides an excellent barrier against water vapour penetration. Shellac-based primers are an effective sealant to control odours associated with fire damage.

Shellac has traditionally been used as a dye for cotton and, especially, silk cloth in Thailand, particularly in the northeastern region.*[17] It yields a range of warm colours from pale yellow through to dark orange-reds and dark ochre.*[18] Naturally dyed silk cloth, including that using shellac, is widely available in the rural northeast, especially in Ban Khwao District, Chaiyaphum province. The Thai name for the insect and the substance is "khrang" (Thai: ครั่ง).

Other

Shellac is used:

- in the tying of artificial flies for trout and salmon where the shellac was used to seal all trimmed materials at the head of the fly.

- in combination with wax for preserving and imparting a shine to citrus fruits, such as lemons.

- in dental technology, where it is occasionally used in the production of custom impression trays and (partial) denture production.

- as a binder in India ink.

- for cycling as a protective and decorative coating for handlebar tape,[19] and as a hard-drying adhesive for tubular cycle tires, particularly for track racing.[20]

- for reattaching ink sacs when restoring vintage fountain pens, the orange variety preferably.

- for fixing pads to the key-cups of woodwind instruments.

- for Luthier applications, to bind wood fibers down and prevent tear out on the soft spruce soundboards.

- to stiffen and impart water-resistance to felt hats, for wood finishing[21] and as a constituent of *gossamer* (or *goss* for short), a cheesecloth fabric coated in shellac and ammonia solution used in the shell of traditional silk top and riding hats.

- to increase the strength and longevity of ballet pointe shoes as a remedy for moisture weakening.[22]

- for mounting insects, in the form of a gel adhesive mixture composed of 75% ethyl alcohol.[23]

- as a binder in the fabrication of abrasive wheels,[24] imparting flexibility and smoothness not found in vitrified (ceramic bond) wheels. 'Elastic' bonded wheels typically contain plaster of paris, yielding a stronger bond when mixed with shellac; the mixture of dry plaster powder, abrasive (e.g. corundum/aluminium oxide Al_2O_3), and shellac are heated and the mixture pressed in a mould.

- in fireworks pyrotechnic compositions as a low-temperature fuel, where it allows the creation of pure 'greens' and 'blues'- colours difficult to achieve with other fuel mixes.

- in watchmaking, due to its low melting temperature (about 80-100 °C), to adjust and adhere pallet stones to the pallet fork. Also for securing small parts to a 'wax chuck' (faceplate) in a watchmakers' lathe.

- in the early 20th century, it was used to protect some military rifle stocks[25]

- in the cosmetic industry, shellac is known as a nail treatment that lasts longer than regular polish. It also gives it a better glossy finish.[26] It is a combination of gel and regular polish and offers a water resistant seal among nail protection. The process consists of three steps and a UV light finish.[27]

- in Jelly Belly jelly beans, in combination with beeswax to give them their final buff and polish.[28]

74.7 Gallery

- Blonde shellac flakes

- Dewaxed Bona (L) and Waxy #1 Orange (R) shellac flakes. The latter orange shellac is the traditional shellac used for decades to finish wooden wall paneling, kitchen cabinets and tool handles.

- Closeup of Waxy #1 Orange (L) and Dewaxed Bona (R) shellac flakes. The former orange shellac is the traditional shellac used for decades to finish wooden wall paneling and kitchen cabinets.

- "Quick and dirty" example of a pine board coated with 1-5 coats of Dewaxed Dark shellac (a darker version of traditional orange shellac)

74.8 References

[1] "shellac" . *Online Etymology Dictionary*. Retrieved 2015-03-17.

[2] Shellac, WoodworkDetails.com: Shellac as a Woodworking Finish

[3] Bangali Baboo; D. N. Goswami (2010). *Processing, Chemistry and Application of Lac*. New Delhi, India: Chandu Press. p. 4.

[4] Naturalhandyman.com : DEFEND, PRESERVE, AND PROTECT WITH SHELLAC : The story of shellac

[5] Yacoubou, Jeanne (30 Nov 2010). "Q & A on Shellac" . *Vegetarian Resource Group*. Retrieved 3 July 2014.

[6] Velji, Vijay (2010). "Shellac Origins and Manufacture" . shellacfinishes.com. Retrieved 3 July 2014.

[7] Berenbaum, May (1993). *Ninety-nine More Maggots, Mites, and Munchers*. University of Illinois Press. p. 27. ISBN 9780252020162.

[8] American Woodworker: Tips for Using Shellac

[9] French polishing tutorial for guitars

[10] Merck Index, 9th Ed. page 8224.

[11] Merrifield, Mary (1849). *Original Treatises on the Art of Painting*. Mineola, N.Y.: Dover Publ. ISBN 0-486-40440-4.

[12] "Furniture finish layer identification by infrared linear mapping microspectroscopy". *JAIC (Journal of the American Institute of Conservation)* **31** (2, Article 6): 225 to 236. 1992.

[13] Woods, C. (1994). "The Nature and Treatment of Wax and Shellac Seals" . *Journal of the Society of Archivists* (15).

[14] "How Shellac Is Manufactured" . The Mail (Adelaide, SA : 1912 - 1954). 18 Dec 1937. Retrieved 3 July 2014.

[15] Shellac film coatings providing release at selected pH and method - US Patent 6620431

[16] US Apple: Consumers - FAQs: Apples and Wax at the Wayback Machine (archived December 3, 2010)

[17] Suanmuang Tulaphan, Phunsap, *Silk Dyeing With Natural Dyestuffs in Northeastern Thailand*, 1999, p. 26-30 (in Thai)

[18] Punyaprasop, Daranee (Ed.)Colour *And Pattern On Native Cloth*, 2001, p. 253, 256 (in Thai)

[19] "Shellac & Twine makes Handlebar fine" . *Out Your Backdoor*. 21 August 2005. Retrieved 2015-03-16.

[20] Mounting Tubular Tires by Jobst Brandt

[21] Jewitt, Jeff. "Shellac: A traditional finish still yields superb results" . *Antique Restorers*. Retrieved 2015-03-16.

[22] Frequently Asked Questions about pointe shoes and ribbons

[23] Fly Times: Shellac gel for insect mounting

[24] Stephen Malkin; Changsheng Guo (2008). *Grinding Technology: Theory and Applications of Machining With Abrasives*. Industrial Press. p. 5.

[25] "What kind of finish is on my stock?". *Russian Mosin Nagant Forum*. Retrieved 2015-03-21.

[26] Pisani, Katrina (5 October 2012). "The Science of Shellac" . *Cool Science*. Retrieved 2015-03-21.

[27] Thompson, Connie (1 June 2011). "Pros and cons of Shellac nail polish" . *Komo News*. Retrieved 2015-03-21.

[28] Q&A - Jelly Belly jelly beans

74.9   External links

- Shellac.net US shellac vendor - properties and uses of dewaxed and non-dewaxed shellac

- The Story of Shellac (history)

- DIYinfo.org's Shellac Wiki, practical information on everything to do with shellac

- Reactive Pyrolysis-Gas Chromatography of Shellac

- Shellac A short introduction to the origin of shellac, the history of Japanning and French polishing, and how to conserve and repair these finishes sympathetically

- Shellac Application By Smith & Rodger

Chapter 75

Tamarind

This article is about the tropical plant. For the South American monkey, see Tamarin.

For other uses see Tamarind (disambiguation) and Tamarindo (disambiguation)

Tamarind (*Tamarindus indica*) (from Arabic: تمر هندي, romanized *tamar hindi*, "Indian date") is a leguminous tree in the family Fabaceae indigenous to tropical Africa. The genus **Tamarindus** is a monotypic taxon, having only a single species.

The tamarind tree produces edible, pod-like fruit which is used extensively in cuisines around the world. Other uses include traditional medicine and metal polish. The wood can be used in carpentry. Because of the tamarind's many uses, cultivation has spread around the world in tropical and subtropical zones.

75.1 Origin

Tamarindus indica is probably indigenous to tropical Africa,[*][2] but has been cultivated for so long on the Indian subcontinent that it is sometimes also reported to be indigenous there.[*][3] It grows wild in Africa in locales as diverse as Sudan, Cameroon, Nigeria and Tanzania. In Arabia, it is found growing wild in Oman, especially Dhofar, where it grows on the sea-facing slopes of mountains. It reached South Asia likely through human transportation and cultivation several thousand years prior to the Common Era.[*][4][*][5] It is widely distributed throughout the tropical belt, from Africa to South Asia, Northern Australia, and throughout Oceania, Southeast Asia, Taiwan and China.

In the 16th century, it was heavily introduced to Mexico, and to a lesser degree to South America, by Spanish and Portuguese colonists, to the degree that it became a staple ingredient in the region's cuisine.[*][6]

Today, India is the largest producer of tamarind.[*][7] The consumption of tamarind is widespread due to its central role in the cuisines of the Indian subcontinent, South East Asia and America, particularly in Mexico.

75.2 Description

A tamarind seedling

The tamarind is a long-lived, medium-growth, bushy tree, which attains a maximum crown height of 12 to 18 metres (39 to 59 ft). The crown has an irregular, vase-shaped outline of dense foliage. The tree grows well in full sun in clay, loam, sandy, and acidic soil types, with a high resistance to drought and aerosol salt (wind-borne salt as found in coastal areas).

The evergreen leaves are alternately arranged and pinnately compound. The leaflets are bright green, elliptical ovular, pinnately veined, and less than 5 cm (2.0 in) in length. The branches droop from a single, central trunk as the tree matures and is often pruned in agriculture to optimize tree density and ease of fruit harvest. At night, the leaflets close up.

The tamarind does flower, though inconspicuously, with red

and yellow elongated flowers. Flowers are 2.5 cm wide (one inch), five-petalled, borne in small racemes, and yellow with orange or red streaks. Buds are pink as the four sepals are pink and are lost when the flower blooms.

Tamarind flowers

The fruit is an indehiscent legume, sometimes called a pod, 12 to 15 cm (4.7 to 5.9 in) in length, with a hard, brown shell.[8][9][10]

The fruit has a fleshy, juicy, acidulous pulp. It is mature when the flesh is coloured brown or reddish-brown. The tamarinds of Asia have longer pods containing six to 12 seeds, whereas African and West Indian varieties have short pods containing one to six seeds. The seeds are somewhat flattened, and glossy brown.

The tamarind is best described as sweet and sour in taste, and is high in tartaric acid, sugar, B vitamins and, oddly for a fruit, calcium.

As a tropical species, it is frost sensitive. The pinnate leaves with opposite leaflets give a billowing effect in the wind. Tamarind timber consists of hard, dark red heartwood and softer, yellowish sapwood.

Tamarindus leaves and pod

It is harvested by pulling the pod from its stalk. A mature tree may be capable of producing up to 175 kg (386 lb) of fruit per year. Veneer grafting, shield (T or inverted T) budding, and air layering may be used to propagate desirable selections. Such trees will usually fruit within three to four years if provided optimum growing conditions.

75.2.1 Etymology

The name derives from Arabic *tamr-hindī*, meaning "date of India". Several early medieval herbalists and physicians wrote *tamar indi*, medieval Latin use was *tamarindus*, and Marco Polo wrote of *tamarandi*.[11]

Raw tamarind fruits

In Colombia, the Dominican Republic, Mexico, Puerto Rico, Venezuela and throughout the Lusosphere, it is called *tamarindo*. In the Caribbean, tamarind is sometimes called *tamón*.[4] Tamarind (*Tamarindus indica*) is sometimes confused with "Manila tamarind" (*Pithecellobium dulce*). While in the same taxonomic family Fabaceae, Manila tamarind is a different plant native to Mexico and known locally as *guamúchil*.

75.3 Cultivation

Seeds can be scarified or briefly boiled to enhance germination. They retain germination capability after several months if kept dry.

The tamarind has also long been naturalized in Indonesia, Malaysia, the Philippines, and the Pacific Islands. Thailand has the largest plantations of the ASEAN nations, followed by Indonesia, Myanmar, and the Philippines. The pulp is marketed in northern Malaya. It is cultivated all over India, especially in the South Indian states of Karnataka, Andhra Pradesh and Tamil Nadu. Extensive tamarind orchards

Three-day-old tamarind seedling

Tamarind sweets in a Mexican candy boutique

Tamarind balls from Trinidad and Tobago

in India produce 275,500 tons (250,000 MT) annually. Commercial plantations throughout tropical Latin America include Brazil, Costa Rica, Colombia, Cuba, Guatemala, Mexico, Nicaragua, Puerto Rico and Venezuela.

In the United States, it is a large-scale crop introduced for commercial use, second in net production quantity to India, in the mainly Southern states due to tropical and semitropical climes, notably South Florida, and as a shade and fruit tree, along roadsides and in dooryards and parks.[12]

75.4 Usage

75.4.1 Culinary uses

The fruit pulp is edible. The hard green pulp of a young fruit is considered by many to be too sour, but is often used as a component of savory dishes, as a pickling agent or as a means of making certain poisonous yams in Ghana safe for human consumption.[13]

The ripened fruit is considered the more palatable, as it becomes sweeter and less sour (acidic) as it matures. It is used in desserts as a jam, blended into juices or sweetened drinks,[14] sorbets, ice creams and other snacks. In Western cuisine, it is found in Worcestershire sauce.[15] In most parts of India, tamarind extract is used to flavor foods ranging from meals to snacks.[16] Across the Middle East, from the Levant to Iran, tamarind is used in savory dishes, notable meat-based stews, and often combined with dried fruits to achieve a sweet-sour tang.[17][18]

A traditional food plant in Africa, tamarind has potential to improve nutrition, boost food security, foster rural development and support sustainable landcare.[19]

In Madagascar, its fruits and leaves are a well-known favorite of the ring-tailed lemurs, providing as much as 50% of their food resources during the year if available.

75.4.2 Traditional medicinal uses

Throughout Southeast Asia, fruit of the tamarind is used as a poultice applied to foreheads of fever sufferers.[8] A 2002 diet control study where subjects were fed tamarind paste, concluded that: "tamarind intake is likely to help in delaying progression of skeletal fluorosis by enhancing urinary excretion of fluoride".[20] Based on a 2012 human study, supplementation of tender tamarind leaves improved disturbances to carbohydrate, lipid and antioxidant metabolism caused by chronic fluoride intake.[21] However, additional research is needed to confirm these results.

Tamarindus indica *tree at Bhopal*

75.4.3 Carpentry uses

Tamarind wood is a bold red color. Due to its density and durability, tamarind heartwood can be used in making furniture and wood flooring.

75.4.4 Metal polish

In homes and temples, especially in Buddhist Asian countries, the fruit pulp is used to polish brass shrine statues and lamps, and copper, brass, and bronze utensils. The copper alone or in brass reacts with moist carbon dioxide to gain a green coat of copper carbonate. Tamarind contains tartaric acid, a weak acid that can remove the coat of copper carbonate. Hence, tarnished copper utensils are cleaned with tamarind or lime, another acidic fruit.[*][4]

75.4.5 Horticultural uses

Throughout Asia and the tropical world, tamarind trees are used as ornamental, garden and cash crop plantings. Commonly used as a bonsai species in many Asian countries, it is also grown as an indoor bonsai in temperate parts of the world.[*][22]

Tamarind on a place of the foundation of city Santa Clara, Cuba

75.5 Research

In hens, tamarind has been found to lower cholesterol in their serum, and in the yolks of the eggs they laid.[*][23][*][24] Due to a lack of available human clinical trials, there is insufficient evidence to recommend tamarind for the treatment of hypercholesterolemia or diabetes.[*][25]

75.6 See also

- Historical tamarind

75.7 References

[1] http://www.theplantlist.org/tpl/record/ild-1720

[2] Diallo, BO; Joly, HI; McKey, D; Hosaert-McKey, M; Chevallier, MH (2007). "Genetic diversity of Tamarindus indica populations: Any clues on the origin from its current distribution?". *African Journal of Biotechnology* **6** (7).

[3] Abukakar, MG; Ukwuani, AN; Shehu, RA (2008). "Phytochemical Screening and Antibacterial Activity of *Tamarindus indica* Pulp Extract". *Asian Journal of Biochemistry* **3** (2): 134–138.

[4] Morton, Julia F. (1987). *Fruits of Warm Climates.* Wipf and Stock Publishers. pp. 115–121. ISBN 0-9653360-7-7.

[5] Popenoe, W. (1974). *Manual of Tropical and Subtropical Fruits.* Hafner Press. pp. 432–436.

[6] Tamale, E.; Jones, N.; Pswarayi-Riddihough, I. (August 1995). *Technologies Related to Participatory Forestry in Tropical and Subtropical Countries.* World Bank Publications. ISBN 978-0-8213-3399-0.

[7] http://www.cropsforthefuture.org/publication/ Monographs/Tamarind%20monograph.pdf

[8] Doughari, J. H. (December 2006). "Antimicrobial Activity of Tamarindus indica". *Tropical Journal of Pharmaceutical Research* **5** (2): 597–603. doi:10.4314/tjpr.v5i2.14637.

[9] "Fact Sheet: Tamarindus indica" (PDF). University of Florida. Retrieved July 22, 2012.

[10] Christman, S. "Tamarindus indica". FloriData. Retrieved January 11, 2010.

[11] Tamarind at the Oxford English dictionary

[12] "Food and Agriculture Organization of the United Nations".

[13] "Tamarind: Tamarindus Indica L.".

[14] Jed Portman. May 12, 2013. Serious Eats: Jarritos Tamarindo. http://drinks.seriouseats.com/2013/03/ essential-sodas-jarritos-tamarindo-mexican-soda.html

[15] "BBC Food:Ingredients—Tamarind recipes". BBC. Retrieved February 23, 2015.

[16] Veg Recipes of India: Tamarind-Date Chutney. http://www.vegrecipesofindia.com/ tamarind-date-chutney-recipe-sweet-chutney-for-chaat/

[17] PRI. Tamarind is the 'sour secret of Syrian cooking'. 2014 July. http://www.pri.org/stories/2014-07-02/ tamarind-sour-secret-syrian-cooking

[18] Joan Nathan. New York Times, 2004. Georgian Chicken in Pomegranate and Tamarind Sauce. http://cooking.nytimes.com/recipes/ 11849-georgian-chicken-in-pomegranate-and-tamarind-sauce

[19] National Research Council (January 25, 2008). "Tamarind". *Lost Crops of Africa: Volume III: Fruits.* Lost Crops of Africa **3**. National Academies Press. ISBN 978-0-309-10596-5. Retrieved July 17, 2008.

[20] "Effect of tamarind ingestion on fluoride excretion in humans.". PubMed. PMID 11840184. Retrieved February 8, 2015.

[21] "Ameliorative effect of tamarind leaf on fluoride-induced metabolic alterations.". PubMed. PMID 22438201. Retrieved July 23, 2013.

[22] D'Cruz, Mark. "Ma-Ke Bonsai Care Guide for Tamarindus indica". Ma-Ke Bonsai. Retrieved August 19, 2011.

[23] Salma, U.; Miah, A. G.; Tareq, K. M. A.; Maki, T.; Tsujii, H. (1 April 2007). "Effect of Dietary Rhodobacter capsulatus on Egg-Yolk Cholesterol and Laying Hen Performance". *Poultry Science* (Oxford University Press) **86** (4): 714–719. doi:10.1093/ps/86.4.714. ISSN 1525-3171. as well as in egg-yolk (13 and 16%)

[24] Chowdhury, SR; Sarker, DK; Chowdhury, SD; Smith, TK; Roy, PK; Wahid, MA (2005). "Effects of dietary tamarind on cholesterol metabolism in laying hens". *Poultry science* **84** (1): 56–60. doi:10.1093/ps/84.1.56. PMID 15685942.

[25] "Tamarindus indica". Health Online. Retrieved January 11, 2010.

75.7.1 Bibliography

- Bhumibhamon, S. 1988. *Multi-purpose trees for small-farm use in the Central Plain of Thailand.* D withington, K MacDicken., CB Sastyr and NR Adams, eds *Multi-purpose trees for small-farm use: Proceedings of an International Workshop* pp. 53–55. November 2–5, 1987, Pattaya Thailand.

- Jean-Marc Boffa, Food and Agriculture Organization of the United Nations Publisher Food & Agriculture Org., 1999. *Agroforestry parklands in Sub-Saharan Africa Volume 34* of *FAO conservation guide Agroforestry Parklands in Sub-Saharan Africa,* ISBN 92-5-104376-0, ISBN 978-92-5-104376-9: 230 pages

- Dassanayake, M. D. & Fosberg, F. R. (Eds.). (1991). *A Revised Handbook to the Flora of Ceylon.* Washington, D. C.: Smithsonian Institution.

- Hooker, Joseph Dalton. (1879). *The Flora of British India*, Vol II. London: L. Reeve & Co.

- Locke J, N Renner: 1991 *Pod Form and Non-Pod Form Variants of Tamarind in Guadelupe* Yaghoubian Agricultural Review 2:122–149

- Michon G, F Mary, J Bopmart: 1986 *Multi-Storied agroforestry Garden System in West Sumatra, Indonesia* Agroforestry Systems 4:315–338

- Narawane SP 1991 *Success stories of* Multi-purpose tree species production by small farmers *in NG Hedge and JN Daniel eds,* Multi-purpose tree species production by small farmers, *proceedings of the National Workshop. January 28–31, 1991 Pune, India.*

- James Rennie: 1834. *Alphabet of medical botany.* Orr and Smith, 1834. 152 page 77. Google Books

- George Spratt, 1830. *Flora Medica*: containing coloured delineations of the various medicinal plants admitted into the London, Edinburgh, and Dublin pharmacopœias; with their natural history, botanical descriptions, medical and chemical properties, Together with a Concise Introduction to Botany; a Copious Glossary of Botanical Terms; and a List of Poisonous Plants. Callow and Wilson, 1830. Google Books.

75.8   External links

- *Tamarindus indica* in Brunken, U., Schmidt, M., Dressler, S., Janssen, T., Thiombiano, A. & Zizka, G. 2008. West African plants – A Photo Guide. www.westafricanplants.senckenberg.de.

- "Tamarind". *Encyclopædia Britannica* (11th ed.). 1911.

- "Tamarind". *The New Student's Reference Work*. 1914.

Chapter 76

Tanbark

For other uses, see Tanbark, Lexington.

Tanbark is the bark of certain species of tree. It is tradi-

Workers peeling hemlock bark for the tannery in Prattsville, New York, United States.

tionally used for tanning hides.[1]

The words "tanning", "tan," and "tawny" are derived from the Medieval Latin *tannare*, "to convert into leather."

Bark mills are horse- or oxen-driven or water powered edge mills[2] and were used in earlier times to shred the tanbark to derive tannins for the leather industry.

A "barker" was a person who stripped bark from trees to supply bark mills.

76.1 Tanbark around the world

In some areas of the United States, such as northern California, tanbark is often called "mulch," even by manufacturers and distributors. In these areas, the word "mulch" may refer to peat moss or to very fine tanbark. In California, *Lithocarpus densiflorus* (commonly known as the *tanoak* or *tanbark-oak*) was used. In New York, on the slopes of Mount Tremper, hemlock bark was a major source of tanbark during the 19th century.

In America, condensed tannins are also present in the bark of blackjack oak (*Quercus marilandica'*).[3]

Around the Mediterranean Sea, sumach (*Rhus coriaria*) leaves and bark are used.

In Africa and Australia, acacia (called "wattle") bark is used by tanners.

76.2 Oak bark

Waterwheel at Combe House Hotel in Holford, Somerset, England. The overshot waterwheel was cast by Bridgwater ironfounder H Culverwell & Co in 1892 to replace an earlier wheel. It was used to grind oak bark for the tannery complex established here in the 1840s by James Hayman. When the tannery closed in 1900 the waterwheel was adapted to other uses such as grinding grain for grist, cutting chaff, chopping apples for the cider press and generating electricity. It also cracked stones in a nearby quarry. The gearing survives too.

In Europe, oak is a common source of tanbark. Quercitannic acid is the chief constituent found in oak barks.[4] The bark is taken from young branches and twigs in oak coppices and can be up to 4 mm thick; it is grayish-brown on the outside and brownish-red on the inner surface.[5]

- Tool to recover bark from oak branches

- Recovery of bark from oak branches

- Another view of the process

- The bark of an oak tree

- Young red oak bark

76.3   See also

- Barkdust

76.4   References

[1] Chapter 8 - Tannins: Major Sources, Properties and Applications. Antonio Pizzi, Monomers, Polymers and Composites from Renewable Resources 2008, Pages 179-199, doi:10.1016/B978-0-08-045316-3.00008-9

[2] cslib.cdmhost.com

[3] Flavan and procyanidin glycosides from the bark of blackjack oak. Young-soo Bae, Johann F.W. Burger, Jan P. Steynberg, Daneel Ferreira and Richard W. Hemingway, Phytochemistry, Volume 35, Issue 2, January 1994, Pages 473-478, doi:10.1016/S0031-9422(00)94785-X

[4] *Quercus* on www.henriettesherbal.com

[5] Oak on www.online-health-care.com

76.5   External links

Chapter 77

Tea tree oil

This article is about essential oil isolated form the leaves of the tea tree, *Melaleuca alternifolia*. For the sweet seasoning oil pressed from *Camellia* seeds, *C. sinensis* or *C. oleifera*, see tea seed oil.

Tea tree oil (TTO), or **melaleuca oil**, is an essential oil

Origin of this essential oil, the tea tree, Melaleuca alternifolia.

Tea tree plantation, Coraki.

with a fresh camphoraceous odor and a color that ranges from pale yellow to nearly colorless and clear.[*][2] It is taken from the leaves of the *Melaleuca alternifolia*, which is native to Southeast Queensland and the Northeast coast of New South Wales, Australia.

Tea tree oil is toxic when taken by mouth,[*][3][*][4] but is widely used in low concentrations in cosmetics and skin washes.[*][1] Tea tree oil has been claimed to be useful for treating a wide variety of medical conditions. It shows some promise as an antimicrobial. Tea tree oil may be effective in a variety of dermatologic conditions including dandruff, acne, lice, herpes, and other skin infections.[*][5]

77.1 History and extraction

The name tea tree is used for several plants, mostly from Australia and New Zealand, from the family Myrtaceae, related to the myrtle. The use of the name probably originated from Captain Cook's description of one of these shrubs, that he used to make an infusion, to drink in place of tea.

The commercial tea tree oil industry originated in the 1920s when Arthur Penfold, an Australian, investigated the business potential of a number of native extracted oils; he reported that tea tree oil had promise as it exhibited powerful antiseptic properties.[*][6]

Tea tree oil was first extracted from *Melaleuca alternifolia* in Australia, and this species remains the most important commercially. Several other species are cultivated for their extracted oil: *Melaleuca armillaris* and *Melaleuca styphelioides* in Tunisia and Egypt; *Melaleuca leucadendron* in Egypt, Malaysia and Vietnam; *Melaleuca acuminata* in Tunisia; *Melaleuca ericifolia* in Egypt; and *Melaleuca quinenervia* in the United States. Similar oils can also be produced by water distillation from *Melaleuca linariifolia* and *Melaleuca dissitiflora*.[*][7]

77.2 Composition and characteristics

Tea tree oil is defined by the International Standard ISO 4730 ("Oil of *Melaleuca*, Terpinen-4-ol type"), which specifies levels of 15 components which are needed to define the oil as "tea tree oil." The oil has been described as having a fresh, camphor-like smell.[8]

Tea tree oils have six types, oils with different chemical compositions. These include a terpinen-4-ol type, a terpinolene type, and four 1,8-cineole types. These various oil types contain over 98 compounds, with terpinen-4-ol the major component responsible for antimicrobial and anti-inflammatory properties.[9] A second component 1,8-cineole, is likely responsible for most allergies in TTO products. Adverse reactions to TTO diminish with minimization of 1,8-cineole content. In commercial production, TTO is prepared as a terpinen-4-ol type.[5]

77.3 Medical use

In vitro studies have shown that tea tree oil kills methicillin-resistant Staphylococcus aureus (MRSA),[10] in nasal or extra-nasal (topical) colonisation studies possibly comparable to treatment with mupirocin,[11] but as of 2005 there appeared to be insufficient evidence to recommend it for clinical use.[10] A 2008 article from the American Cancer Society says that studies have found some promise of a possible role for the topical application of tea tree oil as an antiseptic,[3] but that "despite years of use, available clinical evidence does not support the effectiveness of tea tree oil for treating skin problems and infections in humans" .[3] A 2012 review by the NIH rates tea tree oil as "possibly effective" for three applications, saying that "a 5% tea tree oil gel appears to be as effective as 5% benzoyl peroxide" for treating mild to moderate acne, that "topical application of 100% tea tree oil solution, twice daily for six months, can cure fungal toenail infection in about 18% of people who try it," and that "a 10% tea tree oil cream works about as well as tolnaftate 1% cream" in treating symptoms of athlete's foot, although being less effective than clotrimazole or terbinafine.[12]

A 2006 review of the toxicity of tea tree oil concludes that it may be used externally in its diluted form by the majority of individuals without adverse effect (provided oxidization is avoided).[13] Tea tree oil is poisonous when taken internally.[3] Tea tree oil may be effective in a variety of dermatologic conditions including dandruff, acne, lice, herpes, and other skin infections.[5] A 2012 review of head lice treatment recommended against the use of tea tree oil on children because it could cause skin irritation or allergic re-

actions, because of contraindications, and because of a lack of knowledge about the oil's safety and effectiveness.[14]

77.4 Safety

Tea tree oil is a commercially refined composition of several naturally occurring chemical compounds and is hazardous if misused. Available literature suggests that tea tree oil can be used topically in diluted form by the majority of individuals without adverse effects. Topical application of tea tree oil can cause adverse reactions at high concentration. Adverse effects including skin irritation, allergic contact dermatitis, systemic contact dermatitis, linear immunoglobulin A disease, erythema multiforme like reactions, and systemic hypersensitivity reactions.[5][15]

Tea tree oil is toxic when swallowed.[15] According to the American Cancer Society, ingesting tea tree oil has been reported to cause drowsiness, confusion, hallucinations, coma, unsteadiness, weakness, vomiting, diarrhea, stomach upset, blood cell abnormalities, and severe rashes. It should be kept away from pets and children.[3] Tea tree oil should not be used in or around the mouth.[4] There is at least one case of poisoning reported in medical literature.[16]

Exposure of tea tree oil to air and light results in oxidation of some of its components. Oxidized tea tree oil should not be used.[17] Some people experience allergic contact dermatitis as a reaction to dermal contact with tea tree oil. Allergic reactions may be due to the various oxidation products that are formed by exposure of the oil to light and/or air.[15][18]

In vitro testing of tea tree oil shows that it contains chemicals which are weakly estrogenic, causing particular concern for use with children. However, in tests, the chemicals which show this effect failed to show absorption into the skin, and evidence of a hormonal effect is therefore considered implausible by an EU scientific committee.[1]

In dogs and cats, death[19][20] or transient signs of toxicity (lasting 2 to 3 days), such as depression, weakness, incoordination and muscle tremors, have been reported after external application at high doses.[21] In rats the LD50 is 1.9–2.4 ml/kg.[22]

Undiluted tea tree oil can cause some hearing loss when used in the ears of non-human animals; however, a 2% concentration has not been shown to have any lasting effect. It is not known whether the same is true for humans.[23]

77.5 See also

- Cajuput oil – derived from *Melaleuca leucadendra*

77.6 References

[1] SCCP/1155/08 Scientific Committee on Consumer Products SCCP OPINION ON Tea tree oil – European Union Commission Health and Consumer Union protection director general – adopted 18th plenary of 16 December 2008

[2] "Directory of Essential Oils for Aromatherapy: Tea-Tree Oil (Melaleuca alternifolia)". Holistics Online.

[3] "Tea Tree Oil". American Cancer Society. November 2008. Retrieved September 2013.

[4] "Tea Tree Oil". National Capital Poison Center. Retrieved 4 December 2013.

[5] Pazyar, N; Yaghoobi, R; Bagherani, N; Kazerouni, A (July 2013). "A review of applications of tea tree oil in dermatology". *International Journal of Dermatology* **52** (7): 784–90. doi:10.1111/j.1365-4632.2012.05654.x. PMID 22998411.

[6] Carson, C. F.; Hammer, K. A.; Riley, T. V. (2006). "Melaleuca alternifolia (Tea Tree) Oil: A Review of Antimicrobial and Other Medicinal Properties". *Clinical Microbiology Reviews* **19** (1): 50–62. doi:10.1128/CMR.19.1.50-62.2006. PMC 1360273. PMID 16418522.

[7] Sávia Perina Portilho Falci (2015-07). "Antimicrobial activity of Melaleuca sp. oil against clinical isolates of antibiotics resistant Staphylococcus aureus". *Acta Cirurgica Brasileira* **30** (7). Check date values in: |date= (help)

[8] Billee Sharp (18 September 2013). *Lemons and Lavender: The Eco Guide to Better Homekeeping*. Cleis Press. pp. 43–. ISBN 978-1-936740-11-6.

[9] Hart, P.H.; Brand, C.; Carson, C.F.; Riley, T.V.; Prager, R.H.; Finlay-Jones, J.J. (2000). "Terpinen-4-ol, the main component of the essential oil of Melaleuca alternifolia (tea tree oil), suppresses inflammatory mediator production by activated human monocytes." . *Inflammation Research* **49** (11): 619–26. doi:10.1007/s000110050639. PMID 11131302.

[10] Flaxman, D.; Griffiths, P. (2005). "Is tea tree oil effective at eradicating MRSA colonization? A review". *Br. J. Community Nurs.* **10** (3, March): 123–126. PMID 15824699.

[11] Bradley, Suzanne F (January 2011). "MRSA colonisation (eradicating colonisation in people without active/invasive infection)". *Clinical Evidence* **2011**. PMID 21477403.

[12] "Tea tree oil". U.S. National Library of Medicine. Retrieved 15 December 2014.

[13] Hammer, KA; Carson, CF; Riley, TV; Nielsen, JB (May 2006). "A review of the toxicity of Melaleuca alternifolia (tea tree) oil." . *Food and chemical toxicology : an international journal published for the British Industrial Biological Research Association* **44** (5): 616–25. doi:10.1016/j.fct.2005.09.001. PMID 16243420.

[14] Eisenhower, Christine; Farrington, Elizabeth Anne (2012). "Advancements in the Treatment of Head Lice in Pediatrics". *Journal of Pediatric Health Care* **26** (6): 451–61; quiz 462–4. doi:10.1016/j.pedhc.2012.05.004. PMID 23099312.

[15] Hammer, K; Carson, C; Riley, T; Nielsen, J (2006). "A review of the toxicity of Melaleuca alternifolia (tea tree) oil". *Food and Chemical Toxicology* **44** (5): 616–25. doi:10.1016/j.fct.2005.09.001. PMID 16243420.

[16] "Ingestion of tea tree oil (Melaleucaoil) by a 4-year-old boy" .

[17] "THE EFFECTIVENESS AND SAFETY OF AUSTRALIAN TEA TREE OIL". Australian Government - Rural Industries and Development Corporation. Retrieved 26 February 2014.

[18] Aberer, W (January 2008). "Contact allergy and medicinal herbs". *Journal der Deutschen Dermatologischen Gesellschaft = Journal of the German Society of Dermatology : JDDG* **6** (1): 15–24. doi:10.1111/j.1610-0387.2007.06425.x. PMID 17919303.

[19] "Tea Tree Oil and Dogs, Tea Tree Oil and Cats". Petpoisonhelpline.com. Retrieved December 13, 2012.

[20] "Tea Tree Oil Toxicity". Veterinarywatch. Retrieved December 13, 2012.

[21] Villar, D; Knight, MJ; Hansen, SR; Buck, WB (April 1994). "Toxicity of melaleuca oil and related essential oils applied topically on dogs and cats". *Veterinary and human toxicology* **36** (2): 139–42. PMID 8197716.

[22] Clinical Microbiological Reviews: Melaleuca alternifolia (Tea Tree) Oil: a Review of Antimicrobial and Other Medicinal Properties-C. F. Carson,1 K. A. Hammer,1 and T. V. Riley

[23] "Tea tree oil". Medline Plus, a service of the U.S. National Library of Medicine from the National Institutes of Health. 27 July 2012.

77.7 External links

- "The Marshall Centre: Tea Tree Oil"., research group at the University of Western Australia

- "Tea Tree Oil" . at the American Cancer Society

- "Australian Tea Tree Oil database" ., a searchable abstract database containing 1200+ journal articles on Tea Tree Oil.

Chapter 78

Toddy palm

Toddy palm is a common name for several species of palms used to produce palm wine. Species so used and named include:

- *Borassus flabellifer*, native to South and Southeast Asia, in the Indomalaya ecozone. It is a palm tree, one of the Sugar palm group, found from Indonesia to Pakistan

- *Caryota*, a genus of palm trees

- *Cocos nucifera*

Chapter 79

Turpentine

For other uses, see Turpentine (disambiguation).
 Turpentine (also called **spirit of turpentine**, **oil of tur-**

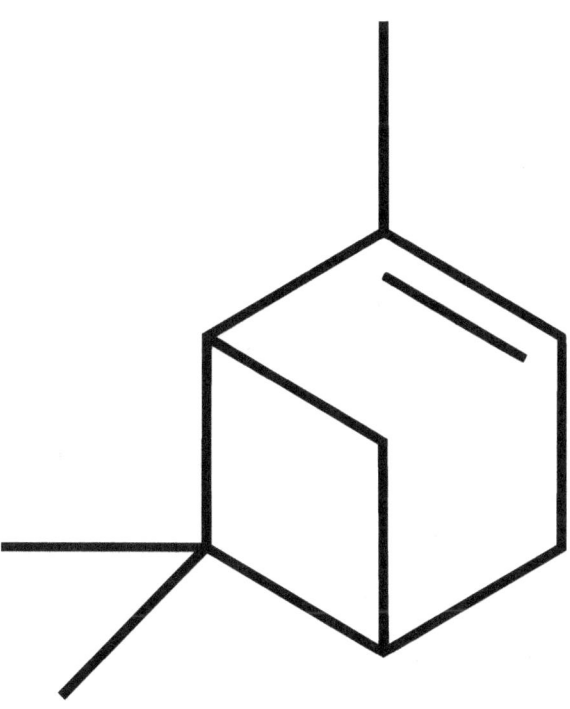

Chemical structure of pinene, a major component of turpentine

pentine, **wood turpentine** and colloquially **turps**[*][1]) is a fluid obtained by the distillation of resin obtained from live trees, mainly pines. It is mainly used as a solvent and as a source of materials for organic synthesis.

Turpentine is composed of terpenes, mainly the monoterpenes alpha-pinene and beta-pinene with lesser amounts of carene, camphene, dipentene, and terpinolene.[*][2]

The word *turpentine* derives (via French and Latin) from the Greek word τερεβινθίνη *terebinthine*, the name of a species of tree, the terebinth tree.[*][3] Mineral turpentine or other petroleum distillates are used to replace turpentine, but they are very different chemically.[*][4]

79.1 Source trees

A 1912 postcard depicting harvesting pine resin for the turpentine industry

One of the earliest sources was the terebinth or turpentine tree (*Pistacia terebinthus*), a Mediterranean tree related to the pistachio. Important pines for turpentine production include: maritime pine (*Pinus pinaster*), Aleppo pine (*Pinus halepensis*), Masson's pine (*Pinus massoniana*), Sumatran pine (*Pinus merkusii*), longleaf pine (*Pinus palustris*), loblolly pine (*Pinus taeda*) and ponderosa pine (*Pinus ponderosa*).

Canada balsam, also called Canada turpentine or balsam of fir, is a turpentine which is made from the oleoresin of the balsam fir. Venice turpentine is produced from the western larch *Larix occidentalis*.

In order to tap into the sap producing layers of the tree, turpentiners used a combination of hacks to remove the pine bark. Once debarked, pine trees secrete oleoresin onto the surface of the wound as a protective measure to seal the opening, resist exposure to micro-organisms and insects and prevent vital sap loss. Turpentiners wounded trees in V-shaped streaks down the length of the trunks so as to channel the oleoresin into containers. It was then collected and processed into spirits of turpentine. Oleoresin yield may be increased by as much as 40% by applying paraquat herbi-

"Herty system" in use on turpentine trees in Northern Florida, circa 1936

Chipping a turpentine tree in Georgia, circa 1906-20

cides to the exposed wood.[*][5]

The V-shaped cuts are called "catfaces" for their resemblance to a cat's whiskers. These marks on a pine tree signify it was used to collect resin for turpentine production.[*][6]

79.2 Converting oleoresin to turpentine

Crude oleoresin collected from wounded trees may be evaporated by steam distillation in a copper still. Molten rosin remains in the still bottoms after turpentine has been evaporated and recovered from a condenser.[*][5] Turpentine may alternatively be condensed from destructive distillation of pine wood.[*][2]

Oleoresin may also be extracted from shredded pine stumps, roots, and slash using the light end of the heavy naphtha fraction (boiling between 90 and 115 °C or 195 and 240 °F) from a crude oil refinery. Multi-stage counter-current extraction is commonly used so fresh naphtha first contacts wood leached in previous stages and naphtha laden with turpentine from previous stages contacts fresh wood before vacuum distillation to recover naphtha from the turpentine. Leached wood is steamed for additional naphtha recovery prior to burning for energy recovery.[*][7]

When producing chemical wood pulp from pines or other coniferous trees, sulfate turpentine may be condensed from the gas generated in Kraft process pulp digesters. The average yield of crude sulfate turpentine is 5–10 kg/t pulp.[*][8] Unless burned at the mill for energy production, sulfate turpentine may require additional treatment measures to remove traces of sulfur compounds.[*][9]

79.3 Industrial and other end uses

79.3.1 Solvent

The two primary uses of turpentine in industry are as a solvent and as a source of materials for organic synthesis. As a solvent, turpentine is used for thinning oil-based paints, for producing varnishes, and as a raw material for the chemical industry. Its industrial use as a solvent in industrialized nations has largely been replaced by the much cheaper turpentine substitutes distilled from crude oil. Turpentine has long been used as a solvent, mixed with beeswax or with carnauba wax, to make fine furniture wax for use as a protective coating over oiled wood finishes (e.g., tung oil).

79.3.2 Source of organic compounds

Turpentine is also used as a source of raw materials in the synthesis of fragrant chemical compounds. Commercially used camphor, linalool, alpha-terpineol, and geraniol are all usually produced from alpha-pinene and beta-pinene, which are two of the chief chemical components of turpentine. These pinenes are separated and purified by distillation. The mixture of diterpenes and triterpenes that is left as residue after turpentine distillation is sold as rosin.

79.3.3 Medicinal elixir

Turpentine and petroleum distillates such as coal oil and kerosene have been used medicinally since ancient times, as topical and sometimes internal home remedies. Topically it has been used for abrasions and wounds, as a treatment for lice, and when mixed with animal fat it has been used as a chest rub, or inhaler for nasal and throat ailments. Many modern chest rubs, such as the Vicks variety, still contain turpentine in their formulations.

Taken internally it was used as treatment for intestinal parasites because of its alleged antiseptic and diuretic properties, and a general cure-all*[10]*[11] as in Hamlin's Wizard Oil. Sugar, molasses or honey were sometimes used to mask the taste. Internal administration of these toxic products is no longer common today.

Turpentine was a common medicine among seamen during the Age of Discovery, and one of several products carried aboard Ferdinand Magellan's fleet in his first circumnavigation of the globe.*[12]

79.3.4 Niche uses

Turpentine is also added to many cleaning and sanitary products due to its antiseptic properties and its "clean scent." In early 19th-century America, turpentine was sometimes burned in lamps as a cheap alternative to whale oil. It was most commonly used for outdoor lighting, due to its strong odour.*[13] A blend of ethanol and turpentine added as an illuminant called burning fluid was also important for several decades. In 1946, Soichiro Honda used turpentine as a fuel for the first Honda motorcycles as gasoline was almost totally unavailable in Japan following World War II.*[14]

Turpentine was a common additive in cheap gin until the 20th century and gave it its characteristic juniper berry flavour without the need for pricier distillations with aromatic spices and berries.*[15]

79.4 Hazards

As an organic solvent, its vapour can irritate the skin and eyes, damage the lungs and respiratory system, as well as the central nervous system when inhaled, and cause renal failure when ingested, among other things. Being combustible, it also poses a fire hazard. Because turpentine can cause spasms of the airways particularly in people with asthma

and whooping cough, it can contribute to a worsening of breathing issues in persons with these diseases if inhaled.

79.5 See also

- Charles Herty

- Galipot

- Naval stores industry

- Patent medicine

- Russia leather, a water-resistant leather, using a birch oil distillate similar to turpentine in its manufacture.

79.6 Sources

- Kent, James A. *Riegel's Handbook of Industrial Chemistry* (Eighth Edition) Van Nostrand Reinhold Company (1983) ISBN 0-442-20164-8

79.7 References

[1] Mayer, Ralph (1991). *The Artist's Handbook of Materials and Techniques* (Fifth ed.). New York: Viking. p. 404. ISBN 0-670-83701-6.

[2] Kent p.569

[3] Barnhart, R.K. (1995). *The Barnhart Concise Dictionary of Etymology*. New York: Harper Collins. ISBN 0-06-270084-7.

[4] Dieter Stoye "Solvents" in *Ullmann's Encyclopedia of Industrial Chemistry*, 2002, Wiley-VCH, Wienheim. doi:10.1002/14356007.a24_437

[5] Kent p.571

[6] Prizer, Tom (June 11, 2010). "Catfaces: Totems of Georgia's Turpentiners | Daily Yonder | Keep It Rural" . *dailyyonder.com*. Retrieved June 5, 2012.

[7] Kent pp.571&572

[8] Stenius, Per, ed. (2000). "2" . *Forest Products Chemistry*. Papermaking Science and Technology **3**. Finland. pp. 73–76. ISBN 952-5216-03-9.

[9] Kent p.572

[10] "Rural Life in the United States: Home Remedies" . *American Memory Timeline*. The Library of Congress. 2002. Retrieved 2008-02-22.

[11] Delbert Trew (15 June 2007). "Coal Oil was Useful All-Purpose Home Remedy". *Texas Escapes*. Blueprints For Travel, LLC. Retrieved 2008-02-22.

[12] Laurence Bergreen (2003). "Over the edge of the world : Magellan's terrifying circumnavigation of the globe". ISBN 0066211735. Retrieved 2009-09-14.

[13] Charles H. Haswell. "Reminiscences of New York By an Octogenarian (1816 - 1860)".

[14] "Honda History". Smokeriders.com.

[15] Patrick Dillon (2002-06-01). "Distil my beating heart". *The Guardian* (London).

79.8   External links

- Inchem.org, IPCS INCHEM Turpentine classification, hazard, and property table.

- CDC - NIOSH Pocket Guide to Chemical Hazards - Turpentine

- FAO.org, Gum naval stores: Turpentine and rosin from pine resin

- FloridaMemory.com, Florida State Archive photographs of turpentine camps and laborers

- HCHSonline.org, Timber and Turpentine Industries

- Distil my beating heart

- Florida's "Turpmtine" Camps

Chapter 80

Urushiol lacquer

For items made with lacquer, see Lacquerware.

Lacquer is a clear or coloured wood finish that dries by

Lacquer box with inlaid mother of pearl peony decor, Ming Dynasty, 16th century

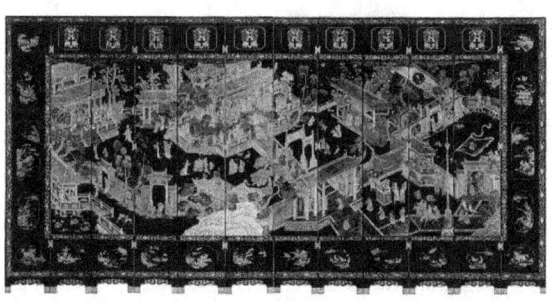

Armorial screen

solvent evaporation or a curing process that produces a hard, durable finish. This finish can be of any sheen level from ultra matte to high gloss, and it can be further polished as required. It is also used for "lacquer paint", which is a paint

that typically dries better on a hard and smooth surface.

The term *lacquer* originates from the Sanskrit word *lāksha'* (लाक्षा) meaning "wax", which was used for both the Lac insect (because of their enormous number) and the scarlet resinous secretion it produces that was used as wood finish in ancient India and neighbouring areas.[*][1] In terms of modern products for coating finishes, lac-based finishes are likely to be referred to as shellac, while lacquer often refers to other polymers dissolved in volatile organic compounds (VOCs), such as nitrocellulose, and later acrylic compounds dissolved in *lacquer thinner*, a mixture of several solvents typically containing butyl acetate and xylene or toluene. Lacquer is more durable than shellac.

In the decorative arts, lacquer or lacquerware refers to a variety of techniques used to decorate wood, metal or other surfaces, especially carving into deep coatings of many layers of lacquer.

80.1 Etymology

The archaic French word *lacre* "a kind of sealing wax", from Portuguese *lacre*, unexplained variant of *lacca* "resinous substance", from Arabic *lakk*, from Telugu 🅰🅰🅰 Persian *lak*, the verb *lac* meaning "to cover or coat with laqueur".[*][2] The root of the word is the Sanskrit word *lāksha'* (लाक्षा), which was used for both the Lac insect (because of their enormous number) and the scarlet resinous secretion it produces that was used as wood finish in ancient India and neighbouring areas.[*][1][*][3] Lac resin was once imported in sizeable quantity into Europe from India along with Eastern woods.[*][4][*][5]

80.2 Sheen measurement

Lacquer sheen is a measurement of the shine for a given lacquer.[*][6] Different manufacturers have their own names and standards for their sheen.[*][6] The most common names from least shiny to most shiny are: flat, matte, egg shell, satin, semi-gloss, and gloss (high).

80.3   Urushiol-based lacquers

A Chinese six-pointed tray, red lacquer over wood, from the Song Dynasty (960–1279), 12th-13th century, Metropolitan Museum of Art.

Ming Dynasty Chinese lacquerware container, dated 16th century.

Urushiol-based lacquers differ from most others, being slow-drying, and set by oxidation and polymerization, rather than by evaporation alone. In order for it to set properly it requires a humid and warm environment. The phenols oxidize and polymerize under the action of an enzyme laccase, yielding a substrate that, upon proper evaporation of its water content, is hard. These lacquers produce very hard, durable finishes that are both beautiful and very resistant to damage by water, acid, alkali or abrasion. The active ingredient of the resin is urushiol, a mixture of various phenols suspended in water, plus a few proteins. The

resin is derived from a tree indigenous to China, species *Toxicodendron vernicifluum*, commonly known as the Lacquer Tree.[7] The fresh resin from the *T. vernicifluum* trees causes urushiol-induced contact dermatitis and great care is required in its use. The Chinese treated the allergic reaction with crushed shellfish, which supposedly prevents lacquer from drying properly.[8] Lacquer skills became very highly developed in Asia, and many highly decorated pieces were produced.

During the Shang Dynasty (1600–1046 B.C.), the sophisticated techniques used in the lacquer process were first developed and it became a highly artistic craft,[9] although various prehistoric lacquerwares have been unearthed in China dating back to the Neolithic period and objects with lacquer coating in Japan from the late Jōmon period.[9] The earliest extant lacquer object, a red wooden bowl,[10] was unearthed at a Hemudu culture (5000-4500 B.C.) site in China.[11] By the Han Dynasty (206 B.C. – 220 a.C), many centres of lacquer production became firmly established.[9] The knowledge of the Chinese methods of the lacquer process spread from China during the Han, Tang and Song dynasties, eventually it was introduced to Korea, Japan, Southeast and South Asia.[12]

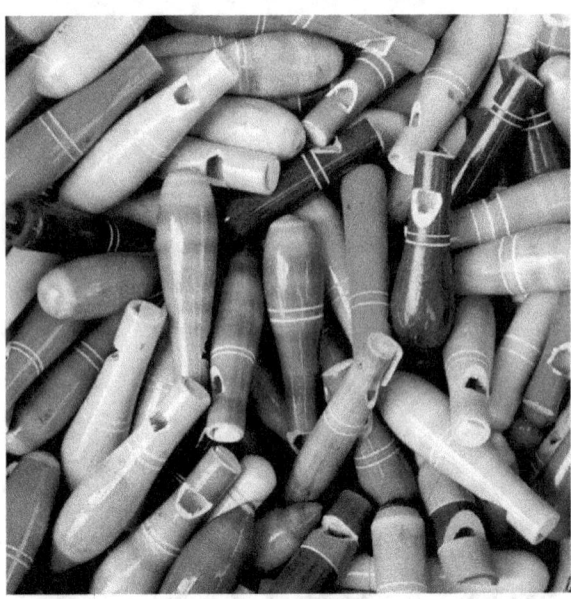

Wooden lacquer-finished whistles made in Channapatna, Karnataka, India

There are two types of lacquer in India: one obtained from the *T. vernicifluum* tree and the other from an insect. In India the insect lac was once used from which a red dye was first extracted, later what was left of the insect was a grease that was used for lacquering objects. Trade of lacquer objects travelled through various routes to the Middle East. Known applications of lacquer in China included coffins, music instruments, furniture, and various household items.*[9] Lacquer mixed with powdered cinnabar is used to produce the traditional red lacquerware from China.

Lacquer mixed with water and turpentine, ready for applying to surface.

The trees must be at least ten years old before cutting to bleed the resin. It sets by a process called "aqua-polymerization", absorbing oxygen to set; placing in a humid environment allows it to absorb more oxygen from the evaporation of the water.

Lacquer-yielding trees in Thailand, Vietnam, Burma and Taiwan, called Thitsi, are slightly different; they do not contain urushiol, but similar substances called "laccol" or "thitsiol". The end result is similar but softer than the Chinese or Japanese lacquer. Unlike Japanese and Chinese *Toxicodendron verniciflua* resin, Burmese lacquer does not cause allergic reactions; it sets slower, and is painted by craftsmen's hands without using brushes.

Raw lacquer can be "coloured" by the addition of small amounts of iron oxides, giving red or black depending on the oxide. There is some evidence that its use is even older than 8,000 years from archaeological digs in China. Later, pigments were added to make colours. It is used not only as a finish, but mixed with ground fired and unfired clays applied to a mould with layers of hemp cloth, it can produce objects without need for another core like wood. The process is called "kanshitsu" in Japan. Advanced decorative techniques using additional materials such as gold and silver powders and flakes ("makie") were refined to very high standards in Japan also after having been introduced from China. In the lacquering of the Chinese musical instrument, the guqin, the lacquer is mixed with deer horn powder (or ceramic powder) to give it more strength so it can stand up to the fingering.

There are more than four forms of urushiol which is written as thus:

A Chinese lacquer coffin decorated with birds and dragons, from the State of Chu, 4th century B.C.

$R = (CH_2)_{14}CH_3$ or
$R = (CH_2)_7CH=CH(CH_2)_5CH_3$ or
$R = (CH_2)_7CH=CHCH_2CH=CH(CH_2)_2CH_3$ or
$R = (CH_2)_7CH=CHCH_2CH=CHCH=CHCH_3$ or
$R = (CH_2)_7CH=CHCH_2CH=CHCH_2CH=CH_2$ and others.

80.3.1 Types of lacquer

Types of lacquer vary from place to place but they can be divided into unprocessed and processed categories.

The basic unprocessed lacquer is called *raw lacquer* (生漆: *ki-urushi* in Japanese, *shengqi* in Chinese). This is directly from the tree itself with some impurities filtered out. Raw lacquer has a water content of around 25% and appears in a light brown colour. This comes in a standard grade made from Chinese lacquer, which is generally used for ground layers by mixing with a powder, and a high quality grade made from Japanese lacquer called *kijomi-urushi* (生正味漆) which is used for the last finishing layers.

The processed form (in which the lacquer is stirred continuously until much of the water content has evaporated) is called *guangqi* (光漆) in Chinese but comes under many different Japanese names depending on the variation, for example, *kijiro-urushi* (木地呂漆) is standard transparent lacquer sometimes used with pigments and *roiro-urushi* (呂色漆) is the same but pre-mixed with iron hydroxide to produce a black coloured lacquer. *Nashiji-urushi* (梨子地漆) is the transparent lacquer but mixed with gamboge to create an even clearer lacquer and is especially used for the sprinkled-gold technique. These lacquers are generally used for the middle layers. Japanese lacquers of this type are generally used for the top layers and are prefixed by the word *jo-* (上) which means 'top (layer)'.

Processed lacquers can have oil added to them to make them glossy, for example, *shuai-urushi* (朱合漆) is mixed with linseed oil. Other specialist lacquers include *ikkake-urushi* (漆) which is thick and used mainly for applying gold or silver leaf.

80.4   Nitrocellulose lacquers

Slow-drying solvent-based lacquers that contain nitrocellulose, a resin obtained from the nitration of cotton and other cellulostic materials, were developed in the early 1920s, and extensively used in the automobile industry for 30 years. Prior to their introduction, mass-produced automotive finishes were limited in colour, with Japan Black being the fastest drying and thus most popular. General Motors Oakland automobile brand automobile was the first (1923) to introduce one of the new fast drying nitrocelluous lacquers, a bright blue, produced by DuPont under their Duco tradename.

These lacquers are also used on wooden products, furniture primarily, and on musical instruments and other objects. Nitrocellulose lacquers are also used to make firework fuses waterproof. The nitrocellulose and other resins and plasticizers are dissolved in the solvent, and each coat of lacquer dissolves some of the previous coat. These lacquers were a huge improvement over earlier automobile and furniture finishes, both in ease of application and in colour retention. The preferred method of applying quick-drying lacquers is by spraying, and the development of nitrocellulose lacquers led to the first extensive use of spray guns. Nitrocellulose lacquers produce a hard yet flexible, durable finish that can be polished to a high sheen. Drawbacks of these lacquers include the hazardous nature of the solvent, which is flammable and toxic, and the hazards of nitrocellulose in the manufacturing process. Lacquer grade of soluble nitrocellulose is closely related to the more highly nitrated form which is used to make explosives. They become relatively non-toxic after approximately a month since at this point, the lacquer has evaporated most of the solvents used in its production.

80.5   Acrylic lacquers

Lacquers using acrylic resin, a synthetic polymer, were developed in the 1950s. Acrylic resin is colourless, transparent thermoplastic, obtained by the polymerization of derivatives of acrylic acid. Acrylic is also used in enamel paints, which have the advantage of not needing to be buffed to obtain a shine. Enamels, however, are slow drying. The advantage of acrylic lacquer is its exceptionally fast drying time. The use of lacquers in automobile finishes was discontinued when tougher, more durable, weather- and chemical-resistant two-component polyurethane coatings were developed. The system usually consists of a primer, colour coat and clear topcoat, commonly known as clear coat finishes.

80.6   Water-based lacquers

Due to health risks and environmental considerations involved in the use of solvent-based lacquers, much work has gone into the development of water-based lacquers. Such lacquers are considerably less toxic and more environmentally friendly, and in many cases, produce acceptable results. More and more water-based coloured lacquers are replacing solvent-based clear and coloured lacquers in underhood and interior applications in the automobile and other similar industrial applications. Water based lacquers are used extensively in wood furniture finishing as well.

80.7   Japanning

Main article: Japanning

Just as *china* is a common name for porcelain, *japanning* is an old name to describe the European technique to imitate Asian lacquerware.*[13] As Asian lacquer work became popular in England, France, the Netherlands, and Spain in the 17th century the Europeans developed imitations that were effectively a different technique of lacquering. The European technique, which is used on furniture and other objects, uses finishes that have a resin base similar to shellac. The technique, which became known as japanning, involves applying several coats of varnish which are each heat-dried and polished. In the 18th century, this type of lacquering gained a large popular following. Although traditionally a pottery and wood coating, japanning was the popular (mostly black) coating of the accelerating metalware industry. By the twentieth century, the term was freely applied to coatings based on various varnishes and lacquers besides the traditional shellac.

80.8   Lakshagraha or The House of Lacquer

Main article: Lakshagraha

The Hindu epic, Mahabharatha, describes the building of a house made of lac. Lakshagraha or Lakshagriha (Sanskrit: लाक्षागृहम्) (The House of Lacquer) is a book or parva from the Mahabharata, one of the two major Sanskrit epics of ancient India, the other being the Ramayana. This house was built under the orders of Duryodhana and his evil uncle and mentor Shakuni in a plot to kill the Pandavas along with their mother Kunti. The architect Purochana was employed in the building of Lakshagraha in the forest of Varnavrat. The house was meant to be a death trap, since lacquer is highly flammable. The plot itself was such that nobody would suspect foul play and the eventual death of the Pandavas would pass off as an accident. In the Mahabharata this incident is considered a major turning point, since the Pandavas were considered dead by their cousins, the Kauravas, which gave them ample opportunity to prepare themselves for an upcoming and unavoidable war. However, an escape route was prepared for the Pandavas who had been warned of the plot.

Lakshgraha Varanavat, is located in modern day Handia in Allahabad in Uttar Pradesh, India. The site at Varnavrat has since become a tourist location.*[14]

80.9 See also

- Lacquerware

- Varnish

- Acetate disc

- Lacquer painting

80.10 References

*[15]

[1] Franco Brunello (1973), *The art of dyeing in the history of mankind*, AATCC, 1973, ... *The word lacquer derives, in fact, from the Sanskrit 'Laksha' and has the same meaning as the Hindi word 'Lakh' which signifies one-hundred thousand ... enormous number of those parasitical insects which infest the plants Acacia catecu, Ficus and Butea frondosa ... great quantity of reddish colored resinous substance ... used in ancient times in India and other parts of Asia ...*

[2] http://www.etymonline.com/index.php?term=lacquer

[3] Ulrich Meier-Westhues (November 2007), *Polyurethanes: coatings, adhesives and sealants*, Vincentz Network GmbH & Co KG, 2007, ISBN 978-3-87870-334-1, ... *Shellac, a natural resin secreted by the scaly lac insect, has been used in India for centuries as a decorative coating for surfaces. The word lacquer in English is derived from the Sanskrit word laksha. which means one hundred thousand ...*

[4] Donald Frederick Lach, Edwin J. Van Kley (1994-02-04), *Asia in the making of Europe, Volume 2, Book 1*, University of Chicago Press, 1971, ISBN 978-0-226-46730-6, ... *Along with valuable woods from the East, the ancients imported lac, a resinous incrustation produced on certain trees by the puncture of the lac insect. In India, lac was used as sealing wax, dye and varnish ... Sanskrit, laksha; Hindi, lakh; Persian, lak; Latin, lacca. The Western word "lacquer" is derived from this term ...*

[5] Thomas Brock, Michael Groteklaes, Peter Mischke (2000), *European coatings handbook*, Vincentz Network GmbH & Co KG, 2000, ISBN 978-3-87870-559-8, ... *The word "lacquer" itself stems from the term "Laksha", from the pre-Christian, sacred Indian language Sanskrit, and originally referred to shellac, a resin produced by special insects ("lac insects") from the sap of an Indian fig tree ...*

[6] Wood Finishers Depot: Lacquer Sheen

[7] Britannica Online Encyclopedia: Oriental lacquer

[8] Major, John S., Sarah Queen, Andrew Meyer, Harold D. Roth, (2010), *The Huainanzi: A Guide to the Theory and Practice of Government in Early Han China*, Columbia University Press, p. 219.

[9] Webb, Marianne (2000). *Lacquer: Technology and conservation*. Oxford: Butterworth-Heinemann. p. 3. ISBN 978-0-7506-4412-9.

[10] Stark, Miriam T. (2005). *Archaeology of Asia*. Malden, MA : Blackwell Pub. Page 30. ISBN 1-4051-0213-6.

[11] Wang, Zhongshu. (1982). *Han Civilization*. Translated by K.C. Chang and Collaborators. New Haven and London: Yale University Press. Page 80. ISBN 0-300-02723-0.

[12] Institute of the History of Natural Sciences and Chinese Academy of Sciences, ed. (1983). *Ancient China's technology and science*. Beijing: Foreign Languages Press. p. 211. ISBN 978-0-8351-1001-3.

[13] Niimura, Noriyasu; Miyakoshi, Tetsuo (2003) Characterization of Natural Resin Films and Identification of Ancient Coating . *J. Mass Spectrom. Soc. Jpn.* 51, 440. JOI:JST.JSTAGE/massspec/51.439

[14] http://www.easternuptourism.com/Lakshagrih.jsp

[15] The Black Lacquer Coffin Black Coffin Treasure, *"The artistic design of the black lacquer coffin reflects the painting skills of the Western Han Dynasty. They often paint mystical and grotesque themes about their myth and legends. But through the technique which involve the use of embossing lacquer application..."*

80.11 Further reading

- Kimes, Beverly R., Editor. Clark, Henry A. (1996), *The Standard Catalog of American Cars 1805–1945*, Kraus Publications, ISBN 0-87341-428-4 p. 1050

- Paolo Nanetti (2006), *Coatings from A to Z*, Vincentz Verlag, Hannover, ISBN 3-87870-173-X - A concise compilation of technical terms. Attached is a register of all German terms with their corresponding English terms and vice versa, in order to facilitate its use as a means for technical translation from one language to the other.

- Webb, Marianne (2000), *Lacquer: Technology and Conservation*, Butterworth Heinemann, ISBN 0-7506-4412-5 A Comprehensive Guide to the Technology and Conservation of Asian and European Lacquer

- Michiko, Suganuma. "Japanese lacquer".

Chapter 81

Vanilla

This article is about the flavoring. For other uses, see Vanilla (disambiguation).

Vanilla is a flavor derived from orchids of the genus *Vanilla*, primarily from the Mexican species, **flat-leaved vanilla** (*V. planifolia*). The word *vanilla*, derived from the diminutive of the Spanish word *vaina* (*vaina* itself meaning sheath or pod), translates simply as "little pod".[1] Pre-Columbian Mesoamerican people cultivated the vine of the vanilla orchid, called *tlilxochitl* by the Aztecs. Spanish conquistador Hernán Cortés is credited with introducing both vanilla and chocolate to Europe in the 1520s.[2]

Initial attempts to cultivate vanilla outside Mexico and Central America proved futile because of the symbiotic relationship between the vanilla orchid and its natural pollinator, the local species of *Melipona* bee.[3] Pollination is required to set the fruit from which the flavoring is derived. In 1837, Belgian botanist Charles François Antoine Morren discovered this fact and pioneered a method of artificially pollinating the plant. The method proved financially unworkable and was not deployed commercially.[4] In 1841, Edmond Albius, a slave who lived on the French island of Réunion in the Indian Ocean, discovered at the age of 12 that the plant could be hand-pollinated. Hand-pollination allowed global cultivation of the plant.[5]

Three major species of vanilla currently are grown globally, all of which derive from a species originally found in Mesoamerica, including parts of modern-day Mexico.[6] The various subspecies are *Vanilla planifolia* (syn. *V. fragrans*), grown on Madagascar, Réunion, and other tropical areas along the Indian Ocean; *V. tahitensis*, grown in the South Pacific; and *V. pompona*, found in the West Indies, and Central and South America.[7] The majority of the world's vanilla is the *V. planifolia* species, more commonly known as **Bourbon vanilla** (after the former name of Réunion, Île Bourbon) or **Madagascar vanilla**, which is produced in Madagascar and neighboring islands in the southwestern Indian Ocean, and in Indonesia.[8][9] *Leptotes bicolor* is used in the same way in South America.

Vanilla is the second most expensive spice after saffron,[10][11] because growing the vanilla seed

Vanilla fruits, dried

395

pods is labor-intensive.[11] Despite the expense, vanilla is highly valued for its flavor, which author Frederic Rosengarten, Jr. described in *The Book of Spices* as "pure, spicy, and delicate"; he called its complex floral aroma a "peculiar bouquet".[12] As a result, vanilla is widely used in both commercial and domestic baking, perfume manufacture and aromatherapy.

81.1 History

The Totonac people, who inhabit the East Coast of Mexico in the present-day state of Veracruz, were the first to cultivate vanilla. According to Totonac mythology, the tropical orchid was born when Princess Xanat, forbidden by her father from marrying a mortal, fled to the forest with her lover. The lovers were captured and beheaded. Where their blood touched the ground, the vine of the tropical orchid grew.[4]

Drawing of Vanilla *from the Florentine Codex* (circa *1580) and description of its use and properties written in the Nahuatl language*

In the 15th century, Aztecs invading from the central highlands of Mexico conquered the Totonacs, and soon developed a taste for the vanilla pods. They named the fruit *tlilxochitl*, or "black flower", after the matured fruit, which shrivels and turns black shortly after it is picked. Subjugated by the Aztecs, the Totonacs paid tribute by sending vanilla fruit to the Aztec capital, Tenochtitlan.

Until the mid-19th century, Mexico was the chief producer of vanilla. In 1819, however, French entrepreneurs shipped vanilla fruits to the islands of Réunion and Mauritius in hopes of producing vanilla there. After Edmond Albius discovered how to pollinate the flowers quickly by hand, the pods began to thrive. Soon, the tropical orchids were sent from Réunion Island to the Comoros Islands, Seychelles and Madagascar, along with instructions for pollinating them. By 1898, Madagascar, Réunion, and the Comoros Islands produced 200 metric tons of vanilla beans, about 80% of world production. According to the United Nations Food and Agriculture Organisation, Indonesia is currently responsible for the vast majority of the world's Bourbon vanilla production[13] and 58% of the world total vanilla fruit production.

Vanilla cultivation

The market price of vanilla rose dramatically in the late 1970s after a tropical cyclone ravaged key croplands. Prices remained high through the early 1980s despite the introduction of Indonesian vanilla. In the mid-1980s, the cartel that had controlled vanilla prices and distribution since its creation in 1930 disbanded.[14] Prices dropped 70% over the next few years, to nearly US$20 per kilogram; prices rose sharply again after tropical cyclone Hudah struck Madagascar in April 2000. The cyclone, political instability, and poor weather in the third year drove vanilla prices to an astonishing US$500 per kilogram in 2004, bringing new countries into the vanilla industry. A good crop, coupled with decreased demand caused by the production of imitation vanilla, pushed the market price down to the $40 per kilogram range in the middle of 2005. By 2010, prices were down to US$20/per kilo.

Madagascar (especially the fertile Sava region) accounts for much of the global production of vanilla. Mexico, once the leading producer of natural vanilla with an annual yield of 500 tons, produced only 10 tons of vanilla in 2006. An estimated 95% of "vanilla" products are artificially flavored with vanillin derived from lignin instead of vanilla fruits.[15]

81.2 Etymology

Vanilla was completely unknown in the Old World before Cortés. Spanish explorers arriving on the Gulf Coast of Mexico in the early 16th century gave vanilla its current name. Spanish and Portuguese sailors and explorers brought vanilla into Africa and Asia later that century. They called it *vainilla*, or "little pod". The word vanilla entered the English language in 1754, when the botanist Philip Miller wrote about the genus in his *Gardener's Dictionary*.[16] *Vainilla* is from the diminutive of *vaina*, from the Latin *vagina* (sheath) to describe the shape of the pods.*[17]

81.3 Biology

81.3.1 Vanilla orchid

Main article: Vanilla planifolia
The main species harvested for vanilla is *Vanilla plani-*

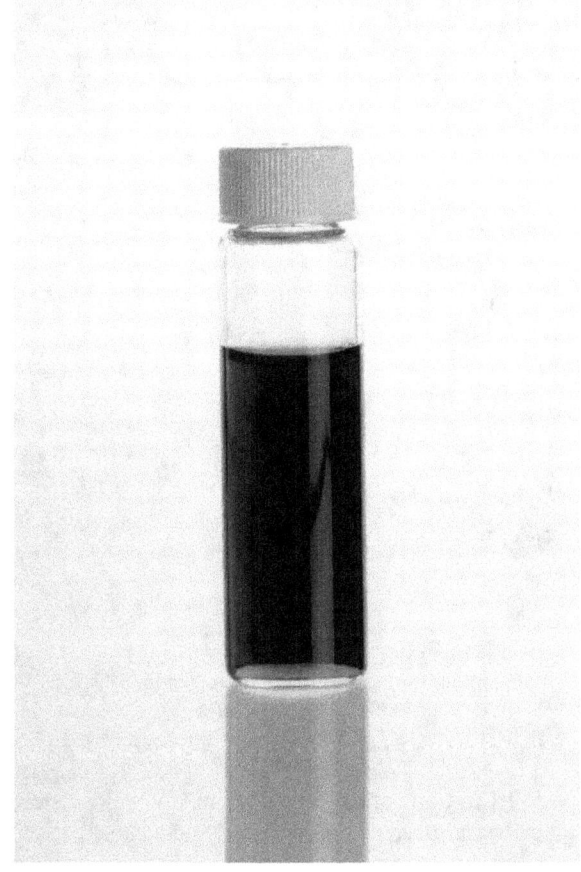

Vanilla extract in a clear glass vial

folia. Although it is native to Mexico, it is now widely grown throughout the tropics. Indonesia and Madagascar are the world's largest producers. Additional sources include *Vanilla pompona* and *Vanilla tahitiensis* (grown in Niue and Tahiti), although the vanillin content of these species is much less than *Vanilla planifolia*.*[18]

Vanilla grows as a vine, climbing up an existing tree (also called a tutor), pole, or other support. It can be grown in a wood (on trees), in a plantation (on trees or poles), or in a "shader", in increasing orders of productivity. Its growth environment is referred to as its *terroir*, and includes not only the adjacent plants, but also the climate, geography,

and local geology. Left alone, it will grow as high as possible on the support, with few flowers. Every year, growers fold the higher parts of the plant downward so the plant stays at heights accessible by a standing human. This also greatly stimulates flowering.

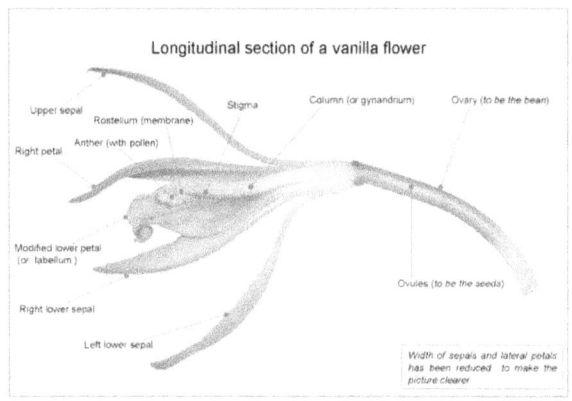

Vanilla planifolia – *flower*

The distinctively flavored compounds are found in the fruit, which results from the pollination of the flower. These seed pods are roughly a third of an inch by six inches, and brownish red to black when ripe. Inside of these pods are an oily liquid full of tiny seeds.*[19] One flower produces one fruit. *V. planifolia* flowers are hermaphroditic: They carry both male (anther) and female (stigma) organs; however, to avoid self-pollination, a membrane separates those organs. The flowers can be naturally pollinated only by bees of the *Melipona* genus found in Mexico (*abeja de monte* or mountain bee). This bee provided Mexico with a 300-year-long monopoly on vanilla production, from the time it was first discovered by Europeans. The first vanilla orchid to flower in Europe was in the London collection of the Honourable Charles Greville in 1806. Cuttings from that plant went to Netherlands and Paris, from which the French first transplanted the vines to their overseas colonies. The vines would grow, but would not fruit outside Mexico. Growers tried to bring this bee into other growing locales, to no avail. The only way to produce fruits without the bees is artificial pollination. And today, even in Mexico, hand pollination is used extensively.

In 1836, botanist Charles François Antoine Morren was drinking coffee on a patio in Papantla (in Veracruz, Mexico) and noticed black bees flying around the vanilla flowers next to his table. He watched their actions closely as they would land and work their way under a flap inside the flower, transferring pollen in the process. Within hours, the flowers closed and several days later, Morren noticed vanilla pods beginning to form. Morren immediately began experimenting with hand pollination. A few years later in 1841, a simple and efficient artificial hand-pollination method was developed by a 12-year-old slave named Edmond Albius on

Réunion, a method still used today. Using a beveled sliver of bamboo,*[20] an agricultural worker lifts the membrane separating the anther and the stigma, then, using the thumb, transfers the pollinia from the anther to the stigma. The flower, self-pollinated, will then produce a fruit. The vanilla flower lasts about one day, sometimes less, so growers have to inspect their plantations every day for open flowers, a labor-intensive task.

The fruit, a seed capsule, if left on the plant, will ripen and open at the end; as it dries, the phenolic compounds crystallize, giving the fruits a diamond-dusted appearance, which the French call *givre* (hoarfrost). It will then release the distinctive vanilla smell. The fruit contains tiny, black seeds. In dishes prepared with whole natural vanilla, these seeds are recognizable as black specks. Both the pod and the seeds are used in cooking.

Like other orchids' seeds, vanilla seeds will not germinate without the presence of certain mycorrhizal fungi. Instead, growers reproduce the plant by cutting: they remove sections of the vine with six or more leaf nodes, a root opposite each leaf. The two lower leaves are removed, and this area is buried in loose soil at the base of a support. The remaining upper roots will cling to the support, and often grow down into the soil. Growth is rapid under good conditions.

81.3.2   Cultivars

- **Bourbon vanilla** or **Bourbon-Madagascar vanilla**, produced from *V. planifolia* plants introduced from the Americas, is the term used for vanilla from Indian Ocean islands such as Madagascar, the Comoros, and Réunion, formerly the *Île Bourbon*. It is also used to describe the distinctive vanilla flavor derived from *V. planifolia* grown successfully in tropical countries such as India.

- **Mexican vanilla**, made from the native *V. planifolia*, is produced in much less quantity and marketed as the vanilla from the land of its origin. Vanilla sold in tourist markets around Mexico is sometimes not actual vanilla extract, but is mixed with an extract of the tonka bean, which contains coumarin. Tonka bean extract smells and tastes like vanilla, but coumarin has been shown to cause liver damage in lab animals and is banned in food in the US by the Food and Drug Administration since 1954.*[21]

- **Tahitian vanilla** is the name for vanilla from French Polynesia, made with the *V. tahitiensis* strain. Genetic analysis shows this species is possibly a cultivar from a hybrid-cross of *V. planifolia* and *V. odorata*. The species was introduced by French Admiral François Alphonse Hamelin to French Polynesia

A bottle of vanilla extract

from the Philippines, where it was introduced from Guatemala by the Manila Galleon trade.*[22]

- **West Indian vanilla** is made from *V. pompona* grown in the Caribbean and Central and South America.*[23]

The term **French vanilla** is often used to designate preparations with a strong vanilla aroma, containing vanilla grains and sometimes also containing eggs (especially egg yolks). The appellation originates from the French style of making vanilla ice cream with a custard base, using vanilla pods, cream, and egg yolks. Inclusion of vanilla varietals from any of the former French dependencies or overseas France noted for their exports may in fact be a part of the flavoring, though it may often be coincidental. Alternatively, French vanilla is taken to refer to a vanilla-custard flavor.*[22] Syrup labeled as French vanilla may include custard, hazelnut, caramel or butterscotch flavors in addition to vanilla.

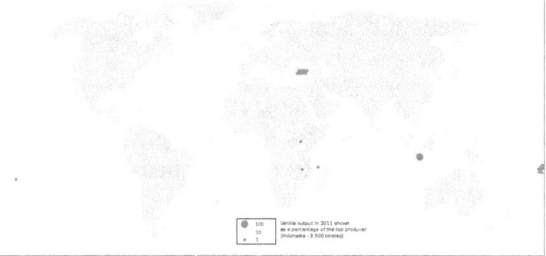

Vanilla output in 2011

81.3.3 Chemistry

Main article: Vanillin

Vanilla essence comes in two forms. Real seedpod extract is an extremely complicated mixture of several hundred different compounds, including vanillin, acetaldehyde, acetic acid, furfural, hexanoic acid, 4-hydroxybenzaldehyde, eugenol, methyl cinnamate, and isobutyric acid. Synthetic essence consists of a solution of synthetic vanillin in ethanol.

The chemical compound vanillin (4-hydroxy-3-methoxybenzaldehyde) is a major contributor to the characteristic flavor and aroma of real vanilla, but hundreds of compounds contribute to a complex flavor that vanillin can only approximate. Another minor component of vanilla extract is piperonal (heliotropin). Vanillin was first isolated from vanilla pods by Gobley in 1858.[*][24] By 1874, it had been obtained from glycosides of pine tree sap, temporarily causing a depression in the natural vanilla industry. Vanillin can be easily synthesized from various raw materials, but the majority of food grade (>99% pure) vanillin is made from guaiacol.

81.4 Production

81.4.1 General guidelines

In general, quality vanilla will only come from good vines and through careful production methods. Commercial vanilla production can be performed under open field and "greenhouse" operations. Both production systems share

the following similarities:

- Plant height and number of years before producing the first grains

- Shade necessities

- Amount of organic matter needed

- A tree or frame to grow around (bamboo, coconut or *Erythrina lanceolata*)

- Labor intensity (pollination and harvest activities)[*][26]

Vanilla grows best in a hot, humid climate from sea level to an elevation of 1500 m. The ideal climate has moderate rainfall, 1500–3000 mm, evenly distributed through 10 months of the year. Optimum temperatures for cultivation are 15–30 °C (59–86 °F) during the day and 15–20 °C (59–68 °F) during the night. Ideal humidity is around 80%, and under normal greenhouse conditions, it can be achieved by an evaporative cooler. However, since greenhouse vanilla is grown near the equator and under polymer (HDPE) netting (shading of 50%), this humidity can be achieved by the environment. Most successful vanilla growing and processing is done in the region within 10 to 20° of the equator.

Soils for vanilla cultivation should be loose, with high organic matter content and loamy texture. They must be well drained, and a slight slope helps in this condition. Soil pH has not been well documented, but some researchers have indicated an optimum soil pH of around 5.3.[*][27] Mulch is very important for proper growth of the vine, and a considerable portion of mulch should be placed in the base of the vine.[*][28] Fertilization varies with soil conditions, but general recommendations are: 40 to 60 g of N, 20 to 30 g of P_2O_5 and 60 to 100 g of K_2O should be applied to each plant per year besides organic manures, such as vermicompost, oil cakes, poultry manure and wood ash. Foliar applications are also good for vanilla, and a solution of 1% NPK (17:17:17) can be sprayed on the plant once a month. Vanilla requires organic matter, so three or four applications of mulch a year are adequate for the plant.

Chemical structure of vanillin

Propagation, preparation and type of stock

Dissemination of vanilla can be achieved either by stem cutting or by tissue culture. For stem cutting, a progeny garden needs to be established. Recommendations for establishing this garden vary, but in general, trenches of 60 cm (24 in) in width, 45 cm (18 in) in depth and 60 cm (24 in) spacing for each plant are necessary. All plants need to grow under 50% shade, as well as the rest of the crop. Mulching the trenches with coconut husk and micro irrigation provide an ideal microclimate for vegetative growth.*[29] Cuttings between 60 and 120 cm (24 and 47 in) should be selected for planting in the field or greenhouse. Cuttings below 60 to 120 cm (24 to 47 in) need to be rooted and raised in a separate nursery before planting. Planting material should always come from unflowered portions of the vine. Wilting of the cuttings before planting provides better conditions for root initiation and establishment.*[26]

Before planting the cuttings, trees to support the vine must be planted at least three months before sowing the cuttings. Pits of 30 x 30 x 30 cm are dug 30 cm (12 in) away from the tree and filled with farm yard manure (vermicompost), sand and top soil mixed well. An average of 2000 cuttings can be planted per hectare (2.5 acres). One important consideration is that when planting the cuttings from the base, four leaves should be pruned and the pruned basal point must be pressed into the soil in a way such that the nodes are in close contact with the soil, and are placed at a depth of 15 to 20 cm (5.9 to 7.9 in).*[28] The top portion of the cutting is tied to the tree using natural fibers such as banana or hemp.

Tissue culture

Tissue culture was first used as a means of creating vanilla plants during the 1980s at Tamil Nadu University. This was the part of the first project to grow *V. planifolia* in India. At that time, a shortage of vanilla planting stock was occurring in India. The approach was inspired by the work going on to tissue culture other flowering plants. Several methods have been proposed for vanilla tissue culture, but all of them begin from axillary buds of the vanilla vine.*[30]*[31] In vitro multiplication has also been achieved through culture of callus masses, protocorns, root tips and stem nodes.*[32] Description of any of these processes can be obtained from the references listed before, but all of them are successful in generation of new vanilla plants that first need to be grown up to a height of at least 30 cm (12 in) before they can be planted in the field or greenhouse.*[26]

Scheduling considerations

In the tropics, the ideal time for planting vanilla is from September to November, when the weather is neither too rainy nor too dry, but this recommendation varies with growing conditions. Cuttings take one to eight weeks to establish roots, and show initial signs of growth from one of the leaf axils. A thick mulch of leaves should be provided immediately after planting as an additional source of organic matter. Three years are required for cuttings to grow enough to produce flowers and subsequent pods. As with most orchids, the blossoms grow along stems branching from the main vine. The buds, growing along the 6 to 10 in (15 to 25 cm) stems, bloom and mature in sequence, each at a different interval.*[29]

Pollination

Flowering normally occurs every spring, and without pollination, the blossom wilts and falls, and no vanilla bean can grow. Each flower must be hand-pollinated within 12 hours of opening. In the wild, very few natural pollinators exist, with most pollination being carried out by bees of the genus *Melipona*. These pollinators do not exist outside the orchid's home range, and even within that range, vanilla orchids have only a 1% chance of successful pollination. As a result, all vanilla grown today is pollinated by hand. A small splinter of wood or a grass stem is used to lift the rostellum or move the flap upward, so the overhanging anther can be pressed against the stigma and self-pollinate the vine. Generally, one flower per raceme opens per day, so the raceme may be in flower for over 20 days. A healthy vine should produce about 50 to 100 beans per year, but growers are careful to pollinate only five or six flowers from the 20 on each raceme. The first flowers that open per vine should be pollinated, so the beans are similar in age. These agronomic practices facilitate harvest and increases bean quality. It takes the fruits five to six weeks to develop, but it takes around six months for the bean to mature. Over-pollination will result in diseases and inferior bean quality.*[28] A vine remains productive between 12 and 14 years.

Pest and disease management

Most diseases come from the uncharacteristic growing conditions of vanilla. Therefore, conditions such as excess water, insufficient drainage, heavy mulch, overpollination and too much shade favor disease development. Vanilla is susceptible to many fungal and viral diseases. *Fusarium*, *Sclerotium*, *Phytophthora*, and *Colletrotrichum* species cause rots of root, stem, leaf, bean and shoot apex. These diseases can be controlled by spraying Bordeaux mixture (1%), carbendazim (0.2%) and copper oxychloride (0.2%).

Biological control of the spread of such diseases can be managed by applying to the soil *Trichoderma* (0.5 kg (1.1 lb) per plant in the rhizosphere) and foliar application of pseudomonads (0.2%). Mosaic virus, leaf curl and cymbidium mosaic potex virus are the common viral diseases. These diseases are transmitted through the sap, so affected plants must be destroyed. The insect pests of vanilla include beetles and weevils that attack the flower, caterpillars, snakes and slugs that damage the tender parts of shoot, flower buds and immature fruit, and grasshoppers that affect cutting shoot tips.[28][29] If organic agriculture is practiced, insecticides are avoided, and mechanical measures are adopted for pest management.[26] Most of these practices are implemented under greenhouse cultivation, since such field conditions are very difficult to achieve.

Artificial vanilla

Most artificial vanilla products contain vanillin, which can be produced synthetically from lignin, a natural polymer found in wood. Most synthetic vanillin is a byproduct from the pulp used in papermaking, in which the lignin is broken down using sulfites or sulfates. However, vanillin is only one of 171 identified aromatic components of real vanilla fruits.[33]

The orchid species *Leptotes bicolor* is used as a natural vanilla replacement in Paraguay and southern Brazil.

Nonplant vanilla flavoring

In the United States, castoreum, the exudate from the castor sacs of mature beavers, has been approved by the Food and Drug Administration (FDA) as a food additive,[34] often referenced simply as a "natural flavoring" in the product's list of ingredients. It is used in both food and beverages, especially as vanilla and raspberry flavoring.[35] It is also used to flavor some cigarettes and in perfume-making.

81.4.2 Stages of production

Harvest

The vanilla fruit grows quickly on the vine, but is not ready for harvest until maturity approximately six months. Harvesting vanilla fruits is as labor-intensive as pollinating the blossoms. Immature dark green pods are not harvested. Pale yellow discoloration that commences at the distal end of the fruits is an indication of the maturity of pods. Each fruit ripens at its own time, requiring a daily harvest. To ensure the finest flavor from every fruit, each individual pod must be picked by hand just as it begins to split on the end. Overmatured fruits are likely to split, causing a reduction

A vanilla plantation in a forest of Réunion Island

in market value. Its commercial value is fixed based on the length and appearance of the pod.

If the fruit is more than 15 cm (5.9 in) in length, it belongs to first-quality product. The largest fruits greater than 16 cm and up to as much as 21 cm are usually reserved for the gourmet vanilla market, for sale to top chefs and restaurants. If the fruits are between 10 and 15 cm long, pods are under the second-quality category, and fruits less than 10 cm in length are under the third-quality category. Each fruit contains thousands of tiny black vanilla seeds. Vanilla fruit yield depends on the care and management given to the hanging and fruiting vines. Any practice directed to stimulate aerial root production has a direct effect on vine productivity. A five-year-old vine can produce between 1.5 and 3 kg (3.3 and 6.6 lb) pods, and this production can increase up to 6 kg (13 lb) after a few years. The harvested green fruit can be commercialized as such or cured to get a better market price.[26][28][29]

Curing

Several methods exist in the market for curing vanilla; nevertheless, all of them consist of four basic steps: killing, sweating, slow-drying, and conditioning of the beans.[36][37]

Killing   The vegetative tissue of the vanilla pod is killed to stop the vegetative growth of the pods and disrupt the cells and tissue of the fruits, which initiates enzymatic reactions responsible for the aroma. The method of killing varies, but may be accomplished by heating in hot water, freezing, or scratching, or killing by heating in an oven or exposing the beans to direct sunlight. The different methods give different profiles of enzymatic activity.[38][39]

Testing has shown mechanical disruption of fruit tissues can cause curing processes,[40] including the degeneration of glucovanillin to vanillin, so the reasoning goes that disrupting the tissues and cells of the fruit allow enzymes and enzyme substrates to interact.[38]

Hot-water killing may consist of dipping the pods in hot water (63–65 °C (145–149 °F)) for three minutes, or at 80 °C (176 °F) for 10 seconds. In scratch killing, fruits are scratched along their length.[39] Frozen or quick-frozen fruits must be thawed again for the subsequent sweating stage. Tied in bundles and rolled in blankets, fruits may be placed in an oven at 60 °C (140 °F) for 36 to 48 hours. Exposing the fruits to sunlight until they turn brown is a method originating in Mexico that was practiced by the Aztecs.[38]

Sweating   Sweating is a hydrolytic and oxidative process. Traditionally, it consists of keeping fruits, for seven to 10 days, densely stacked and insulated in wool or other cloth. This retains a temperature of 45–65 °C (113–149 °F) and high humidity. Daily exposure to the sun may also be used, or dipping the fruits in hot water. The fruits are brown and have attained much of the characteristic vanilla flavor and aroma by the end of this process, but still retain a 60-70% moisture content by weight.[38]

Drying   Reduction of the beans to 25–30% moisture by weight, to prevent rotting and to lock the aroma in the pods, is always achieved by some exposure of the beans to air, and usually (and traditionally) intermittent shade and sunlight. Fruits may be laid out in the sun during the mornings and returned to their boxes in the afternoons, or spread on a wooden rack in a room for three to four weeks, sometimes with periods of sun exposure. Drying is the most problematic of the curing stages; unevenness in the drying process can lead to the loss of vanillin content of some fruits by the

time the others are cured.[38]

Grading vanilla beans at Sambava, Madagascar

Conditioning   Conditioning is performed by storing the pods for five to six months in closed boxes, where the fragrance develops. The processed fruits are sorted, graded, bundled, and wrapped in paraffin paper and preserved for the development of desired bean qualities, especially flavor and aroma. The cured vanilla fruits contain an average of 2.5% vanillin.

Grading

See also: Food grading

Once fully cured, the vanilla fruits are sorted by quality and graded.

Several vanilla fruit grading systems are in use. Each country which produces vanilla has its own grading system,[41] and individual vendors, in turn, sometimes use their own criteria for describing the quality of the fruits they offer for sale.[42]

In general, vanilla fruit grade is based on the length, appearance (color, sheen, presence of any splits, presence of blemishes), and moisture content of the fruit.[41][43] Whole, dark, plump and oily pods that are visually attractive, with no blemishes, and that have a higher moisture content are graded most highly.[44] Such pods are particularly prized by chefs for their appearance and can be featured in gourmet dishes.[42] Beans that show localized signs of disease or other physical defects are cut to remove the blemishes; the shorter fragments left are called "cuts" and are assigned lower grades, as are fruits with lower moisture contents.[43] Lower-grade fruits tend to be favored for uses in which the appearance is not as important, such as in the production of vanilla flavoring extract and in the fragrance industry.

Higher-grade fruits command higher prices in the market.[41][43] However, because grade is so dependent on visual appearance and moisture content, fruits with the highest grade do not necessarily contain the highest concentration of characteristic flavor molecules such as vanillin,[45] and are not necessarily the most flavorful.[42]

† *moisture content varies among sources cited*

A simplified, alternative grading system has been proposed for classifying vanilla fruits suitable for use in cooking:[42]

Under this scheme, vanilla extract is normally made from Grade B fruits.[42]

81.5 Usage

81.5.1 Culinary uses

Vanilla rum, Madagascar

Pure Vanilla Powder

There are four main commercial preparations of natural vanilla:

- whole pod

- powder (ground pods, kept pure or blended with sugar, starch, or other ingredients)[50]

- extract (in alcoholic or occasionally glycerol solution; both pure and imitation forms of vanilla contain at least 35% alcohol)[51]

- vanilla sugar, a pre-packaged mix of sugar and vanilla extract

Vanilla flavoring in food may be achieved by adding vanilla extract or by cooking vanilla pods in the liquid preparation. A stronger aroma may be attained if the pods are split in two, exposing more of a pod's surface area to the liquid. In this case, the pods' seeds are mixed into the preparation. Natural vanilla gives a brown or yellow color to preparations, depending on the concentration. Good-quality vanilla has a strong aromatic flavor, but food with small amounts of low-quality vanilla or artificial vanilla-like flavorings are far more common, since true vanilla is much more expensive.

A major use of vanilla is in flavoring ice cream. The most common flavor of ice cream is vanilla, and thus most people consider it to be the "default" flavor. By analogy, the term "vanilla" is sometimes used as a synonym for "plain". Although vanilla is a prized flavoring agent on its own, it is also used to enhance the flavor of other substances, to which its own flavor is often complementary, such as chocolate, custard, caramel, coffee, cakes, and others.

The food industry uses methyl and ethyl vanillin. Ethyl vanillin is more expensive, but has a stronger note. *Cook's Illustrated* ran several taste tests pitting vanilla against vanillin in baked goods and other applications, and, to the consternation of the magazine editors, tasters could not differentiate the flavor of vanillin from vanilla;[52] however, for the case of vanilla ice cream, natural vanilla won out.[53] A more recent and thorough test by the same group produced a more interesting variety of results; namely, high-quality artificial vanilla flavoring is best for cookies, while high-quality real vanilla is very slightly better for cakes and significantly better for unheated or lightly heated foods.[54]

It was once believed that the liquid extracted from vanilla pods had medical properties, helping with various stomach ailments.[55]

81.6   Gallery

- A vanilla plantation in open field on Réunion

- A vanilla plantation in a "shader" (*ombrière*) on Réunion

- Flower

- Green fruits

81.7   Notes

[1] James D. Ackerman (June 2003). "Vanilla". *Flora of South America* **26** (4): 507. Retrieved 22 July 2008. Spanish vainilla, little pod or capsule, referring to long, podlike fruits

[2] The Herb Society of Nashville. "The Life of Spice". The Herb Society of Nashville. Archived from the original on 20 September 2011. Following Montezuma's capture, one of Cortés' officers saw him drinking "chocolatl" (made of powdered cocoa beans and ground corn flavored with ground vanilla pods and honey). The Spanish tried this drink themselves and were so impressed by this new taste sensation that they took samples back to Spain.' and 'Actually it was vanilla rather than the chocolate that made a bigger hit and by 1700 the use of vanilla was spread over all of Europe. Mexico became the leading producer of vanilla for three centuries. – Excerpted from 'Spices of the World Cookbook' by McCormick and 'The Book of Spices' by Frederic Rosengarten, Jr

[3] Lubinsky P., Van Dam M.Y. H., Van Dam A.R. Lindleyana 2006 Pollination of Vanilla and evolution in Orchidaceae

[4] Janet Hazen (1995). *Vanilla*. Chronicle Books.

[5] Silver Cloud Estates. "History of Vanilla". Silver Cloud Estates. Retrieved 23 July 2008. In 1837 the Belgian botanist Morren succeeded in artificially pollinating the vanilla flower. On Reunion, Morren's process was attempted, but failed. It was not until 1841 that a 12-year-old slave by the name of Edmond Albius discovered the correct technique of hand-pollinating the flowers.

[6] Lubinsky, Pesach; Bory, Séverine; Hernández Hernández, Juan; Kim, Seung-Chul; Gómez-Pompa, Arturo (2008). "Origins and Dispersal of Cultivated Vanilla (*Vanilla planifolia* Jacks. [Orchidaceae])". *Economic Botany* **62** (2): 127–38. doi:10.1007/s12231-008-9014-y.

[7] Besse, Pascale; Silva, Denis Da; Bory, Séverine; Grisoni, Michel; Le Bellec, Fabrice; Duval, Marie-France (2004). "RAPD genetic diversity in cultivated vanilla: *Vanilla planifolia*, and relationships with *V. Tahitensis* and *V. Pompona*". *Plant Science* **167** (2): 379–85. doi:10.1016/j.plantsci.2004.04.007.

[8] "Vanilla growing regions". The Rodell Company. 7 January 2008. Archived from the original on 10 June 2008. Retrieved 22 July 2008. ...Madagascar is the world's primary growing region, cured vanilla pods are produced in the Comoros Islands, French Polynesia, Guatemala, India, Indonesia, Mexico, Sri Lanka, Tonga and Uganda.

[9] The Nielsen-Massey Company. "History of vanilla". The Nielsen-Massey Company. Archived from the original on 1 March 2012. Madagascar and Indonesia produce 90 percent of the world's vanilla bean crop.

[10] Le Cordon Bleu (2009). *Le Cordon Bleu Cuisine Foundations*. Cengage learning. p. 213. ISBN 978-1-4354-8137-4.

[11] Parthasarathy, V. A.; Chempakam, Bhageerathy; Zachariah, T. John (2008). *Chemistry of Spices*. CABI. p. 2. ISBN 978-1-84593-405-7.

[12] Rosengarten, Frederic (1973). *The Book of Spices*. Pyramid Books. ISBN 978-0-515-03220-8.

[13] "FAO's Statistical Database - FAOSTAT". 2011.

[14] Le Cordon Bleu Cuisine Foundations

[15] "Rainforest Vanilla Conservation Association". RVCA. Archived from the original on 24 June 2009. Retrieved 16 June 2011.

[16] Correll D (1953) Vanilla: its botany, history, cultivation and economic importance. Econ Bo 7(4): 291–358.

[17] "Online Etymology Dictionary". Etymonline.com. Retrieved 1 May 2010.

[18] "Brockman, Terra *Types of Vanilla* June 11, 2008 Chicago Tribune". Chicagotribune.com. 11 June 2008. Retrieved 1 May 2010.

[19] Diderot, Denis. "Vanilla". *The Encyclopedia of Diderot & d'Alembert: Collaborative Translations Project*. Retrieved 1 April 2015.

[20] "Flower with money power". The Hindu. 10 May 2004. Retrieved 1 May 2010.

[21] "IMPORT ALERT IA2807: "DETENTION WITHOUT PHYSICAL EXAMINATION OF COUMARIN IN VANILLA PRODUCTS (EXTRACTS – FLAVORINGS – IMITATIONS)"". U.S. Food and Drug Administration Office of Regulatory Affairs. January 1998. Retrieved 21 December 2007.

[22] "Tahitian vanilla originated in Maya forests, says botanist". GeneticArchaeology.com. 22 August 2008.

[23] USDA publication. "*Vanilla pompona* Schiede/West Indian vanilla". United Dept. of Agriculture. Retrieved 24 July 2008.

[24] Gobley, N.-T. (1858). "Recherches sur le principe odorant de la vanilla" [Research on the fragrant substance of vanilla]. *Journal de Pharmacie et de Chimie, Series 3* **34**: 401–405.

[25] "Faostat". Faostat.fao.org. 16 December 2009. Retrieved 19 January 2013.

[26] Anilkumar, A. S. (February 2004). "Vanilla cultivation: A profitable agri-based enterprise" (PDF). *Kerala Calling*: 26–30.

[27] Berninger, F., Salas, E., 2003. Biomass dynamics of Erythrina lanceolata as influenced by shoot-pruning intensity in Costa Rica. Agro-forestry Systems, 57:19–28.

[28] Davis, Elmo W. (1983). "Experiences with growing vanilla (Vanilla planifolia)". *Acta Horticulturae* **132**: 23–9.

[29] Elizabeth, K. G. (2002). "Vanilla: an orchid spice". *Indian Journal of Arecanut Spices and Medicinal Plants* **4** (2): 96–8.

[30] George, P. S.; Ravishankar, G. A. (1997). "In vitro multiplication of *Vanilla planifolia* using axillary bud explants". *Plant Cell Reports* **16** (7): 490–4. doi:10.1007/BF01092772.

[31] Kononowicz, H.; Janick, J. (1984). "In vitro propagation of Vanilla planifolia". *HortScience* **19** (1): 58–9.

[32] Giridhar P, Ravishankar GA (2004). "Efficient micropropagation of *Vanilla planifolia* Andr. under influence of thidiazuron, zeatin and coconut milk". *Indian Journal of Biotechnology* **3** (1): 113–8.

[33] "About Vanilla – Vanilla imitations". Cook Flavoring Company. 2011. Retrieved 22 June 2011.

[34] Burdock GA (2007). "Safety assessment of castoreum extract as a food ingredient". *Int. J. Toxicol.* **26** (1): 51–5. doi:10.1080/10915810601120145. PMID 17365147.

[35] Burdock, George A., Fenaroli's handbook of flavor ingredients. CRC Press, 2005. p. 277.

[36] Havkin-Frenkel D, French JC, Graft NM (2004). "Interrelation of curing and botany in vanilla (vanilla planifolia) bean". *Acta Horticulturae* **629**: 93–102.

[37] Havkin-Frenkel, D.; French, J. C.; Pak, F. E.; Frenkel, C. (2003). "Botany and curing of vanilla". *Journal of Aromatic medicinal plants*.

[38] Frenkel, Chaim; Ranadive, Arvind S.; Vázquez, Javier Tochihuitl; Havkin-Frenkel, Daphna (2010). "Curing of Vanilla". In Havkin-Frenkel, Daphna; Belanger, Faith. *Handbook of Vanilla Science and Technology*. John Wiley & Sons. pp. 79–106 [87]. ISBN 978-1-4443-2937-7.

[39] Arana, Francisca E. (October 1944). "Vanilla curing and its chemistry". *Bulletin* (Federal Experiment Station of the United States Department of Agriculture in Mayaguez, Puerto Rico) (42): 1–17.

[40] Methods of dehydrating and curing vanilla fruit US Patent 2,621,127

[41] Havkin-Frenkel, Daphna; Belanger, Faith C. (2011). *Handbook of Vanilla Science and Technology*. Chichester, UK: Wiley-Blackwell. pp. 142–145. ISBN 978-1-4051-9325-2.

[42] "Vanilla". VanillaReview.com. Retrieved 15 January 2012.

[43] Nielsen, Jr., Chat (1985). *The Story of Vanilla*. Chicago: Nielsen-Massey Vanillas.

[44] "Vanilla". *Spices Board of India*. Ministry of Commerce & Industry, Government of India. Retrieved 16 January 2012.

[45] K. Gassenheimer; E. Binggeli (2008). Imre Blank, Matthias Wüst, Chahan Yeretzian, ed. "Vanilla Bean Quality - A Flavour Industry View" in Expression of Multidisciplinary Flavour Science: Proceedings of the 12th Weurman Symposium (Interlaken, Switzerland 2008). Wädensil, Switzerland: Zürich University of Applied Sciences. pp. 203–206. ISBN 978-3-905745-19-1.

[46] "LFIE Vanilla Products". *Lopat Frederic Import Export*. Retrieved 16 January 2012.

[47] "Vanilla Bourbon". SA. VA. Import - Export. Retrieved 16 January 2012.

[48] "Vanilla Products". Gascar Trading Company. Retrieved 16 January 2012.

[49] "Vanilla Bean Products". Vanexco. Retrieved 16 January 2012.

[50] The U.S. Food and Drug Administration requires at least 12.5% of pure vanilla (ground pods or oleoresin) in the mixture

[51] The U.S. Food and Drug Administration requires at least 35% vol. of alcohol and 13.35 ounces of pod per gallon

[52] "Pure versus Imitation Vanilla Extract". Cooks Illustrated. 1 March 2009. Retrieved 30 April 2013.

[53] "Tasting lab: Vanilla Ice Cream". Cooks Illustrated. 1 May 2010. Retrieved 30 April 2013.

[54] "Vanilla Extract". 1 March 2009. Retrieved 19 June 2012.

[55] Jaucourt, Louis (1765). "Vanilla". *Encyclopédie ou Dictionnaire raisonné des sciences, des arts et des métiers*. Retrieved 31 March 2015.

81.8   Further reading

- Ecott, Tim (2004). *Vanilla: Travels in Search of the Luscious Substance*. London: Penguin, New York: Grove Atlantic

- Rain, Patricia (2004). *Vanilla: The Cultural History of the World's Favorite Flavor and Fragrance*. New York: J. P. Tarcher/Penguin.

81.9   External links

- "Vanilla". *Encyclopedia Americana*. 1920.

- Kew Species Profile: *Vanilla planifolia* (vanilla)

- History, Classification and Lifecycle of Vanilla planifolia

- *Spices* at UCLA History & Special Collections

- Vanilla and Extracts at DMOZ

- "The Present State of the West-Indies: Containing an Accurate Description of What Parts Are Possessed by the Several Powers in Europe" by Thomas Kitchin, 1778, in which Kitchin discusses vanilla

Chapter 82

Wildcrafting

Wildcrafting is the practice of harvesting plants from their natural, or "wild" habitat, for food or medicinal purposes. It applies to uncultivated plants wherever they may be found, and is not necessarily limited to wilderness areas. Ethical considerations are often involved, such as protecting endangered species.

When wildcrafting is done sustainably with proper respect, generally only the fruit, flowers or branches from plants are taken and the living plant is left, or if it is necessary to take the whole plant, seeds of the plant are placed in the empty hole from which the plant was taken. Care is taken to only remove a few plants, flowers, or branches, so plenty remains to continue the supply.*[1]

82.1   See also

- Agroforestry

- Biomass (ecology)

- Biomass

- Bioproducts

- Permaforestry

- World Forestry Congress

82.2   References

[1] Buren, Bruce. "Wildcrafting: A "simple" life fraught with a host of complex ethical and practical considerations". *The Rodale Institute*. Retrieved 12 October 2010.

Chapter 83

Willow

For other uses, see Willow (disambiguation).
"Willow tree" redirects here. For Willow tree figurines, see Willow Tree (figurines).
"Salix" redirects here. For other uses, see Salix (disambiguation).

Willows, also called **sallows**, and **osiers**, form the genus *Salix*, around 400 species*[2] of deciduous trees and shrubs, found primarily on moist soils in cold and temperate regions of the Northern Hemisphere. Most species are known as willow, but some narrow-leaved shrub species are called **osier**, and some broader-leaved species are referred to as **sallow** (from Old English *sealh*, related to the Latin word *salix*, willow). Some willows (particularly arctic and alpine species) are low-growing or creeping shrubs; for example, the dwarf willow (*Salix herbacea*) rarely exceeds 6 cm (2.4 in) in height, though it spreads widely across the ground.

83.1 Description

At the base of the petiole a pair of stipules form. These may fall in spring, or last for much of the summer or even for more than one year (marcescence).

Willows all have abundant watery bark sap, which is heavily charged with salicylic acid, soft, usually pliant, tough wood, slender branches, and large, fibrous, often stoloniferous roots. The roots are remarkable for their toughness, size, and tenacity to life, and roots readily grow from aerial parts of the plant.

The leaves are typically elongated, but may also be round to oval, frequently with serrated edges. Most species are deciduous; semievergreen willows; coriaceous leaves are rare, e.g. *Salix micans* and *S. australior* in the eastern Mediterranean. All the buds are lateral; no absolutely terminal bud is ever formed. The buds are covered by a single scale. Usually, the bud scale is fused into a cap-like shape, but in some species it wraps around and the edges overlap.*[3] The leaves are simple, feather-veined, and typically linear-lanceolate. Usually they are serrate, rounded at base, acute or acuminate. The leaf petioles are short, the stipules often very conspicuous, resembling tiny, round leaves, and sometimes remaining for half the summer. On some species, however, they are small, inconspicuous, and caducous (soon falling). In color, the leaves show a great variety of greens, ranging from yellowish to bluish.

83.1.1 Flowers

Willows are dioecious, with male and female flowers appearing as catkins on different plants; the catkins are produced early in the spring, often before the leaves, or as the new leaves open.

The staminate (male) flowers are without either calyx or corolla; they consist simply of stamens, varying in number from two to 10, accompanied by a nectariferous gland and inserted on the base of a scale which is itself borne on the rachis of a drooping raceme called a catkin, or ament. This scale is square, entire, and very hairy. The anthers are rose-colored in the bud, but orange or purple after the flower opens; they are two-celled and the cells open longitudinally. The filaments are threadlike, usually pale brown, and often bald.

Young male catkin

83.4 Ecological issues

Knotted willow and woodpile in the Bourgoyen-Ossemeersen, Ghent, Belgium

The pistillate (female) flowers are also without calyx or corolla, and consist of a single ovary accompanied by a small, flat nectar gland and inserted on the base of a scale which is likewise borne on the rachis of a catkin. The ovary is one-celled, the style two-lobed, and the ovules numerous.

83.2 Cultivation

Almost all willows take root very readily from cuttings or where broken branches lie on the ground. The few exceptions include the goat willow (*Salix caprea*) and peachleaf willow (*Salix amygdaloides*). One famous example of such growth from cuttings involves the poet Alexander Pope, who begged a twig from a parcel tied with twigs sent from Spain to Lady Suffolk. This twig was planted and thrived, and legend has it that all of England's weeping willows are descended from this first one.*[4]*[5]

Willows are often planted on the borders of streams so their interlacing roots may protect the bank against the action of the water. Frequently, the roots are much larger than the stem which grows from them.

83.3 Hybrids

Willows are very cross-fertile, and numerous hybrids occur, both naturally and in cultivation. A well-known ornamental example is the weeping willow (*Salix × sepulcralis*), which is a hybrid of Peking willow (*Salix babylonica*) from China and white willow (*Salix alba*) from Europe.

The hybrid cultivar 'Boydii' has gained the Royal Horticultural Society's Award of Garden Merit.*[6]

Willows are used as food plants by the larvae of some Lepidoptera species, such as the mourning cloak butterfly.*[7] Ants, such as wood ants, are common on willows inhabited by aphids, coming to collect aphid honeydew, as sometimes do wasps.

A small number of willow species were widely planted in Australia, notably as erosion-control measures along watercourses. They are now regarded as invasive weeds, and many catchment management authorities are removing and replacing them with native trees.*[8]*[9]

Willow roots spread widely and are very aggressive in seeking out moisture; for this reason, they can become problematic when planted in residential areas, where the roots are notorious for clogging French drains, drainage systems, weeping tiles, septic systems, storm drains, and sewer systems, particularly older, tile, concrete, or ceramic pipes. Newer, PVC sewer pipes are much less leaky at the joints, and are therefore less susceptible to problems from willow roots; the same is true of water supply piping.*[10]*[11]

83.5 Pests and diseases

Willow species are hosts to more than a hundred aphid species, belonging to *Chaitophorus* and other genera,[12] forming large colonies to feed on plant juices, on the underside of leaves in particular.[13] Rust, caused by fungi of genus *Melampsora*, is known to damage leaves of willows, covering them with orange spots.[14]

83.6 Uses

83.6.1 Medicine

The leaves and bark of the willow tree have been mentioned in ancient texts from Assyria, Sumer and Egypt[15] as a remedy for aches and fever,[16] and in Ancient Greece the physician Hippocrates wrote about its medicinal properties in the fifth century BC. Native Americans across the Americas relied on it as a staple of their medical treatments. It provides temporary pain relief. Salicin is metabolized into salicylic acid in the human body, and is a precursor of aspirin.[17] In 1763, its medicinal properties were observed by the Reverend Edward Stone in England. He notified the Royal Society, which published his findings. The active extract of the bark, called salicin, was isolated to its crystalline form in 1828 by Henri Leroux, a French pharmacist, and Raffaele Piria, an Italian chemist, who then succeeded in separating out the compound in its pure state. In 1897, Felix Hoffmann created a synthetically altered version of salicin (in his case derived from the *Spiraea* plant), which caused less digestive upset than pure salicylic acid. The new drug, formally acetylsalicylic acid, was named Aspirin by Hoffmann's employer Bayer AG. This gave rise to the hugely important class of drugs known as nonsteroidal anti-inflammatory drugs (NSAIDs).

Willow has been listed as one of the 38 substances used to prepare Bach flower remedies,[18] a kind of alternative medicine promoted for its effect on health. However, according to Cancer Research UK, "there is no scientific evidence to prove that flower remedies can control, cure or prevent any type of disease, including cancer" .[19]

83.6.2 Manufacturing

Some of humans' earliest manufactured items may have been made from willow. A fishing net made from willow dates back to 8300 BC.[20] Basic crafts, such as baskets, fish traps, wattle fences and wattle and daub house walls, were often woven from osiers or withies (rod-like willow shoots, often grown in coppices). One of the forms of Welsh coracle traditionally uses willow in the 'lats'. Thin or split willow rods can be woven into wicker, which also has a long history. The relatively pliable willow is less likely to split while being woven than many other woods, and can be bent around sharp corners in basketry. Willow wood is also used in the manufacture of boxes, brooms, cricket bats (grown from certain strains of white willow), cradle boards, chairs and other furniture, dolls, flutes, poles, sweat lodges, toys, turnery, tool handles, veneer, wands and whistles. In addition, tannin, fibre, paper, rope and string can be produced from the wood. Willow is also used in the manufacture of double basses for backs, sides and linings, and in making splines and blocks for bass repair.

83.6.3 Other

Male catkin of Salix cinerea *with bee*

Willow tree in spring, England

- **agriculture**: Willows produce a modest amount of nectar from which bees can make honey, and are especially valued as a source of early pollen for bees. Poor people at one time often ate willow catkins that had been cooked to form a mash.[21]

Willow tree with woodbine honeysuckle

The willow tree as seen as the main part of an heraldic escutcheon over the main portal of a patrician house belonging to the Salis family in Chur, Switzerland, circa 1750

Environmental art installation "Sandworm" in the Wenduine Dunes, Belgium, made entirely out of willow

- **Art**: Willow is used to make charcoal (for drawing) and in living sculptures. Living sculptures are created from live willow rods planted in the ground and wo-

ven into shapes such as domes and tunnels. Willow stems are used to weave baskets and three-dimensional sculptures, such as animals and figures. Willow stems are also used to create garden features, such as decorative panels and obelisks.

- **Energy**: Willow is grown for biomass or biofuel, in energy forestry systems, as a consequence of its high energy in-energy out ratio, large carbon mitigation potential and fast growth.[*][22] Large-scale projects to support willow as an energy crop are already at commercial scale in Sweden,[*][23] and in other countries, others are being developed through initiatives such as the Willow Biomass Project in the US and the Energy Coppice Project in the UK.[*][24] Willow may also be grown to produce charcoal.

- **Environment**: As a plant, willow is used for biofiltration, constructed wetlands, ecological wastewater treatment systems, hedges, land reclamation, landscaping, phytoremediation, streambank stabilisation (bioengineering), slope stabilisation, soil erosion control, shelterbelt and windbreak, soil building, soil reclamation, tree bog compost toilet, and wildlife habitat.

- **Religion**: Willow is one of the "Four Species" used ritually during the Jewish holiday of Sukkot. In Buddhism, a willow branch is one of the chief attributes of Kwan Yin, the *bodhisattva* of compassion. Christian churches in northwestern Europe and Ukraine often used willow branches in place of palms in the ceremonies on Palm Sunday.[*][25]

83.7 Culture

In China, some people carry willow branches with them on the day of their Tomb Sweeping or Qingming Festival. Willow branches are also put up on gates and/or front doors, which they believe help ward off the evil spirits that wander on Qingming. Legend states that on Qingming Festival, the ruler of the underworld allows the spirits of the dead to return to earth. Since their presence may not always be welcome, willow branches keep them away.[*][26] In traditional pictures of the Goddess of Mercy Guanyin, she is often shown seated on a rock with a willow branch in a vase of water at her side. The Goddess employs this mysterious water and the branch for putting demons to flight. Taoist witches also use a small carving made from willow wood for communicating with the spirits of the dead. The image is sent to the nether world, where the disembodied spirit is deemed to enter it, and give the desired information to surviving relatives on its return.[*][27] The willow is a famous subject in many East Asian nations' cultures, particularly in pen and ink paintings from China and Japan.

A *gisaeng* (Korean geisha) named Hongrang, who lived in the middle of the Joseon Dynasty, wrote the poem "By the willow in the rain in the evening", which she gave to her parting lover (Choi Gyeong-chang).[28] Hongrang wrote:

> "...I will be the willow on your bedside."

In Japanese tradition, the willow is associated with ghosts. It is popularly supposed that a ghost will appear where a willow grows. Willow trees are also quite prevalent in folklore and myths.[29][30]

In English folklore, a willow tree is believed to be quite sinister, capable of uprooting itself and stalking travellers. The Viminal Hill, one of the Seven Hills Of Rome, derives it name from the Latin word for osier, *viminia* (pl.).

- Hans Christian Andersen wrote a story called "Under the Willow Tree" (1853) in which children ask questions of a tree they call "willow-father", paired with another entity called "elder-mother".[31]

- Old Man Willow in J. R. R. Tolkien's legendarium, appearing in *The Lord of the Rings*.

- "Green Willow" is a Japanese ghost story in which a young samurai falls in love with a woman called Green Willow who has a close spiritual connection with a willow tree.[32] "The Willow Wife" is another, not dissimilar tale.[33] "Wisdom of the Willow Tree" is an Osage Nation story in which a young man seeks answers from a willow tree, addressing the tree in conversation as 'Grandfather'.[34]

- In "The Secret of Salix Babylonicus," set in the Babylonian exile, the willow is portrayed benignly as a symbol of sympathy in grief and perseverance in suffering, an upright character who shows solidarity with the exiles and comes to express the spirit of the psalms.

- The "Whomping Willow" is a feature of the grounds of Hogwarts School for Witchcraft and Wizardry in the Harry Potter stories.

83.8 Selected species

Main article: List of Salix species

The genus *Salix* is made up of around 400 species[2] of deciduous trees and shrubs:

- *Salix acutifolia* Willd. – long-leaved violet willow

- *Salix alaxensis* (Andersson) Coville

- *Salix alba* L. – white willow

- *Salix amygdaloides* Andersson – peachleaf willow

- *Salix arbuscula* L.

- *Salix arbusculoides* – littletree willow

- *Salix arctica* Pall. – Arctic willow

- *Salix arizonica* Dorn

- *Salix atrocinerea* Brot. – grey willow

- *Salix aurita* L. – eared willow

- *Salix babylonica* L. – Babylon willow, Peking willow or weeping willow

- *Salix bakko*

- *Salix barclayi* Andersson

- *Salix barrattiana* – Barratt's willow

- *Salix bebbiana* Sarg. – beaked willow, long-beaked willow, and Bebb's willow

- *Salix bicolor*

- *Salix bonplandiana* Kunth – Bonpland willow

- *Salix boothii* Dorn – Booth's willow

- *Salix brachycarpa* Nutt.

- *Salix breweri* Bebb – Brewer's willow

- *Salix canariensis* Chr. Sm.

- *Salix candida* Flüggé ex Willd. – sageleaf willow

- *Salix caprea* L. – goat willow, or pussy willow

- *Salix caroliniana* Michx. – coastal plain willow

- *Salix chaenomeloides* Kimura

- *Salix cinerea* L. – grey willow

- *Salix cordata* Michx. – sand dune willow, furry willow, or heartleaf willow

- *Salix delnortensis* C.K.Schneid. – Del Norte willow

- *Salix discolor* Muhl. – American willow

- *Salix drummondiana* Barratt ex Hook. – Drummond's willow

- *Salix eastwoodiae* Cockerell ex A.Heller – Eastwood's willow, mountain willow, or Sierra willow

- *Salix eleagnos* Scop. - olive willow

- *Salix eriocarpa*

- *Salix exigua* Nutt. – sandbar willow, narrowleaf willow, or coyote willow

- *Salix floridana*

- *Salix fragilis* L. – crack willow

- *Salix fuscescens* - Alaska bog willow

- *Salix futura*

- *Salix geyeriana* Andersson – Geyer's willow

- *Salix gilgiana* Seemen

- *Salix glauca* L.

- *Salix glaucosericea*

- *Salix gooddingii* C. R. Ball – Goodding's willow, or Goodding's black willow

- *Salix gracilistyla* Miq.

- *Salix hastata* L.

- *Salix herbacea* L. – dwarf willow, least willow or snowbed willow

- *Salix hookeriana* Barratt ex Hook. – dune willow, coastal willow, or Hooker's willow

- *Salix hultenii*

- *Salix integra* Thunb.

- *Salix interior*

- *Salix japonica* Thunb.

- *Salix jepsonii* C.K.Schneid. – Jepson's willow

- *Salix jessoensis* Seemen

- *Salix koriyanagi* Kimura ex Goerz

- *Salix kusanoi*

- *Salix laevigata* Bebb – red willow or polished willow

- *Salix lanata* L. – woolly willow

- *Salix lapponum* L. - downy willow

- *Salix lasiolepis* Benth. – arroyo willow

- *Salix lemmonii* Bebb – Lemmon's willow

- *Salix libani* – Lebanese willow

- *Salix ligulifolia* C.R.Ball – strapleaf willow

- *Salix lucida* Muhl. – shining willow, Pacific willow, or whiplash willow

- *Salix lutea* Nutt. – yellow willow

- *Salix magnifica* Hemsl.

- *Salix matsudana* Koidz. – Chinese willow or twisted willow, variant *Corkscrew*

- *Salix melanopsis* Nutt. – dusky willow

- *Salix miyabeana* Seemen

- *Salix monticola*

- *Salix mucronata* - Cape silver willow

- *Salix myrsinifolia* Salisb.

- *Salix myrtillifolia*

- *Salix myrtilloides* L. – swamp willow

- *Salix nakamurana*

- *Salix nigra* Marshall – black willow

- *Salix orestera* C.K.Schneid. – Sierra willow or gray-leafed Sierra willow

- *Salix pentandra* L. – bay willow

- *Salix phylicifolia* L.

- *Salix planifolia* Pursh. – diamondleaf willow or tea-leafed willow

- *Salix polaris* Wahlenb. – polar willow

- *Salix prolixa* Andersson – MacKenzie's willow

- *Salix pulchra*

- *Salix purpurea* L. – purple willow or purple osier

- *Salix reinii*

- *Salix reticulata* L. – net-veined willow

- *Salix retusa*

- *Salix richardsonii*

- *Salix rorida* Lacksch.

- *Salix rupifraga*

- *Salix schwerinii* E. L. Wolf

- *Salix scouleriana* Barratt ex Hook. – Scouler's willow

- *Salix sepulcralis* group – hybrid willows

- *Salix sericea* Marshall – silky willow

- *Salix serissaefolia*

- *Salix serissima* (L. H. Bailey) Fernald autumn willow or fall willow

- *Salix serpyllifolia*

- *Salix sessilifolia* Nutt. – northwest sandbar willow

- *Salix shiraii*

- *Salix sieboldiana*

- *Salix sitchensis* C. A. Sanson ex Bong. – Sitka willow

- *Salix subfragilis*

- *Salix subopposita* Miq.

- *Salix taraikensis*

- *Salix tarraconensis*

- *Salix taxifolia* Kunth – yew-leaf willow

- *Salix tetrasperma* Roxb. – Indian willow

- *Salix triandra* L. – almond willow or almond-leaved willow

- *Salix udensis* Trautv. & C. A. Mey.

- *Salix viminalis* L. – common osier

- *Salix vulpina* Andersson

- *Salix yezoalpina* Koidz.

- *Salix yoshinoi*

83.9   See also

- *Aravah*, the Hebrew name of the willow, for its ritual use during the Jewish Feast of Tabernacles;

- List of Lepidoptera that feed on willows

- *Rhabdophaga rosaria*, a willow gall;

- Willow Biomass Project;

- Willow water, using the biological rooting hormones indolebutyric acid and salicylic acid from willow branches to stimulate root growth in new cuttings;

- Sail ogham letter meaning willow.

83.10   References

[1] "Genus **Salix (willows)**". *Taxonomy*. UniProt. Retrieved 2010-02-04.

[2] Mabberley, D.J. 1997. The Plant Book, Cambridge University Press #2: Cambridge.

[3] George W. Argus. *The Genus* Salix *(Salicaceae) in the Southeastern United States*. Systematic Botany Monographs **9**.

[4] Leland, John (2005) *Aliens in the Backyard: Plant and Animal Imports into America*, p. 70. Univ of South Carolina Press. At Google Books. Retrieved 11 August 2013.

[5] Laird, Mark (1999) *The Flowering of the Landscape Garden: English Pleasure Grounds, 1720-1800*, p. 403. University of Pennsylvania Press At Google Books. Retrieved 11 August 2013.

[6] "RHS Plant Selector - *Salix* 'Boydii'". Retrieved 2 June 2013.

[7] "Mourning Cloak" . Study of Northern Virginia Ecology. Fairfax County Public Schools.

[8] Albury/Wodonga Willow Management Working Group (December 1998). "Willows along watercourses: managing, removing and replacing" . Department of Primary Industries, State Government of Victoria.

[9] Cremer, Kurt W. (2003). "Introduced willows can become invasive pests in Australia" (PDF).

[10] Salix spp. UFL/edu, *Weeping Willow* Fact Sheet ST-576, Edward F. Gilman and Dennis G. Watson, United States Forest Service

[11] "Rooting Around: Tree Roots", Dave Hanson, *Yard & Garden Line News* Volume 5 Number 15, University of Minnesota Extension, October 1, 2003

[12] Blackman, R. L.; Eastop, V. F. (1994). *Aphids on the World's Trees*. CABI. ISBN 9780851988771.

[13] David V. Alford (2012). *Pests of Ornamental Trees, Shrubs and Flowers*. p. 78.

[14] Kenaley, Shawn C.; et al. (2010). "Leaf Rust" (PDF).

[15] James Breasted (English translation). "The Edwin Smith Papyrus" . Retrieved 2007-06-09.

[16] "An aspirin a day keeps the doctor at bay: The world's first blockbuster drug is a hundred years old this week" . Retrieved 2007-06-09.

[17] W. Hale White. "Materia Medica Pharmacy, Pharmacology and Therapeutics" . Retrieved 2011-04-02.

[18] D. S. Vohra (1 June 2004). *Bach Flower Remedies: A Comprehensive Study*. B. Jain Publishers. p. 3. ISBN 978-81-7021-271-3. Retrieved 2 September 2013.

[19] "Flower remedies". Cancer Research UK. Retrieved September 2013.

[20] The palaeoenvironment of the Antrea Net Find The Department of Geography, University of Helsinki

[21] Hageneder, Fred (2001). *The Heritage of Trees.* Edinburgh : Floris. ISBN 0-86315-359-3. p.172

[22] Aylott, Matthew J.; Casella, E; Tubby, I; Street, NR; Smith, P; Taylor, G (2008). "Yield and spatial supply of bioenergy poplar and willow short-rotation coppice in the UK" (PDF). *New Phytologist* **178** (2): 358–370. doi:10.1111/j.1469-8137.2008.02396.x. PMID 18331429. Retrieved 2008-10-22.

[23] Mola-Yudego, Blas; Aronsson, Pär. (2008). "Yield models for commercial willow biomass plantations in Sweden" (PDF). *Biomass and Bioenergy* **32** (9): 829–837. doi:10.1016/j.biombioe.2008.01.002. Retrieved 2009-11-20.

[24] "Forestresearch.gov.uk". Forestresearch.gov.uk. Retrieved 2011-12-18.

[25] "ChurchYear.net". ChurchYear.net. Retrieved 2011-12-18.

[26] Doolittle, Justus (2002) [1876]. *Social Life of the Chinese.* Routledge. ISBN 978-0-7103-0753-8.

[27] Doré S.J., Henry; Kennelly, S.J. (Translator), M. (1914). *Researches into Chinese Superstitions.* Tusewei Press, Shanghai. Vol I p. 2

[28] "The Forest of Willows in Our Minds". Arirang TV. August 20, 2007. Retrieved September 10, 2007. Check date values in: |date= (help)

[29] "In Worship of Trees by George Knowles: Willow".

[30] "Mythology and Folklore of the Willow".

[31] "Under The Willow Tree". Hca.gilead.org.il. 2007-12-13. Retrieved 2011-12-18.

[32] "Green Willow". Spiritoftrees.org. Retrieved 2011-12-18.

[33] The Willow Wife Archived May 18, 2008 at the Wayback Machine

[34] "Wisdom of the Willow Tree". Tweedsblues.net. Archived from the original on September 29, 2011. Retrieved 2011-12-18.

83.11 Bibliography

- Keeler, Harriet L. (1990). *Our Native Trees and How to Identify Them.* New York: Charles Scriber's Sons. Pages 393–395. ISBN 0-87338-838-0.

- Newsholme, C. (1992). *Willows: The Genus Salix.* ISBN 0-88192-565-9

- Warren-Wren, S.C. (1992). *The Complete Book of Willows.* ISBN 0-498-01262-X

- Sviatlana Trybush, Šárka Jahodová, William Macalpine and Angela Karp. (2008). *A genetic study of a* Salix *germplasm resource reveals new insights into relationships among subgenera, sections and species* BioEnergy Research. 1(1):67 – 79.

83.12 External links

- 1911 Encyclopaedia Britannica

- "Willow". *The New Student's Reference Work.* 1914.

83.13 Text and image sources, contributors, and licenses

83.13.1 Text

- **Non-timber forest product** *Source:* https://en.wikipedia.org/wiki/Non-timber_forest_product?oldid=654693781 *Contributors:* Samsara, Utcursch, Rich Farmbrough, Nigelj, Gary, Woohookitty, Rjwilmsi, Vegaswikian, Malcolma, SmackBot, Derek R Bullamore, Gnome (Bot), BetacommandBot, JamesAM, Thijs!bot, PierceG, Uruiamme, Aprawa, Naniwako, Lady Dreamgirl, DASonnenfeld, VolkovBot, Why Not A Duck, COBot, A21sauce, ClueBot, Auntof6, Dthomsen8, Addbot, Favonian, Lightbot, Yobot, AnomieBOT, Minnecologies, Anna Frodesiak, FrescoBot, Remotelysensed, Bioguru1, Pat604, FoxBot, EmausBot, Look2See1, ClueBot NG, Meredithmartin, BG19bot, Declangi, BattyBot, Pankaj Oudhia, Rjsshawty, Macaulay27, Cuteyasaswini, Troubleinhighheels, Eight2twelve, Babitaarora, Superhomosapien, LucianHelzapoppin, Cameron Ehteshami, Lilwaveyre, Geohutt1, Thecoolguy1233333, Henri Trudeau, Hitherehello102, Ryanckulp, Mirahlucas, Ss557, I94barman, Aaykmr, Shajia Khan, Nainsal, Asykes3, Sangoku09, QueenBoopse, Idkbreanna, Muralidharan24, JoeyJoeyJoeMC, TIMETODREAM, Halomace, Selom10, SteveTheGreat1997, ArchonMarcus, Zachh87, Keegster200034, Monkbot and Anonymous: 14

- **Akpeteshie** *Source:* https://en.wikipedia.org/wiki/Akpeteshie?oldid=671401427 *Contributors:* Excalibur, Dr.frog, Grutness, BD2412, Rjwilmsi, SmackBot, Centrepull, TAnthony, PMG, DASonnenfeld, Funandtrvl, Behemothing, Msrasnw, Mild Bill Hiccup, Arjayay, Dthomsen8, Addbot, AnomieBOT, DSisyphBot, Steven.mccurdy, Jesse V., BattyBot, Nkansahrexford, Yarrmatey and Anonymous: 4

- **Allspice** *Source:* https://en.wikipedia.org/wiki/Allspice?oldid=676916296 *Contributors:* Tarquin, Rmhermen, Heron, Ewen, Edward, Dgrant, Jacquez, Victor Engel, Nohat, Wik, Joy, Wetman, Slawojarek, Peak, Naddy, Modulatum, SchmuckyTheCat, Bkell, Modeha, Cyrius, Joconnor, Macrakis, JoJan, Burschik, Timothy Campbell, EugeneZelenko, Mwmnp, LuciferBlack, Aquillion, Hesperian, Jonathunder, Jumbuck, Hugowolf, Ctande, JK the unwise, Shoefly, Axeman89, Kazvorpal, Feezo, Sandover, ChrisJMoor, Spettro9, Bunchofgrapes, Loniceas, Ricardo Carneiro Pires, FlaBot, Lemmikkipuu, Ronebofh, Gdrbot, WriterHound, YurikBot, Martnik, AVM, Gaius Cornelius, Curtis Clark, Badagnani, Burman, TDogg310, GeorgeC, SameerKhan, DeadEyeArrow, Asarelah, Trainra, IceCreamAntisocial, Where next Columbus?, Mmcannis, GrinBot~enwiki, Schizobullet, SmackBot, Krychek, Kintetsubuffalo, Straitgate, Gilliam, Ohnoitsjamie, Hmains, Salvor, Deli nk, Uthbrian, Rrburke, Smokefoot, Bejnar, Euchiasmus, Johanna-Hypatia, Maksim L., Courcelles, Tau'olunga, ZsinjBot, Ibadibam, Cydebot, Miclwilson, Nick Y., JamesLucas, Ebyabe, Thijs!bot, J. W. Love, Escarbot, Luna Santin, Glennwells, Aholdhusen01, TuvicBot, JAnDbot, MER-C, Quentar~enwiki, Yowzaboodle, Jerem43, Yenisey, Keith D, CommonsDelinker, DBlomgren, Trusilver, Acalamari, Thesis4Eva, STBotD, DASonnenfeld, Idioma-bot, Redwood57, Tumblingsky, Paul Tomlinson, AlleborgoBot, Angelastic, SieBot, OlliffeObscurity, Jtonsing, Rob.bastholm, Flyer22, ClueBot, PipepBot, Jmgarg1, Johnfredx, TReubens, Jeremiestrother, Gnome de plume, EE4E, Gwellesley, Snacks, Slidinandridin06, Thingg, XLinkBot, Truetom, Addbot, Cuaxdon, GyroMagician, CanadianLinuxUser, Ebudan, Kk sze, Tassedethe, Luckas-bot, WikiDan61, KamikazeBot, AnomieBOT, Rubinbot, Piano non troppo, RayvnEQ, Tiggsy, Materialscientist, Addihockey10, Gigemag76, TheGunn, Bear8it, Shirik, Leno4ka~enwiki, Bornnablesme, Some standardized rigour, FrescoBot, Recognizance, Rgvis, Xxglennxx, Abductive, AmphBot, Dosinovsky, RedBot, Peace and Passion, Boricuamark, RjwilmsiBot, Pigletcrenn, Kamran the Great, DASHBot, Ebe123, Thaumaturgist, ZéroBot, Ὁ οἶστρος, Donner60, Babapa, ClueBot NG, Plantdrew, Hadashi Black, BG19bot, Duckbilldanny, ChrisGualtieri, Aliwal2012, Sminthopsis84, Jamermell, Eyesnore, Onuphriate, Birdfrancis21, SovalValtos, Julietdeltalima, ScrapIronIV, Denniscabrams, Pimentowoodproducts and Anonymous: 165

- **Bay leaf** *Source:* https://en.wikipedia.org/wiki/Bay_leaf?oldid=679789560 *Contributors:* Brion VIBBER, Andre Engels, Kowloonese, Unukorno, PierreAbbat, Frecklefoot, Dominus, Mkweise, Ellywa, Jengod, Loren Rosen, Savirr, Wiwaxia, KeithH, WormRunner, Rfc1394, Tom Radulovich, Agentseven, Robert Weemeyer, Burschik, Guanabot, Circeus, Longhair, RaffiKojian~enwiki, Jonathunder, Hugowolf, Eleland, Kfitzner, Edgewise, Grika, Georgelazenby, Eubot, Alphachimp, CJLL Wright, Chris Capoccia, Grubber, Dforest, Badagnani, Smartyhall, Superluser, Seb1982, Reyk, Katieh5584, SmackBot, Istrebitjel, Brianski, VMS Mosaic, Cybercobra, Sparky-sama, Markj99, Ben Moore, BillFlis, Waggers, Iridescent, Sameboat, Janus303, Robfwoods, Beznas, CmdrObot, Bonás, CWY2190, Epbr123, IvanStepaniuk, Headbomb, Marek69, A3RO, Touko vk, Majorly, Nipisiquit, JAnDbot, Husond, MER-C, Ericoides, Hut 8.5, VoABot II, Ali, Benjamint444, Vanished user g454XxNpUVWvxzlr, KylieTastic, DASonnenfeld, JochemKaas, Part Deux, Fences and windows, Vincent Lextrait, Drussaxe, Martin451, Brian Huffman, Malick78, Burntsauce, Vikrant42, Wikineer, SieBot, N2thai, Palachandra, WTucker, Dawn Bard, Serag4000, Cacycle test, Smf5000, Dillard421, Q1969NL, ClueBot, Pressforaction, Jian Heseri, Tanglewood4, Mezigue, Kitsunegami, Jusdafax, Lobsternar, NuclearWarfare, Hhhggg111, ProSvet, Fastily, JonhySK, Addbot, Cuaxdon, Saurabh1981, Deciarea, Supersqr, Numbo3-bot, Lightbot, Delta 51, Yobot, Tempodivalse, Jychemist, AnomieBOT, Citation bot, Wherrelz, Xqbot, Gigemag76, Nezinau, Mayor mt, Ellenois, Getspaper, VI, Mfwitten, Lilaac, DrilBot, Psionic Fighter, Deagle AP, DASHBot, DrJGMD, Az jor, Jkadavoor, MarinaMichaels, Breiz~enwiki, Rcsprinter123, Donner60, JonRichfield, ClueBot NG, Chester Markel, Macdicilla, ScottSteiner, Clescuyer, Widr, Novusuna, Plantdrew, Emayv, NotWith, ChrisGualtieri, Sminthopsis84, RandomFixer and Anonymous: 205

- **Benzoin resin** *Source:* https://en.wikipedia.org/wiki/Benzoin_resin?oldid=653763675 *Contributors:* Stone, Dogface, Francs2000, MPF, Rich Farmbrough, CanisRufus, Longhair, Hooperbloob, Anthony Appleyard, Sjschen, Andrewpmk, Sl, Dryman, Allen3, V8rik, Anarchivist, Wiserd911, SkyCaptain~enwiki, RussBot, Shaddack, Prime Entelechy, El Cazangero, Dysmorodrepanis~enwiki, Trovatore, DAJF, IceCreamAntisocial, NielsenGW, The Famous Movie Director, Sei Shonagon~enwiki, Drphilharmonic, Ligulembot, Snowgrouse, SilkTork, Christian75, JAnDbot, Epeefleche, Magioladitis, Maproom, DASonnenfeld, Idioma-bot, Almazi, Wie146, Italtrav, Wikiisawesome, SieBot, Calliopejen1, LarryMorseDCOhio, Jht4060, Addbot, Mr0t1633, DOI bot, Panfily, SamatBot, Yobot, PMLawrence, Wisg, LilHelpa, Janmeut~enwiki, Dinamik-bot, Gixie, Kamran the Great, EmausBot, WikitanvirBot, Tmrdean, BG19bot, PhnomPencil, Bangalorius, Solar Dynasty, Mpiva, Lemnaminor, Arun Skeet, Alluppity, Karish10, P. S. Sena and Anonymous: 23

- **Berry** *Source:* https://en.wikipedia.org/wiki/Berry?oldid=684703256 *Contributors:* Vicki Rosenzweig, Mav, Tarquin, Stephen Gilbert, Rmhermen, PierreAbbat, Ellmist, Montrealais, Michael Hardy, Tompagenet, Egil, Ahoerstemeier, Ronz, Angela, Julesd, Glenn, Andres, Steve nova, Hike395, Timwi, WhisperToMe, Zoicon5, Tpbradbury, Marshman, Grendelkhan, Ed g2s, Topbanana, Pakaran, Jusjih, Finlay McWalter, Owen, Gromlakh, Robbot, Wjhonson, Clarkk, Lupo, Marc Venot, MPF, Yekrats, Solipsist, Chaerani, Pgan002, Alexf, Knutux, Sonjaaa, Phe, MisfitToys, Flow~enwiki, Jossi, SimonLyall, Neutrality, Mike Rosoft, Discospinster, GoD, Bishonen, H0tte, Jnestorius, MisterSheik, MBisanz, El C, Galf, RoyBoy, Dannown, Circeus, Meggar, Viriditas, Giraffedata, Man vyi, Alansohn, Keenan Pepper, Hippophaë~enwiki, AzaToth, Wtmitchell, Rebroad, RJFJR, Gene Nygaard, Kazvorpal, Angr, Pekinensis, Camw, James Kemp, BillC, WadeSimMiser, MGTom, Burgher, Isnow, Macaddct1984, Blackcats, Mendaliv, Rjwilmsi, BlueMoonlet, Oblivious, Dolphonia, FlaBot, RexNL, Czar, Ronebofh, Slow Graffiti,

Kbdank71, Bunchofgrapes, FreplySpang, Pmj, Rjwilmsi, NatusRoma, Jweiss11, Harro5, Mork the delayer, Oblivious, CQJ, Ricardo Carneiro Pires, Brighterorange, Yug, Panterka, Fish and karate, Naraht, Eubot, RobertG, Awotter, Chanting Fox, Meeve, Spudtater, RexNL, Kolbasz, OpenToppedBus, Ronebofh, Le Anh-Huy, Mongreilf, Gdrbot, Bgwhite, NSR, WriterHound, The Rambling Man, YurikBot, Sceptre, SkyCaptain~enwiki, Red Slash, Kvuo, IBook of the Revolution, Stephenb, Mithridates, Gaius Cornelius, Wimt, Curtis Clark, Dysmorodrepanis~enwiki, Nirvana2013, Veledan, Badagnani, Geeksquad, Joelr31, Mccready, Ragesoss, Banes, Coderzombie, Marknesbitt, Syrthiss, Gadget850, Asarelah, Tachs, 1978~enwiki, Crisco 1492, Leptictidium, Codrinb, Theda, SMcCandlish, Canley, BorgQueen, MrHen, Katieh5584, Kungfuadam, DVD R W, Jaysscholar, A bit iffy, SmackBot, YellowMonkey, Bouette, Erwinrossen, Ma8thew, Gubby, Olorin28, Hydrogen Iodide, Sanjay ach, Blue520, Thunderboltz, Grey Shadow, Eskimbot, Anomaly2002, Stevegallery, TypoDotOrg, Gilliam, Ohnoitsjamie, Cibyd, KaragouniS, Smalltowngirl, CrookedAsterisk, Thumperward, HartzR, Hollow Wilerding, Deli nk, ImpuMozhi, Uthbrian, Octahedron80, Junius49, Royboycrashfan, Scray, Rfwoolf, Flibbert, Andrewin, Yidisheryid, 66664yyiotjwoier, CJ666, Neonow, Jkirish5, Rrburke, Kingdon, TedE, VegaDark, BryanG, Ligulembot, Mu2, Tjrichter, Bejnar, Apalaria, DrJoe, Eliyak, Geach, Teneriff, Kuru, Ian Spackman, Joffeloff, IronGargoyle, A. Parrot, Eternal Equinox, NcSchu, SQGibbon, SandyGeorgia, Ryulong, M855GT, Peyre, OnBeyondZebrax, Tina Brooks, Blehfu, Tawkerbot2, Mschroebel, JForget, KNM, Code E, Ale jrb, Albert.white, IP Address, GargoyleMT, Pgr94, Tomw91, Doctormatt, Cydebot, Webaware, Anthonyhcole, Tawkerbot4, Damianrafferty, Zalgo, Calvero JP, Casliber, SummonerMarc, Epbr123, Devadaskrishnan, Bryanwake, Hammerhorn~enwiki, Dgies, AntiVandalBot, Widefox, CLSwiki, Esesk, Danger, Mutt Lunker, NinaSpeaking, Skarkkai, StrawberryClock, JAnDbot, Tigga, Altairisfar, MER-C, Ericoides, Awien, Colotfox, KaUni, Magioladitis, WolfmanSF, Bongwarrior, VoABot II, JNW, Rivertorch, Trugster, Rimibchatterjee, Jessicapierce, Robotman1974, Rhalden, 28421u2232nfenfcenc, Spellmaster, The cattr, DerHexer, Philg88, Jerem43, MartinBot, Grandia01, R'n'B, CommonsDelinker, Fconaway, Wlodzimierz, J.delanoy, Pharaoh of the Wizards, Thegreenj, Whitebox, Rstevec, Salih, Jigesh, Anonywiki, AntiSpamBot, NewEnglandYankee, Fakewcfrog, Biglovinb, Juliancolton, DorganBot, Treisijs, PurelyNina, Nashville Monkey, DASonnenfeld, Xiahou, Idioma-bot, Redtigerxyz, Valugi, Sniper1rfa, VolkovBot, Saddy Dumpington, GimmeBot, Abtinb, Fcb981, Tusbra, Someguy1221, Naohiro19 revertvandal, Seraphim, BotKung, Pigslookfunny, Spicedoctor, Fotek, Jaguarlaser, Falcon8765, Spinningspark, GoddersUK, SieBot, Ivan Štambuk, Prakash Nadkarni, Dawn Bard, Ghaag, WestCoastMusketeer, Lydiafiedler, Flyer22, Jojalozzo, Carnun, Arknascar44, Ftindia, Oxymoron83, Goustien, Lightmouse, Gordonofcartoon, Moletrouser, Afernand74, Dabomb87, Escape Orbit, ClueBot, Ian S. Richards, Abee60, Jmgarg1, The Thing That Should Not Be, Kallidaimaniac, EoGuy, Av99, Mild Bill Hiccup, Niceguyedc, Another Matt, Auntof6, Paulcmnt, Akk7a, Excirial, Yossman97, PixelBot, Vivio Testarossa, NuclearWarfare, Georgiamonet, Lalitstar, SoHome, Slidinandridin06, JasonAQuest, Versus22, MuckFizzou, DumZiBoT, XLinkBot, Drhealthnutty, Vanished 45kd09la13, Wikiuser100, AndreNatas, Senzuri, ZooFari, Airplaneman, HexaChord, Johnny apples, GDibyendu, Addbot, DOI bot, Tcncv, Rav314, Annielogue, Nath1991, CanadianLinuxUser, Bazza1971, Glane23, Bassbonerocks, Chzz, Zanbuist, Ahmad.ghamdi.24, CarTick, Americanfreedom, Kommus, Tide rolls, Lightbot, Luckas-bot, Yobot, TaBOT-zerem, Choosetocount, Legobot II, Sanyi4, KamikazeBot, IW.HG, AnomieBOT, Cicero in utero, Qwertyuiopasdfghjklñ, Piano non troppo, Golb12, Zeisterre, Sz-iwbot, Citation bot, Lightningman26, Quebec99, Capricorn42, Astudent1, ChildofMidnight, Vanished user xlkvmskgm4k, GrouchoBot, Jhbdel, Itineranttrader, Brandon5485, Zefr, Dr.Dunn, Bonerbiter69, Doulos Christos, Grammarfixeruper, Dasaradhawiki, Hamamelis, Dan Wylie-Sears 2, Starwars1791...continued, FrescoBot, XXeggsexXx, Pepper, Racingstripes, Kites11, PiperNigrum, Citation bot 1, Pinethicket, HRoestBot, Abductive, Faerra, Dosinovsky, Wikitza, STEV56, My05hammer, Tyler-willard, Zhonghuo~enwiki, IJBall, Trappist the monk, Xlxfjh, FlamingMoonsOfSaturn, Diannaa, Firebeastm, Tbhotch, Bariapepper, Obsidian Soul, Servbot777, Ripchip Bot, The Stick Man, Sajoshthambi, Binoyjsdk, EmausBot, Acather96, WikitanvirBot, Dewritech, Jkadavoor, Rajkiandris, Slightsmile, Wikipelli, Djembayz, Urbain23, Balisong5, Comesturnruler, ZéroBot, John Cline, Fæ, Schubert80, Traxs7, Access Denied, H3llBot, Wayne Slam, Erianna, Rcsprinter123, Anglais1, Donner60, Phanjuy, Uziel302, Mikhail Ryazanov, ClueBot NG, Mr. Glengarry Glen Ross, Rich Smith, Ishfaq Khan8, Gareth Griffith-Jones, Servbot0303, Jiwa Matahari, Movses-bot, A.D.Balasubramaniyan, Samalambam1, Dr.Cena, Wangond, Diyasp, Widr, Joelcostanzo21, Ante Vranković, Seemerock, Helpful Pixie Bot, Bobby chauhan, ?oygul, Calidum, DBigXray, Plantdrew, Gomada, Blake Burba, Northamerica1000, Cold Season, Jahnavisatyan, OttawaAC, Yowanvista, MsBatfish, YVSREDDY, Vrraybadboy3013, Pratap.sps, Crabapple44, Vanished user lt94ma34le12, Nguyễn Quốc Việt, Sanjaya weerasinghe, ChrisGualtieri, Ancienzus, TylerDurden8823, One-ply, Sumimary, Dexbot, Jesse sidhu, Hmainsbot1, Webclient101, Tommy Pinball, Frosty, Corn cheese, Ihatelettuce, Rocketman2699, Aftabbanoori, The Ajan, Jamie LaDawn, Rockonomics, Ministar Nesigurnosti, JamesMoose, Rybec, Gunny777, Mantedayya, Ugog Nizdast, NottNott, Chindukulkarni, Kind Tennis Fan, Kochay1, Stamptrader, JaconaFrere, Safarigal11, Gbrager, Sanskari, Oiyarbepsy, Einskisson, Julietdeltalima, Aristo Class, Keerthijagannath and Anonymous: 535

- **Boscia senegalensis** *Source:* https://en.wikipedia.org/wiki/Boscia_senegalensis?oldid=663685890 *Contributors:* Tpbradbury, Alan Liefting, Hesperian, Stemonitis, Rjwilmsi, ENeville, Asarelah, RDBrown, OrphanBot, Cydebot, Dancter, Marco Schmidt, Idioma-bot, Hippo99, Not aplikabul, Addbot, Flakinho, Pganas, GrouchoBot, Trappist the monk, RjwilmsiBot, John of Reading, Bamyers99, Plantdrew, MKwek, Ariditeprospere, Donnem, Paulvandam22 and Anonymous: 3

- **Camphor** *Source:* https://en.wikipedia.org/wiki/Camphor?oldid=684993798 *Contributors:* Liftarn, Minesweeper, Narrowhouse, Snoyes, Evercat, Iorsh, Humehwy, Maximus Rex, Carlossuarez46, Riddley, Stewartadcock, Hadal, MPF, Kbahey, Mintleaf~enwiki, Gilgamesh~enwiki, Jason Quinn, Foobar, MacGyverMagic, Iantresman, Cacycle, Van Flamm~enwiki, Mani1, DcoetzeeBot~enwiki, Bender235, Robotje, MPerel, Free Bear, Tek022, Wiki-uk, Benjah-bmm27, Walkerma, Vuo, Gene Nygaard, Kazvorpal, Saxifrage, Firsfron, Julo, Mb1000, V8rik, De-Piep, Rjwilmsi, Yamamoto Ichiro, Vuong Ngan Ha, FlaBot, TSamuel, WriterHound, EamonnPKeane, Deeptrivia, Jurijbavdaz, Xihr, Chris Capoccia, El Cazangero, Dialectric, Grafen, Badagnani, Mardochaios, Chewyrunt, Elkman, Wikicheng, Kriscotta, Some guy, GrinBot~enwiki, SmackBot, Jagged 85, Kintetsubuffalo, Edgar181, RDBrown, MalafayaBot, Hgrosser, VMS Mosaic, Kingdon, Jcgarcow, Bejnar, Cesium 133, Writtenonsand, A. Parrot, Cowbert, BillFlis, Beetstra, Ewulp, Thricecube, Tawkerbot2, Bubbha, CRGreathouse, Neelix, Cydebot, Rifleman 82, Christian75, Casliber, Thijs!bot, Kablammo, Missvain, Heroeswithmetaphors, Escarbot, AntiVandalBot, Deflective, Ekabhishek, MER-C, Patrus, Lord antares, Albmont, Ling.Nut, MartinBot, NReitzel, Gunkarta, DBlomgren, Tgeairn, Trusilver, Rdhinakar, Nemo bis, Alphapeta, Chemartist, DASonnenfeld, Mutlay, Idioma-bot, VolkovBot, TXiKiBoT, Rvraman0, Anna Lincoln, Steven J. Anderson, Martin451, Milkbreath, Hassan1357, Lamro, Alcmaeonid, Wavehunter, AlleborgoBot, Shanmugammpl, SieBot, Gobugmom, Phe-bot, Дарко Максимовић, Sean.hoyland, Zwanzig~enwiki, Chem-awb, ClueBot, Gaia Octavia Agrippa, Mild Bill Hiccup, Thegeneralguy, Shriniwaskashalikar, Sensonet, Lessogg, Seanwal111111, DragonBot, Anon lynx, Iohannes Animosus, Slidinandridin06, Plasmic Physics, MystBot, Samin096, Addbot, Gxcf1, SpBot, Tide rolls, Lightbot, ScAvenger, Luckas-bot, Yobot, CheMoBot, Dmarquard, AnomieBOT, Casforty, IRP, Sabata, Materialscientist, Citation bot, غامد.أحم.غ.م24, ي عليﺍﻟﺒﻂ حسن, Programming gecko, J04n, GrouchoBot, Omnipaedista, Siddaarth.s, FK1954, ISamratC, Ellenois, FrescoBot, Tobby72, Riventree, Saiarcot895, Jrew86, BogBot, Trappist the monk, Sheogorath, Vrenator, Brumon, Difu Wu, RjwilmsiBot, Beyond My Ken, Dg harini, WikitanvirBot, Jnanadevm, Lucas Dziesinski, Jrfep,

NeRF 342951, Pinethicket, I dream of horses, LittleWink, 10metreh, Jonesey95, A8UDI, Achaemenes, Ssone, Myxsix, Raamah, Koakhtzvigad, Paragrapher, Pouyakhani, Thrissel, Trappist the monk, Sheogorath, Kalaiarasy, Jonkerz, Vrenator, Blind cyclist, Aoidh, David Hedlund, Kidsnaturalnews, RjwilmsiBot, PipingHotSoup, Timothyandrewelliott, Anand.athikesavan, WildBot, Autumnalmonk, Skamecrazy123, Kamran the Great, DrJGMD, EmausBot, Wiroshermes, Arineat, Nuujinn, Abby 92, Emabartolomucci, Oncology group, GoingBatty, Slightsmile, Tommy2010, Wikipelli, Sepguilherme, Kavaliltt, PBS-AWB, Liquidmetalrob, Matthewcgirling, Onlytock, H3llBot, SporkBot, Stainless steel cat, Prabinepali, Wayne Slam, Erianna, Petropetro, Jay-Sebastos, TyA, IGeMiNix, Brandmeister, Krissco, Motopower, TYelliot, Mephisto spa, JonRichfield, Michael Bailes, ClueBot NG, Mechanical digger, WIERDGREENMAN, This lousy T-shirt, Lahedoniste, ةيذغتـلا ةيواه, O.Koslowski, U.Steele, Widr, Modamin sticks, Traveletti, MerIlwBot, BayaniMills, Helpful Pixie Bot, Bobby chauhan, Vincent Marcus, Titodutta, DBigXray, Plantdrew, Jmreese, Agus stiw, Lacannelle, Northamerica1000, Shivendu ranjan, AvocatoBot, TheMan4000, Hud0007, Jahnavisatyan, ZarinTheZarin, Aubrey.knight73, Willpolydna, NotWith, Ohhey123, Dreambeaver, Fylbecatulous, BattyBot, Bmo27, Jimw338, Curritocurrito, BoyagamaLasal, Mikeyfitz, Dexbot, JurgenNL, Sminthopsis84, Ocotea, Lugia2453, Corinne, Epicgenius, AmaryllisGardener, JakeWi, Ghernzy, Vskthas, Sotkil, Thats ma name, Kingofaces43, Ugog Nizdast, Mihajilla, Saehry2, Kind Tennis Fan, HalfGig, Spirited-Michelle, Ethically Yours, CogitoErgoSum14, Technotoad, Longseeyes, Craigrottman, Writers Bond, Drjbdc, Monkbot, Drummerdude17065, KevinThomas69, Almorell, Sapientiaeetveritatis, Sanskari, Crowmed, ChamithN, Woderman, Jakebeegen, Redrightguy1, Charithonline, Kate jonson, Mahajandeepakv, Dil jimboy, Susquianna, Nomisel, Drkawatra8587, Kodilla, Jeendanie, Atudu, Ashlynsimpson, Syedkasfur9, Bornofsithis and Anonymous: 742

- **Clove** *Source:* https://en.wikipedia.org/wiki/Clove?oldid=685330858 *Contributors:* WojPob, PierreAbbat, Karen Johnson, Zadcat, Montrealais, DopefishJustin, Liftarn, Mkweise, Ronz, Darkwind, Emperorbma, Marshman, Furrykef, Topbanana, Robbot, Yosri, Rfc1394, Bkell, Modeha, Anthony, Adhib, MPF, Mark Richards, Tom Radulovich, DO'Neil, Mboverload, Matt Crypto, Chameleon, Pne, LiDaobing, Quadell, Williamb, Ary29, Zfr, Burschik, Quota, Ukexpat, Eyrian, DanielCD, Rich Farmbrough, Cacycle, Bender235, Mcpusc, MisterSheik, Pjf, *drew, Kwamikagami, Renice, Olve Utne, Viriditas, Physicistjedi, Hesperian, Idleguy, Haham hanuka, Alansohn, Gary, Hugowolf, Anthony Appleyard, Sabine's Sunbird, Sjschen, Ducttapeavenger, Clarahamster, Ghirlandajo, Dismas, Ari x, Stemonitis, Angr, Velho, Woohookitty, WadeSimMiser, John Hill, Male1979, Allen3, Bunchofgrapes, Rjwilmsi, Feydey, Rui Silva, Vuong Ngan Ha, Eubot, Dvortygirl, Karelj, TheDJ, Gdrbot, Cactus.man, WriterHound, Gwernol, Roboto de Ajvol, Gaius Cornelius, Rsrikanth05, Grafen, Badagnani, Tetsuo, SivaKumar, Lockesdonkey, Asarelah, Tachs, Wknight94, FF2010, BorgQueen, Pádraic MacUidhir, Maxsupereme, Alexandrov, SmackBot, Bobet, Sanjay ach, Delldot, Edgar181, Brianski, Ohnoitsjamie, JAn Dudík, Anwar saadat, Davea0511, Rkitko, Moshe Constantine Hassan Al-Silverburg, Deli nk, Bazonka, PureRED, Foxjwill, Egsan Bacon, TheGerm, Downtown dan seattle, Vprajkumar, Smokefoot, BullRangifer, Chaitanya ch, Zzorse, Lambiam, SilkTork, Aljullu, Mgiganteus1, Hargle, Hartono citra setiawan, Dalstadt, Chathurank, Marenf, Kranthikiranakula, Igoldste, Courcelles, Firewall62, AbsolutDan, Illegal.person, Danberbro, Srini.ms, CmdrObot, Binky The WonderSkull, Cydebot, Dominicanpapi82, JamesAM, Thijs!bot, Nalvage, John254, Merbabu, Heroeswithmetaphors, Niduzzi, Widefox, Seaphoto, AlexandriNo, Postlewaight, John Moss, Storkk, JAnDbot, Davewho2, Koibeatu, MER-C, SAMbo, Some thing, PhilKnight, WildlifeAnalysis, Joelfoss, VoABot II, Dufko, JNW, TheEsb, Catgut, Fuzzyllama, SlamDiego, Ksvaughan2, MartinBot, STBot, ChemNerd, CommonsDelinker, Nono64, JAGO, Tgeairn, Huzzlet the bot, J.delanoy, Maproom, Acalamari, Alphapeta, Floaterfluss, Belovedfreak, Zl.xyz.lz, Jester7777, Nvram, DASonnenfeld, Idioma-bot, VolkovBot, AlnoktaBOT, TXiKiBoT, Oshwah, A4bot, Uch, ElinorD, Chhoton, Modal Jig, BotKung, TheLoverly, Ian Glenn, SieBot, Coffee, Grant.Alpaugh, BBKurt, Benea, Bill9933, Denisarona, Escape Orbit, Fangjian, ClueBot, Shan iuctt, Infoeco, Icarusgeek, EoGuy, CyrilThePig4, Drmies, Piledhigheranddeeper, Neverquick, Niaaz, Jimmy Hammerfist, Akk7a, Kitsunegami, Francisco Herrero, Vanisheduser12345, Jotterbot, Ngebendi, Tuchomator, Iohannes Animosus, ChrisHodgesUK, Thingg, DumZiBoT, Funnychanges, Rickremember, HawaiiHangin10, Mavigogun, SilvonenBot, Solmefun, Addbot, Kk sze, Glane23, Debresser, SpBot, VASANTH S.N., Tide rolls, Lightbot, Gail, Zorrobot, Legobot, Luckas-bot, Bjoertvedt, Fallen.cze, AnomieBOT, Götz, Coopkev2, Galoubet, JackieBot, Piano non troppo, Sz-iwbot, Materialscientist, Info-farmer, Citation bot, Xqbot, TinucherianBot II, Cinagua, JimVC3, Capricorn42, Gigemag76, Maddie!, Squeegie25, Ita140188, Zefr, Mayor mt, Legato33, Ellenois, Drunauthorized, Citation bot 1, Krish Dulal, Pinethicket, Melba1, RedBot, Rholme, FoxBot, UranianPoet, Lotje, Boricuamark, Trnesjevski, Obsidian Soul, RjwilmsiBot, TjBot, Alph Bot, DrJGMD, EmausBot, WikitanvirBot, Amudha1, GoingBatty, Cladogrammatic, Abazub, Kavaliltt, ZéroBot, Shuipzv3, Борунс, Erianna, Donner60, Manytexts, Michael Bailes, ClueBot NG, Peter James, Primergrey, Mannanan51, Widr, Bobby chauhan, Tdimhcs, Calabe1992, Mouchumi, Plantdrew, BG19bot, Ahura21, Agus stiw, MKar, Declangi, Metricopolus, Jahnavisatyan, BlastodermMan, Kimimila58, NotWith, Geegad, Kobbersmed, Eslamaher, NewzealanderA, Hmainsbot1, Aidanf41, Flatspinjim, Poonam tasare, 11alfa, Kind Tennis Fan, LadyCailin, Onuphriate, CodyHofstetter, Dr.Farhat Naz, Sanskari, EllaJameson, Julietdeltalima, Jmconeby, Mdfoard34, KasparBot, Twajanak, Shion~dewiki, BU Rob13, Atrihemtn and Anonymous: 302

- **Cocoa bean** *Source:* https://en.wikipedia.org/wiki/Cocoa_bean?oldid=686100492 *Contributors:* Paul Drye, The Epopt, Vicki Rosenzweig, Jeronimo, Andre Engels, Rmhermen, Anthere, Heron, Jaknouse, Hephaestos, Chuq, Bdesham, Infrogmation, Pit~enwiki, Gabbe, Zanimum, Sannse, Delirium, Gbleem, Looxix~enwiki, Ahoerstemeier, Ronz, Den fjättrade ankan~enwiki, Schneelocke, JidGom, Timwi, Nohat, Bemoeial, JonMoore, Andrewman327, Kierant, Spikey, David.Monniaux, Anthony Fok, Phil Boswell, Gentgeen, Sanders muc, Altenmann, Halibutt, Sunray, Bkell, J.Rohrer, Cyrius, Solver, Marc Venot, Giftlite, DocWatson42, MPF, Barbara Shack, Mintleaf~enwiki, BenFrantzDale, Ds13, Tecpaocelotl, Bkonrad, DO'Neil, Finn-Zoltan, AlistairMcMillan, Eequor, Gzornenplatz, Bobblewik, Tagishsimon, JRR Trollkien, Golbez, Pgan002, Williamb, Rdsmith4, Sam Hocevar, Quota, Vsb, Trevor MacInnis, Discospinster, Solitude, Rich Farmbrough, Xezbeth, Stereotek, Bender235, ESkog, Kbh3rd, Elwikipedista~enwiki, DamianFinol, Kwamikagami, Aude, Xed, CDN99, Bobo192, Smalljim, Electrolito, Nivagh, R. S. Shaw, Adrian~enwiki, Geocachernemesis~enwiki, TheProject, Ranveig, Alansohn, Gary, Chino, Sjschen, Xanxz, Thoric, Sligocki, Muugokszhiion, Bart133, DreamGuy, Wtmitchell, Bucephalus, Velella, SidP, Tony Sidaway, BlastOButter42, Samsoncity, Benbest, Jugger90, Rotten, MarkusHagenlocher, Wayward, Prashanthns, Ineko, IIBewegung, Imersion, Ando228, Phillipedison1891, Rjwilmsi, Heah, Alan J Shea, Gsp, Nihiltres, David H Braun (1964), Angus Lepper, Kollision, Pigman, Hydrargyrum, Stephenb, Gaius Cornelius, Dijan, Eleassar, Stassats, NawlinWiki, ENeville, Janke, DAJF, Danlaycock, Dbfirs, Psy guy, Maunus, Super Rad!, E Wing, Sean Whitton, BorgQueen, AnimeJanai, GraemeL, Shyam, Kevin, Allens, Bluezy, Glorgana, CIreland, SmackBot, Brian1979, Astavrou, Honza Záruba, Peloneous, Frymaster, Edgar181, Cazort, Yamaguchi 先生, Gilliam, Hmains, Anwar saadat, Dycotiles, RDBrown, Anchoress, SchfiftyThree, Hibernian, Deli nk, Sadads, Epastore, ACupOfCoffee, Oatmeal batman, Peter Campbell, Can't sleep, clown will eat me, Frap, Geoffrey Gibson, JohnHR11, Reclarke, VMS Mosaic, Abrahami, Jiddisch~enwiki, BiggKwell, Daniel.Cardenas, Gnome of Fury, Lambiam, U-571, Kuru, John, Microchip08, Jidanni, Oscark, Buchanan-Hemit, Gobonobo, NeantHumain, Ninnnu~enwiki, Ben Moore, Munita Prasad, Meco, Sijo Ripa, Skinsmoke, Peyre, Caiaffa, Hu12, Iridescent, Phoenix2275, Missionary, Joseph Solis in Australia, Tmangray, Nhinchey, Splitpeasoup, Puffin10, Tawkerbot2, Wolfdog, Van helsing, Woudloper, Ibadibam, Green caterpillar, Nasorenga, Themightyquill, Cydebot, Eroach, Naturalhomes, Gogo Dodo, Cancun771, Jon C.,

Thebotanist, Bongwarrior, VoABot II, MartinDK, Spikeyone, MJD86, JNW, JamesBWatson, Economicprof, Praveenp, CeeJay.dk, Faizhaider, Think outside the box, Cadsuane Melaidhrin, Ling.Nut, Confiteordeo, Skew-t, Prestonmcconkie, Neweco, Catgut, Indon, Jessicapierce, Animum, Kiko76, Shauny bhoy, Sugarcaddy, Allstarecho, Robert Wetzlmayr, Elowan, Talon Artaine, DerHexer, Felisopus, Lenticel, Psym, Dkriegls, WLU, Patstuart, Digi23azlan, Yobol, Hdt83, MartinBot, PAK Man, Intesvensk, Shanomack66, Faizalkc, Personstuff18, Oncamera, Rockrangoon, R'n'B, CommonsDelinker, AlexiusHoratius, PrestonH, Tgeairn, Jeroldc, Wlodzimierz, J.delanoy, Pharaoh of the Wizards, Mloughran69, Try0yrt, Fowler&fowler, Ali, Adavidb, Bogey97, Uncle Dick, Ginsengbomb, Ashcraft, Ear32, Cheeriokole, Ron mathwizard, Balsa10, Acalamari, IdLoveOne, Katalaveno, Thomas Larsen, (jarbarf), Sweethome101, SJP, Toon05, Mufka, Rumpelstiltskin223, Ayyah tubby, Juliancolton, Tangmo~enwiki, Jevansen, Vanished user 39948282, Treisijs, Grendlefuzz, Useight, Koeho, DASonnenfeld, Idioma-bot, Paddycomeback, Funandtrvl, Braidedheadman, Caribbean H.Q., Deor, Netmonger, Caspian blue, Hale storm93, Fishnoodler, ABF, DSRH, Flyingidiot, Nullx42, Bobmuffin, Ryan032, Ampzapper, JBazuzi, Philip Trueman, Somanypeople, TXiKiBoT, Serg!o, A4bot, FlagSteward, Anonymous Dissident, GcSwRhIc, Jobey3, Sankalpdravid, Charlesdrakew, Chapala, Anna Lincoln, Una Smith, Jhkey, Saint Invective, Clarince63, Martin451, Leafyplant, Iikiikii, ^demonBot2, Iaminfo, Muhammad Mahdi Karim, Seb az86556, Elchip, Quis sugar, Pod Bay Door, ACEOREVIVED, Madhero88, Tanner-Christopher, Kaggar, Davetrotter, Rhopkins8, Jaguarlaser, Falcon8765, Kuabt, Monty845, Froto79, AlleborgoBot, Mdotdot, Arjun024, SD Martin61, Avinesh, Botev, Pare Mo, SieBot, Dusti, Tiddly Tom, Caulde, Peopledowhattheyoughttodo, StopPickingOnMe, M00cherman, Gerakibot, Claus Ableiter, Yohlanduh, Mbz1, Dawn Bard, Caltas, Matthew Yeager, Mahudhy, Yintan, Whodunnit-notme, Got.rice.biarch, Sreeji maxima, Keilana, Flyer22, Mocktard, Uwmad, Alexbrn, Sballal, Oda Mari, Gpics, Arknascar44, JSpung, Sbowers3, Allmightyduck, Hzh, Sjwells53, Wmpearl, Oxymoron83, Nuttycoconut, Baseball Bugs, Veronica.eyebrow, Cbl62, BIGShorts, 123ilikecheese, Khvalamde, Rekk, AniChai, Fratrep, Lauracs, Jongleur100, Geodanny, Sean.hoyland, Mygerardromance, The second fugue, Danieltiger45, BackToThePast, Rewang67, Runner5k, Florentino floro, Wahrmund, Denisarona, Soyseñorsnibbles, Escape Orbit, Thorncrag, Maxschmelling, Beemer69, Ecp3, Kinkyturnip, Beeblebrox, Soph983, Elassint, ClueBot, ICAPTCHA, GorillaWarfare, Cpljwlusmc, Snigbrook, The Thing That Should Not Be, DeadManTyping, TinyMark, Plastikspork, Abhinav, Ishida639, Jonbe, Ndenison, GreenSpigot, Babando, Drmies, Mild Bill Hiccup, RRS Trojan, Sickbrah, Niceguyedc, Blanchardb, LizardJr8, Neumeiko~enwiki, Colossus888, Auntof6, Jimmy Hammerfist, Fat Tummy, Harrieshc, Aua, Trav100, Pondos, Takeaway, Excirial, Jusdafax, Darthdinn, Adu65, Fasttimes68, Mindcry, Watermelo, Children.of.the.Kron, Lartoven, Ngebendi, Iohannes Animosus, Singhalawap, Razorflame, Percy s a carballo, UltimateDestroyerOfWorlds, Thehelpfulone, Kakofonous, Rui Gabriel Correia, Robfbms, Thingg, Vegetator, Versus22, Berean Hunter, Qwfp, Mythdon, SoxBot III, Ginbot86, DumZiBoT, Lpjames, XLinkBot, Hotcrocodile, ZicoCoconutWater, Medifix, Spitfire, Adeuss, Stickee, Jovianeye, Onehundredtrillion, Johntabs, Mjpresson, Feinoha, Nepenthes, WikHead, NellieBly, Mifter, Wjw0111, Zlzpqx, Mareechikaa, Kvkpanniyur, WikiDao, Wowzer224, Paulmoody14, Boobiess, Yes.aravind, Bobbarker97, PurrfectPeach, HexaChord, GDibyendu, Addbot, Gmsoccer18, Ryankunz21, Darius Sinclair, TheGeekHead, Kerina yin, Landon1980, Luis2134, Ronhjones, Movingboxes, Davrosuk, Reddevil1213, Shirtwaist, Vishnava, Paul Yeratz, Si Gam Acèh, Download, HackerNOT, Glane23, Coconutballs, Favonian, Dnadan319, 5 albert square, Tyw7, Tassedethe, Flakinho, SK 1993, Peridon, Theking17825, Hello2000, Mr.Xp, Rehman, Tide rolls, William S. Saturn, Lightbot, Tekstman, Tel'Quess, Ahmedcyad, VP-bot, Legobot, Luckas-bot, Yobot, Tohd8BohaithuGh1, Fraggle81, Cflm001, Donthereal, Kevinbrogers, THEN WHO WAS PHONE?, Cottonshirt, MacTire02, Dmarquard, Kulmalukko, AnomieBOT, Mnewhous, DemocraticLuntz, Floquenbeam, Oskulo, Hairhorn, Jim1138, Piano non troppo, Xufanc, LlywelynII, Kingpin13, Jo3sampl, Crecy99, Materialscientist, Hbkrishnan, The High Fin Sperm Whale, Aff123a, Citation bot, Lauralongtime, Reginarivera, JohnnyB256, Durgeshbonde, Gossipguy, Neurolysis, ‏مدي.غ.محمد‏24, Quebec99, LilHelpa, Jay L09, Xqbot, Jake7890, Cake1039, Belasted, Character.assassin, Weepingraf, Capricorn42, Gigemag76, Jeffrey Mall, Nasnema, Rai nitk, Anna Frodesiak, Maddie!, Ijustwishtohelp, Streed51, Shazine43, Imfreakintired, Arafitos, Pmlineditor, GrouchoBot, AVBOT, George94, Illini407, Zefr, RibotBOT, The Interior, Amaury, Mollisande, Brutaldeluxe, Cyfraw, Richard BB, Verbum Veritas, Algysea, Nutlover, Shadowjams, Hamamelis, Bloobedyblahblah, Iamstillandalwayswillbethewalrus, Ben1488002, Gamespirit, Gtimm13, Jakekerrr, Løken, Akathetruth111, Prari, FrescoBot, Magicalme 2, Axmann8, Patchy1, Kotekaiga, Katrina133113, Asianloverx0x0, AlexanderKaras, MusicInTheHouse, Maruchan1, Blackout2009, VS6507, Qwert987654321, Limpato, Dizymauler, RoyGoldsmith, Willard84, Haeinous, MGA73bot, Oldmanwearingawig, Flareflame93, Craig Pemberton, Citation bot 2, HamburgerRadio, Citation bot 1, Avyn2, Drasek Riven, Srisez, Javert, Dtfman, WQUlrich, Cdm1991, Pinethicket, I dream of horses, Edderso, Tlusta, Calmer Waters, Drkvmallesh, Bunnyboy6, Btlm, MastiBot, Phearson, Nijgoykar, Metalmaddog, Vencel, Brookbond, Midnight Comet, Fumitol, Robo Cop, Maceis, Alnachiappa, Lipidresearch, Zeeth, Tim1357, FoxBot, Jedi94, Yunshui, Tydude187, Sheogorath, LogAntiLog, Spartan0007, Theoffice89, BPK, Lotje, Fox Wilson, GregKaye, Kolobochek, TRYPPN, Reaper Eternal, Aiken drum, Boricuamark, Lambanog, AARRIISSHHAA, LuftWaffle0, Sugarysweetflowerchildxoxo, Cocktail-media, Toreydag26, Peacelovedance, PleaseStand, Tbhotch, Reach Out to the Truth, Bleachitblue, Obsidian Soul, Mean as custard, Neinsun, DexDor, Alph Bot, Ripchip Bot, Auntypog, DRAGON BOOSTER, Mangotango4ever, Trejo98, In ictu oculi, Rakeshmallick27, Salvio giuliano, Bennyhui, Hahahamuamua112, Giri1234, EmausBot, CR1928, TreeXpert, Orphan Wiki, Immunize, Look2See1, Zollerriia, Ibbn, Racerx11, Newo67, Ramon FVelasquez, RA0808, Philipp Wetzlar, Jkadavoor, RenamedUser01302013, Kalinjar, AndrewStar, J-james1234, Tommy2010, Landbridge1, Wikipelli, Doctor Boboshango, K6ka, Etan Doronne, Tyranny Sue, Circabro2, Lucas Thoms, Hype5ocity, Subin.a.mathew, Sherwin ph, AvicBot, Comesturnruler, ZéroBot, John Cline, Pauldr713, Bollyjeff, Érico Júnior Wouters, Magicfiji, Podruzny, Juandel, Jay2980, Mpathma, A930913, Manyoudude, EJavanainen, SelvanathanPerumal, MajorVariola, AManWithNoPlan, Wayne Slam, Mcmatter, Tolly4bolly, Jankdog, Erianna, Noescapefromreality, Anglais1, Mitchellellis, Cameron11598, L Kensington, Donner60, Puffin, CoCoNuts26, Bill william compton, ChuispastonBot, Coconut tree dude, Utharamalabar, Nayansatya, Lguipontes, El Whizzo, Swathivijay, Lodoicea, Rocketrod1960, Delegance, Michael Bailes, E. Fokker, Helpsome, ClueBot NG, Marrier & Fruitful., Jack Greenmaven, Reji Jacob, Mariagomez39, Jofsig, This lousy T-shirt, Jiwa Matahari, Satellizer, Hooverbahz, Crazyman9898, Bruce.rowe, Doh5678, Yucatan111, Birdlass, Snotbot, Hindustanilanguage, Stormodboyo, Braincricket, Aksamary, Widr, GlassLadyBug, Cprees112, Theopolisme, IgnorantArmies, Coconuts11, Math cat`ing, Helpful Pixie Bot, Sherwm1, CarTowner, Candleabracadabra, Plantdrew, Enchantedlandscapes, BG19bot, Ilikesivan2, CoconutFreak, Mangalgi, Rolandkeeps, Northamerica1000, Declangi, Cyberpower678, Wiki13, MusikAnimal, Jahnavisatyan, Just another guy, Amr20, Mrchickenpop, Aranea Mortem, Sadfghj, The masterbadder, The 4567890, NotWith, Rsamahamed, DdraconiandevilL, TurkeyBurgers, Zujua, Pikachu Bros., Elevation ripped, Minsbot, Mosnoph, Dhess13, MrCallus97, Anbu121, Wikipeddler001, Bonkers The Clown, Muffin Wizard, Mikeshinobi1, EricEnfermero, Simeondahl, Balachand, Barefootwriter, HueSatLum, Nannerguy7, Karthiksabaasha, Plutoniumjesus, Cv45319, Nguyễn Quốc Việt, Hsuadre, Cyberbot II, Duakembang, Mediran, Drewster137, Tiato, Its-lyke-summer, Miguel raul, JYBot, Musiclanka, Damndembeaverflaps, Dexbot, Vwalvekar, Wgracen, Wow its a tree shocker carrot head, Coolieconch, Webclient101, Nouniquenames, Mogism, Geremy.Hebert, Kjbp, 88saGekiK, Lugia2453, Bmillicent, Graphium, Sriharsh1234, Cutieitalian, Vegetarianve2975, Athomeinkobe, DanielTom, Howtobuttrape4u, Sebidya, Youtnsnjufn, Lopipoo, Mydreamsparrow, Collin123321, PinkAmpersand, Epicgenius, Jwoodward48wiki, Godot13, Vanamonde93, Gh326312738, Camyoung54, Masterjediwiki, AmaryllisGardener, Megayofish, Everymorning, Jakec, Pineapple coconut cheeseball, Sonuistom, Jhipad, Barfolemew, Asdfjkl;714,

Tide rolls, Lightbot, Jarble, AnomieBOT, Materialscientist, LilHelpa, Tomdo08, JascalX, حسن علي البط, Champlax, J04n, Shirik, RibotBOT, Chloralhydrat, FrescoBot, Xxxcolexxx19, Chenopodiaceous, Pinethicket, I dream of horses, Daveandkay, Trappist the monk, Lotje, RjwilmsiBot, EmausBot, John of Reading, Penom, ZéroBot, JonRichfield, ClueBot NG, Astatine211, Dougmcdonell, Lifeformnoho, Gorthian, NotWith, Brummyjim, Gordon Davy, Leprof 7272, Adbar, Monkbot, Epigogue, Xycomgroup, Pezuszw, Devcrate and Anonymous: 97

- **Cycas circinalis** *Source:* https://en.wikipedia.org/wiki/Cycas_circinalis?oldid=665043933 *Contributors:* Raul654, Hesperian, Alai, Stemonitis, Miss Madeline, Eyu100, Eubot, Snek01, Masparasol, Mgiganteus1, Alaibot, Thijs!bot, Maias, Magioladitis, A Nobody, Rufous-crowned Sparrow, Chiswick Chap, DASonnenfeld, VolkovBot, Vinayaraj, QualiaBot, Fuzzonian, DragonBot, Jotterbot, Addbot, Flakinho, Gigemag76, RedBot, TobeBot, EmausBot, Jkadavoor, ZéroBot, Floscuculi, NotWith, YVSREDDY, Reynoldjose, Joseph Laferriere, Jaehaeron and Anonymous: 8

- **Dehesa (pastoral management)** *Source:* https://en.wikipedia.org/wiki/Dehesa_(pastoral_management)?oldid=685998265 *Contributors:* William Avery, Jmabel, Alan Liefting, Eggstasy, Chanheigeorge, SmackBot, Zeorymer, Alanmaher, Nick Number, Escarbot, Steven Walling, DASonnenfeld, Richard New Forest, Hugo999, Pare Mo, Lightmouse, TubularWorld, No such user, SchreiberBike, MatthewVanitas, Addbot, HerculeBot, Yobot, AnomieBOT, Xufanc, Minnecologies, Apothecia, Madrid Tiger, Daouuud, Lynnstarrs, Imaginibus, Rowan Adams, Khazar2, Robert4565, Edwinhere, Dehesas ibericas and Anonymous: 12

- **Diospyros melanoxylon** *Source:* https://en.wikipedia.org/wiki/Diospyros_melanoxylon?oldid=671209039 *Contributors:* Florian Blaschke, Kwamikagami, Gdrbot, Butsuri, Dysmorodrepanis~enwiki, TDogg310, McGeddon, Snori, Skinsmoke, Dl2000, Paul venter, DASonnenfeld, WereSpielChequers, SilvonenBot, Addbot, Martin-vogel, JackieBot, Jo3sampl, Xqbot, Abajo estaba el pez, Akkida, Drajay2010, Puffin, Neha.Vindhya, Declangi, YVSREDDY, Sminthopsis84, Pankaj Oudhia, Jaytwist, CastielTheSexAngel, Gihan Jayaweera and Anonymous: 7

- **Durian** *Source:* https://en.wikipedia.org/wiki/Durian?oldid=686311148 *Contributors:* AxelBoldt, Tobias Hoevekamp, Eloquence, Malcolm Farmer, Alex.tan, Danny, Rmhermen, Heron, JakeVortex, Earth, Liftarn, Gabbe, Seav, Ahoerstemeier, Ronz, Jpatokal, Ijon, Bogdangiusca, Ike9898, Dysprosia, WhisperToMe, Peregrine981, Tpbradbury, Samsara, Jeeves, Wetman, Secretlondon, David.Monniaux, THSlone, Jeffq, Donarreiskoffer, Robbot, Hankwang, Tomchiukc, RedWolf, Kowey, Yosri, DHN, Kamakura, Mandel, Xanzzibar, The Fellowship of the Troll, MPF, Nickdc, Michael Devore, Henry Flower, Malbear, Gilgamesh~enwiki, Bobblewik, Ragib, PeterC, James Crippen, Andycjp, Yath, PDH, Yik Lin Khoo, Jeremykemp, Beginning, Neutrality, Squash, Rickvaughn, Dr.frog, Ouro, Discospinster, Rich Farmbrough, Guanabot, Drano, Xezbeth, Bender235, Aqua008, CanisRufus, Kwamikagami, Art LaPella, Aaron D. Ball, Bobo192, Circeus, Smalljim, Viriditas, La goutte de pluie, Jojit fb, Hesperian, Jonathunder, Quaternion, Orangemarlin, Alansohn, Jeltz, Munchkinguy, Mysdaao, Wtmitchell, Eli the Bearded, Zantastik, Garzo, Tony Sidaway, Mikeo, T1980, Galaxiaad, Stemonitis, Velho, Richard Arthur Norton (1958-), Femmy~enwiki, Lochaber, AshishG, Terence, TotoBaggins, Fooby, Sengkang, Prashanthns, MarcoTolo, Greenmars, Graham87, BD2412, Kbdank71, Bunchofgrapes, Krymson, Rjwilmsi, Koavf, Matt.whitby, Aldeyan, Dzhango, Seraphimblade, Ligulem, Ricardo Carneiro Pires, Brighterorange, Maitch, Eubot, RobertG, PlatypeanArchcow, Celestianpower, RexNL, Srleffler, Stakeda, Socalmikey84, Chobot, Gdrbot, Bgwhite, Vmenkov, YurikBot, Wavelength, Jimp, JarrahTree, Reo On, Wisekwai, Killervogel5, Ivirivi00, Qwertzy2, IBook of the Revolution, SpuriousQ, Chaser, Hydrargyrum, Cplbeaudoin, Gaius Cornelius, Sacre, Joncolvin, Dysmorodrepanis~enwiki, Dat789, Badagnani, Janet13, Bukhrin, Sethos, Haoie, TDogg310, Gadget850, IainDavidson, AjaxSmack, Eurosong, Crisco 1492, Gsherry, Zargulon, Zzuuzz, TheMadBaron, Closedmouth, MarsJenkar, Sarefo, BorgQueen, Ekeb, LeonardoRob0t, Jaranda, Kaicarver, Bly1993, Fitness~enwiki, JDspeeder1, Alsharine, Cmglee, SpLoT, SmackBot, Saravask, KnowledgeOfSelf, Lawrencekhoo, Stifle, Eskimbot, Chych, HalfShadow, Peter Isotalo, Gilliam, Pstraten, Thesparrows, Kurykh, Rkitko, Thumperward, Jon513, ABC1357, Berton, Stevage, Silverelf, Tripledot, CSWarren, DHN-bot~enwiki, Colonies Chris, Rlevse, Verrai, Ekrenor, Wilybadger, Can't sleep, clown will eat me, Xiner, Arabiafish, Kingdon, Tapered, Bonecrushah, ColumbiaKid, Dcamp314, SeanAhern, DMacks, Powelldinho, Zzorse, Mion, Samuel Sol, Pilotguy, Qmwne235, Andrew Dalby, Dinkybarrel, Paul 012, HighwayCello, Euchiasmus, Khim1, Tomorts, RomanSpa, BillFlis, Wander apr, Yvesnimmo, PRRfan, Ryulong, Paukrus, Norm mit, Iridescent, Kencf0618, Daniel5127, SkyWalker, RSido, CmdrObot, Porterjoh, Paperdoll51, John Riemann Soong, Some P. Erson, Whereizben, Felixboy, Nilfanion, Cydebot, Hebrides, Daniel J. Leivick, Reijiro, Asenine, GangstaEB, Vibodha, Scarpy, Gimmetrow, Casliber, Thijs!bot, Epbr123, Dasani, GentlemanGhost, Sry85, Headbomb, Tonyle, RevolverOcelotX, Biyan-to, AlexanderM, Merbabu, Ongjyhseng, Dfrg.msc, Natalie Erin, Visik, Niduzzi, AntiVandalBot, Gioto, Luna Santin, Val2397~enwiki, Quintote, Adam11600, Leafeater, LegitimateAndEvenCompelling, Minun, JAnDbot, Deflective, MER-C, 1 Cent In Mind, Areaseven, Sophie means wisdom, Krasanen, TheEditrix2, Rothorpe, Kerotan, Naval Scene, Penubag, WolfmanSF, Murgh, Bongwarrior, Ling.Nut, Sodabottle, Andryono, IP Singh, Indon, Chinese.fatty, Torchiest, EquatorialSky, Johnkoh, SKULLSPLITTER, Ilovelucyblue24, Digi23azlan, Juansidious, Anaxial, Sport woman, Bus stop, CommonsDelinker, AlexiusHoratius, Gunkarta, Jarhed, Drfoop, J.delanoy, DrKay, Trusilver, Bongomatic, SiliconDioxide, MTLskyline, HiLo48, Erest, Zishanallibhai, MKoltnow, Danaidae, Juliancolton, Tangmo~enwiki, STBotD, Treisijs, Sypherin, S, Useight, DASonnenfeld, Idioma-bot, Chedz, Roddersg, Richardinho~enwiki, PeaceNT, Caspian blue, Lebart, VolkovBot, CWii, AlnoktaBOT, TXiKiBoT, Kraikk, Kww, Clarince63, GlobeGores, LeaveSleaves, Raryel, Wandering canadian, Yaan, Tanner-Christopher, Billgordon1099, Maladroitmortal, Tikuko, Enviroboy, WatermelonPotion, Insanity Incarnate, AaronD12, Oregoniansdoitintherain, Piekfrosch, MuzikJunky, Work permit, Peopledowhattheyoughttodo, DavisGL, Triwbe, Keilana, Flyer22, Gr33nmachin3, Jotracy21, Mimihitam, Hzh, Oxymoron83, Lightmouse, TheDailyKimchi, Efe, BorgKingIV, Genya Avocado, Tatterfly, Invertzoo, SallyForth123, Sfan00 IMG, De728631, ClueBot, Hippo99, The Thing That Should Not Be, Hongthay, Northdot9, Drmies, Gr0ff, Altenhofen, Othmanskn, Wikijens, Blanchardb, Takeaway, Axxand, Lartoven, Chezumar, Carl Francis, Saralonde, Rui Gabriel Correia, Mrjmcneil, Thingg, Aitias, NJGW, Qing Yi G, Vanished user uih38riiw4hjlsd, DumZiBoT, Wikicentral, XLinkBot, Johntabs, Durzatwink, Ariconte, Xxca, Addbot, DOI bot, Jaypark97, Herm555, Nukeguy04, Sidorak900, Ronhjones, Martindo, Vishnava, Jim10701, Download, AchromatReader, Ld100, Favonian, Blaylockjam10, Lime82, Pietrow, مانع, Rojypala, 白布 , Bessiefu, Darielquiogue, Nxtcrazy, Luckas-bot, Yobot, Chùn-hiàn, Yngvadottir, KamikazeBot, Mfbz78, Langthorne, 虞海, Diádoco, Jim1138, JackieBot, Xufanc, Materialscientist, Citation bot, Pomeapplepome, Haleyga, Frankenpuppy, Xqbot, ChildofMidnight, DSisyphBot, Abyssquick, Barriofiesta, Brandon5485, Shadowjams, Hamamelis, Weslwh, Citation bot 1, HRoestBot, Hamtechperson, SpaceFlight89, Tanzania, Mikespedia, Full-date unlinking bot, Banej, Turian, PacificWarrior101, FoxBot, Trappist the monk, Wotnow, Hannysog, Vrenator, Josegeographic, Victor Pogadaev, Leonster, Miracle Pen, Davidisere, Jynto, Raintreer, RjwilmsiBot, TjBot, EmausBot, Look2See1, Ajraddatz, Dewritech, مانی فا, Somemadman, Brendan42, Durianlover, Kmoksy, Maglame, Daonguyen95, Miras.ravi, Fæ, Ὁ οἶστρος, H3llBot, Bo1234567890, Frugivore, SporkBot, Surya Prakash.S.A., Music Sorter, Mayur, ChuispastonBot, Kleopatra, AlleinStein, ClueBot NG, Braincricket, Churchhands, Anakuching, Dean Turbo, Jacgyc, Kristendv, Helpful Pixie Bot, Vball199, Plantdrew, Alexis4444, Victorianarsyawondergirl2, AshunGurl, Northamerica1000, Declangi, MusikAnimal, Badon, Duriancrepe, Mark Arsten, CutieEzha, Ktitimbo, YVSREDDY, Whiteboardcheese, MHugh, MKwek, Aisteco, Muffin Wizard, BattyBot, Jxh298, Khazar2,

Midtownmonkey55, Cwobeel, Sminthopsis84, Jeremy Kuan Chern Eu, Numerousflaws, Furry tin cans, Graphium, Mr Teamo, Corinne, SON-NYPW, Faizan, Lightsaber45, Jodibus, 0987654321sd, Factfindest, Tentinator, Backendgaming, Alfkja, Ugog Nizdast, Casheymay, CarterRios, FDMS4, AfadsBad, Jackmcbarn, WPGA2345, Albertleeisqueenie, Sportsguy17, PRanXz, Allanhenderson1998, Magicloveisintheair, Monkbot, MrRainbows, Owais Khursheed, Junchuann, King of all fruit, Path2zero, Yuel0022, Hertz107, Debtang1019, Applezap4, Nurhayati1990, Bakebread and Anonymous: 594

- **Durio zibethinus** *Source:* https://en.wikipedia.org/wiki/Durio_zibethinus?oldid=672376737 *Contributors:* BorgQueen, Raz1el, Phil Bridger, 白布   , Trappist the monk, EmausBot, Plantdrew, Declangi, Dexbot, Sminthopsis84, KasparBot and Anonymous: 2

- **Eucalyptus oil** *Source:* https://en.wikipedia.org/wiki/Eucalyptus_oil?oldid=678133500 *Contributors:* Fubar Obfusco, Jpatokal, PFHLai, Rich Farmbrough, Bender235, Axl, Rjwilmsi, Waitak, JarrahTree, RussBot, Hydrargyrum, DAJF, Zagalejo, Asarelah, Rathfelder, Groyolo, Smack-Bot, Melchoir, Chris the speller, JDCMAN, Stepho-wrs, Zearin, Liam Skoda, CmdrObot, I.M.S., John Moss, DuncanHill, RebelRobot, Prestonmcconkie, WhatamIdoing, Enquire, Th84, DASonnenfeld, VolkovBot, Ktalon, Una Smith, Petteri Aimonen, Logan, Roo1812, Hordaland, Neznanec, Mild Bill Hiccup, SchreiberBike, GKantaris, Dthomsen8, Doug butler, Addbot, Colibri37, Luckas-bot, Bunnyhop11, Evilyoshimax, EryZ, Obersachsebot, Borys bond, FrescoBot, LucienBOT, Citation bot 4, Aareo, Zzzomg2, Tea with toast, Trappist the monk, TjBot, MrFawwaz, GoingBatty, Klbrain, John Cline, AManWithNoPlan, ClueBot NG, Ehrjej, TwoTwoHello, Binko100, SantiLak, P. S. Sena and Anonymous: 45

- **Fern** *Source:* https://en.wikipedia.org/wiki/Fern?oldid=685204316 *Contributors:* Magnus Manske, Josh Grosse, Nonenmac, Jaknouse, Liftarn, Gabbe, Menchi, Ixfd64, Lquilter, Ellywa, Pratyeka, Glenn, Cimon Avaro, Cratbro, Jallan, JCarriker, Tpbradbury, Marshman, Renato Caniatti~enwiki, Pakaran, Shafei, Robbot, Astronautics~enwiki, Altenmann, Arkuat, Hadal, UtherSRG, Wikibot, JerryFriedman, Pengo, Alan Liefting, MPF, Orangemike, Tom Radulovich, Mboverload, Just Another Dan, Wmahan, Gadfium, Utcursch, Alexf, Gdr, Antandrus, Onco p53, Evertype, Talrias, Neale Monks, Ngc1976, Mike Rosoft, Stupid girl, DanielCD, Kbh3rd, Vzb83~enwiki, El C, -jkb-, Art LaPella, Roy-Boy, Bobo192, Vervin, Fir0002, Arcadian, ParticleMan, Nk, Microtony, Alphax, Obradovic Goran, Jumbuck, Zachlipton, Qwe, AnnaP, Eric Kvaalen, Plumbago, Pion, Snowolf, Wtmitchell, Bsadowski1, Ghirlandajo, Zntrip, Stemonitis, Angr, Woohookitty, Mondhir~enwiki, Mrs Trellis, Tkessler, TheAlphaWolf, Mandarax, BD2412, Rjwilmsi, Pabix, SMC, Ucucha, MWAK, Mike Dallwitz~enwiki, Daderot, Latka, RexNL, Gurch, Jrtayloriv, Choess, M7bot, Chobot, Antilived, Gdrbot, Bgwhite, Dj Capricorn, Banaticus, Roboto de Ajvol, YurikBot, Wavelength, Huw Powell, Arjuna909, Splash, Aaron Walden, Hellbus, Stephenb, Thane, Anomalocaris, ENeville, Curtis Clark, Dysmorodrepanis~enwiki, Dforest, Thiseye, E rulez, Moe Epsilon, Dbfirs, WAS 4.250, Theodolite, Wildjourne, CWenger, Anclation~enwiki, Emc2, Kungfuadam, True Pagan Warrior, SmackBot, MicruBot, Brian1979, KnowledgeOfSelf, Sanjay ach, Vald, Bomac, EncycloPetey, Hardyplants, Canthusus, Zephyris, Gilliam, Skizzik, Jon513, Berton, DARQ MX, Paalexan, A. B., Rlevse, Can't sleep, clown will eat me, Abyssal, Sahmeditor, Rrburke, Sundar-Bot, Phaedriel, Illiquent, Dasunt, Smooth O, Kingdon, Bigturtle, Bansp, Richard001, Illnab1024, Cephal-odd, Kukini, Clicketyclack, Jomegat, MrDarwin, Ram32110, J. Finkelstein, Lapaz, This user has left wikipedia, Mgiganteus1, Smith609, Makyen, Ulysses Fiuza, Thrindel, Iridescent, JMK, Aperium, Newone, Bobim, Courcelles, Tawkerbot2, Dto, Rawling, Dr.Bastedo, Lmcelhiney, Philiptdotcom, Baskaufs, Brownlee, Gogo Dodo, Clovis Sangrail, DumbBOT, Olegivvit~enwiki, Rosser1954, Thijs!bot, Epbr123, Kilva, CopperKettle, Nudgnudge, Marek69, Doyley, RFerreira, Zachary, FERN EU, Dawnseeker2000, Escarbot, Majorly, Plantguy, KP Botany, Ste4k, Smartse, JAnDbot, Husond, Samar, Barek, Plantsurfer, PhilKnight, Affinis, Maias, SteveSims, Bongwarrior, VoABot II, Think outside the box, Michael Goodyear, Lucyin, Crazytonyi, Wertloo, PIrish, Animum, Kotoreru, DerHexer, Pyrochem, Nopira, Peter coxhead, NatureA16, MartinBot, Anaxial, CommonsDelinker, Boston, J.delanoy, Frinkus, Maurice Carbonaro, Captain Infinity, Lemur12, Charlez123, McSly, Betswiki, Coppertwig, Chiswick Chap, NewEnglandYankee, Shoessss, Cometstyles, ACBest, Treisijs, Pdcook, DASonnenfeld, Idioma-bot, Lights, Deor, VolkovBot, CWii, Thedjatclubrock, Chienlit, Vlmastra, Philip Trueman, TXiKiBoT, Tom612pl, Antoni Barau, Technopat, Miranda, Dendodge, Philli100, Jackfork, LeaveSleaves, BotKung, Krzysfr, Ferrari2503, Enviroboy, Burntsauce, Dessymona, Sue Rangell, Jimmi Hugh, PGWG, Ojoe2000, Gaelen S., Britzingen, Jauerback, Callipides~enwiki, Viskonsas, Caltas, RoRo, Happysailor, Flyer22, Radon210, Wilson44691, Oxymoron83, Nuttycoconut, Aspects, StaticGull, Troop350, Denisarona, Furado, Forest Ash, Faithlessthewonderboy, Ratemonth, ClueBot, Botanybob, NickCT, Jmgarg1, The Thing That Should Not Be, EnterpriseCrew, Cygnis insignis, Rotational, Neverquick, Excirial, Eeekster, Katiesolly, Sun Creator, Sbfw, Coccyx Bloccyx, La Pianista, Thingg, Versus22, Dana boomer, SoxBot III, Party, DumZiBoT, Abhijit borah, Fastily, Rror, Little Mountain 5, Avoided, Addbot, DOI bot, Jennabcool, DaughterofSun, Monanobi, Ronhjones, CanadianLinuxUser, Fluffernutter, BepBot, Moonsrock13, LinkFA-Bot, Numbo3-bot, Tide rolls, ماني, Middayexpress, Luckas-bot, Yobot, Fraggle81, TaBOT-zerem, THEN WHO WAS PHONE?, Tearanz, Eric-Wester, AnomieBOT, IRP, Ryanjoyce11, Carolina wren, Ulric1313, Milkdudman, Materialscientist, Citation bot, OllieFury, Bob Burkhardt, Grantabb, Elucidate, Sionus, Capricorn42, Shimmin Beg, Hannah 50, Mathonius, Geopersona, Doulos Christos, Eyepodd, Manuelt15, Shadowjams, Santăr, Dougofborg, Stirred-not-shaken, FrescoBot, VI, Schnobby, Citation bot 1, Cubs197, Pinethicket, I dream of horses, Comppsi, Jonesey95, Impala2009, Serols, Foobarnix, Bgpaulus, Transope, AGiorgio08, SkyMachine, FoxBot, Cark bakerclarkbaker, LogAntiLog, Miracle Pen, Guyonstreets, DARTH SIDIOUS 2, Dlyanelza, HOMER6969, Mean as custard, RjwilmsiBot, Noommos, DSP-user, EmausBot, Orphan Wiki, Look2See1, Roastedpepper, Assamsefood, Rajkiandris, Tommy2010, Persianseeker, Wikipelli, K6ka, Thecheesykid, Pipipoo231, Josve05a, Traxs7, Editing19473, Donner60, Keidax, TYelliot, David 5252, Mattmaul1, Kleopatra, Xanchester, ClueBot NG, Satellizer, A520, Wornwinter11, O.Koslowski, Asukite, Widr, GlassLadyBug, Calabe1992, Ramaksoud2000, Nas119, WNYY98, BG19bot, Mnbappy, Rijinatwiki, MusikAnimal, Ushakaron, CimanyD, Astpurcell, Writ Keeper, Tu7uh, MrBill3, Qwekiop147, Achowat, Cricetus, Lokinorsemyth, Jml3, T.seppelt, Darorcilmir, Cyberbot II, GoShow, YFdyh-bot, Dexbot, TheIrishWarden, Sidelight12, Yahye123123, VanishedUser 2313214sad1, Thisboylikeschicken, Epicgenius, Michipedian, Chemistryg112, DavidLeighEllis, ButterNuggets777, GetGarrisonned, Shearflyer, Jiyeonsuh, Νημινυλι, HalfGig, Coreyemotela, Mrasta4836, YOLOSWAGKING, Garrisonn, Chaya5260, Harrisonn, Monkbot, Gmoran182, Peter Smyth, Bob123543, Biblioworm, Grayz654, Taze x panda, Nartman99, Dylandass, Dylandass1, Lefinnen, MacPoli1, Stepbang, Ashraf.shopnil, Anoakes11 and Anonymous: 543

- **Forage** *Source:* https://en.wikipedia.org/wiki/Forage?oldid=683509440 *Contributors:* Rmhermen, Rbrwr, Docu, Bogdangiusca, Wik, Imc, Pollinator, Robbot, Nurg, H-2-O, Just Another Dan, Onco p53, DanielCD, Longhair, Remuel, Cohesion, Carbon Caryatid, Shoefly, Ricardo Carneiro Pires, Common Man, Wavelength, Epipelagic, Allens, Asterion, Myrabella, SmackBot, Nick Dillinger, Amatulic, Chris the speller, Elagatis, Snowmanradio, Abbott75, Bdiscoe, Montanabw, Burningflag, JAnDbot, Daniel Cordoba-Bahle, Rich257, Cgingold, TDCunm16, Ciotog, DA-Sonnenfeld, Richard New Forest, Philip Trueman, Supfatty, LeaveSleaves, Atubeileh, Ethel Aardvark, Twinsday, Gabacho2, Kembangraps, Wyatt915, Addbot, Lucian Sunday, Luckas-bot, Xufanc, Citation bot, Pkravchenko, Dehaan, Gumruch, Joxemai, BenzolBot, I dream of horses, HRoestBot, Eva Ekeblad, Glorioussandwich, PBS-AWB, ClueBot NG, Brunswick Dude, Helpful Pixie Bot, Mark Marathon, CitationCleanerBot, Milad, Anonymous1466, ChrisGualtieri, Yamaha5, Tortie tude and Anonymous: 37

- **Forest farming** *Source:* https://en.wikipedia.org/wiki/Forest_farming?oldid=625111178 *Contributors:* Skysmith, Lumos3, Alan Liefting, Bobo192, RJFJR, RHaworth, Nirvana2013, SmackBot, Bwpach, Bddmagic, Alaibot, Richhoncho, Fabrictramp, Gabriel Kielland, Gomm, Jeannie kendrick, DASonnenfeld, JL-Bot, Occur Curve, ClueBot, Auntof6, Aprock, MrOllie, AnomieBOT, Citation bot, Xqbot, Anna Frodesiak, Jonesey95, Isiaunia, The Ent, Tcazes, NGPriest, EdoBot, ClueBot NG, Northamerica1000, SFK2, Cjbukows, Laurieds, Harmoniclag, Olenyash, Monkbot, Jbanegas and Anonymous: 13

- **Forest gardening** *Source:* https://en.wikipedia.org/wiki/Forest_gardening?oldid=684764019 *Contributors:* Ray Van De Walker, Anthere, Quercusrobur, Lquilter, Stan Shebs, Artost, Glenn, Marshman, Vardion, Alan Liefting, Everyking, Bobblewik, Serendeva, Pgan002, Mike Rosoft, Chris j wood, Guanabot, Bender235, Eadmund~enwiki, Erauch, Sumalsn, Anthony Appleyard, Velella, Kazvorpal, Bobrayner, Rtdrury, Benjitz, Salix alba, Gaius Cornelius, Dialectric, Nirvana2013, Kevin, SmackBot, Cacuija, Lotusduck, Chris the speller, Bluebot, Brimba, Abrahami, Byelf2007, Dandelion1, SilkTork, Gobonobo, Rkmlai, DabMachine, Lograph, Doug Weller, Marek69, Ingolfson, Daniel Cordoba-Bahle, Sustainableyes, Skier Dude, Madbishop, Jorfer, Woodsguy, Scott Roy Atwood, DASonnenfeld, Lightmouse, Der Golem, Mild Bill Hiccup, XLinkBot, Edibleforests, Addbot, Granitethighs, Jarble, Bermicourt, Luckas-bot, AnomieBOT, Rubinbot, Citation bot, Anna Frodesiak, Legion23, BoundaryRider, Citation bot 1, Lotje, Vrenator, RjwilmsiBot, Look2See1, EME44, Mmeijeri, Lexandalf, ZéroBot, Popok75, Walter Ralt, ClueBot NG, PaleCloudedWhite, Helpful Pixie Bot, Philospelunk, Wbm1058, Gob Lofa, Lavenderdawn, Northamerica1000, Mr. Joca, Rowan Adams, Dexbot, Sminthopsis84, Lisamd, Bleu8, Yackityyack, Ginsuloft, EChastain, JoannaHoman and Anonymous: 51

- **Forest produce (India)** *Source:* https://en.wikipedia.org/wiki/Forest_produce_(India)?oldid=668868224 *Contributors:* Booyabazooka, Utcursch, Bobo192, RussBot, SmackBot, Gilliam, Shyamsunder, Skapur, Woodshed, Tawkerbot2, MoogleDan, Loaa~enwiki, Mkyadava, DASonnenfeld, Marcus334, Pradtke, KnowledgeHegemonyPart2, Sadie82, AnomieBOT, Minnecologies, Materialscientist, Anna Frodesiak, Erik9bot, ChrisGualtieri, Pankaj Oudhia, Main priyanshi and Anonymous: 11

- **Fur** *Source:* https://en.wikipedia.org/wiki/Fur?oldid=683572672 *Contributors:* Magnus Manske, Bryan Derksen, Rmhermen, Heron, Leandrod, Patrick, Fred Bauder, Menchi, Eurleif, Ahoerstemeier, Julesd, Glenn, Cimon Avaro, Timwi, Pedant17, Furrykef, Rossumcapek, Gentgeen, Robbot, Baldhur, Altenmann, Alan Liefting, Matt Gies, Gwalla, Elf, Vfp15, Ferkelparade, Peruvianllama, Alison, Bobblewik, Pehrs, Pgan002, Andycjp, Toytoy, Rosemaryamey, Scott Burley, Karl-Henner, Adashiel, Grunt, SYSS Mouse, Quill, Jiy, Kathar, Discospinster, Rhobite, Smyth, ESkog, El C, Femto, Smalljim, Viriditas, Pschemp, Pharos, Rolfmueller, Alansohn, Trysha, PatrickFisher, AzaToth, SlimVirgin, Patangay, Kesh, BanyanTree, Super-Magician, Heida Maria, Versageek, Japanese Searobin, -oo0(GoldTrader)0oo-, Chrissymatt, KaurJmeb, Marudubshinki, Ashmoo, BD2412, Vegaswikian, Feco, Fuzzyeric, Marquee~enwiki, Isotope23, Nimur, WikiWikiPhil, Schwern, Bgwhite, Lar, Chunitaku, NawlinWiki, Voyevoda, XamiXiarus, CarlFink, Iancarter, Aaron Schulz, Caerwine, Eli lilly, AmyBeth, FF2010, Wolle212, Chase me ladies, I'm the Cavalry, Tsunaminoai, ChemGardener, SmackBot, Ashenai, Moeron, Od Mishehu, Gjs238, Gaff, Donama, Ohnoitsjamie, Cgoodwin, Octahedron80, Colonies Chris, Foxjwill, Kotra, Can't sleep, clown will eat me, EuroTrash, VMS Mosaic, Bardsandwarriors, Brainshmain, Lila13~enwiki, BullRangifer, Will Beback, Xerocs, General Ization, Ian Spackman, Scientizzle, Phajj, Superbootneck, Bjankuloski06en~enwiki, MarkSutton, TastyPoutine, Danilot, Kaarel, Courcelles, Dlohcierekim, Dycedarg, KyraVixen, Ballista, Neelix, Coderoyal, Peridot9tikal, Gogo Dodo, Maximilian Schönherr, X201, Escarbot, Tangerines, Egpetersen, PhJ, JAnDbot, Xeno, The elephant, Petersonpet, Acroterion, Bongwarrior, Carlwev, Pvmoutside, Nyttend, Mwalimu59, Catgut, NatureA16, Hdt83, MartinBot, Xand3r, Uncle Dick, Sidhekin, Aqwis, Ignatzmice, Koven.rm, Povertycat, Tomy45, DASonnenfeld, Xwwxw, Almiospacio, Bovineboy2008, Philip Trueman, JuneGloom07, Fastback68, Benthackray~enwiki, Technopat, Una Smith, Martin451, Deranged bulbasaur, Fatty911, Munci, Trackinfo, Euryalus, Caltas, Bentogoa, Android Mouse, Oda Mari, Lmc169, Bob98133, Nuttycoconut, Hobartimus, Dillard421, Improvements For All, Habbo raider 4 life, Fangjian, ClueBot, Polentario, Rumping, Piledhigheranddeeper, Jusdafax, Kuerschner, Cenarium, JamieS93, Dekisugi, Thehelpfulone, Kakofonous, Lx 121, Olaf.gustafson, XLinkBot, Roxy the dog, Maky, Fede.Campana, Bogylux, Glassdooru, Doc9871, Noctibus, TravelinSista, Addbot, Boomur, Usadrumhead, Zacrayniak, Snapplewhitetea, LaaknorBot, Glane23, RussianFox, Panther4517, ILOVEPINK2008, Tide rolls, Jarble, Putnik, Legobot, Luckas-bot, Yobot, TaBOT-zerem, BuckwikiPDa535, Nallimbot, Maxí, Kürschner, Itwastrees, Piano non troppo, Wolvenmoon, Materialscientist, Bob Burkhardt, GB fan, Quebec99, Mariomassone, Omnipaedista, Kyng, Waldo1961, Shadowjams, Schekinov Alexey Victorovich, Joaquin008, Dougofborg, FrescoBot, Recognizance, Booppoop234, Tintenfischlein, Pinethicket, I dream of horses, Hamtechperson, Jschnur, Impala2009, White Shadows, P00p1515, Kelly2357, Vrenator, TBloemink, Reaper Eternal, Diannaa, Suffusion of Yellow, Blueseadream, Zachzolty, Itrucid, Immunize, I love animals62, Benisacat, Heyheyheyhey4, Wikipelli, PBS-AWB, Mh7kJ, Brandmeister, Coasterlover1994, ChuispastonBot, JonRichfield, Editor3445, ClueBot NG, Sw2nd, Widr, DrChrissy, Theopolisme, Jk2q3jrklse, MerlIwBot, Furkhaocean, AvocatoBot, Mark Arsten, Fvrconstant, Nuaryolcay, Medieval furrier, GoShow, Timelezz, Hwqutree, WookieDaisy, Dexbot, Barrrrrr, Infopediagirl, Reatlas, Nuclear Optimism, ZGorilla, Roberta jr., Ylyudmil, Hellogangstagangsta, 19sweibler, Belchior90, KasparBot, Catsssssss and Anonymous: 233

- **Game (hunting)** *Source:* https://en.wikipedia.org/wiki/Game_(hunting)?oldid=673697168 *Contributors:* Rmhermen, Edward, Lexor, Norm, Jimfbleak, David Thrale, Marshman, Gentgeen, Robbot, Meelar, Darrien, Zondor, Dr.frog, AliveFreeHappy, EugeneZelenko, Guanabot, Dbachmann, Harriv, Janderk, Redlentil, Quintucket, Carbon Caryatid, Cdc, Wdfarmer, Gene Nygaard, Mindmatrix, MONGO, Miss Madeline, Tabletop, Graham87, BD2412, Kanadier~enwiki, I know who my sister is., Sjö, Dosowski, Ucucha, Syced, Astatine, RexNL, Quuxplusone, McDogm, Benjwong, Ariele, YurikBot, RussBot, Pigman, IanManka, TheGrappler, Shanel, R'son-W, Epipelagic, Cinik, GraemeL, Nessuno834, Kungfuadam, Rooivalk, That Guy, From That Show!, Yvwv, SmackBot, Wcquidditch, McGeddon, Eskimbot, Rojomoke, Yellowbounder, Sloman, Hmains, Thumperward, Ghemesh, Rorybowman, Bazonka, Yaf, Muckapedia, Smooth O, Abbott75, SirIsaacBrock, Gobonobo, Mgiganteus1, Bjankuloski06en~enwiki, Seanoquinn, Maksim L., Texas Dervish, Floridan, DabMachine, MikeHobday, Courcelles, Steve64, Greenfinch100, ShelfSkewed, Gogo Dodo, Ebyabe, Marek69, Wikidenizen, KrakatoaKatie, Jayron32, Nipisiquit, Glacierfairy, JAnDbot, QuantumEngineer, Lucyin, SwiftBot, Thernlund, Cpl Syx, Atulsnischal, Timmah48, Damien Shiest, Fleebo, Greatestrowerever, Ja 62, DASonnenfeld, Squids and Chips, VolkovBot, TheMindsEye, Naysie, Anonymous Dissident, Aymatth2, Mardhil, InMooseWeTrust, CMBJ, Cransdell, Mycomp, BotMultichill, Bob98133, Jóhann Heiðar Árnason, Sugarcubez, Dust Filter, Tony9wuty, Faithlessthewonderboy, Myth010101, ClueBot, TableManners, RafaAzevedo, UlitmateLoserNo1, Elephantissimo, Arjayay, SchreiberBike, ChrisHodgesUK, DumZiBoT, Maky, Ost316, MystBot, Addbot, Mortense, LaaknorBot, ChenzwBot, Lemonade100, The PeteMan, Zorrobot, Pointer1, Luckas-bot, Yobot, AnomieBOT, Andrewrp, Dinesh smita, Frscght, Anna Frodesiak, Haselbrunner278, FrescoBot, Tobby72, ReigneBOT, Timmyfreaker, CarstenHermann, Jamesslong, MastiBot, Vinay84, EmausBot, WikitanvirBot, RenamedUser01302013, Dcirovic, ZéroBot, H3llBot, Erianna, Rcsprinter123, L Kensington, Carsten R D, Donner60, Il Saggiatore, ClueBot NG, CopperSquare, Morgan Riley, Gob Lofa, Plantdrew, Northamerica1000, MrBill3, ChrisGualtieri, Aliwal2012, Kevinzhang27, SFK2, Lupus Bellator, Epicgenius, Bahooka, Mat-Moel, Comp.arch, Ilabsboy, Wes52342, Riot25, Editor abcdef, 字板 and Anonymous: 157

Sir Jamie2, WerdeMikan, Pinethicket, I dream of horses, Rushbugled13, A8UDI, RedBot, Ongar the World-Weary, FormerIP, Wikihenna, Masyaina, Smhenna2004, Jikybebna, FoxBot, Trappist the monk, Umlfl, Lotje, Callanecc, MarkGT, Vrenator, Ivanvector, Between My Ken, RjwilmsiBot, DexDor, Regancy42, Donnet info, Rajettan, EmausBot, John of Reading, Super48paul, AlphaGamma1991, Capostrophe Jones, Soulspothenna, AsceticRose, AvicBot, MithrandirAgain, The Nut, Azuris, Myhaseeb1, Makecat, Furries, Gz33, Wayne Slam, Tolly4bolly, L Kensington, Donner60, PunitSep3097, Puffin, Dineshkumar Ponnusamy, ChuispastonBot, Hina.sabreen, ClueBot NG, Horoporo, Pebble101, LA Henna, Rtucker913, Hurremyamakoglu, Aksamary, Widr, VLewis1025, Pluma, Helpful Pixie Bot, Plantdrew, Dead Mars, BG19bot, Caitsmith21, Mdpurple, TiagoEspinha, Vincent Liu, OttawaAC, Tu7uh, BattyBot, Prof. Squirrel, Khazar2, HelenWA4711, Mogism, Makecatbot, Lugia2453, Graphium, Jonhope123, Fmc47, Bluemaruti, Epicgenius, Mmm999~enwiki, Latestlifestyles, ADZQ90, Pooo666, Melaphyre, Janellwashere, Hennasaif4lyf, Musicnotes2001, SJ Defender, Liz, Bodroom, Akappe, Deepcruze, HennaServices, JaconaFrere, 7Sidz, Monkbot, Ifaddish, Rohitchhibber420, MariaRodrgz, Vieque, Rosie227, Josephmaroon, ContentKing01, Secretkeeper12, Shivani Henna Art, Sandipmsu, Zakariabenx, Cakepoop, Vaqas Ahmed, Josh9007, FreeatlastChitchat, Rubbish computer, Twinkle Parekh, Joseph2302, Jasminelehal, BBStijn, HennaAddict, Exacrion, Deepa Topiwala, Elioun, Endless90*, Alyssac210, Kalyanijainapally, Keyan890, A4Aarthi and Anonymous: 453

- **Honey** *Source:* https://en.wikipedia.org/wiki/Honey?oldid=686416460 *Contributors:* WojPob, Bryan Derksen, Ap, Alex.tan, Brant Boucher, Karen Johnson, Ortolan88, Roadrunner, SimonP, Heron, Montrealais, Sfdan, Fxmastermind, Someone else, Chuq, Spiff~enwiki, Edward, D, JohnOwens, Michael Hardy, Stormwriter, Llywrch, Dominus, Menchi, Ixfd64, Frank Shearar, Ahoerstemeier, Lovely Chris, Ronz, William M. Connolley, Theresa knott, Jebba, Александър, Pratyeka, Rossami, Andres, Samw, Focus mankind~enwiki, Dyss, RodC, Gutza, Markhurd, Tpbradbury, Taxman, Jimbreed, Ed g2s, Bevo, Shizhao, Wetman, Ortonmc, Jerzy, Pollinator, JorgeGG, Rossumcapek, Shantavira, Phil Boswell, Gentgeen, Robbot, Ee00224, Postdlf, Pingveno, Rfc1394, Auric, Rhombus, Hadal, Wayland, Znode, Dina, Argasp~enwiki, Parasite, DocWatson42, MPF, Kbahey, Dinomite, Mintleaf~enwiki, Drewzhrodague, Tom harrison, Martijn faassen, Lupin, Peruvianllama, Everyking, Moyogo, Jfdwolff, T0m, Dawidl, Sundar, Moogle10000, Gyrofrog, R. fiend, Yath, Lode~enwiki, Antandrus, Beland, Piotrus, GeoGreg, Karl-Henner, Blue387, Neutrality, Dcandeto, Kasreyn, Adashiel, Thorwald, Jfpierce, Mike Rosoft, Dr.frog, Heegoop, Eyrian, DanielCD, Dcfleck, Jiy, EugeneZelenko, Discospinster, Rich Farmbrough, Chammy Koala, Roybb95~enwiki, Alistair1978, Night Gyr, Bender235, ESkog, MisterSheik, CanisRufus, FirstPrinciples, MBisanz, Hayabusa future, Cafzal, Svdmolen, Fir0002, Petronivs, Palmiro, Scott Moore, Brutulf, Justinsomnia, Darwinek, Eritain, Sukiari, AppleJuggler, Jonathunder, Pogo747, Ranveig, Jumbuck, Alansohn, Gerweck, EvanGrim, Arthena, GrantNeufeld, Revmachine21, Fuzlogic, Iris lorain, Celzrro, Tchalvak, Mbloore, Wtmitchell, Melaen, Velella, Danntm, Shoefly, Sumergocognito, Wadems, Versageek, Gene Nygaard, Alai, Ghirlandajo, Dan East, Kouban, Netkinetic, HenryLi, Kazvorpal, Yurivict, Jimgeorge, Feezo, Simetrical, OwenX, Woohookitty, Mindmatrix, TigerShark, Hunding, StradivariusTV, Pol098, Amigadave, Matijap, Zach Alexander, AshishG, Tbc2, Clemmy, Bbatsell, GregorB, Evilmoo, Apavlova, MarcoTolo, DavidFarmbrough, GraemeLeggett, SausMeester, Allen3, Mandarax, Sherpajohn, Ashmoo, Graham87, BD2412, FreplySpang, Davogones, Pmj, Kane5187, Sjö, Rjwilmsi, Coemgenus, Fieari, Koavf, Ian Page, Ae77, Quiddity, Linuxbeak, Fish-Face, OmegaWikipedia, Ricardo Carneiro Pires, Brighterorange, The wub, Bhadani, MarnetteD, Exeunt, FayssalF, FlaBot, AED, Nihiltres, Isotope23, Sean WI, Ppk80, RexNL, Gurch, Mitsukai, Bigdottawa, Penguin, Cyko149, Thndr333, Zotel, King of Hearts, Jared Preston, DVdm, 334a, Bgwhite, Digitalme, Vmenkov, PointedEars, EamonnPKeane, Roboto de Ajvol, The Rambling Man, YurikBot, Wavelength, Sortan, Jimp, RussBot, RadioFan2 (usurped), Stephenb, Manop, Perodicticus, Gaius Cornelius, Shaddack, Lovesick, Thane, Anomalocaris, K.C. Tang, NawlinWiki, Shreshth91, SEWilcoBot, Dysmorodrepanis~enwiki, Naushad~enwiki, Wiki alf, Bloodofox, Welsh, Mhartl, The Grot, Cleared as filed, Renata3, THB, Abexy, Silvery, Larsinio, Haoie, LexieM, SameerKhan, Squally, Zephalis, DeadEyeArrow, Bota47, Ke5crz, Elemtilas, Igiffin, Silverchemist, Closedmouth, Dale662, Josh3580, BorgQueen, Diddims, LeonardoRob0t, Natgoo, Anclation~enwiki, Allens, Appleseed, Moomoomoo, Meegs, Stepan~enwiki, NeilN, Luk, Av.P, Attilios, Kenji Toyama, SmackBot, FocalPoint, Nicolas Barbier, Fireworks, Hanchi, McGeddon, Allixpeeke, Jagged 85, Bwithh, Anastrophe, WayneConrad, Kintetsubuffalo, Mintpieman, Joshfriel, Peter Isotalo, Gilliam, Ohnoitsjamie, Kaiwen1, Dingar, Jonathonbarton, Dark jedi requiem, Durova, Thebrainkid, Anwar saadat, Tyciol, Smileyborg, Persian Poet Gal, Dwinetsk, DennisTheTiger, JennyRad, Jprg1966, Jeekc, Elagatis, Raasnoerd, MalafayaBot, SchfiftyThree, Goldfinger820, Mrarfarf, Deli nk, Uthbrian, WeniWidiWiki, Szepattila, Epastore, Oatmeal batman, Scwlong, Can't sleep, clown will eat me, Frap, Onorem, KevM, Rrburke, Wes!, VMS Mosaic, ARA, Celarnor, Krich, Flyguy649, Gabi S., Model Citizen, Cloud02, Jaibe, Nakon, TedE, DChetkovich, Jiddisch~enwiki, Blake-, WookMuff, Yulia Romero, Tomtefarbror, Rockrapdude, Zexarious, Honey101, Kahuroa, Ck lostsword, Bezapt, GameKeeper, Keyesc, Ohconfucius, SashatoBot, Lambiam, ArglebargleIV, Rklawton, Zearin, Dbtfz, The idiot, Kipala, SilkTork, Jakobat, Hemmingsen, Minna Sora no Shita, Peterlewis, Bilby, Stoa, LebanonChild, A. Parrot, Slakr, Boomshadow, Renwick, Martinp23, Malachite36, Godfrey Daniel, Bendzh, Dicklyon, MrArt, Mathsci, Supacabs, Jrt989, Jose77, Dodo bird, Peyre, Aloysius XII, Wikixoox, Hu12, Japoniano, Alan.ca, Iridescent, Pwforaker, Lakers, T.O. Rainy Day, Shoeofdeath, Evildoctorbluetooth, Baloglu, DavidOaks, NoQuarter, Tawkerbot2, Ellin Beltz, Chetvorno, Fvasconcellos, HDCase, JForget, Rambam rashi, Wafulz, Dycedarg, Chrumps, Brainslug, Rwflammang, Dgw, ShelfSkewed, Meodipt, Pablo180, Cydebot, NealIRC, Vanished user vjhsduheuiui4t5hjri, Gogo Dodo, JFreeman, Rishodi, Elustran, Zeno Izen, Tletnes, Odie5533, Shirulashem, Doug Weller, Roberta F., Loki125, Dooly00000, Vanished User jdksfajlasd, Dyanega, Gimmetrow, LukeS, Thijs!bot, Epbr123, Bigwyrm, Tomcwheeler, Emily~enwiki, Mojo Hand, Marek69, Second Quantization, Jacobshaven3, Bksimonb, Sivazh, BalfourCentre, Davidhorman, Aquilosion, Benqish, Natalie Erin, Jpsaleeby, RoboServien, Escarbot, Hmrox, AntiVandalBot, The Obento Musubi, Majorly, Luna Santin, Seaphoto, SummerPhD, Nelsonc5, Tchoutoye, Shui9, Whytecypress, Kristoferb, Usman.shaheen, Lfstevens, Canadian-Bacon, Res2216firestar, .alyn.post., JAnDbot, Deflective, Husond, Bundas, MER-C, Dream Focus, Aviatophobiac, Y2kcrazyjoker4, Stardotboy, .anacondabot, Parsecboy, The Norse, Bongwarrior, VoABot II, Fusionmix, J.P.Lon, GearedBull, Kajasudhakarababu, Mbc362, Decembermouse, CameronB, Twsx, Brusegadi, Catgut, Jimjamjak, Theroadislong, Jessicapierce, Sgr927, Loonymonkey, 28421u2232nfenfcenc, Boffob, Hamiltonstone, Cpl Syx, Thibbs, DerHexer, Edward321, Garik 11, Ryandsmith, Squarefaced, NatureA16, B9 hummingbird hovering, S3000, Yobol, Hdt83, MartinBot, Grandia01, BetBot~enwiki, Ariel., MikePhobos~enwiki, 1staroundtheworld, Numero4, Jpyeron, Gidip, ColorOfSuffering, Erikun, R'n'B, Verdatum, Fconaway, Thewallowmaker, Proabivouac, Tgeairn, AlphaEta, J.delanoy, Pharaoh of the Wizards, Trusilver, Bongomatic, Arrow740, Rbrewer42, Numbo3, Keithkml, Boghog, Piercetheorganist, Uncle Dick, Extransit, Floros, Derwig, Acalamari, 999mal, Bot-Schafter, Squad51, McSly, Clerks, Naniwako, Davandron, Toothrees, Laura H S, Twyman1988, Kelvin599, AntiSpamBot, Spinach Dip, Richard D. LeCour, Ljgua124, Pundit, Hanacy, Sunderland06, Juliancolton, Entropy, Jamesontai, Atama, Vanished user 39948282, DorganBot, Zzzronnyzzz, DASonnenfeld, RjCan, Xiahou, Idioma-bot, Vranak, Meiskam, VolkovBot, Catwhoorg, TreasuryTag, Johan1298~enwiki, ICE77, Carter, Leebo, Jeff G., JoeDeRose, AlnoktaBOT, Kansaisamurai, QuackGuru, Wolfnix, Philip Trueman, Stealthbreed, Martinevans123, Alkanbalkaya, Summersg, TXiKiBoT, Joopercoopers, Davehi1, Ederny, OShunKarma, Hypnopomp, Someguy1221, Muleattack, Patrikk07, Gekritzl, Digby Tantrum, Dlae, MDSL2005, Raymondwinn, Elchip, Ilyushka88, Trilokeshwar Shonku, Deranged bulbasaur, RiverStyx23, Goodbye 2, Nazar, Uannis~enwiki, Moshi1618, Weather333, Andy Dingley, Finngall, AronR, Wolfrock, Mspritch, Enviroboy, Robatwikipedia, Why Not A Duck, Bobo The Ninja, Doc James, AlleborgoBot, Jungegift, CMBJ, Mehmet Karatay, PeterBFZ, Charles noooo, WassermeloneMan, SieBot, Love-

Gigs, Rojomoke, Aaronproot, Kintetsubuffalo, Virdi, Ohnoitsjamie, Rkitko, Deli nk, Shane.julian, Cjk91, MrPMonday, Smokefoot, BiggKwell, Vina-iwbot~enwiki, Takowl, U-571, Tktktk, Myopic Bookworm, Paul venter, WeggeBot, Cydebot, Ntsimp, Damianrafferty, Dasani, Pepperbeast, Calathan, JHFTC, JAnDbot, 100110100, Zack2007, LittleOldMe, Magioladitis, Faizhaider, Lucyin, Lenticel, Jerem43, EricTheise, David matthews, CommonsDelinker, Maproom, Acalamari, Benjamint444, DadaNeem, DASonnenfeld, Ultimate Death, Mirrordor, AlnoktaBOT, Feroshki, Matahari Pagi, Derekawesome, Atubeileh, SieBot, Sylverfysh, Vodnokon4e, Billyg, Animeronin, Ottawahitech, GUYTONIAN, IthinkIwannaLeia, Lalitstar, Magnetic Rag, Ginbot86, EdChem, Dthomsen8, Snowmonster, Marchije, Addbot, Jacopo Werther, Ben10027, DOI bot, Anishviswa, Morning277, Debresser, Numbo3-bot, Flakinho, Bff, Luckas-bot, Yobot, Ptbotgourou, Gabe brady, AnomieBOT, Piano non troppo, Materialscientist, Citation bot, Xqbot, Ignusb, Gigemag76, Programming gecko, GrouchoBot, RibotBOT, Bigger digger, Migpi, GreenZmiy, Recognizance, Spindocter123, DrilBot, Micromesistius, Sct3030, Boricuamark, EmausBot, Ajay Dotar Sojat, Ebe123, ZxxZxxZ, Slightsmile, Erpert, AshrafQuraishi, Medeis, Erianna, ChuispastonBot, Hungda, Kleopatra, Michael Bailes, ClueBot NG, Benzshkh, Widr, Beorhtwulf, Plantdrew, RobMarvin, BG19bot, MKar, Declangi, Weblars, Kurt91k, TheNuszAbides, ChrisGualtieri, Kpmsrikanth, Dexbot, Sminthopsis84, Ellis Novak, Vieque, Napijsc, Sanskari and Anonymous: 112

- **Jackfruit** *Source:* https://en.wikipedia.org/wiki/Jackfruit?oldid=686084461 *Contributors:* Shyamal, Ixfd64, Ahoerstemeier, Jimfbleak, KayEss, Mxn, Xam~enwiki, RodC, Zoicon5, IceKarma, Tpbradbury, Marshman, Imc, Hyacinth, C Fenijn, Karukera, Jose Ramos, Samsara, Bwmodular, THSlone, Altenmann, Kowey, Yosri, Davidcannon, MPF, Nickdc, Everyking, ManicParroT, Jorge Stolfi, Ragib, Pinnecco, Utcursch, Andycjp, Creidieki, Neutrality, Wadsworth, Squash, Dr.frog, Random contributor, Discospinster, Xezbeth, Bender235, CanisRufus, Kwamikagami, Hayabusa future, Guettarda, Bobo192, MANOJTV, Reinyday, Gingko, Nesnad, RaffiKojian~enwiki, Espoo, Ixfalia, Alansohn, Gary, Ibn zareena, Guy Harris, The.lawrd, LRBurdak, Snowolf, Steveo83, Yuckfoo, Lerdsuwa, Pauli133, Axeman89, Woohookitty, MGTom, Bluesleeper, Sengkang, Isnow, SDC, Fxer, Sikandarji, BD2412, Mancunius, Rjwilmsi, Angusmclellan, Ricardo Carneiro Pires, Eleazar~enwiki, SanGatiche, Vuong Ngan Ha, Avocado, Xero, Eubot, Harmil, Sahana, Prophet121, Alphachimp, Gdrbot, Bgwhite, Vmenkov, Wavelength, Rejith, Gaius Cornelius, Rsrikanth05, NawlinWiki, Seb35, Voyevoda, Badagnani, Welsh, E rulez, JPMcGrath, SameerKhan, Skbhat, BOT-Superzerocool, Wujastyk, Tux the penguin, Dspradau, Sanyarajan, JoanneB, Chriswaterguy, Katieh5584, Jade Knight, Tom Morris, KasugaHuang, Junyi, Parvez gsm, SmackBot, Thierry Caro, Cdogsimmons, Sssatyan, Od Mishehu, Kintetsubuffalo, Tomh009, Skizzik, Carl.bunderson, Chris the speller, Rkitko, T3mujin, DHN-bot~enwiki, Hongooi, Mike hayes, Brimba, OrphanBot, GRuban, SundarBot, Caniago, Peaceduck, Andrew c, Paul 012, Titus III, Jidanni, Ortho, Saigon punkid, Miltonzs, Kaviswarna, Vikasapte, Pjrm, Manojb, Sinaloa, PhilipDM, Courcelles, Radiant chains, Tau'olunga, Daniel5127, Vanisaac, CmdrObot, Pramesh.poudel, Zureks, Rafael Archuleta, InspiredLight, Sreekumarr, Corp1117, CumbiaDude, Cydebot, Eroach, Ntsimp, Gogo Dodo, MysticMetal, Jeff forssell, JohnInDC, הספרן, Casliber, Thijs!bot, Epbr123, TonyTheTiger, Pepperbeast, Marek69, Basu69, Som123, Geordles, Tellyaddict, Rajaramraok, Marco Schmidt, Nick Number, Erik Neves, Escarbot, Val2397~enwiki, Julia Rossi, Trakesht, Markusbradley, Hazel Grace, Leuko, Husond, Davewho2, Barek, MER-C, Plantsurfer, Kaidenx, Roidroid, Acroterion, Anoop anooprs, EvilPizza, VoABot II, Mrund, JamesBWatson, Lelkesa, Jarene, RetypePassword, Ravindiran, Salsa man, Quanticle, R'n'B, Nono64, Ashwan, Gunkarta, Schlumpff, Tgeairn, Artaxiad, Thaurisil, Wikramadithya, Btouburg, Maproom, Notreallydavid, Kesal, NS Zakeruga, Belovedfreak, Nadiatalent, KylieTastic, Treisijs, DASonnenfeld, Xiahou, Levydav, Idioma-bot, Funandtrvl, Caspian blue, VolkovBot, Sporti, Chango369w, TXiKiBoT, Hqb, T.sujatha, Treeears, Jmac0585, Suniltg, Wiikipedian, Jack Garfield, Ajursha, Maboughey, Jackfork, LeaveSleaves, Barquentine, McM.bot, Suriel1981, Jaguarlaser, Sentryman101, Entirelybs, AlleborgoBot, SieBot, Chintancd, Psbsub, Lucasbfrbot, Xenobiologista, Kmhasanat, Alexbrn, Jvs, MaynardClark, Oda Mari, Hzh, Mankar Camoran, Helikophis, Manway, Gordonofcartoon, Aravind V R, Reubentg, Superbeecat, Cookmughlai, Kanonkas, Forest Ash, RockyAlley, ClueBot, TwoHundredOk, The Thing That Should Not Be, Awesomebitch, Zfc89, Av99, WDavis1911, Mild Bill Hiccup, Aggarsandeep, Hafspajen, CounterVandalismBot, Niceguyedc, Crazypersonbb, StigBot, Reeloo, DragonBot, Takeaway, Opcassio~enwiki, Gateaudenoel, Jotterbot, Carriearchdale, Vtalbot, Briarfallen, SoxBot III, Vanished user uih38riiw4hjlsd, Operationrandomfix, Kinglaw, BarretB, Jytdog, Sophie1979, Ost316, Skarebo, GayanRS, Thatguyflint, Kembangraps, Wyatt915, Kajabla, Addbot, Proofreader77, JBsupreme, Adjaq, Tcncv, Srimanta.Bhuyan, AkhtaBot, Donwen, Yohannvt, Btheb, NjardarBot, Download, Musicollector, Lihaas, Favonian, SpBot, TheHamburger, Numbo3-bot, Ben Ben, Galanga, Luckas-bot, Nyanatusita, Yobot, Worldbruce, Legobot II, Amirobot, Annatanama, KamikazeBot, 虞海, AnomieBOT, Sanfy, Xufanc, Kingpin13, Algorithme, Cerme, Mahmudmasri, Materialscientist, KevinBaulch, ArthurBot, Xqbot, TinucherianBot II, Frank juhas, Anon423, Ryangiggs69, Ani medjool, J04n, Jayachandrank, RibotBOT, RomanHunt, Sasikiran 10, Eugene-elgato, Hamamelis, Griffinofwales, Green Cardamom, FrescoBot, Denatured Alcohol, Hasiru, Hridayalu, HamburgerRadio, Krish Dulal, Pinethicket, HRoestBot, Faerra, Jonesey95, King Zebu, Btilm, Nijgoykar, Raamah, Chethankg, Mauritianpride, Lotje, Zvn, Albiorix123, Emjaymem, Umesh119, ZhBot, Cowlibob, Codobai, Ivanvector, Avedeus, RjwilmsiBot, Alph Bot, Abbasjnr, Dr PL Narayanan, EmausBot, WikitanvirBot, T3dkjn89q00vl02Cxp1kqs3x7, Heracles31, Ramon FVelasquez, GoingBatty, Winner 42, Kmoksy, ZéroBot, Platypus333, Langra, Muneerpnm, Lateg, Anir1uph, Hassan usmanipak, Prabinepali, Staszek Lem, Jsayre64, EdwinAmi, Kratvej4, Nayansatya, Aznaturalist, ClueBot NG, Subhash ict003, Getauvi, Euglossine, FusionLord, Lolo Sambinho, Widr, Beayeem, Theopolisme, Oddbodz, Healthoz, Titodutta, Plantdrew, Jack nageon, BG19bot, NusHub, M0rphzone, PhnomPencil, Jackfruitlove, AndiNell, Jahnavisatyan, Yowanvista, Innertruth108, Trevayne08, CitationCleanerBot, ASCIIn2Bme, Krupasindhu Muduli, Sbblr geervaanee, Tu7uh, Lekro, Cottonflop, MKwek, Vanischenu, Chandana12, Retnuh66, Jewel457, Ayanwiki, Bij021, Keshav08, BattyBot, Justincheng12345-bot, ChrisGualtieri, Ancienzus, TylerDurden8823, Jethro B, Winarto km, Earth100, Mogism, Manishteo, Ricardopena15, Sa.feroz, Corinne, Syedshahidabbas, Rheetuz, Mydreamsparrow, Ramcram, Howicus, Dinesh Poudel, Sakthiprasanna, Tentinator, DavidLeighEllis, Himali liyanage, Gihan Jayaweera, Pratibha pandit, AfadsBad, NutrientGirl, Cboning, Trần Anh Mỹ, Lakun.patra, Kaushalie dharmarathna, Magicloveisintheair, Monkbot, Filedelinkerbot, Amit Diwakar Pandey, Mmukndh, Vinegarymass911, Vijayganesh.s1996, Oiyarbepsy, UPKS, Maheshwar7, Ashwinkudkuli, Debtang1019, The best ever386, NuclearNakedMoleRat, Afizere7, Tormipro, Equinox, FRUIT, Deshitech, HelloDaally, Tusiel, Rumi Senthil and Anonymous: 510

- **Japan wax** *Source:* https://en.wikipedia.org/wiki/Japan_wax?oldid=673014456 *Contributors:* Quartertone, TheBlunderbuss, Shaddack, SmackBot, Stevegallery, Smokefoot, Plantsurfer, STBot, DASonnenfeld, Tonkawa68, Addbot, Erik9bot, Pinethicket, Plantdrew, NotWith, Scienceenforcer and Anonymous: 9

- **Juniper berry** *Source:* https://en.wikipedia.org/wiki/Juniper_berry?oldid=667614412 *Contributors:* Enchanter, Ixfd64, Ebruchez, Wetman, HaeB, MPF, ComaVN, Jason Quinn, OldakQuill, RetiredUser2, Rich Farmbrough, Bobo192, Circeus, Alansohn, Keenan Pepper, Ish ishwar, Walshga, Sandover, Bunchofgrapes, Ttwaring, WriterHound, Noclador, Wavelength, SkyCaptain~enwiki, Pigman, Calicore, Curtis Clark, Astral, Veledan, DAJF, Lockesdonkey, Asarelah, Smaines, BorgQueen, DVD R W, SmackBot, Melchoir, Kintetsubuffalo, Here.it.comes.again, Nakon, BullRangifer, MrDarwin, Iridescent, Pukkie, ShelfSkewed, Cydebot, Kozuch, JorgonQ, Antique Rose, JAnDbot, Jdcook, Natureguy1980, Fradlin, VoABot II, Rarian rakista, McSly, DASonnenfeld, Salvar, Lamro, BotanyBot, Mr.Sourcebook, Radon210, ClueBot, Justin W Smith,

Benwedge, Myrtone86, Raj Fra, Tawkerbot2, Jh12, Winston Spencer, HDCase, Mapsax, Sakurambo, CmdrObot, FunPika, Scohoust, Iced Kola, Dgw, ShelfSkewed, Moreschi, Ken Gallager, Maxxicum, Slazenger, Cydebot, Samuell, Lesqual, Ryan, Carboncopy, Balrog30, Nick Wilson, Gogo Dodo, Boardhead, Bazzargh, Moonslight, Doug Weller, DumbBOT, Septagram, Brad101, Greenboxed, Aldis90, Saintrain, Cuziyq, Casliber, LilDice, JamesAM, Thijs!bot, Epbr123, Mushoku, Andyjsmith, Oliver202, John254, VinnieF, AgentPeppermint, Blathnaid, Ericjs, Oreo Priest, Mentifisto, AntiVandalBot, Majorly, Pfist3r, Seaphoto, Leena, Spencer, Gökhan, Canadian-Bacon, Leuko, Ronjamin, Ericoides, Katiemur, Fweezle, GurchBot, Denimadept, Connormah, WolfmanSF, Celithemis, Bongwarrior, VoABot II, Nyq, Brianjames12890, JamesB-Watson, Gezellig, GearedBull, Cadsuane Melaidhrin, Tedickey, Objectivesea, JMBryant, Rusty Cashman, Albert85~enwiki, Dirac66, Boffob, JMyrleFuller, Spellmaster, WLU, Drm310, Jerem43, MartinBot, Knightskye, TechnoFaye, R'n'B, Nono64, Richardphythian, Cbmsu01, Tgeairn, J.delanoy, Pharaoh of the Wizards, DrKay, Tasarin, Richiekim, Captain Infinity, Acalamari, DanielEng, Shawn in Montreal, Katalaveno, Mc-Sly, PhilDNC, Betswiki, Richard D. LeCour, Student7, Largoplazo, Mr.friend, Deathinfrench, Hbayat, Yodler, Belegur, Andy Marchbanks, DASonnenfeld, RjCan, Mdee, X!, TreasuryTag, Thedjatclubrock, Bentonia School, Jeff G., Soliloquial, TheOtherJesse, Yoho2001, Oshwah, GimmeBot, Cosmic Latte, MooseKin, Qxz, Oxfordwang, Una Smith, Yilloslime, JhsBot, Vtliving, Guest9999, Maxim, Joeldl, Doug, Meters, CephasE, Why Not A Duck, Logan, Stomme, Ellomate, Hazel77, NHRHS2010, Dlcantrell, Bfpage, Acer saccharum, StAnselm, WTucker, Richyrich460, Bielle, Keilana, Happysailor, Toddst1, Oxymoron83, Antonio Lopez, Drewsachs, Nancy, Dodger67, Faithlessthewonderboy, Clue-Bot, Trfasulo, Fyyer, The Thing That Should Not Be, DbelangeA, Andyikea, Dirtdog52658, Cahabon, Mild Bill Hiccup, DanielDeibler, Kroyw, Niceguyedc, Richerman, Piledhigheranddeeper, Neverquick, Sensonet, Auntof6, Excirial, Alexbot, Jusdafax, Eeekster, Zaharous, Estirabot, LarryMorseDCOhio, ErinSzeto, Strangecow, ChrisHodgesUK, Another Believer, Thingg, Versus22, Pzoxicuvybtnrm, Dana boomer, Wynford, Lx 121, Atalide, DumZiBoT, Fastily, MadPoster, Rreagan007, Mifter, Sweedishman666, Klandeen, Mm40, RyanCross, Sabasaba~enwiki, Addbot, Cxz111, JBsupreme, Some jerk on the Internet, Tcncv, ConicProjection, Blethering Scot, Ronhjones, Fieldday-sunday, Bootboy41, Ronkonkaman, AliceCake, Cst17, Download, Toastermaster, Glane23, AndersBot, ChenzwBot, Ryan Goldschlager, Rockport1019, 5 albert square, Aardnavark, Shortymama, Dbroer, Tide rolls, Ettrig, LuK3, Luckas-bot, 2D, MountainMan1983, Berkay0652, Evans1982, Elmridge, ArchonMagnus, ZEKKELZ, Keep your fork, there's pie, Irolpat, Rubinbot, 1exec1, Jim1138, Accuruss, Piano non troppo, Dweeebis, Kingpin13, Hmvont, Ulric1313, Flewis, Materialscientist, Tapmytrees, Aurorated, Citation bot, Basilisk4u, ‏حم.غ.ع.امدي.‏124, LovesMacs, Capricorn42, .45Colt, Hitchin' A Ride, Acebulf, Nickkid5, AbigailAbernathy, Click This, no contest, Abce2, Sugarmaker, Ed8r, Earlypsychosis, Zefr, Amaury, Brutaldeluxe, Moxy, Tipherf, Shadowjams, Joaquin008, Green Cardamom, BoomerAB, Captain-n00dle, Captain Weirdo the Great, Surv1v4l1st, Lozada.8, Crisisdebt, Dogposter, Qwell the pell, HJ Mitchell, Tj8805, Neko 105, HamiltonLake, OgreBot, Jakob Russian, Pinethicket, Rhianjones, Jschnur, Gurwlsww, SpaceFlight89, Σ, Emika22, M̧, FoxBot, Trappist the monk, Tubby23, Jambonilton, Vrenator, Bluefist, Creatoroffacts, Bluerasberryrocker, Lammidhania, Kathleen5454, Trentwolodko, Mapleysyrup, Innotata, Tbhotch, Hwd345247, Minimac, E3man, Andrea105, Carbooty, RjwilmsiBot, Hajatvrc, InstantOwn, Slon02, Maxkreusen, EmausBot, WikitanvirBot, Leech44, Philippe (WMF), Minimac's Clone, RenamedUser01302013, Tempestz, Tommy2010, Wikipelli, Djembayz, Cptbutt, Physcovideos, Uvmcdi, John Cline, Liquidmetalrob, Infamouspineapple, Parsonscat, Jeanpetr, Fishing Chimp, A930913, H3llBot, Murmuration, MJ for U, Grahamharris, Erianna, Jsayre64, Demiurge1000, Jay-Sebastos, Brandmeister, Larkinf, Donner60, Chickagirl1321, Rocketrod1960, Creepelectronics, Autodidact1, ClueBot NG, Namelessblackcat, Mr. Berty, Lindapeck, This lousy T-shirt, VanishedUser sdu8asdasd, O.Koslowski, Khakhalin, Marechal Ney, Go Phightins!, CopperSquare, Widr, Danim, Abagaba, Crosstemplejay, Helpful Pixie Bot, Accedie, Tholme, Calidum, Gob Lofa, Lowercase sigmabot, Arnavchaudhary, Vagobot, Mapledesk1234567890, Ceradon, George Ponderevo, Northamerica1000, Saraghhoey, Hurricanefan25, Interchangeable, Dan653, Logandibble, Toccata quarta, Harizotoh9, Khagerty13, Bryantqb21, Cray54, Ryanw023, MsBatfish, Oneluffy, Dr.fluff, Loriendrew, Dr.c8dc, Chiefrunningbear, MarcosFus, Michael James Owen, Chief O Medicine, TerryTweel, YUMA928, Jake Greenberg, Luiscrawford, Joeydog2000, BattyBot, Holmenat, Janerosi, Bremwinston, Quant18, Randomcat311, Webclient101, Mogism, Skykabob,   , Lugia2453, RotlinkBot, Bananasoldier, Decaract10, Aemc412, No1inparticularhere, Babitaarora, Lizzy Kitty, Kind Tennis Fan, Stroumel, Tristan.andrade.136, Flyguy2040, MegaMind69, EnergySta5, Shoogg, Sexbug, Da salt, WorldOfSchlong, Monkbot, BethNaught, IP.303, We r fun people, John Jacob junglehiemer Schmidt, Julietdeltalima, Noahpearlman123, Andreeaep10, Tropicalkitty, Maplesyrupzoey, ArmanEbrahimmmi, Cuunntmuffin and Anonymous: 738

- **Matsutake** *Source:* https://en.wikipedia.org/wiki/Matsutake?oldid=674292323 *Contributors:* Yas~enwiki, Chris 73, DocWatson42, Marcus2, Jpgordon, Sjschen, Hoary, Supergloom, Duff, BlueCanoe, Tokek, Sin-man, Kbdank71, Makaristos, Ericsteinert~enwiki, Eubot, DannyWilde, Chobot, Gdrbot, Debivort, Vmenkov, Calicore, ENeville, Badagnani, Tomomarusan, Dogcow, Shyam, SmackBot, Cla68, Kintetsubuffalo, Skizzik, Tyciol, Bluebot, MalafayaBot, Wikipediatrix, Nbarth, Midori, Memming, Sasata, Jjok, J Milburn, Myasuda, Cydebot, Pihka, Casliber, Thijs!bot, Angelofdeath275, Pedro, JamesBWatson, Neweco, Trusilver, Adavidb, Naniwako, DASonnenfeld, Caspian blue, CWii, Alan Rockefeller, Fifi1314, Staka, Lamro, Azukimonaka, Mycomp, Phe-bot, LeadSongDog, Oxymoron83, InMemoriamLuangPu, Addbot, Ka Faraq Gatri, Аимаина хикари, Luckas-bot, Nallimbot, AnomieBOT, Xqbot, GrouchoBot, Jatlas, Phoenix7777, Shanmugamp7, Hauva, Cresus22, Emaus-Bot, WikitanvirBot, Ὁ οἶστρος, Alexpmuller, SporkBot, ChuispastonBot, Helpful Pixie Bot, Plantdrew, Tomohiro HIRAO, Anguson o Keg and Anonymous: 57

- **Metroxylon sagu** *Source:* https://en.wikipedia.org/wiki/Metroxylon_sagu?oldid=644934797 *Contributors:* MPF, Guettarda, Hesperian, Stemonitis, Toksave, Eubot, Arjuna909, Rkitko, Kingdon, Voceditenore, JAnDbot, Chiswick Chap, TottyBot, DASonnenfeld, TXiKiBoT, Flag-Steward, Una Smith, Jaguarlaser, SieBot, Nagatang, Mild Bill Hiccup, Addbot, Flakinho, Yobot, Ptbotgourou, Rubinbot, Walrus heart, Xqbot, RedBot, Gulbenk, Ripchip Bot, EmausBot, HiW-Bot, ZéroBot, ChuispastonBot, Spicemix, ClueBot NG, Declangi, LylaGirly, RikSchuiling, Ducknish, Makecat-bot, Luxure, Joseph Laferriere and Anonymous: 3

- **Mushroom** *Source:* https://en.wikipedia.org/wiki/Mushroom?oldid=685229420 *Contributors:* Damian Yerrick, Magnus Manske, Vicki Rosenzweig, Tarquin, Mark, Enchanter, Kip~enwiki, Merphant, Ellmist, Heron, Patrick, DopefishJustin, Dante Alighieri, Shyamal, Menchi, Wapcaplet, Ixfd64, Tango, Cyde, Gbleem, Ahoerstemeier, Mac, Andrewa, Julesd, Ugen64, Tristanb, Jeandré du Toit, Evercat, Smack, Nikola Smolenski, GeShane, Marknen, Timwi, Janko, Dino, David Latapie, ARog, Tpbradbury, Marshman, Taxman, Rei, Phoebe, Xevi~enwiki, Sandman~enwiki, Wetman, David.Monniaux, Pollinator, Garo, PuzzletChung, Chuunen Baka, Donarreiskoffer, Bearcat, Astronautics~enwiki, YahoKa, Goethean, Altenmann, Naddy, Lowellian, Rfc1394, Academic Challenger, Texture, Ojigiri~enwiki, Andrew Levine, Sunray, UtherSRG, JesseW, Wikibot, Avij, Mushroom, Cordell, Carnildo, Enochlau, Fabiform, Giftlite, Wolfkeeper, Abigail-II, Tom harrison, Fastfission, Ausir, Everyking, Moyogo, Wouterhagens, Curps, Sik0fewl, Pparadox, FriedMilk, Summerbellrc, Horatio, Solipsist, Foobar, SoWhy, Traumerei, Antandrus, MisfitToys, Annom, Rdsmith4, Mzajac, Kesac, Vbs, Bumm13, TonyW, Aramgutang, Urhixidur, Joyous!, Ukexpat, Ratiocinate, Adashiel, ELApro, Mike Rosoft, Dr.frog, Haiduc, Discospinster, Pak21, Cacycle, Vsmith, Smyth, Swid, MBisanz, Pjrich, Simonfairfax, Shanes, Zegoma beach, EmilJ, Mairi, Adambro, Bobo192, Circeus, Meggar, Smalljim, Viriditas, Elipongo, Chirag, Forteanajones, Bdamokos, Ben@liddicott.com,

Lulubell77, FrescoBot, TeeJay2009, Surv1v4l1st, ع‍بـد ال‍مؤمن, Limbodog, Jatlas, Pepper, Wikipe-tan, Sky Attacker, Xscenexemox, Jason-jambalya, Jleer1, Recognizance, Vicharam, Suki77, HJ Mitchell, Jamesooders, Alysdream, A little insignificant, Dramartistic, Challenger 2, Hockeydude622, Citation bot 1, Biodina87, Ilovejoe13, Bonginhaler, Pinethicket, I dream of horses, Edderso, SoccerMan2009, Bejinhan, Hamtechperson, Vahn dole, Jschnur, Jatlas2, BubikolRamios, Zhonghuo~enwiki, ActivExpression, FoxBot, Jonkerz, Lotje, Fox Mcloud, Mab-sal, Vrenator, Leondumontfollower, Reaper Eternal, Dtierney786, Bkunesh, Adi4094, Countskull, Tbhotch, Stroppolo, DARTH SIDIOUS 2, Mean as custard, RjwilmsiBot, Уральский Кот, Baklap123, KathieBanks1982, Skamecrazy123, EmausBot, John of Reading, Orphan Wiki, Acather96, WikitanvirBot, Mythic Dawn Agent, Look2See1, Qaedtgujol, Jamie99999, Bluejellyisawesome, RA0808, Tacomaster196, Brittany Ringer, Slightsmile, Holyhellshrooms, Booknotes, Wikipelli, PAcmanJones123, JimbobGuy, Comesturnruler, Mh7kJ, Traxs7, Killr96, Dekkun, Dken, Letsnothavefun, Alpha Quadrant (alt), GZ-Bot, Oscarta, Wayne Slam, Erianna, Isarra, TyA, L Kensington, Dickslikesuger1, Hecnevill, Donner60, Carmichael, Subrata Roy, ChuispastonBot, Peter Karlsen, Emcee69, RwAr2012, Roosterlover, TYelliot, Akeenley, AMD, Xgenz6, Petrb, Optobume, ClueBot NG, Unknown652, Jaklaz, Anagogist, This lousy T-shirt, Superlegoman100, Neilz N13, Kpsimoulis, Satellizer, Gavincnom, Jcgoble3, PedR, Cntras, Kevin Gorman, Gooselover12, Widr, GlassLadyBug, JaxHawk, Cloudymushroom, Mtking, Helpful Pixie Bot, Kaiman999, Soroplebo, Calabe1992, Bob12345swn, Qwerty1234567890ooO, Lowercase sigmabot, Agshsgfhxhgfh, Billygardell, DuoMind, Userwhosit3, Wiki13, Lozzieloler, MusikAnimal, Jahnavisatyan, Mark Arsten, Shoefrog11, Xxmorgzxx, Snow Blizzard, YVSREDDY, Ez-Jay, Klilidiplomus, Triangles of the square, Senic, Mateuy, TheBaur, Vishal cherrian, Cimorcus, Mush zombie, Cyberbot II, Bunnyfist, Taiya26, Snowyrancis, Aliwal2012, MrGreendew2, CrunchySkies, Greenleafnetwork, Ekren, Ducknish, Dexbot, Wikiwahtyourpedia, LacrosseExpert, Webclient101, Hfortuin, Arunsivan, Frosty, SFK2, Chance2497, Zagmore, Deraileddeath, 069952497a, Helium105, Maniesansdelire, Mmaprofan, Camilla2002, Jct123, I am One of Many, AmaryllisGardener, JoeMeas, Setery, Silver213p, Dustin V. S., Irg1969, CensoredScribe, Buffbills7701, MushyMushtip, Robert Quillen, Jeffjohnjeff, Sef599, Tjm16313, 00dcrotzer, Oldmcdonald123, Noamilanidema, Amrtahio1, Derekmagalong17, TuxLibNit, Liar123456, Arvelers, Goldi.negi, Manliestmushroom, Rabbismile, Jehdc, UnpluggedAbroad, Rdrafter01, Patrictia, Monkeydoodle10, Djkjdk, Learnerktm, Jeh10, Mouseythemouse, 69zepo, SuperTeacherAfro, Slasher7291, Spaetzel02, Random369, Qwehfh-fjgjfjfjvivuvic, Joxon10, Chase Swag, Thetrollofalltrolls, Michael Palmer, KasparBot, Williditor and Anonymous: 1433

- **Mushroom hunting** *Source:* https://en.wikipedia.org/wiki/Mushroom_hunting?oldid=684560065 *Contributors:* Mav, Mintguy, Dante Alighieri, Menchi, CesarB, Haakon, Angela, Andrewa, Julesd, Alex756, Dysprosia, IceKarma, Marshman, Maximus Rex, Taxman, Rei, Wetman, David.Monniaux, UninvitedCompany, Hankwang, Dusik, Geeoharee, Kpalion, Mboverload, Bobblewik, CryptoDerk, Mzajac, Vbs, Vishahu, Aramgutang, Quasistoic, Jbinder, Poccil, Jkl, Deirdre~enwiki, Bender235, Nabla, Keno, Jpgordon, ToastieIL, Alansohn, Zyqqh, Blaxthos, Kenyon, Dismas, Tariqabjotu, BlueCanoe, Stemonitis, Woohookitty, LOL, XIcy, Jeff3000, Kbdank71, Strobilomyces, Peter Bladwell, Kmorozov, Scoo, WriterHound, Debivort, YurikBot, Wavelength, Peter G Werner, Diliff, Gaius Cornelius, Marcus Cyron, ENeville, Joncolvin, Renata3, JPMcGrath, Cinik, Whitejay251, Modify, Petri Krohn, SmackBot, TestPilot, KocjoBot~enwiki, PizzaMargherita, Dims, Michaelll, PiKeeper, Mycota, Toddlisonbee, Colonies Chris, Huwmanbeing, LBM, Mion, Scharks, ComSpex, T3hZ10n, Cortezz~enwiki, Rainwarrior, Bendzh, Pseudoanonymous, Beefyt, Alan.ca, Iridescent, J Milburn, Laonikoss, CmdrObot, Neelix, Myasuda, Kirkesque, Lugnuts, Daniel J. Leivick, Revdrace, Kozuch, Cs california, Casliber, RSIferd, Ante Aikio, Hugo.arg, Ruber chiken, Just Chilling, Marokwitz, Alphachimpbot, Larrybot3000, Pedro, Pixel ;-), Galileo01, Cgingold, Dunham, Lost tourist, Guyfrompc19, Wiki wiki1, J.delanoy, Myredroom, Majamin, Qatter, Colchicum, NewEnglandYankee, Jmcw37, Zara1709, DASonnenfeld, Steel1943, Idioma-bot, Roaring phoenix, VolkovBot, Alan Rockefeller, Technopat, Z.E.R.O., Maxim, Malick78, Twooars, Cosprings, SieBot, YAYsocialism, Twinkliest, Nzfiend, Danvandan, Miotch711, Judicatus, Peterthinks, Crealizate, The Thing That Should Not Be, Mycologyauthor, Auntof6, Excirial, Chrismatherly, Tnxman307, Djk3, Anticipation of a New Lover's Arrival, The, Thebestofall007, Addbot, Sillyfolkboy, Ehrenkater, Аимаина хикари, Bunnyhop11, Tirithel, AnomieBOT, FadulJoseA, Jim1138, Xu-fanc, Liopac, Tom87020, Xqbot, UMDorota, FadulJoseArabe, Prezbo, Belsavis, Josef Papi, FrescoBot, Jatlas, D'ohBot, HJ Mitchell, George Chernilevsky, Pinethicket, Patspacersfan, IKILLEDmufasa, Jikyebbna, Jonkerz, Aisadore, Mabsal, Connelly90, Orphan Wiki, WikitanvirBot, Tommy2010, Bongoramsey, Coryv1985, Bamyers99, Brandmeister, Donner60, Sol Goldstone, Chester704, ClueBot NG, Frerin, Snotbot, Billodea, Primergrey, Helpful Pixie Bot, BG19bot, Jimbobhawk, Bondaruk85, Hillcrest98, WikiHannibal, SatenikTamar, Pc-pdx, Mogism, B14709, Ada564, AMartiniouk, Lisatort, Therewillbesixmushrooms, Formycat and Anonymous: 148

- **Myrobalans** *Source:* https://en.wikipedia.org/wiki/Myrobalans?oldid=575632447 *Contributors:* DanielCD, Murty, Ksvaughan2, DASonnenfeld, Addbot, Xufanc, Analphabot, RibotBOT, Ansumang, Plantdrew, Sminthopsis84 and Anonymous: 3

- **Natural dye** *Source:* https://en.wikipedia.org/wiki/Natural_dye?oldid=681805192 *Contributors:* IceKarma, Wetman, Mike Rosoft, Rjwilmsi, PKM, Wavelength, Fnorp, Pyrotec, Sasata, James086, Dr. Blofeld, Sluzzelin, Bongwarrior, Sustainableyes, Keith D, R'n'B, Adavidb, Paris1127, Johnbod, DASonnenfeld, Uyvsdi, Mercurywoodrose, Blahmos, Brenont, Ketone16, Flyer22, Correogsk, Thelmadatter, Sfan00 IMG, Polyamorph, Rui Gabriel Correia, Felix Folio Secundus, Addbot, Annielogue, West.andrew.g, Guy1890, KamikazeBot, AnomieBOT, Bluerasberry, Materialscientist, FrescoBot, ManfromButtonwillow, Madison60, RjwilmsiBot, WikitanvirBot, Erianna, ClueBot NG, Widr, Helpful Pixie Bot, Wbm1058, KLBot2, KimS012, OttawaAC, Rjlanc, Ices2Csharp, Toxophilus, Anoronha, Bsfs and Anonymous: 32

- **Natural rubber** *Source:* https://en.wikipedia.org/wiki/Natural_rubber?oldid=685544480 *Contributors:* AxelBoldt, LC~enwiki, Bryan Derksen, Timo Honkasalo, The Anome, Malcolm Farmer, Mark Ryan, Rjstott, Mirwin, Shsilver, Vanderesch, PierreAbbat, Heron, Hephaestos, Olivier, Edward, Patrick, Infrogmation, D, Michael Hardy, Earth, Mac, CatherineMunro, Julesd, Ineuw, Mxn, Emperorbma, Charles Matthews, Jay, Thomasgl, Dogface, Moros~enwiki, Finlay McWalter, Anthony Fok, Rogper~enwiki, Kristof vt, Securiger, Miles, Ambarish, Isopropyl, Lupo, Giftlite, DocWatson42, Lupin, Brian Kendig, Tubular, Mark Richards, Monedula, Everyking, Solipsist, Tagishsimon, R. fiend, Antandrus, Rdsmith4, Togo~enwiki, H Padleckas, Neutrality, Jh51681, Miguel3000, Corti, Mike Rosoft, Discospinster, FiP, Mani1, Pavel Vozenilek, Patrickov, Darren Foong, Pjrich, Imoen, Jpgordon, Bill Thayer, Bobo192, DaveGorman, Kjkolb, Nk, Idleguy, Haham hanuka, Polylerus, Hooperbloob, Klafubra, Alansohn, Anthony Appleyard, Walter Görlitz, Jamyskis, Dowcet, Malo, Wtmitchell, Stephan Leeds, Leoadec, RainbowOfLight, LFaraone, Henry W. Schmitt, Drbreznjev, Luigizanasi, Brookie, Crosbiesmith, Dejvid, CNRNair, Gmaxwell, Nuno Tavares, Woohookitty, Deeahbz, Kenneth Stephen, MONGO, Nklatt, GregorB, SCEhardt, Wayward, Wgsimon, Dysepsion, Magister Mathematicae, BD2412, Susten.biz, Rjwilmsi, Gudeldar, Matt Deres, Latka, Margosbot~enwiki, Nihiltres, RexNL, Chooyimooyi, Physchim62, Imnotminkus, Butros, King of Hearts, Chobot, Bornhj, Phantom Thief, YurikBot, Wavelength, Tom Barnwell 0, Phantomsteve, RussBot, Icarus3, Hydrargyrum, Stephenb, CambridgeBayWeather, Shaddack, Eleassar, Cryptic, Wimt, NawlinWiki, Wiki alf, Thatdog, Bukhrin, Cleared as filed, Dmaestoso, TDogg310, Rwalker, DeadEyeArrow, Cavan, Xabian40409, Light current, Josh3580, Canley, BorgQueen, Kevin, ArielGold, David Biddulph, Allens, DVD R W, The Yeti, Eog1916, Luk, ChemGardener, Snalwibma, Attilios, Yakudza, SmackBot, Amcbride, Oxford Comma, C.Fred, PRA, WookieInHeat, Delldot, PJM, Ilikeeatingwaffles, Edgar181, Gilliam, Brianski, Ohnoitsjamie, Skizzik, Carl.bunderson, JAn Dudík,

Chris the speller, Bluebot, MikeParker, Persian Poet Gal, JDowning, NCurse, SchfiftyThree, Deli nk, Vampyrecat, Dlohcierekim's sock, Octahedron80, Darth Panda, Gracenotes, Can't sleep, clown will eat me, Милан Јелисавчић, DeFacto, Yidisheryid, Rubber~enwiki, Xiner, Prmacn, Rrburke, CorbinSimpson, Bardsandwarriors, Stevenmitchell, Flyguy649, Kthxhax, Kingdon, MichaelBillington, M.Kris, Mistress Selina Kyle, Smokefoot, PStatic, The PIPE, DMacks, Salamurai, Risker, Sadi Carnot, Kukini, G4sxe, Rklawton, DO11.10, Turtleflipper, John, Dog Eat Dog World, Kuksul, KarlM, Peterlewis, IronGargoyle, Ckatz, 16@r, Twalls, TastyPoutine, EEPROM Eagle, Dik~enwiki, Iridescent, Father Time89, Shoeofdeath, Courcelles, Bassclef, Linkspamremover, Phasmatisnox, Toomers57, Tawkerbot2, Dlohcierekim, Lahiru k, SkyWalker, CRGreathouse, Ale jrb, Liammars, Benwildeboer, Dgw, NickW557, PlanetEric, Casper2k3, Funnyfarmofdoom, Kupirijo, Pais, Abeg92, Jefferyseow, MC10, Gogo Dodo, Travelbird, JFreeman, Llort, Vegfarandi, Tawkerbot4, Karuna8, Smeazel, Malleus Fatuorum, Epbr123, Ishdarian, Headbomb, Sinn, Philippe, Michael A. White, Ohaider, Escarbot, Temers, Babudaniel, AntiVandalBot, Prolog, KP Botany, RapidR, Jj137, Pro crast in a tor, Cinnamon42, LibLord, Spencer, Ingolfson, JAnDbot, Husond, Wiki0709, Harryzilber, MER-C, Bob the Dino, Instinct, Sitethief, RicardoFachada, PhilKnight, Cynwolfe, Bongwarrior, VoABot II, AuburnPilot, Jéské Couriano, Brother Francis, Prestonmcconkie, Arz1969, Bubba hotep, Panser Born, Dirac66, 28421u2232nfenfcenc, Francescobrisa, Glen, DerHexer, Mikewhitcombe, Mmustafa~enwiki, ClubOranje, MartinBot, T.c.newman, Kostisl, CommonsDelinker, AlexiusHoratius, Verdatum, Tgeairn, J.delanoy, Trusilver, Ginsengbomb, Thaurisil, OfficeGirl, Newpaltzbob, McSly, AntiSpamBot, Recurve7, NewEnglandYankee, Trilobitealive, Gregfitzy, Sam Paris, Dacrycarpus, Jamesontai, Vanished user 39948282, Exelby, DASonnenfeld, Martial75, Idioma-bot, Signalhead, Deor, VolkovBot, Cireshoe, Richardwhitelaw, Jmrowland, Rmstorey, LaborRightsNow, LeilaniLad, Philip Trueman, Konfucious, TXiKiBoT, Alesnormales, Rotor DB, Dobrinia, Tnewman.ILRF, Ann Stouter, JhsBot, Tpk5010, Monkeynoze, Tri400, Havlat195, Billinghurst, Andy Dingley, Omulazimoglu, Jaguarlaser, Vector Potential, Dockmno, Locke9k, HeirloomGardener, Skarz, Symane, Cvf-ps, Deconstructhis, Avinesh, MiddleNation101, SieBot, Speed Air Man, Krawi, Caltas, Tomlet12481632, Brando55555, Yea booooooooooiiiiii, Harold91, Francish7, Sunny910910, Happysailor, Tiptoety, Radon210, Report5rose, Oda Mari, Jojalozzo, Chickenwing.69, JSpung, Crvpicture, Oxymoron83, Lightmouse, Steve the e-o, Arahar, Jacob.jose, PabloStraub, Efe, Church, Tanvir Ahmmed, ClueBot, Rayyung, The Thing That Should Not Be, Permanerd, Kafka Liz, Techdawg667, Iulian28ti, Garyzx, Ajdehoog, Blanchardb, Khateeb88, DragonBot, Ice scream123, Robert Skyhawk, Excirial, Pumpmeup, TonyBallioni, AlphaAqua, Spike-Toronto, Lostpostcards, The Founders Intent, Arjayay, Bald Zebra, Thingg, Aitias, Spongeguart, Versus22, SoxBot III, DumZiBoT, Daven brown, XLinkBot, Figgalicous, HarlandQPitt, JinJian, Joemcauley, Angelfirelol123, Tedelex06, Addbot, Xp54321, Proofreader77, Aqua785, Succu, C6541, Element16, WPHyundai, Tcncv, Non-dropframe, Theleftorium, Kongr43gpen, DougsTech, Rubberducky3201, Fieldday-sunday, Mr. Wheely Guy, Aboctok, Fet'our, Vishnava, CanadianLinuxUser, Leszek Jańczuk, Aransil, SaishankerV, MrOllie, Download, Favonian, Tassedethe, Shekure, Gooberck, Tide rolls, Gumshoea, Lightbot, Teles, Jarble, Luckas-bot, Yobot, 2D, Kartano, Gobbleswoggler, Maxí, Језка, Cnt-make-a-name, Tempodivalse, Synchronism, AnomieBOT, DemocraticLuntz, 1exec1, AdjustShift, Fahadsadah, Ulric1313, RandomAct, HdeK, Giants27, Materialscientist, Citation bot, Eumolpo, Holyname, Xqbot, Waynekhue, S h i v a (Visnu), Transity, Apothecia, Capricorn42, Gigemag76, Umoveudie1, Corruptcopper, Shirik, Aphid974, Sahehco, Rainald62, Ll1324, Rupunkel, Luigi mad, FalconL, Contino, General-Septem, Pepper, Ace of Spades, Patronus789, Just Say No, Theballoonwentbooom, HJ Mitchell, BenzolBot, Sanoopj, Owen21, Pinethicket, Elockid, Edderso, Tom.Reding, Calmer Waters, Jschnur, Impala2009, Jaybird vt, Footwarrior, Hellzies, Kgrad, Ashisfishy, Inbamkumar86, Leeatcookerly, Lotje, Callanecc, Extra999, Defender of torch, Inferior Olive, Vanished user aoiowaiuyr894isdik43, Tbhotch, Mean as custard, RjwilmsiBot, Amieab, Fratter, Wiesner1252, Slon02, DASHBot, Oliverlyc, Orphan Wiki, Efcmagnew, ScottyBerg, Racerx11, GoingBatty, RenamedUser01302013, Wikipelli, Dcirovic, K6ka, Gadget142, Thecheesykid, Sherwin ph, Brothernight, Sunnyn88, Liquidmetalrob, Fæ, ElationAviation, Hazard-SJ, DihanPerera, Char466, Wagino 20100516, Naveenzcherian, MonoAV, Donner60, Rejinp, Blu3skys, GrayFullbuster, DASHBotAV, JonRichfield, Edgar181 is dumb, ClueBot NG, This lousy T-shirt, Raghith, BarrelProof, HovhannesKarapetyan, Skoot13, Primergrey, Alefeb, Widr, ?oygul, JohnSRoberts99, Strike Eagle, DBigXray, Plantdrew, Kinaro, Mickey big dig mouse, BG19bot, Lowercase Sigma, Declangi, MusikAnimal, Erica.largen, GreenJAMO, Altaïr, Sean-is-here, Hamish59, Glacialfox, Vanischenu, Mastergontoj, Shayanmardanpour, Anbu121, BattyBot, Tutelary, Suresh Baddika, Mrt3366, Tonyxc600, Cyberbot II, ChrisGualtieri, Ducknish, Ncellett1, Eoktay, Reatlas, NewNew22, Epicgenius, Vanamonde93, Forevs, I am One of Many, Eyesnore, Tonymena1, Zack from 9c, Manhire7, Xuanmingzi, Jt940, MUnsell, CensoredScribe, Serten, AresLiam, Ugog Nizdast, BatMan (M), Thailandfish, Lreuber, Jianhui67, TCMemoire, Sprocketmama, Boot Blues, Davidjohn11, CDLatexLover, Paul Badger Brown, Howunusual, Hawkfolkin54, Nickid12, JenmartNJ, Anonomus03, Unnontoxic, Bobbyjoecheese, Krieaersaeh, Quickway, Stefanut6060, Angelinadamico, Crackstack22, Cybersmart123, KasparBot, 123456789rocks, Ipadcase, Naturalrubberpro and Anonymous: 862

- **Naval stores** *Source:* https://en.wikipedia.org/wiki/Naval_stores?oldid=684294562 *Contributors:* Taxman, Qutezuce, YUL89YYZ, Alansohn, Sjschen, Canderson7, Frekja, Alex20850, Welsh, Hmains, Matchups, Bill.albing, Sopoforic, Cydebot, Mary Mark Ockerbloom, Cfdbowens, Textorus, DASonnenfeld, Brewcrewer, Thewellman, Minnecologies, Materialscientist, Purplebackpack89, Shadowjams, Colinr9, DexDor, John of Reading, Dewritech and Anonymous: 15

- **Nutmeg** *Source:* https://en.wikipedia.org/wiki/Nutmeg?oldid=681269892 *Contributors:* Zundark, Timo Honkasalo, Espen, Ted Longstaffe, Edolin, Danny, Darius Bacon, Robert Foley, Arj, Jaknouse, Kwertii, Jketola, Tannin, Ixfd64, Mkweise, Darkwind, Andrewa, BenKovitz, David Stewart, Ehn, Phr, Wik, Jessel, Ann O'nyme, HarryHenryGebel, Spinster, David.Monniaux, Francs2000, Robbot, Tomchiukc, Sanders muc, WormRunner, Naddy, Rfc1394, Academic Challenger, Auric, Moink, Mervyn, Fuelbottle, Exploding Boy, MPF, Cantara, Tom harrison, Orangemike, Wyss, Eequor, Dumbo1, Golbez, Manuel Anastácio, Utcursch, Andycjp, Quadell, Williamb, Beland, Kvasir, JoJan, Sh~enwiki, Rdsmith4, DragonflySixtyseven, Sharavanabhava, Mike Rosoft, DanielCD, Jhwelsch, Discospinster, Hydrox, Arthur Holland, Gonzalo Diethelm, Bender235, Android79, STGM, Pjf, Lankiveil, Bobo192, Clawson, WikiLeon, Hesperian, Pacula, Haham hanuka, Jonathunder, Jhfrontz, Alansohn, Anthony Appleyard, LtNOWIS, Ctande, Atlant, Rd232, Reprah, Kz9dsr0t387346, MattWade, Hohum, Marianocecowski, Mlutfi, RainbowOfLight, Gene Nygaard, Markaci, Killing Vector, Mcsee, Stemonitis, Richard Arthur Norton (1958-), Alvis, Woohookitty, Richard Barlow, Pol098, JeremyA, Exxolon, MrDarcy, Fxer, Allen3, Dysepsion, Mandarax, Graham87, RxS, Adamccl, Rjwilmsi, Ncc1701zzz, Jake Wartenberg, Heah, Matt Deres, X1987x, Eubot, Windchaser, Gurch, Preslethe, Chobot, Gdrbot, WriterHound, YurikBot, Wavelength, Jimp, Pburka, Muchness, Lexi Marie, Rsrikanth05, Anomalocaris, NawlinWiki, Seanlavelle, Dysmorodrepanis~enwiki, Noddycr, Axgoss, Buster79, Dforest, Badagnani, Awiseman, Irishguy, Catharticflux, Terpdx, Tumey, EEMIV, IceCreamAntisocial, FF2010, Scott Adler, JoanneB, Chriswaterguy, Digfarenough, Pinothyj, GrinBot~enwiki, Snowboardpunk, DVD R W, Kf4bdy, Wolf1728, SmackBot, FocalPoint, GoldenXuniversity, McGeddon, KocjoBot~enwiki, Stephensuleeman, Edgar181, Eloil, Gilliam, Skizzik, Chris the speller, TimBentley, E.T.Smith, Cews, Jprg1966, Salvor, Melburnian, Bazonka, JONJONAUG, Polyhedron, A. B., Verrai, Scwlong, KaiserbBot, Juandev, VMS Mosaic, Xyzzyplugh, Addshore, Kcordina, NoIdeaNick, Kingdon, Cybercobra, Nakon, Aelffin, Hans Frandsen, Sharp962, Kukini, Andrei Stroe, Secret7000, Lisapollison, Anjow, Mr. Lefty, IronGargoyle, Beetstra, Mr Stephen, Bendzh, Peyre, JaymzSpyhunter, Iridescent, JoeBot, Tau'olunga, Yashgaroth, Nutster, Earthlyreason, BoH, Bonás, Lentower, Matthew Treder, Jane023, Cydebot, Gogo Dodo, A Softer Answer, Pascal.Tesson, Tkynerd, Tawkerbot4, DumbBOT, Nooby aok, Karuna8, JodyB, JamesAM, Epbr123, KnightValor, Bigwyrm, N5iln, Wickerprints, PierceG, Superbeef, John254, A3RO, Merbabu, Windi, CharlotteWebb, Holocam, Escarbot, DewiMorgan, Niduzzi, AntiVandalBot, Seaphoto, Halvorf~enwiki, Nipisiquit, Dmethoxibit, Random user 8384993, Noremacmada, Edwardtbabinski, Klow, Ioeth, Deadbeef, JAnDbot, Andonic, Hut 8.5, Some thing, Rothorpe, Acroterion, FaerieInGrey, Pedro, VoABot II, Kaivosukeltaja, Sparisi1122, Aka042, Fabrictramp, Catgut, Jessicapierce, Frotz, Halogenated, Philg88, Johnbrownsbody, Peter coxhead, Oroso, Erpbridge, Jerem43, MartinBot, Vijendrapal, Kateshortforbob, CommonsDelinker, Nono64, J.delanoy, Trusilver, GSEkng, Uncle Dick, Meta-Physician, Moaks9, DoubleD17, Jcwf, Qbot (renamed), Truthandcoffee, Alnokta, Ken Dailey, Duellist, DASonnenfeld, Segilla, RjCan, Richard New Forest, Idioma-bot, Lights, Derekbd, Mogk, Mrh30, Driehoek, Soliloquial, LeilaniLad, Philip Trueman, TXiKiBoT, Jkeene, Moogwrench, Abtinb, Sankalpdravid, Tsm17, Sintaku, Arunrj, Martin451, Sniperz11, Beatle2102, Cremepuff222, Maxim, Madhero88, Ameilio, Luminum, HarmonyRocket, Falcon8765, Freeet, Justinleif, Snoopy321, Alfrodull, Balloon6, Samdira, Sumo180, Pollyclass, EJF, SieBot, Mbz1, Mentalmoose, Emoxstoned, Crazykid4u, GlassCobra, Flyer22, Slugfilm, Damniggarsumgoodbud, Antonio Lopez, Billyk63, KPH2293, Obesehawk676, Superbeecat, Joesteels phd, Denisarona, Invertzoo, ClueBot, Binksternet, Justin W Smith, Jmgarg1, Manggo, Rjd0060, RYNORT, Boing! said Zebedee, Niceguyedc, Blanchardb, Neverquick, Wspr81, Excirial, GngstrMNKY, Kjramesh, Maddieroth1, Buchem, Jotterbot, Tuchomator, La Pianista, Jacknjill123, Scottcal, XLinkBot, Strdst grl, Jovianeye, Vanished 45kd09la13, Dthomsen8, WikHead, Addbot, Bad words suc, Grey Geezer, C6541, Jojhutton, Wsvlqc, Tcncv, Novalia, Cuaxdon, Ronkonkaman, NjardarBot, Kk sze, Glane23, Chzz, Norman21, Tyw7, EdPeggJr, Shekure, Rediretihw, Tide rolls, SDJ, Lightbot, Mirza Barlas, Bff, Luckas-bot, Fraggle81, Legobot II, II MusLiM HyBRiD II, Gobbleswoggler, Becky Sayles, English spy, Brougham96, IW.HG, Eric-Wester, Soaringhawk21, AnomieBOT, Götz, Adenosine Triphosphate, Killiondude, Jim1138, Palace of the Purple Emperor, Piano non troppo, Sven70, Admch4, Capricorn42, Termininja, Gigemag76, Anna Frodesiak, Mcoupal, Uxbona, Brandon5485, Mark Schierbecker, Amit 9b, Mayor mt, 78.26, Kylekieran, AntiAbuseBot, Mattis, Hamamelis, E0steven, Ellenois, Wiki User 68, Custoo, Macruzq, The Nerd from Earth, Newbie82, Whatinthewampa, OgreBot, DrilBot, Pinethicket, Moonraker, HowardJWilk, Midnight Comet, Jujutacular, Zairatool, Ashokpmeena, LogAntiLog, Vrenator, Singlemaltscotch, Reaper Eternal, Boricuamark, TjBot, VernoWhitney, Beyond My Ken, Lopifalko, Mrcopyfighter, EmausBot, Jackmingo, Fransouski, Detrickm, Bassistdakota, Xxxterminatorxxx, XxGirly Girl14xX, Slightsmile, K6ka, Bravo.dexter, Maglame, ZéroBot, John Cline, Fæ, Duvallrobbie, E557, Halouu, H3llBot, Erianna, TyA, L Kensington, Killerwinky, Bob1033, Petrb, Piex17, ClueBot NG, Anudada, This lousy T-shirt, CyberknightMK, O.Koslowski, Breogan2008, Pushthatrock, Thewranglerstrangler, Vhkdhg, Mannanan51, Widr, Jklein1692, DanKosher, Plantdrew, BG19bot, Krenair, Agus stiw, Chrysalifourfour, Declangi, AvocatoBot, Azhar feder, Jahnavisatyan, Belliesmom, CitationCleanerBot, Alexei Zverev, Sensesfail123, AntanO, Bus-token, IsraphelMac, Dexbot, Sminthopsis84, Frosty, Corinne, Fun&games, 71Gabi, AsianGeographer, Blah90, EvergreenFir, Clr324, Avi8tor, Tortie tude, HalfGig, Lfsean, Markcalba, Sanskari, Superdude255, Lwilson262, Lunastar321, Atudu, Scourge of Trumpton and Anonymous: 595

- **Oak** *Source:* https://en.wikipedia.org/wiki/Oak?oldid=686199212 *Contributors:* ClaudeMuncey, Mav, Manning Bartlett, Malcolm Farmer, Rizniz, Rgamble, Enchanter, PierreAbbat, William Avery, TomCerul, Zoe, Azhyd, Heron, Jaknouse, Quercusrobur, Olivier, Infrogmation, Dan Koehl, Liftarn, Deadstar, Ixfd64, Ellywa, Ahoerstemeier, Glenn, Bogdangiusca, Wnissen, Dmsar, Phr, Tjunier, Radiojon, Tpbradbury, Imc, Nv8200pa, SEWilco, Roehsler, David.Monniaux, Shafei, Palefire, Owen, Carlossuarez46, Robbot, Kristof vt, Moncrief, Altenmann, Romanm, Arkuat, Caknuck, UtherSRG, Srtxg, Fabiform, Giftlite, DocWatson42, MPF, Abigail-II, Tom harrison, Tom Radulovich, Ds13, Dick Bos, Guanaco, Critto~enwiki, Chowbok, Andycjp, Antandrus, Williamb, Kesac, Maximaximax, Icairns, Rellis1067, Kevin Rector, Canterbury Tail, Andete, DanielCD, Discospinster, LegCircus, Guanabot, Vsmith, Bender235, Kwamikagami, Mwanner, QuartierLatin1968, Susvolans, RoyBoy, Stesmo, Sortior, Smalljim, Dungodung, Nk, Troels Nybo~enwiki, Hesperian, MPerel, HasharBot~enwiki, JustJuthan, Alansohn, Quatermass, Algirdas, Theodore Kloba, Snowolf, Wtmitchell, Djlayton4, Garzo, HenryLi, Kazvorpal, Feezo, Pekinensis, Woohookitty, Richard Barlow, Georgia guy, Jeff3000, Pixeltoo, Miss Madeline, Jean-Pol Grandmont, Kelisi, Schzmo, Wiggy!, Blisco, Jobnikon, BD2412, FreplySpang, Rjwilmsi, Astropithicus, Vagab, Ian Lancaster, Yamamoto Ichiro, FayssalF, Eubot, Andymadigan, Crazycomputers, AJR, Hottentot, Gemmabass, RexNL, Gurch, Schmerguls, CiaPan, CJLL Wright, Chobot, DaGizza, Gdrbot, VolatileChemical, YurikBot, Wavelength, Hairy Dude, FrenchIsAwesome, Pigman, Eupator, Jon Peli Oleaga, Igo4U, Aaron Walden, Stephenb, Wimt, Tavilis, ENeville, Grafen, Seegoon, Misza13, Zwobot, Aaron Schulz, Xompanthy, Lockesdonkey, Samir, Amphis, Asarelah, Tigershrike, Zzuuzz, Argo Navis, Alureiter, Katieh5584, Appleseed, Ajdebre, DVD R W, Myrabella, SmackBot, Amcbride, Unschool, Ex0pos, KnowledgeOfSelf, Pgk, Eventer, Eskimbot, Hardyplants, Onebravemonkey, MelancholieBot, HalfShadow, Yamaguchi 先生, Gilliam, Ohnoitsjamie, Kaiwen1, Skizzik, Rkitko, Ian13, Bjmullan, Melburnian, I7s, Deli nk, Andrej Šalov, Gsp8181, Can't sleep, clown will eat me, Shalom Yechiel, OrphanBot, Sommers, Snowmanradio, Lesnail, Addshore, Amazins490, Mermoz~enwiki, Wirbelwind, Lcarscad, Kotjze, SirIsaacBrock, The undertow, Scientizzle, J 1982, ML5, Mgiganteus1, Bjankuloski06, Archangel127, A. Parrot, Smith609, Bendzh, Maksim L., Waggers, Peter Horn, MTSbot~enwiki, Ginkgo100, Pauric, Nehrams2020, Iridescent, LadyofShalott, AGK, Maelor, Civil Engineer III, Tawkerbot2, Enginear, Alecsescu, JForget, Dgw, NickW557, Moreschi, Davnor, ArgentTurquoise, Gogo Dodo, Agne27, Optimist on the run, Throquzum, Daven200520, Victoriaedwards, Daniel Olsen, Rosser1954, Oxonhutch, Thijs!bot, Epbr123, Pstanton, Kablammo, Sagaciousuk, N5iln, Mojo Hand, Oliver202, Horologium, Edal, Z10x, Big Bird, RoboServien, Escarbot, Dainis, Ericjs, Darekun, AntiVandalBot, Luna Santin, Seaphoto, Emeraldcityserendipity, QuiteUnusual, Tchoutoye, Smartse, Harborsparrow, Nipisiquit, Storkk, AubreyEllenShomo, Helot, BeefRendang, Sluzzelin, Parande, JAnDbot, Deflective, Cantabwarrior, Barek, MER-C, Plantsurfer, Lionchow, Krasanen, 100110100, Snowolfd4, WolfmanSF, Bongwarrior, VoABot II, JNW, JamesBWatson, Steven Walling, 0x0n, Catgut, Daarznies, 28421u2232nfenfcenc, Nat, Cpl Syx, DerHexer, Daemonic Kangaroo, Patstuart, Pere prlpz, MartinBot, Krosth~enwiki, Glossando, R'n'B, Vox Rationis, Alexnevzorov, Nono64, Brunch vs. breakfast, Fconaway, Tgeairn, J.delanoy, Trusilver, Love Krittaya, Uncle Dick, Acalamari, DarkFalls, Gordonmonaghan, Novis-M, Cognita, Bumper12, Plasticup, NewEnglandYankee, Hello i am a guy, Juliancolton, Kmanblue, Ginga123, Bonadea, Kinigi, Christopher.wright.2006, Rasgfsadgas, Xiahou, ThePointblank, CardinalDan, Richard New Forest, Idioma-bot, Spellcast, Ottershrew, Wordreader, VolkovBot, CWii, ABF, AlnoktaBOT, Philip Trueman, Yomammammammamifjuyigb, Nypr1nc355, TXiKiBoT, VivekVish, Carlangas, Anonymous Dissident, Historyexpert, Denseatoms, Hrothberht, Clarince63, Martin451, Leafyplant, LeaveSleaves, Snowonweb, Andrewrost3241981, Jeremy Bolwell, Maxim, Car utas, Falcon8765, Andygharvey, TommyMac71, Monty845, Nagy, Logan, Pruxo, Legoktm, NHRHS2010, Matt.t.martin, SkyViewOrphanage, Bfpage, SieBot, Tiddly Tom, Ellbeecee, Jauerback, Kodd, Camaranj, Krawi, Winchelsea, Dawn Bard, Caltas, Geniuskyle, Triwbe, SuzanneIAM, Matthew Brandon Yeager, Nelsondh, Happysailor, Flyer22, Alexbrn, Hobojoebob, Oda Mari, Buttons, Kay Körner~enwiki, JSpung, Allmightyduck, Koi-chan, Oxymoron83, Smilesfozwood, ShadowShift, Pymouss, Gift Of Ireland, CObot, Diego Grez-Cañete, Privilegg, Ascidian, Denisarona, Richard David Ramsey, Lascorz, Martarius, ClueBot, Jchatter, GorillaWarfare, 4everboy, Jeffy666, Iliaorkin, Fyyer, The Thing That Should Not Be, Drmies, SuperHamster, AtSwimTwoBirds, Neverquick, Puchiko, Excirial, Filippakos, Northernhenge, Naleh, Lucas hersey9, Gulmammad, Vivio Testarossa, Lartoven, Arjayay, Razorflame, Starrystar849, RegalStar, Mikaey, SchreiberBike, Thingg, Aitias, Gejan, PCHS-NJROTC, Amaltheus, Columbiabotany,

Mzmadmike, LeaveSleaves, Yeah568, Natg 19, Wikiisawesome, Lamro, Synthebot, KjellG, Biscuittin, SieBot, Calliopejen1, YonaBot, Caltas, Yintan, MudMonster, Happysailor, Flyer22, Oxymoron83, Kentynet, Pnelnik, Mrfebruary, Sfan00 IMG, TonyDodson, ClueBot, The Thing That Should Not Be, Watti Renew, Boing! said Zebedee, Niceguyedc, Ibormeith, Auntof6, Magnificascriptor, Alexbot, Jusdafax, Zaharous, Sun Creator, Mhockey, DumZiBoT, XLinkBot, HexaChord, Addbot, Narayansg, Some jerk on the Internet, Ronhjones, Pelex, CanadianLinuxUser, Groundsquirrel13, Grubel, BepBot, Glane23, AndersBot, Farmercarlos, LinkFA-Bot, Dyuku, Ejjjido., Rehman, Tide rolls, Function95, Gail, Jarble, Greyhood, Ettrig, Legobot, Luckas-bot, Yobot, Fraggle81, Geshrwh, Buddy431, Aboalbiss, Magog the Ogre, AnomieBOT, KDS4444, ThaddeusB, Collieuk, Apau98, World affairs, Bluerasberry, Materialscientist, A333, Beenturns22, ArthurBot, Xqbot, Cureden, Jonoaf, Sylwia Ufnalska, Anna Frodesiak, GrouchoBott, Mattis, GNRY09, Praisesouljaboy, ჯაყუ ქ, Mokopuna, FrescoBot, Jamesybhoycfc, Beaber, Tranletuhan, I.Sáček, senior, Lastlaff, Rekordrivne, K'Anpo, Lilaac, AstaBOTh15, Pinethicket, LittleWink, Jonesey95, Hamtechperson, Jschnur, PerV, TobeBot, كاشف عقيل, Lotje, Easttyrone, Specs112, Minimac, RjwilmsiBot, Alph Bot, Eekerz, Yca.zuback, Ajraddatz, Boundarylayer, Dewritech, K6ka, Chiton magnificus, John Cline, Matthewcgirling, Erianna, Augurar, Gsarwa, Uthican, Cd-Rom-1387, ChuispastonBot, 22tho, ClueBot NG, Wetlands International, Qarakesek, Clearlyfakeusername, Kasirbot, Bblawsonnn, Helpful Pixie Bot, Tholme, Gob Lofa, BG19bot, Jamie.mcarthur, Bcoultrip, AvocatoBot, Wikih101, Kvnchris, Cengime, Zackmann08, DignitySun53, R7500, IkbenFrank, Cyberbot II, YFdyhbot, Dexbot, Nonovyurbizniz, Corinne, Todaloo potato, NGREU, EMCKR, Tlambf,gh,gh, Njol, Library Guy, Skr15081997, St170e, Monkbot, Pokeball120, Kristof Jacunski, Johnjog, Macbane86, KasparBot, Luke jones3131 and Anonymous: 332

- **Permaforestry** *Source:* https://en.wikipedia.org/wiki/Permaforestry?oldid=612288110 *Contributors:* Edward, Alan Liefting, Pmetzger, Woohookitty, Nirvana2013, Gobonobo, DASonnenfeld, Rstafursky, Otisjimmy1, Themfromspace, AnomieBOT, Minnecologies, LilHelpa, Anna Frodesiak, Pat604, Isiaunia, Look2See1, Hamiltha, The Banner Turbo, Northamerica1000 and Anonymous: 7

- **Pine tar** *Source:* https://en.wikipedia.org/wiki/Pine_tar?oldid=685130117 *Contributors:* Cimon Avaro, Richard Avery, Zoicon5, AaronSw, Oaktree b, Auric, MPF, Sj, HangingCurve, Kpalion, MisfitToys, Rwv37, DragonflySixtyseven, Oknazevad, Mattingly23, Thuresson, Leadingbrand, Denniss, Avenue, Vuo, RHaworth, Zzyzx11, Banpei~enwiki, DePiep, Alyosha8, Koavf, Jweiss11, Kazrak, Sango123, Butsuri, Hydrargyrum, CrazyTalk, The JumpStation, Happydrifter, Wknight94, Conman33, SmackBot, Bluebot, Deli nk, Tsca.bot, Muboshgu, AdamWeeden, Sirgregmac, NaySay, Wizardman, Spentangeli, Cosmoline, Khono, IronGargoyle, 16@r, Aarktica, Dfred, PaulGS, Acsails, Zachary, AntiVandalBot, Brandonsmith, Epsoul, Father Goose, Anemte, Zetterberg40, Wdflake, BMRR, Ekotekk, R'n'B, J.delanoy, BigrTex, Khathi, AAA!, Jesant13, Ahhhsean, BrettAllen, DASonnenfeld, Oshwah, Paulburnett, Brianga, Rlendog, Flyer22, Ronny22, Andysjs, Xenon54, Addbot, Glane23, Lightbot, OlEnglish, Yobot, AnomieBOT, Bagumba, Goudron92, Jibjab1, Pinethicket, Aerolin55, Epsteamboat, MrX, OnePt618, GermanJoe, ClueBot NG, Chitt66, BG19bot, BattyBot, Foreverest97, Zenibus, Tymon.r, LaserShark, Freebie1492, ThaiWood and Anonymous: 114

- **Pinyon pine** *Source:* https://en.wikipedia.org/wiki/Pinyon_pine?oldid=685349398 *Contributors:* Malcolm Farmer, Marshman, JerryFriedman, DocWatson42, MPF, Abigail-II, Kate, Frank101, BryanD, Fledgeling, Schzmo, Rjwilmsi, Eubot, DVdm, Gdrbot, RussBot, Pigman, Hydrargyrum, Neilbeach, SmackBot, Thaagenson, Brossow, Kintetsubuffalo, Rkitko, Drphilharmonic, Thijs!bot, Peter coxhead, Grandia01, BeadleB, Acalamari, Sylfred1977, Colchicum, DASonnenfeld, Seb az86556, RW Marloe, Pinon~enwiki, Saint Aardvark, Addbot, OlEnglish, Yobot, Ptbotgourou, AnomieBOT, Plumpurple, DSisyphBot, LucienBOT, Zencowboy27, Wantonlife, DexDor, John of Reading, Look2See1, Smallchief, ClueBot NG, Alvisa88, Widr, Plantdrew, BG19bot, Kehkou, FoCuSandLeArN, Monkbot and Anonymous: 27

- **Pitch (resin)** *Source:* https://en.wikipedia.org/wiki/Pitch_(resin)?oldid=685882243 *Contributors:* Bryan Derksen, Montrealais, Hephaestos, Edward, Axlrosen, LMB, Drew3D, Kurt Eichenberger, Yath, Deglr6328, Rich Farmbrough, Mattisgoo, Eleland, Geo Swan, Stephan Leeds, Gpvos, Angr, Wafry, YurikBot, RussBot, AirLiner, Mike Dillon, Yvwv, SmackBot, Portillo, Ohnoitsjamie, NickGarvey, Agateller, Just plain Bill, The Man in Question, 10014derek, Angelpeream, Ale jrb, Ibadibam, The Ultimate Koopa, Thijs!bot, Moondigger, Saibo, AntiVandalBot, Dyolf~enwiki, Dougher, JAnDbot, 100110100, JamesBWatson, Swpb, TheEsb, Ectonic, Thehalfone, CommonsDelinker, Tgeairn, Pharaoh of the Wizards, Ayecee, Afluegel, DadaNeem, DASonnenfeld, WarddrBOT, TXiKiBoT, Perohanych, Lamro, AlleborgoBot, Biscuittin, SieBot, LeadSongDog, TX55, Ronny22, ClueBot, Followingjoshua, PixelBot, DumZiBoT, Freetrashbox, Addbot, MrVanBot, Mpfiz, Lightbot, PM-Lawrence, SlickRIFF, SaaHc2B, FrescoBot, RicHard-59, HRoestBot, Tom.Reding, Christopher chalmers, Alarichus, Vrenator, WCCasey, EmausBot, Wikipelli, ClueBot NG, Alex Nico, Badon, BattyBot, Jimw338, Frosty, KimMarie787, Thobis91, Berginsd, Monkbot, Shawtaroh, Thuyhuyhr634ujhws3ujhtw3y8o4t and Anonymous: 63

- **Putu (mushroom)** *Source:* https://en.wikipedia.org/wiki/Putu_(mushroom)?oldid=670813942 *Contributors:* KylieTastic, Dthomsen8, AnomieBOT, K6ka, Plantdrew, Cpt.a.haddock, Randykitty, Eman235, Deskolinsore and Leelambars

- **Rattan** *Source:* https://en.wikipedia.org/wiki/Rattan?oldid=684614351 *Contributors:* Miguel~enwiki, Notheruser, Mxn, WhisperToMe, Wik, Eugene van der Pijll, Tomchiukc, Chris 73, DocWatson42, MPF, Jason Quinn, Kudz75, Pgan002, Andycjp, CryptoDerk, Necrothesp, DanielCD, FT2, Xezbeth, Hesperian, Haham hanuka, Sjschen, Dowcet, Proski, LFaraone, IJzeren Jan, Lemworld, Woohookitty, Dhasenan, BD2412, FlaBot, Eubot, Margosbot~enwiki, JdforresterBot, Mathiastck, Gdrbot, YurikBot, Huw Powell, Gaius Cornelius, Lockesdonkey, Asarelah, Raijinili, Eno-ja, CharlieHuang, Thomas Blomberg, Nekura, SmackBot, Lawrencekhoo, EncycloPetey, Wikipedian c, Thumperward, Snori, Xx236, Shunpiker, Bardsandwarriors, RossF18, Lambiam, Snow cat, 2T, Dr.K., Iridescent, Paul venter, Sharp11, Phauly, Madun, Acs4b, Luccas, Thijs!bot, Missvain, Mmcknight4, Rautheyellavar, Seaphoto, Magioladitis, WLU, FisherQueen, R'n'B, CommonsDelinker, Kguirnela, J.delanoy, CSchoenberger, DASonnenfeld, Wilhelm meis, Idioma-bot, VolkovBot, Milnivlek, Lova Falk, SieBot, Johanges, Walls bridges, Flyer22, Gliu, Almufasa, Cyfal, ClueBot, Boing! said Zebedee, WDM27, Tryptamine dreamer, Zippyweirdo, Dekisugi, Ost316, Jakswan, Addbot, Grayfell, Ronhjones, Aamrs, CarsracBot, Nerdypunkkid, Tassedethe, Tide rolls, Bff, Yobot, Captain Spleen, Xufanc, Xvlcm12, Object404, Bob Burkhardt, ArthurBot, LilHelpa, Anna Frodesiak, William branston, DeNoel, FrescoBot, Alarics, Tim1357, FoxBot, Obsidian Soul, RjwilmsiBot, EmausBot, Mixedkitty, Tommy2010, PBS-AWB, Standupstriker9, Rickyhandicraft, ClueBot NG, CocuBot, Widr, Bl2nk0r, Richfrost, Plantdrew, Cerabot~enwiki, Whitestore, Hubertl, Monkbot, RiyadMA, TheQ Editor and Anonymous: 76

- **Resin** *Source:* https://en.wikipedia.org/wiki/Resin?oldid=682540243 *Contributors:* Vicki Rosenzweig, Ellmist, Heron, Vkem~enwiki, Lisiate, Lexor, Stan Shebs, Rl, Ehn, Sehrgut, Furrykef, Nv8200pa, Spikey, PuzzletChung, Robbot, Kristof vt, Dina, Alexwcovington, Giftlite, DocWatson42, MPF, Everyking, Solipsist, Bumm13, MattKingston, Discospinster, Darkone, Brian0918, Mickeymousechen~enwiki, Bob the Cannibal, Haham hanuka, Jordan117, Arthena, Sjschen, Walkerma, Wtmitchell, HenryLi, Cédric, Novacatz, TigerShark, Mandarax, Old Moonraker, Srleffler, Dunemaire, BMF81, Chobot, YurikBot, Tadanisakari, RattusMaximus, Shaddack, NawlinWiki, Tearlach, DAJF, LeonardoRob0t, Slipdisk101, SmackBot, Kilo-Lima, AndreasJS, Elwood j blues, Edgar181, Gilliam, Chris the speller, MalafayaBot, Da Vynci, Gruzd, Sbharris, Langbein Rise, Can't sleep, clown will eat me, Silverjonny, Kingdon, Smokefoot, DMacks, Just plain Bill, Clicketyclack, Geoffrey Wickham, AThing, Yakob~enwiki, Peterlewis, Majorclanger, Ginkgo100, Iridescent, Ü, Sinse~enwiki, Kanjilearner, Lord Warlock, BethHonan,

Erik Kennedy, Jibi44, GHe, Pewwer42, Gogo Dodo, Khatru2, Room101, Rosser1954, Thijs!bot, Marek69, Lethargy, Savatar, Siawase, Nick Number, Dr. Blofeld, TuvicBot, Plantsurfer, Epeefleche, Nodoremi, Jonemerson, Hdmodi, CTF83!, Jim Douglas, Euhedral, Wikiak, Philander, Ksero, Stealthound, CHENCHICHEN, Jager1224~enwiki, Bissinger, Icorey, Nono64, Slash, Hans Dunkelberg, Uncle Dick, Eliz81, Mikael Häggström, NewEnglandYankee, Touch Of Light, DorganBot, DASonnenfeld, Idioma-bot, VolkovBot, TXiKiBoT, Vipinhari, JhsBot, Falcon8765, Moral Simplex, AlleborgoBot, Solicitr, Vincevance, Samuelmbaskin, Roshav ohluv, Jamiehook01, The Wolfgang, Martarius, ClueBot, Ferred, Dpachterberg~enwiki, Matdrodes, JesSiiCaaH, SuperHamster, Auntof6, Takeaway, Muhandes, Kauri Gumdigger, Gupta02kanishk, Eliezerh, ChrisHodgesUK, Schatten213, Widenercontent, Vojtěch Dostál, Alexius08, Kaiwhakahaere, Addbot, Some jerk on the Internet, Daughterof-Sun, Tide rolls, Lightbot, OlEnglish, ChemProf101, Middayexpress, Luckas-bot, Yobot, AnomieBOT, Walrus heart, AdjustShift, Koshelyev, Derkernsting, Hunnjazal, Kerryw85, ArthurBot, Xqbot, Anna Frodesiak, GrouchoBot, Joey2201, RibotBOT, Riventree, חנש1979, Cannolis, Dneyder, Pinethicket, HRoestBot, RedBot, Inbamkumar86, TobeBot, Lotje, Callanecc, Dinamik-bot, Amyisthis, Problems123, EmausBot, Craxyxarc, Wikiuser34, Wikipelli, ZéroBot, Westley Turner, SporkBot, Mjbmrbot, ClueBot NG, Uacs451, Mannanan51, Widr, BG19bot, PhnomPencil, NotWith, Henry McClean, Calathea, Mr. Guye, Sminthopsis84, WolfShadow9192, 93, Ugog Nizdast, Glaisher, Ghost Lourde, P. S. Sena, KasparBot and Anonymous: 215

- **Root beer** *Source:* https://en.wikipedia.org/wiki/Root_beer?oldid=686296957 *Contributors:* Magnus Manske, Mav, MarkAtwood, Rmhermen, DavidLevinson, Olivier, RTC, Infrogmation, Dominus, Ixfd64, Gbleem, Minesweeper, Dgrant, Kingturtle, Sugarfish, Vzbs34, Cimon Avaro, Palfrey, Lukobe, Mxn, Fuzheado, Furrykef, Nv8200pa, Secretlondon, EdwinHJ, KnightofNEE, LGagnon, Andrew Levine, MarkMcDermott, Lupo, Cyberia23, Cecropia, Xanzzibar, Centrx, DocWatson42, MPF, Fennec, Philwelch, BenFrantzDale, Abigail-II, Zigger, Cool Hand Luke, Xerxes314, Everyking, Niteowlneils, Yekrats, Macrakis, SWAdair, Pne, Gyrofrog, Uranographer, Alexf, Mr d logan, Beland, Piotrus, Polyfrog, Latitude0116, Yossarian, Tooki, JulieADriver, Adrian727, Oknazevad, Safety Cap, Flex, Grstain, Dr.frog, Jayjg, Freakofnurture, Discospinster, Ahkond, Eric Shalov, AlexKepler, Duemellon, ESkog, Nharmon, Yamavu, Neko-chan, MisterSheik, Mr. Billion, Workster, Aaronbrick, Adambro, Bobo192, Smalljim, TheProject, NickSchweitzer, Mh26, Sam Korn, Polylerus, Jonathunder, Merope, Alansohn, Retran, Eric Kvaalen, Munchkinguy, Toon81, Snowolf, Wtmitchell, Velella, Ronark, SidP, KingTT, Zantastik, Cburnett, Yuckfoo, Cyraan, RainbowOfLight, Guthrie, Werty8472, Saxifrage, Dismas, The JPS, Minivet, Starblind, Woohookitty, Grillo, Akira625, Agaran, Zzyzx11, Kesla, Yoghurt, Graham87, Deltabeignet, Deadcorpse, FreplySpang, Rjwilmsi, Nightscream, Jweiss11, Ian Lancaster, SeanMack, Rangek, FlaBot, Nivix, Gurch, DevastatorIIC, VolatileChemical, Gwernol, Measure, YurikBot, Jimp, Priest4hire, FrenchIsAwesome, Bhny, Lord Voldemort, Akhristov, Philopedia, Bovineone, Thane, Friday, NawlinWiki, Astral, Cryptoid, AJHalliwell, Trovatore, SigPig, Haoie, Raven4x4x, Nut-meg, KARENJC, Vlad, Dddstone, MacMog, Mistercow, Superiority, Varano, Super jedi droid, Emana, Open2universe, Chase me ladies, I'm the Cavalry, Theda, Fourohfour, Ilmari Karonen, Mrblondnyc, Junglecat, Captain Cornflake, Airconswitch, SmackBot, Not cat, Bigbluefish, C.Fred, Strabismus, Michael Dorosh, Antrophica, Jab843, Piper108, Kintetsubuffalo, Edgar181, Srnec, David Fuchs, Yamaguchi 先生, Aksi great, Gilliam, Ohnoitsjamie, Hmains, Smartducky, DanMonkey, Chris the speller, Bidgee, Keegan, TimBentley, Doctor Love, Jnelson09, Jprg1966, Barthalamule, Fuzzform, Mike1, SchfiftyThree, Moshe Constantine Hassan Al-Silverburg, Fishhead2100, Nbarth, Colonies Chris, Oatmeal batman, Gracenotes, Gsp8181, KittyKat, Fiddling Frog, Can't sleep, clown will eat me, Tamfang, Djido, Typofixer76, Mr.bonus, JonHarder, EvelinaB, Captbob007, Addshore, Squamate, EI at10s, PrometheusX303, Khukri, Nibuod, JGGardiner, Kneale, Doberman Pharaoh, Only, Zzorse, Kalathalan, Marcus Brute, Textor, Pilotguy, Ged UK, Kilonum, Valfontis, JzG, John, SilkTork, LWF, Narmical, Robofish, Benesch, Gonioul, Minna Sora no Shita, Ztras, Mxreb0, JHoltzman, LorD, Oughgh, Scoty6776, CPAScott, A. Parrot, WhiteHatLurker, Skaroly, Doczilla, Synergism, Mountainmage, Jack O'Neill, DarkCow, Iridescent, J Di, UncleDouggie, Quixotequest, Drvanthorp, Courcelles, JayHenry, Tawkerbot2, Billastro, IronChris, Amniarix, CalebNoble, JForget, Unidyne, Wizardimps, Timothy Chavis, Skabat169, PuerExMachina, Wafulz, Roseynose, Makeemlighter, Jsmaye, Matthew Auger, Mauvais, Bennyp77, Aaronpark, Moreschi, Liu Bei, Slazenger, Scottiscool, JonnyLightning, Gogo Dodo, Forest guy, Adolphus79, Darkerzen, Tawkerbot4, Doug Weller, DumbBOT, Chrislk02, Northwest, Daven200520, Em-jay-es, John Lake, Thijs!bot, Epbr123, Chitomcgee, Orlyorlyorly, Beerman~enwiki, Marek69, Tapir Terrific, Ufwuct, X96lee15, Floridasand, Scottandrewhutchins, Visik, Jackftwist, KrakatoaKatie, AntiVandalBot, Colo1115, Seaphoto, NeilEvans, Doc halidai, Jj137, Tmopkisn, RobJ1981, LibLord, Farosdaughter, Pixelface, Altamel, Fireice, Pailman, Apavlo, Canadian-Bacon, Billertl, Agentrootbeer007, Davewho2, Tymeflame, InsaneShiyn, PubliusFL, Mrwhizzard, Suede~enwiki, Acroterion, Freedomlinux, ZPM, Bongwarrior, VoABot II, Thedoorhinge, GameMasterCor, JNW, Fordsfords, Jay Gatsby, -Kerplunk-, J.P.Lon, Unholy Enkidu, Think outside the box, Cadsuane Melaidhrin, Srbs, Tedickey, Eldude611, Satch234, SparrowsWing, Midgrid, Allstarecho, XMog, LorenzoB, $yD!, JaGa, Edward321, WLU, TheManStan, AdmRiley, FilmFemme, Theupsguy, MartinBot, Anaxial, CommonsDelinker, J.delanoy, BeTheB, Rgoodermote, SimpsonDG, Eliz81, Extransit, T2t2, Whitebox, Theeurocrat, Maproom, Gman124, Woodega, Pyrospirit, RenniePet, Comp25, Belovedfreak, NewEnglandYankee, Hennessey, Patrick, Witeandnerdy, Lightningbug, Nadiatalent, Omegamormegil, LordCo Centre, Useight, Fankhadb, DASonnenfeld, Upex, Danpskinner, Joharri, Deor, CWii, Jmrowland, Philip Trueman, Damonshutak, Somvilay, TXiKiBoT, Someguy06, Oshwah, Wikidemon, Finniksa, Monkey Bounce, LeaveSleaves, Aloha27, RiverStyx23, Billybobmoose, Bgf456, DeathNomad, Meters, Wolfrock, Faa Q, Ilovepierap, Falcon8765, Magiclite, Insanity Incarnate, Mark164, Cenzo 3, Eloc Jcg, Zxyggrhyn, Vampire wings252, StAnselm, Mikemoral, Upex1, Yintan, Johnagraham, Gravitan, Aaronh7, Keilana, Flyer22, Tiptoety, Longshotz, Kguske, Editore99, ColaDude, Snevetsm, Antonio Lopez, Faradayplank, Nuttycoconut, Smhaus, Techman224, Sean.hoyland, Mygerardromance, DRTllbrg, Escape Orbit, Randy Kryn, Blahblah5555, ClueBot, AndrePeltier, Snigbrook, The Thing That Should Not Be, Chad01, Pi zero, Jimmyjuan, Arakunem, Adventhesis, Tubetendsd, Knowledgepower23, Foofbun, Blanchardb, Stylteralmaldo, Piledhigherand-deeper, Breulib, Excirial, Danielabt, Rhododendrites, JamieS93, Eastcote, Finfid, Carawombat, Bald Zebra, Возрождение, Aitias, Shava23, Rolosrevenge, Berean Hunter, Canihaveacookie, Apparition11, Vanished User 1004, DumZiBoT, Linkofspades, XLinkBot, Facts707, Mifter, Taylorbanana, Deaniew557, MatthewVanitas, Captain Package, Addbot, Willking1979, Slayoxus, Some jerk on the Internet, Jojhutton, Captaintucker, Atethnekos, Blethering Scot, Jb0007, Pinbob87, Cammmmmmmmeron, Fluffernutter, Dreambig125, Cst17, Download, Glane23, Jomunro, Jasper Deng, West.andrew.g, Rawr5656, Peridon, Tide rolls, Lesbrown99, Ochib, Root beer, Luckas-bot, Yobot, 2D, RockfangBot, II MusLiM HyBRiD II, Gobbleswoggler, AnomieBOT, Samterch, Rubinbot, Piano non troppo, Sodajerk66, Materialscientist, Jaycor77, Clark89, MaksimBurnin, Obersachsebot, Poop1994, Darth Beppo, Yomama127, The sock that should not be, Capricorn42, Ryomaandres, ChildofMidnight, Lsevedge, Purplebackpack89, NateY94, Hobosteeth, Nataliejinlubber, AbigailAbernathy, Sabritones, Ruy Pugliesi, GrouchoBot, Tschingeld, Mechla, Asoccer11, Rootbeerjunky, Thaflinger, Paytokay, E0steven, Griffinofwales, FrescoBot, Sweetestmalefica, Pepper, Sky Attacker, Sean4131987, Johnnyfuga, Weetoddid, Simpsons1994, Pinethicket, HRoestBot, Abductive, Emo-kid-99, Jschnur, Bigal2rin, Serols, SkyMachine, Irbisgreif, Jerksoda66, Deeminus, Yojoemcfee, GregWooledge, Bento00, AgentSG, Salvio giuliano, Skamecrazy123, EmausBot, Orphan Wiki, Sm123rocks, RA0808, Gogophergo, Smappy, Wikipelli, AsceticRose, K1113k21k7, Imperial Monarch, Juneeluv, Ὁ οἶστρος, AvicAWB, UrbanNerd, Christina Silverman, FinalRapture, Wingman4l7, 11sharpshot, Erianna, Rickinasia, L1A1 FAL, AlexPressey, Jp98745615555478, Traversablehen, Thesmartalec, Kilroywasnthere, Cgt, ClueBot NG, Sprecher, This lousy T-shirt, Catlemur, Danmorcos, Joefromrandb, Frashizzletrainer, Angelinarox, Twillisjr, CopperSquare, Widr, Condor66, Helpful Pixie Bot, Lawnbon, Ashokhatt, Roberticus, Northamerica1000, Cthulhu Rising, AvocatoBot, ErikBly, Dipankan001, Owhatn0wo, Amandazu, Wombatcombat11, DeltaCommand, Whitenigar360, Mediran, NFLisAwesome, Lugia2453, Frosty, Jamesx12345, Little green rosetta, Hairynuts, The Green Dragons with Hats, Corn cheese, GWPabst, MichealfosterSTI, Nessabug714, Jhscarborough, JakeWi, Guysthatarereallycool1s2, Anthony rootbeer, PurserSmith, Welderhelper, Mange Bjuder på kladdkaka, SL1358, Gknotts, Somchai Sun, StanleyGrohl, Manul, AddWittyNameHere, Truenorth069, DudeWithAFeud, Hushed-scream, Oficialeditor, JaconaFrere, Nyoung914, CyberXRef, Pencilcity, Zesterer, Undentman, AwwwYeahRulerz, Danpburton, Ghost 1231, 1234567890098dhfjksdhfsjkhfkshfks, Rkon3318, Ian Doyle Capita, Straightouttabrazil, Hellodolly3, Skullpig, JJMC89, Tropicalkitty, Lluuk-keeppaayynnee and Anonymous: 922

- **Rubia** *Source:* https://en.wikipedia.org/wiki/Rubia?oldid=680179719 *Contributors:* WojPob, WormRunner, Ashley Y, DocWatson42, MPF, Yak, Brona, Yath, JoJan, Jossi, Rellis1067, DanielCD, Mani1, Jonsafari, Hesperian, Stemonitis, Richard Barlow, Ketiltrout, Vberger~enwiki, Kugamazog, FlaBot, Duomillia, PKM, Gdrbot, Vmenkov, Gaius Cornelius, Shaddack, TDogg310, Asarelah, Emijrp, SmackBot, Brossow, Kintetsubuffalo, Durova, MalafayaBot, ChazYork, Juhachi, Neelix, Arb, Rosarinagazo, Escarbot, JAnDbot, Hydro, Maias, Klausok, Lackwit, CommonsDelinker, Nono64, Wlodzimierz, Kloisiie, DASonnenfeld, Steel1943, VolkovBot, Rei-bot, Taranah, Keilana, Moonraker12, Dstlas-caux, Jordi Roqué, CarolSpears, Wildcolour, Cygnis insignis, HumphUK, PixelBot, SchreiberBike, Andieluijk, Addbot, LatitudeBot, Spilling-Bot, JSR, Yobot, Choij, Daniele Pugliesi, ArthurBot, LilHelpa, Weepingraf, Noder4, Manjishtha, Dechog, Orbicule, Balablitz, RjwilmsiBot, Skamecrazy123, EmausBot, Robiquetgobley, MerlIwBot, Plantdrew, Rjlanc, NotWith, DaltonCastle, KasparBot and Anonymous: 29

- **Sassafras** *Source:* https://en.wikipedia.org/wiki/Sassafras?oldid=683355565 *Contributors:* The Anome, Danny, Roadrunner, Heron, Chuq, Ram-Man, Infrogmation, Delirium, CatherineMunro, Kimiko, Chrysalis, Arteitle, SEWilco, Mp~enwiki, Nixve, Pollinator, Tomchiukc, Worm-Runner, Nurg, Rfc1394, MPF, Guanaco, Jorge Stolfi, Foobar, Ryanaxp, SoWhy, Bumm13, Aramgutang, Oknazevad, Ratiocinate, DanielCD, Discospinster, Guanabot, Vsmith, CanisRufus, Sole Soul, Cmdrjameson, Mh26, Hesperian, Nsaa, Alansohn, Pen1234567, Velella, Axeman89, Djsasso, RM, Ipeterson, Pixeltoo, Mandarax, BD2412, Rjwilmsi, Ittiz, JHMM13, Heah, FlaBot, TrinkerBell, Eubot, Wowbobwow12, Gdr-bot, The Rambling Man, YurikBot, Jhbeck23, Hellbus, Gaius Cornelius, Bovineone, Alynna Kasmira, ENeville, Joshdboz, Grafen, Ezeu, TDogg310, JPMcGrath, Davidbessler, Josh3580, Katieh5584, Alexandrov, Sacxpert, SmackBot, Mangoe, Kintetsubuffalo, Edgar181, Yam-aguchi 先生, Rkitko, Fuzzform, Ankurjain, Berton, Deli nk, Bazonka, Wikipediatrix, Nbarth, Can't sleep, clown will eat me, TheKMan, DAMD, Cunners, Kingdon, Cybercobra, Nakon, NaySay, DMacks, BiggKwell, AThing, U-571, Vanished user 9i39j3, Zaphraud, Heimstern, Yfoiler, Kevmin, BillFlis, Martinp23, Jgrahn, Shoeofdeath, Courcelles, Enock1970, Tawkerbot2, Tanthalas39, Egoadvocate, Exhummerdude, Baskaufs, NickW557, Mlwilson, Shawnzhao, Epbr123, Typing monkey, Gerard c, PHaze, Uruiamme, Shirt58, Kbthompson, Smartse, Fxnick, Altamel, Res2216firestar, Erxnmedia, JAnDbot, Altairisfar, Plm209, Meeples, VoABot II, CTF83!, Nyttend, Allstarecho, Kjmoran, Carstor, Peter coxhead, Rickard Vogelberg, Jerem43, MartinBot, J.delanoy, Vanished user 342562, Jameschristopher, Chiswick Chap, Knulclunk, Na-diatalent, Juliancolton, Mirithing, Use the force, DASonnenfeld, Jmrowland, Philip Trueman, Kuyan, LeaveSleaves, RiverStyx23, January2007, Doug, Falcon8765, Burntsauce, Socksysquirrel, Brianga, SieBot, Calliopejen1, Denimskater, TheAngriestPharmacist, Keilana, Oxymoron83, Diyforlife, Rfstell, The Wolfgang, KBYU, ClueBot, LAX, Avenged Eightfold, Waxsin, Clairelawt, Parkwells, Jeffmizzohr, LarryMorseDCO-hio, Versus22, Katanada, OldOrnerySnitch, Addbot, DOI bot, Friginator, Terrillja, Luckas-bot, Yobot, Jason Recliner, Esq., Intothewoods29, THEN WHO WAS PHONE?, Tempodivalse, AnomieBOT, Usetica, Jim1138, Xufanc, Citation bot, Bob Burkhardt, Stanislao Avogadro, Lil-Helpa, Xqbot, Darth Beppo, JimVC3, Gigemag76, Renaissancee, Patsfball, Phn229, RibotBOT, Brutaldeluxe, Bshell, Slowart, Hamamelis, Andromeas, FrescoBot, Cranhandler, BlueCap88, Rocazii, Citation bot 1, Pinethicket, HRoestBot, Jonesey95, Evenrød, 村木荒茂, System-atic review, Karen1monger, Tbhotch, Obsidian Soul, EmausBot, Acather96, Dewritech, Racerx11, Slightswikie, Wikipelli, Glavindref, Fæ, Ὁ οἶστρος, Tolly4bolly, Removingext, Donner60, Ego White Tray, Potatoepoptarts, Architect7, Johnny chedda, ClueBot NG, Neux-Neux, Eocene guy, Theopolisme, Helpful Pixie Bot, BG19bot, Dylanhaynes, Altaïr, Shisha-Tom, BattyBot, Sminthopsis84, Makecat-bot, Morfusmax, Every-morning, PurserSmith, MAXISWRONG, Manul, Jachoox, TF92, Monkbot, Thunderfinger, Jarrodhill, Jazmin7778, Betyeyanna, Cayton47 and Anonymous: 270

- **Serenoa** *Source:* https://en.wikipedia.org/wiki/Serenoa?oldid=683584013 *Contributors:* Jaknouse, Tedernst, Stan Shebs, Jdforrester, Fredrik, Eric Yarnell, Seglea, Texture, MPF, Beland, D6, Bastique, Arcadian, Dejitarob, MARQUIS111, Hesperian, A2Kafir, MikeMaughan, Ish ishwar, Stemonitis, Richard Barlow, Rjwilmsi, Nneonneo, Eubot, Gdrbot, Debivort, Wiserd911, YurikBot, JoeMystical, RussBot, Alynna Kasmira, Psi-kat, TDogg310, Asarelah, IceCreamAntisocial, 2over0, Jm546, Chriswaterguy, F. Cosoleto, Mmcannis, DearPrudence, John Broughton, SmackBot, Slashme, EncycloPetey, Edgar181, Apers0n, Deli nk, Balin42632003, Uthbrian, DMacks, Ligulembot, InNuce, MrDarwin, Pondle, Beetstra, SQGibbon, LAlawMedMBA, Ulefoss300, MayaSimFan, FrancoGG, Thijs!bot, Mmcknight4, Emeraldcityserendipity, Tachyonwaves, Altairisfar, MER-C, Hydro, Alexander Domanda, Thunderhead~enwiki, Johnshopkinshealthalerts, Icenine378, J.delanoy, Bonadea, DASon-nenfeld, Idioma-bot, Jamiegreen19, VolkovBot, Philip Trueman, TXiKiBoT, FlorisGr, Malljaja, Someguy1221, Flopster2, Fbs. 13, Moretto, Alexbrn, Fbguzz36, HighInBC, IvanTortuga, Emokillerjord, Cephacles, PixelBot, Prof Health, Sallicio, LizGere, Addbot, DOI bot, Rorycalvin2, SamatBot, Yobot, AnomieBOT, Materialscientist, Citation bot, حمد.غ.مد.يديماح24, LilHelpa, GrouchoBot, Mathonius, Hamamelis, Andromeas, Dan Wylie-Sears 2, Citation bot 1, Pinethicket, Mrernesto, RjwilmsiBot, 7mike5000, Peterusso, Djhomeworks, ZéroBot, TheoriginalNothingNow, Ihardlythinkso, Michael Bailes, Rsc227, Helpful Pixie Bot, Jamespenn1234, NotWith, Catlith, BattyBot, LieberPeppi, Everymorning, Joseph Laferriere, Monkbot, SilverBlogs, Bfxieucd and Anonymous: 45

- **Shellac** *Source:* https://en.wikipedia.org/wiki/Shellac?oldid=681766700 *Contributors:* Bryan Derksen, KF, Dante Alighieri, GTBacchus, Ahoer-stemeier, Jimregan, Schneelocke, Sanders muc, Xanzzibar, Pengo, Alan Liefting, BenFrantzDale, Tsca, Saxsux, Beland, Joyous!, Moxfyre, N-k, Discospinster, Vsmith, Horsten, Bender235, Felagund, Bobo192, Jxn, Pharos, Pearle, Mareino, Abstraktn, Arlosuave, Walter Görlitz, Arthena, Philip Cross, Sjschen, Hohum, Luigizanasi, Woohookitty, Qanuta, Travis Emergency, Palica, Deltabeignet, V8rik, Rjwilmsi, Nightscream, Quiddity, MWAK, Hannu83, Galessa, YurikBot, Jimp, Gilesgoat, Awave, Shaddack, Veledan, Janke, Irishguy, Neil Hooey, Bota47, Kkmurray, Scheinwerfermann, Where next Columbus?, CharlieHuang, NeilN, SmackBot, Lilliwan, Slashme, David Shear, Stepa, Anastrophe, Kintetsubuf-falo, Hmains, Krnntp, Chris the speller, Avin, Columba livia, SchfiftyThree, RayAYang, Oni Ookami Alfador, Epastore, Sbharris, Kotra, Qabach, Wes!, Just plain Bill, LDHan, Carnby, JohnI, Peterlewis, Anderh, Bendzh, A.kruse, Mvsrhollywood, DouglasCalvert, Tmangray, Haus, Fanacek, Fumblebruschi, PKT, AndrewDressel, James086, Dekker, Nick Number, Chubbles, DarkAudit, CForrester, DagosNavy, Belg4mit, Attarparn, Rlramirez, Oskay, Alvian, Frotz, Philg88, Blacksqr, Rettetast, R'n'B, Verdatum, Millwark, Rod57, Foetusized, AntiSpamBot, Chiswick Chap, Group29, DorganBot, Inwind, DASonnenfeld, Kyle the bot, Newtown11, Burpen, Rei-bot, Slysplace, BotKung, Modocc, Nedrutland, Andy Din-gley, Falcon8765, Woodturner9, Ajrocke, Gaelen S., Bdentremont, Scalebug, Pteronura brasiliensis, Imcaslan, Lizardo tx, BayReckoner, Chump Manbear, ClueBot, Meltingfeather, Samba pa ti, DragonBot, AndrzejGlo, Doprendek, SchreiberBike, DumZiBoT, Londonsista, Wikiuser100,

Durden8823, TheAnonymousRabbit, EncycloCritique, FriendlyParasprite, Clr324, Dough34, Mogs03, RYasmeen (WMF), Anamikagioiaxo, Monkbot, ACathe-Leigh, Apkurnia1, Markie1992, P. S. Sena, Requiem for a Daydream, Blascelle, אילן שמעוני and Anonymous: 249

- **Toddy palm** *Source:* https://en.wikipedia.org/wiki/Toddy_palm?oldid=665019485 *Contributors:* Badagnani, Closedmouth, Cencurut~enwiki, Ronhjones, Josve05a, Minerv, Plantdrew, Aisteco, Vladimir Alexiev and Anonymous: 1

- **Turpentine** *Source:* https://en.wikipedia.org/wiki/Turpentine?oldid=683574413 *Contributors:* The Anome, Kowloonese, Rmhermen, Heron, Hephaestos, Nealmcb, Lexor, Kku, Mkweise, Ellywa, Glenn, Crusadeonilliteracy, Marshman, Imc, Secretlondon, Owen, Klang, Pingveno, MPF, Fennec, Luis Dantas, BenFrantzDale, Timpo, Fak119, Solipsist, Bobblewik, Andycjp, Publunch, Ukexpat, Picapica, Eyrian, Luqui, Flapdragon, Billlion, Phoenix Hacker, Shenme, Cohesion, PaulHanson, Jezmck, Sjschen, AndreasPraefcke, Fledgeling, Vuo, Gene Nygaard, Kazvorpal, Brycebenton en, CharlesC, Andrev, Emerson7, Scottanon, Rjwilmsi, Bubba73, Alex20850, FlaBot, Margosbot~enwiki, Antrax, Srleffler, Chobot, WriterHound, Kummi, YurikBot, Jimp, DMahalko, Okedem, Manop, Shaddack, Matnkat, Inhighspeed, Snagglepuss, Syrthiss, Hosterweis, Kirper, Ninly, BorgQueen, JLaTondre, AssistantX, SaulPerdomo, SmackBot, Kintetsubuffalo, Edgar181, Bluebot, Sadads, Langbein Rise, Yaf, Mosca, Abrahami, Blake-, Ilovebees, Smokefoot, Just plain Bill, Kc12286, Bezapt, Lambiam, Turbothy, John, Žiga, Beetstra, Dicklyon, Rubikfreak, Wwallacee, Dani 7C3, MightyWarrior, FlyingToaster, Rifleman 82, Bilal's dog, Sagaciousuk, Oreo Priest, Whiteknox, EdgarCarpenter, Nelsonc5, Noroton, Golgofrinchian, Sluzzelin, JuanSacco, JAnDbot, VoABot II, Bertramanda, Senthryl, TimBuck2, It Is Me Here, Arrgh406, Balthazarduju, Colchicum, Cadwaladr, DadaNeem, DASonnenfeld, Chris™, Anonymous Dissident, Andy Dingley, Prakash Nadkarni, WacoJacko, Richard David Ramsey, Escape Orbit, Chem-awb, ClueBot, EoGuy, Crywalt, Thewellman, Johnuniq, Erodium, Ost316, Wiki-Dao, MystBot, Curtlee2002, Addbot, RobinClay, 安德烈烈斯, QuadrivialMind, OrlandoVenice, Luckas-bot, Yobot, CheMoBot, AnomieBOT, Jim1138, Bob Burkhardt, Neurolysis, Xqbot, TheAMmollusc, K731, Omnipaedista, Amaury, Riventree, HamburgerRadio, Lilaac, HRoestBot, A412, Abc518, Lotje, Medchemct, RjwilmsiBot, Aethalides, DASHBot, Muchie11791, Gfoley4, Milibethjida, SirSirrrr, ResearchRave, ClueBot NG, Joefromrandb, Castncoot, Theopolisme, Helpful Pixie Bot, Addihockey10 (automated), Explodo-nerd, Ugncreative Usergname, Blakpik, BattyBot, Springmata, JZNIOSH, Tentinator, Chemieexperte12, Astro interest, Bobpussy, KasparBot and Anonymous: 142

- **Urushiol lacquer** *Source:* https://en.wikipedia.org/wiki/Lacquer?oldid=686413579 *Contributors:* Kowloonese, Dominus, Stan Shebs, Bluelion, Fuzheado, Lkesteloot, SEWilco, Xuanwu, Knobunc, DocWatson42, MPF, Ahltorp, Michael Devore, Revth, Mboverload, Chowbok, Zeimusu, Beland, Kusunose, Sam Hocevar, Sonett72, Maikel, Discospinster, Bender235, Kbh3rd, Aranel, Kwamikagami, Aaronbrick, Stude62, Sorent, Eric Kvaalen, Sjschen, Velella, LordAmeth, Prattflora~enwiki, HenryLi, Ceyockey, Luigizanasi, Kznf, Brookie, Rvlaw, Woohookitty, Jean-Pol Grandmont, TotoBaggins, SDC, Sparkit, Biriwilg, Ian Lancaster, SMC, Benjwong, Narvalo, YurikBot, Wavelength, Grifter84, Bhny, Gaius Cornelius, Shaddack, ENeville, The Ogre, Dforest, Froth, Zephalis, Tachs, Vivafelis, Ash Crow, CharlieHuang, That Guy, From That Show!, SmackBot, Vorophobe, Edgar181, Hmains, Hyper megaman, Chris the speller, Thumperward, TheLeopard, Trekphiler, KaiserbBot, Scooterman, Soumyasch, Niczar, HappyVR, Keahapana, Norm mit, JoeBot, Tmangray, Tawkerbot2, CmdrObot, MiShogun, Cydebot, Luvlymish, Mato, Gogo Dodo, Zeno Izen, Trident13, Rajesh dangi, AntiVandalBot, Goldenrowley, Austinmurphy, Ipthief, الدبوني, LorenzoB, DerHexer, Truthordare, Arjun01, Red Sunset, R'n'B, CommonsDelinker, Verdatum, Mrgr4nt, Silverxxx, Johnbod, Balthazarduju, Bushcarrot, Cadwaladr, Gavinmason, DASonnenfeld, Iwavns, Jeff G., Andy Dingley, Geanixx, Galamazoo12, Northfox, Michael Frind, PericlesofAthens, Botev, GeiwTeol, Keilana, Permacultura, ImageRemovalBot, ClueBot, Himansud, Obelix83, Seanwal111111, Tonkawa68, Hwalee76, Vilkapi, Johnuniq, DumZiBoT, Wikiuser100, Addbot, Heavenlyblue, Hattar393, New-deal2007, Kman543210, Chamal N, Ws227, Math Champion, Unara, Materialscientist, Hunnjazal, Citation bot, حسن علي البط, Byassine52, Snehaparmekar, Ganeshjdev, Banhtrung1, Lotje, John of Reading, AManWithNoPlan, Deutschgirl, Tomásdearg92, Ebehn, ClueBot NG, Widr, Rurik the Varangian, MerlIwBot, Helpful Pixie Bot, BG19bot, Cold Season, BattyBot, ChrisGualtieri, SherabPuntso, Frosty, Rchadel~enwiki, SamoaBot, IQ125, Marigold100, YiFeiBot, Writers Bond, Boo2blue, Poultice, Niyayeye, KasparBot, Archaeo2, Theeditor500000 and Anonymous: 144

- **Vanilla** *Source:* https://en.wikipedia.org/wiki/Vanilla?oldid=685756766 *Contributors:* 0, Brion VIBBER, Vicki Rosenzweig, Malcolm Farmer, Rmhermen, Novalis, Heron, Edward, Bdesham, RTC, Infrogmation, Michael Hardy, Dominus, Ixfd64, Tomi, Pcb21, Ahoerstemeier, Stan Shebs, Jimfbleak, Docu, Snoyes, Suisui, Julesd, BenKovitz, SeeSchloss, Pascal, JonMoore, Dysprosia, Tpbradbury, Big Bob the Finder, E23~enwiki, Topbanana, David.Monniaux, Pollinator, Rhys~enwiki, Gentgeen, Robbot, WormRunner, Rfc1394, Fuelbottle, Oobopshark, Dina, Lysy, DocWatson42, BenFrantzDale, Lode Runner, Dratman, Gilgamesh~enwiki, Ceejayoz, Mboverload, Eequor, JoJan, Hans castorp81~enwiki, Arsene, Adamsan, DragonflySixtyseven, Deleting Unnecessary Words, B.d.mills, Burschik, Acad Ronin, Mike Rosoft, Rich Farmbrough, Francis Davey, Xezbeth, Darren Olivier, ESkog, Andrejj, Neko-chan, MisterSheik, Goto, Svdmolen, Bobo192, Longhair, Geocachernemesis~enwiki, ArkansasTraveler, Cayte, Nk, Physicistjedi, LostLeviathan, MPerel, Vizcarra, Grutness, Alansohn, Gary, Ctande, Philip Cross, Riana, Derumi, Nighthawk4211, Suruena, Angelo, Fdewaele, Sandover, Mindmatrix, Daniel Case, Lifung, SCEhardt, J3bus, Mandarax, Graham87, Bunchofgrapes, Jclemens, Yardensachs~enwiki, Rjwilmsi, JHMM13, Soakologist, Fred Bradstadt, Orchi, Gringer, FayssalF, RexNL, Gurch, Alphachimp, King of Hearts, Metropolitan90, DVdm, Gdrbot, Bgwhite, Roboto de Ajvol, The Rambling Man, Wavelength, Peter G Werner, Pip2andahalf, RussBot, Chris Capoccia, Epolk, Elephant53, Hellbus, Gaius Cornelius, Shaddack, Emiellaiendiay, Rsrikanth05, GeeJo, Anomalocaris, ENeville, Shikan, Apokryltaros, PeterDelmonte, THB, Maunus, Miraculouschaos, Emijrp, Ninly, Abu adam~enwiki, Yaztromo, CWenger, Jeffreymcmanus, The Way, Alexandrov, Mouseboks, Jade Knight, ChemGardener, Pouchkidium, SmackBot, F, Griot~enwiki, Zen00, Delldot, StefanoC, Aaronproot, Canthusus, Kintetsubuffalo, TheFourthWay, Edgar181, Yamaguchi 先生, Peter Isotalo, Anwar saadat, Bluebot, Rajanm13, Moshe Constantine Hassan Al-Silverburg, Deli nk, Fluggo~enwiki, H Bruthzoo, Can't sleep, clown will eat me, Nitchawan~enwiki, Kerecsen~enwiki, VMS Mosaic, Addshore, Flyguy649, Cybercobra, Nakon, MichaelBillington, Lpgeffen, Drphilharmonic, Muslimguy 77, Ck lostsword, Kukini, Rdavout, AThing, Mangiraikos, Rousse, Iglew, Bob the Hamster, NeantHumain, IronGargoyle, Bendzh, Waggers, Purenoni, Gandalf1491, JoeBot, Tmangray, Igoldste, Suboptimal Username, Ewulp, Tawkerbot2, Daniel5127, JForget, Sadalmelik, Randhirreddy, Shandris, Cydebot, MC10, Mato, Gogo Dodo, Zgystardst, Channer, Moonslight, Mr Hap0, Sioneholof, Thijs!bot, Marek69, Anskrev, Missvain, John254, WillMak050389, Escarbot, Glegoo, Majorly, Mandra Oleka, SgtDrini, Tjmayerinsf, Fayenatic london, Spencer, Ingolfson, JAnDbot, Plantsurfer, Supertheman, Warthog32, Bongwarrior, VoABot II, Yandman, WhatamIdoing, Theroadislong, Vssun, Lenticel, Rickard Vogelberg, Oroso, Jerem43, MartinBot, FlieGerFaUstMe262, Joie de Vivre, Mustard Pot, VirtualDelight, Uncle Dick, Whitebox, Don Cuan, Acalamari, Dskluz, Erik the guy, Belovedfreak, NewEnglandYankee, Flylady, Woood, Gwen Gale, Gemini1980, DASonnenfeld, Idioma-bot, Oaxaca dan, Jeff G., QuackGuru, Sześćsetsześćdziesiątsześć, TXiKiBoT, Vipinhari, Abtinb, ElinorD, Logicology69, DennyColt, Anarchangel, CK7, Jaguarlaser, Sexmaster6969, Banjo Fraser, Ambrosechukakiu, Sonicology, Harpyionycteris, Legion fi, Cwkmail, France3470, Flyer22, Musicandnintendo, Tiptoety, Nuttycoconut, Samatarou, Pittsburghmuggle, The Dharmatist, StaticGull, Mike2vil, Nn123645, Laurentrouble, Explicit, Starrychloe, ClueBot, GorillaWarfare, Purpleishtay, Watkinsln, Helenabella, DreamStar05, CounterVandalismBot, Fuglymugly, Rotational, Piledhigheranddeeper, Oriolpont, Riptiki, Excirial, Canis Lupus, Carsrac, Wiki libs, Laurakatie, The Red, Doprendek, Thingg, PeterNuernberg, Aitias, Plasmic

83.13.2 Images

biodivlibrary/6280048728/in/set-72157627975114672 *Original artist:* Harold Maxwell-Lefroy

- **File:066a_Josepinism_Style_Furniture.jpg** *Source:* https://upload.wikimedia.org/wikipedia/commons/8/8f/066a_Josepinism_Style_Furniture.jpg *License:* Public domain *Contributors:* Map of Furnitures of Theresianism and Josephinism in Vienna, Scholl Verlag 1912 *Original artist:* Scan and postprocessing by Hubertl

- **File:1-Arraiolos-0050.jpg** *Source:* https://upload.wikimedia.org/wikipedia/commons/5/57/1-Arraiolos-0050.jpg *License:* CC BY 3.0 *Contributors:* Own work *Original artist:* User:Sallyofmayflower

- **File:16th_century_brewer_80px.png** *Source:* https://upload.wikimedia.org/wikipedia/commons/6/6c/16th_century_brewer_80px.png *License:* Public domain *Contributors:*

- The_Brewer_designed_and_engraved_in_the_Sixteenth._Century_by_J_Amman.png   *Original   artist:* The_Brewer_designed_and_engraved_in_the_Sixteenth._Century_by_J_Amman.png: J. Amman

- **File:1890_newspaper_advertisement_showing_tin_of_desiccated_cocoanut.jpg** *Source:* https://upload.wikimedia.org/wikipedia/commons/f/f4/1890_newspaper_advertisement_showing_tin_of_desiccated_cocoanut.jpg *License:* Public domain *Contributors:* **Original publication**: Unknown original pub.
  Immediate source: http://chroniclingamerica.loc.gov/lccn/sn84025968/1890-04-25/ed-1/seq-6.pdf Library of Congress Chronicling America *Original artist:* Unknown.
  (Life time: Unknown.)

- **File:2005clove.PNG** *Source:* https://upload.wikimedia.org/wikipedia/commons/3/36/2005clove.PNG *License:* CC BY-SA 3.0 *Contributors:* Transferred from en.wikipedia to Commons by Stefan4 using CommonsHelper. *Original artist:* Anwar saadat at English Wikipedia

- **File:2005mushroom_and_truffle.PNG** *Source:* https://upload.wikimedia.org/wikipedia/commons/8/82/2005mushroom_and_truffle.PNG *License:* CC BY-SA 3.0 *Contributors:* Transferred from en.wikipedia to Commons by Stefan4 using CommonsHelper. *Original artist:* Anwar saadat at English Wikipedia

- **File:2011vanilla.svg** *Source:* https://upload.wikimedia.org/wikipedia/commons/4/42/2011vanilla.svg *License:* CC BY-SA 3.0 *Contributors:* Based on BlankMap-World6, compact.svg *Original artist:* SeeSchloss

- **File:2012cocoa.png** *Source:* https://upload.wikimedia.org/wikipedia/commons/b/b1/2012cocoa.png *License:* CC BY-SA 4.0 *Contributors:* Own work *Original artist:* Swidran

- **File:3_day_old_tamarind_sprout.jpg** *Source:* https://upload.wikimedia.org/wikipedia/commons/6/6c/3_day_old_tamarind_sprout.jpg *License:* CC BY-SA 3.0 *Contributors:* Own work *Original artist:* Davidals

- **File:4628_-_Bacche_di_ginepro_al_mercato_di_Ortigia,_Siracusa_-_Foto_Giovanni_Dall'Orto,_20_marzo_2014.jpg** *Source:* https://upload.wikimedia.org/wikipedia/commons/f/fa/4628_-_Bacche_di_ginepro_al_mercato_di_Ortigia%2C_Siracusa_-_Foto_Giovanni_Dall%27Orto%2C_20_marzo_2014.jpg *License:* Attribution *Contributors:* Self-photographed *Original artist:* Giovanni Dall'Orto

- **File:4_color_mix_of_peppercorns.jpg** *Source:* https://upload.wikimedia.org/wikipedia/commons/f/f6/4_color_mix_of_peppercorns.jpg *License:* CC BY-SA 3.0 *Contributors:* Own work *Original artist:* Ragesoss

- **File:ABS-5439.0-InternationalMerchandiseImportsAustralia-StandardInternationalTradeClassificationCustomsValue-633CorkManufactures-A18 svg** *Source:* https://upload.wikimedia.org/wikipedia/commons/3/37/ABS-5439.0-InternationalMerchandiseImportsAustralia-StandardInternationalTradeClassi svg *License:* CC BY-SA 3.0 au *Contributors:* This image is based on Australian Bureau of Statistics data. *Original artist:* Toby Hudson

- **File:A_cut_coconut_shell.JPG** *Source:* https://upload.wikimedia.org/wikipedia/commons/3/32/A_cut_coconut_shell.JPG *License:* CC BY-SA 3.0 *Contributors:* Own work *Original artist:* Aravind Sivaraj

- **File:AbbotV1Tab02A.jpg** *Source:* https://upload.wikimedia.org/wikipedia/commons/1/12/AbbotV1Tab02A.jpg *License:* Public domain *Contributors:* *The natural history of the rarer lepidopterous insects of Georgia*, volume 1, Tab 2 (modified from Biodiversity Heritage Library) *Original artist:* Smith, James Edward, Sir, 1759-1828 (text), Abbot, John, 1751-1840 (artist)

- **File:Acer_saccharum.jpg** *Source:* https://upload.wikimedia.org/wikipedia/commons/2/26/Acer_saccharum.jpg *License:* CC BY-SA 2.5 *Contributors:* Own work http://www.cirrusimage.com/tree_maple_sugar.htm *Original artist:* Bruce Marlin

- **File:Adnexed_gills_icon2.svg** *Source:* https://upload.wikimedia.org/wikipedia/commons/5/5c/Adnexed_gills_icon2.svg *License:* CC-BY-SA-3.0 *Contributors:* en:File:Adnexed_gills_icon2.png *Original artist:* User:Bryan Derksen

- **File:Aegopodium_podagraria1_ies.jpg** *Source:* https://upload.wikimedia.org/wikipedia/commons/b/bf/Aegopodium_podagraria1_ies.jpg *License:* CC-BY-SA-3.0 *Contributors:* Own work *Original artist:* Frank Vincentz

- **File:African_Buffalo.JPG** *Source:* https://upload.wikimedia.org/wikipedia/commons/c/cb/African_Buffalo.JPG *License:* CC BY 2.5 *Contributors:* ? *Original artist:* ?

- **File:Akpeteshie_local_distillation_process.jpg** *Source:* https://upload.wikimedia.org/wikipedia/commons/b/b3/Akpeteshie_local_distillation_process.jpg *License:* CC BY-SA 3.0 *Contributors:* Own work *Original artist:* Nkansahrexford

- **File:Al_hafa_corniche.jpg** *Source:* https://upload.wikimedia.org/wikipedia/commons/a/ae/Al_hafa_corniche.jpg *License:* CC BY 2.0 *Contributors:* http://www.flickr.com/photos/50755773@N06/5471194938/in/photostream/ *Original artist:* billy1847

- **File:Alambic.svg** *Source:* https://upload.wikimedia.org/wikipedia/commons/2/2a/Alambic.svg *License:* CC BY-SA 3.0 *Contributors:* Own work *Original artist:* Ayack

- **File:Alcoholic_beverages.jpg** *Source:* https://upload.wikimedia.org/wikipedia/commons/7/7a/Alcoholic_beverages.jpg *License:* Public domain *Contributors:* Own work *Original artist:* Or17

- **File:Alice_par_John_Tenniel_21.png** *Source:* https://upload.wikimedia.org/wikipedia/commons/a/a4/Alice_par_John_Tenniel_21.png *License:* Public domain *Contributors:* Scanned from "The collected verse of Lewis Carroll". New York, The Macmillan Co., 1933. ([1]) *Original artist:* Sir John Tenniel

- **File:Dehesa_cabaneros.jpg** *Source:* https://upload.wikimedia.org/wikipedia/commons/b/b1/Dehesa_cabaneros.jpg *License:* Public domain *Contributors:* ? *Original artist:* ?

- **File:Derivation_of_coal-tar_creosote.svg** *Source:* https://upload.wikimedia.org/wikipedia/en/a/ab/Derivation_of_coal-tar_creosote.svg *License:* CC0 *Contributors:*
  Adapted from source, p.12: Price, Overton W.; Kellogg, R.S.; Cox, W.T. (1909). Forests of the United States: Their Use. Government printing office. http://books.google.com/books?id=vGxGAQAAIAAJ *Original artist:*
  Brian Shapiro

- **File:Derivation_of_water-gas-tar_creosote.svg** *Source:* https://upload.wikimedia.org/wikipedia/en/1/1b/Derivation_of_water-gas-tar_creosote.svg *License:* CC0 *Contributors:*
  Adapted from source, p.12: Price, Overton W.; Kellogg, R.S.; Cox, W.T. (1909). Forests of the United States: Their Use. Government printing office. http://books.google.com/books?id=vGxGAQAAIAAJ *Original artist:*
  Brian Shapiro

- **File:Derivation_of_wood-tar_creosote.svg** *Source:* https://upload.wikimedia.org/wikipedia/en/2/23/Derivation_of_wood-tar_creosote.svg *License:* CC0 *Contributors:*
  Adapted from source, p.13: Price, Overton W.; Kellogg, R.S.; Cox, W.T. (1909). Forests of the United States: Their Use. Government printing office. http://books.google.com/books?id=vGxGAQAAIAAJ *Original artist:*
  Brian Shapiro

- **File:Diospyros_melanoxylon_Tendu.jpg** *Source:* https://upload.wikimedia.org/wikipedia/commons/c/c2/Diospyros_melanoxylon_Tendu.jpg *License:* CC BY-SA 3.0 *Contributors:* Own work *Original artist:* Pankaj Oudhia

- **File:Down_Awn_and_guard_hairs_of_cat_2012_11_13_9203r.JPG** *Source:* https://upload.wikimedia.org/wikipedia/commons/d/de/Down_Awn_and_guard_hairs_of_cat_2012_11_13_9203r.JPG *License:* CC BY-SA 3.0 *Contributors:* Own work *Original artist:* JonRichfield

- **File:Dried_Peppercorns.jpg** *Source:* https://upload.wikimedia.org/wikipedia/commons/a/a8/Dried_Peppercorns.jpg *License:* CC-BY-SA-3.0 *Contributors:* en:Image:Dried Peppercorns.jpg *Original artist:* en:User:Bunchofgrapes

- **File:Dried_berries.jpg** *Source:* https://upload.wikimedia.org/wikipedia/commons/5/57/Dried_berries.jpg *License:* CC BY-SA 3.0 *Contributors:* Own work *Original artist:* WhistlingBird

- **File:Drinking_coconut.jpg** *Source:* https://upload.wikimedia.org/wikipedia/commons/6/6f/Drinking_coconut.jpg *License:* CC0 *Contributors:* Own work *Original artist:* Dtfman

- **File:Durian.ogg** *Source:* https://upload.wikimedia.org/wikipedia/commons/d/d9/Durian.ogg *License:* CC-BY-SA-3.0 *Contributors:* Transferred from en.wikipedia; transferred to Commons by User:Sfan00_IMG using CommonsHelper. *Original artist:* Original uploader was Tonyle at en.wikipedia

- **File:Durian_customer.jpg** *Source:* https://upload.wikimedia.org/wikipedia/commons/7/71/Durian_customer.jpg *License:* CC BY 2.0 *Contributors:* another durian man *Original artist:* Stories Told By My Zadeva

- **File:Durian_flower.jpg** *Source:* https://upload.wikimedia.org/wikipedia/commons/5/58/Durian_flower.jpg *License:* CC BY-SA 2.0 *Contributors:* Durian King of Fuits is coming *Original artist:* bRYAN lOH

- **File:Durian_native_and_exotic_range_map.svg** *Source:* https://upload.wikimedia.org/wikipedia/commons/9/94/Durian_native_and_exotic_range_map.svg *License:* CC0 *Contributors:* Made from File:Location_Map_Asia.svg and File:Asia-Durian.png *Original artist:* Blank map by Haha169

- **File:Durian_on_sale_near_Cirebon.jpg** *Source:* https://upload.wikimedia.org/wikipedia/commons/9/9d/Durian_on_sale_near_Cirebon.jpg *License:* CC BY-SA 3.0 *Contributors:* via email *Original artist:* Michael Hermann

- **File:Durian_stall.JPG** *Source:* https://upload.wikimedia.org/wikipedia/commons/b/b9/Durian_stall.JPG *License:* CC BY-SA 1.0 *Contributors:* Photograph taken 2002 November 26 in Singapore by Eli the Bearded. *Original artist:* User Eli the Bearded on en.wikipedia

- **File:Durian_tree_in_malaysia.jpg** *Source:* https://upload.wikimedia.org/wikipedia/commons/6/61/Durian_tree_in_malaysia.jpg *License:* CC BY-SA 2.0 *Contributors:* Durian trees are quite tall *Original artist:* Yun Huang Yong from Harbord, Australia

- **File:Durian_with_sharp_thorns.jpg** *Source:* https://upload.wikimedia.org/wikipedia/commons/c/c6/Durian_with_sharp_thorns.jpg *License:* CC BY-SA 2.0 *Contributors:* durian *Original artist:* takato marui from Osaka, Japan

- **File:Durianpack01.JPG** *Source:* https://upload.wikimedia.org/wikipedia/commons/4/40/Durianpack01.JPG *License:* CC-BY-SA-3.0 *Contributors:* ? *Original artist:* ?

- **File:Durians_in_mesh_bags.jpg** *Source:* https://upload.wikimedia.org/wikipedia/commons/4/47/Durians_in_mesh_bags.jpg *License:* CC-BY-SA-3.0 *Contributors:* English wikipedia *Original artist:* ChildofMidnight

- **File:Durio_Zibethinus_Van_Nooten.jpg** *Source:* https://upload.wikimedia.org/wikipedia/commons/8/84/Durio_Zibethinus_Van_Nooten.jpg *License:* Public domain *Contributors:* ? *Original artist:* ?

- **File:Dyeing_British_Library_Royal_MS_15.E.iii,_f._269_1482.jpg** *Source:* https://upload.wikimedia.org/wikipedia/commons/6/6b/Dyeing_British_Library_Royal_MS_15.E.iii%2C_f._269_1482.jpg *License:* Public domain *Contributors:* http://www.nationalarchives.gov.uk/utk/england/popup/wool.htm *Original artist:* Unknown

- **File:Earth_Day_Flag.png** *Source:* https://upload.wikimedia.org/wikipedia/commons/6/6a/Earth_Day_Flag.png *License:* Public domain *Contributors:* File:Earth flag PD.jpg, File:The Earth seen from Apollo 17 with transparent background.png *Original artist:* NASA (Earth photograph) SiBr4 (flag image)

- **File:Edible_fungi_in_basket_2012_G1.jpg** *Source:* https://upload.wikimedia.org/wikipedia/commons/2/26/Edible_fungi_in_basket_2012_G1.jpg *License:* Public domain *Contributors:* Own work *Original artist:* George Chernilevsky

83.13.3 Content license